HANDBOOK OF NEURAL ACTIVITY MEASUREMENT

Neuroscientists employ many different techniques to observe the activity of the brain, from single-channel recording to functional imaging (fMRI). Many practical books explain how to use these techniques, but in order to extract meaningful information from the results it is necessary to understand the physical and mathematical principles underlying each measurement. This book covers an exhaustive range of techniques, with each chapter focusing on one in particular. Each author, a leading expert, explains exactly which quantity is being measured, the underlying principles at work, and most importantly the precise relationship between the signals measured and neural activity.

The book is an important reference for neuroscientists who use these techniques in their own experimental protocols and need to interpret their results precisely, for computational neuroscientists who use such experimental results in their models, and for scientists who want to develop new measurement techniques or enhance existing ones.

ROMAIN BRETTE is Associate Professor in the Cognitive Science Department at Ecole Normale Supérieure, Paris.

ALAIN DESTEXHE is CNRS Research Director in the Unit for Neuroscience, Information and Complexity, Gif-sur-Yvette.

HANDBOOK OF NEURAL ACTIVITY MEASUREMENT

Edited by

ROMAIN BRETTE

Ecole Normale Supérieure, Paris

ALAIN DESTEXHE

CNRS, Unit for Neuroscience, Information and Complexity, Gif-sur-Yvette

CAMBRIDGE
UNIVERSITY PRESS

CAMBRIDGE
UNIVERSITY PRESS

University Printing House, Cambridge CB2 8BS, United Kingdom

One Liberty Plaza, 20th Floor, New York, NY 10006, USA

477 Williamstown Road, Port Melbourne, VIC 3207, Australia

314-321, 3rd Floor, Plot 3, Splendor Forum, Jasola District Centre, New Delhi - 110025, India

103 Penang Road, #05-06/07, Visioncrest Commercial, Singapore 238467

Cambridge University Press is part of the University of Cambridge.

It furthers the University's mission by disseminating knowledge in the pursuit of education, learning and research at the highest international levels of excellence.

www.cambridge.org
Information on this title: www.cambridge.org/9780521516228

First published 2012

A catalogue record for this publication is available from the British Library

Library of Congress Cataloging in Publication data
Handbook of neural activity measurement / edited by Romain Brette, Alain Destexhe.
p. ; cm.
Includes bibliographical references and index.
ISBN 978-0-521-51622-8 (hardback)
I. Brette, Romain, 1977– II. Destexhe, Alain, 1962–
[DNLM: 1. Neurons – physiology. 2. Electroencephalography – methods.
3. Models, Neurological. 4. Nerve Net. 5. Neuroimaging – methods.
6. Signal Transduction. WL 102.5]
616.8′047547–dc23
2012021709

ISBN 978-0-521-51622-8 Hardback

Contents

The color plates are situated between pages 248 and 249.

Contributors

Seppo P. Ahlfors
Athinoula A. Martinos Centre for Biomedical Imaging, Department of Radiology, Massachusetts General Hospital, and Harvard MIT Division of Health Sciences and Technology, Charlestown, MA, USA

Andreas Bartels
Department of Physiology of Cognitive Processes, Max Planck Institute for Biological Cybernetics, Tübingen, Germany

Claude Bédard
Unit for Neuroscience, Information and Complexity (UNIC), CNRS, Gif-sur-Yvette, France

Romain Brette
Department of Cognitive Science, Ecole Normale Supérieure, Paris, France

Frédéric Chavane
Institut de Neurosciences Cognitives de la Méditerranée, CNRS, Aix-Marseille Université, Marseille, France

Sandrine Chemla
Canadian Centre for Behavioural Neuroscience, University of Lethbridge, Lethbridge, Alberta, Canada

Cynthia H. Chen-Bee
Department of Neurobiology and Behavior, Department of Biomedical Engineering, and the Center for the Neurobiology of Learning and Memory, University of California Irvine, Irvine, CA, USA

Maureen Clerc
Athena Project Team, INRIA Sophia Antipolis Méditerranée, France

Anders M. Dale
Departments of Radiology and Neurosciences, University of California San
Diego, La Jolla, CA, USA

Jan C. De Munck
Department of Physics and Medical Technology, VU University Medical Centre,
Amsterdam, The Netherlands

Alain Destexhe
Unit for Neuroscience, Information and Complexity (UNIC), CNRS,
Gif-sur-Yvette, France

Gaute T. Einevoll
Department of Mathematical Sciences and Technology, and Center for Integrative
Genetics (CIGENE), Norwegian University of Life Sciences, Ås, Norway

Ron D. Frostig
Department of Neurobiology and Behavior, Department of Biomedical
Engineering, and the Center for the Neurobiology of Learning and Memory,
University of California Irvine, Irvine, CA, USA

Jozien Goense
Department of Physiology of Cognitive Processes, Max Planck Institute for
Biological Cybernetics, Tübingen, Germany

Matti S. Hämäläinen
Athinoula A. Martinos Center for Biomedical Imaging, Department of Radiology,
Massachusetts General Hospital, and Harvard MIT Division of Health Sciences
and Technology, Charlestown, MA, USA

Fritjof Helmchen
Brain Research Institute, University of Zürich, Zürich, Switzerland

Henrik Lindén
Department of Mathematical Sciences and Technology, Norwegian University of
Life Sciences, Ås, Norway

Nikos Logothetis
Department of Physiology of Cognitive Processes, Max Planck Institute for
Biological Cybernetics, Tübingen, Germany

Klas H. Pettersen
Department of Mathematical Sciences and Technology, Norwegian University of
Life Sciences, Ås, Norway

Thomas Stieglitz
Laboratory for Biomedical Microtechnology, Department of Microsystems Engineering IMTEK, Faculty of Engineering, and Bernstein Center Freiburg, Albert-Ludwig-University of Freiburg, Freiburg, Germany

Carsten H. Wolters
Institute for Biomagnetism and Biosignal Analysis, University of Münster, Münster, Germany

1

Introduction

ROMAIN BRETTE AND ALAIN DESTEXHE

Most of what we know about the biology of the brain has been obtained using a large variety of measurement techniques, from the intracellular electrode recordings used by Hodgkin and Huxley to understand the initiation of action potentials in squid axons to functional magnetic resonance imaging (fMRI), used to explore higher cognitive functions. To extract meaningful information from these measurements, one needs to relate them to neural activity, but this relationship is usually not trivial. For example, electroencephalograms (EEG) measure the summed electrical activity of many neurons, and relating the electrical signals of the electrodes to neural activity in specific brain areas requires a deep understanding of how these signals are formed. Therefore, the interpretation of measurements relies not only on an understanding of the physical measurement devices (what physical quantity is measured), but also on our current understanding of the brain (the relationship between the measured quantity and neural activity).

The biophysics of neurons is explained in great detail in a number of books. This book deals with the biophysical and mathematical principles of neural activity measurement, and provides models of experimental measures. We believe this should be useful for at least three broad categories of scientists: (1) neuroscientists who use these techniques in their own experimental protocols and need to interpret the results precisely, (2) computational neuroscientists who use the experimental results for their models, (3) scientists who want to develop new techniques or enhance existing techniques. Chapters in this book cover an exhaustive range of techniques used to measure neuronal activity, from intracellular recording to imaging techniques. Each chapter explains precisely what physical quantity the technique actually measures and how it relates quantitatively to neural activity.

In the remainder of this introduction, we will give a brief overview of neuronal biophysics, in relation with the different measurement techniques that are covered in this book. More detailed accounts can be found in several books (Tuckwell, 1988; Koch, 1999; Dayan and Abbott, 2001; Hille, 2001; Gerstner and Kistler, 2002).

Handbook of Neural Activity Measurement, ed. Romain Brette and Alain Destexhe. Published by Cambridge University Press. © Cambridge University Press 2012.

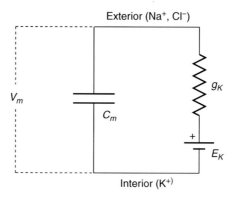

Figure 1.1 Equivalent electrical circuit of a patch of neuronal membrane: V is the membrane potential, g_K is the conductance of ionic channels permeable to K^+ ions, and E_K is the corresponding reversal potential.

The membrane of a neuron is a bipilid layer, which is an electrical insulator (see Figure 1.1). It separates the interior and the exterior of the cell, which contain ions in different proportions: sodium (Na^+) and chloride (Cl^-) ions outside the cell, potassium (K^+) ions inside the cell. These ions can enter or leave the cell through proteins in the membrane named *ionic channels*. An ionic channel forms a tiny hole in the membrane, that specific types of ions (e.g. K^+) can cross. At rest, the membrane is mostly permeable to K^+ (and a bit less to Na^+ and Cl^-). By diffusion, K^+ ions will tend to move from the interior of the cell, where the concentration is high, to the exterior of the cell, where it is lower. This phenomenon moves positive charges outside the cell, which creates an electrical field across the membrane. This field produces a movement of ions in the other direction (positive charges move to where there are fewer positive charges), and therefore an equilibrium is reached when the outward ion flux due to diffusion exactly matches the inward flux due to the electrical field. These are the basic principles of electrodiffusion. At equilibrium, there are more positive charges outside than inside the cell, and therefore the electrical potential is higher outside than inside. The *membrane potential* is defined as the difference $V_m = V_{in} - V_{out}$, and is thus negative at rest (typically around -70 mV): the membrane is *polarized*. The membrane potential at equilibrium for a given ionic channel is named the *equilibrium potential*, the *Nernst potential* or the *reversal potential*. The latter denomination means that the equilibrium potential is the membrane potential E at which the transmembrane current I changes sign: when $V_m = E_K$ no current passes through the channels (by definition), when $V_m > E_K$ positive charges exit the cell and therefore $I_K > 0$ (where the current is defined from inside to outside), and when $V_m < E_K$ positive charges enter the cell, $I_K < 0$.

Thus the transmembrane current I has the sign of $V_m - E_K$. A linear approximation yields $I = g_K(V_m - E_K)$. The parameter g has the dimensions of a conductance and is therefore called the *channel conductance*. It is the inverse of a resistance: $g = 1/R$. Electrically, this is equivalent to a resistor in series with a battery (see Figure 1.1). If we assume that all points inside the cell have the same potential, and in the same way that the potential outside the cell is constant, then the membrane is equivalent to a capacitor, representing the bilipid layer, in parallel with a resistor in series with a battery, representing the ionic channels. Using Kirchhoff's law, we can describe the temporal evolution of the membrane potential with the following *membrane equation*:

$$C\frac{dV_m}{dt} + g_K(V_m - E_K) = 0$$

where the first term is the capacitive current, and the second term is the ionic channel current. This equation is often rewritten as follows:

$$\tau\frac{dV_m}{dt} + V_m - E_K = 0$$

where $\tau = C/g_K$ is the membrane time constant. To these currents, we should add many others: currents from synapses and from the dendritic tree, and currents through other types of ionic channels. In particular, action potentials are produced by currents through voltage-dependent ionic channels, i.e. channels with a conductance that depends on the membrane potential. We also assumed that ionic channels were permeable to a single type of ion (here K^+), while in reality they are permeable to several types. These finer aspects are described in more detail in all the books mentioned above.

The membrane potential can be recorded by inserting an electrode in the cell (Figure 1.2). Electrodes for neural recording are described in Chapter 2, and intracellular recording is described in Chapter 3.

In our description of neuronal electricity above, we assumed that the potentials inside and outside the cell were spatially uniform. This is only true for a small patch of membrane or for the soma of the neuron. Along a dendrite, the potential can vary because such small processes have an electrical resistance in the longitudinal direction (quantified by the *intracellular resistivity*). This longitudinal current can be added to the membrane equation, defined at a specific point along the dendrite:

$$\tau\frac{\partial V_m(x, t)}{\partial t} + V_m(x, t) - E_K - \lambda^2\frac{\partial^2 V_m(x, t)}{\partial x^2} = 0$$

where $V_m(x, t)$ is the membrane potential at a specific point of the dendrite, and λ is the space or length constant, also called *electrotonic length*. The last term is

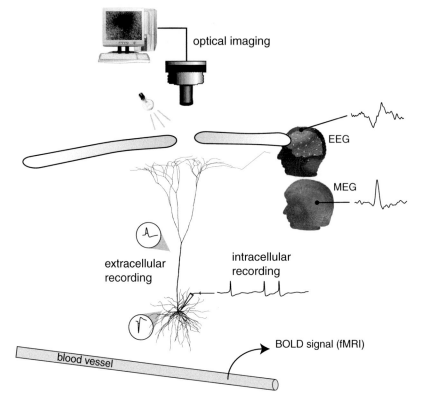

Figure 1.2 (See plate section for color version.) Correlates of neural activity and their measurement. A pyramidal cortical cell is displayed in the middle. Its membrane potential can be recorded with an intracellular electrode (*intracellular recording*, Chapter 3). Current flowing through the neuronal membrane creates extracellular potentials, which can be measured with an extracellular electrode (*extracellular recording*, Chapters 4 and 5). These potentials can also be measured with electrodes on the scalp (*EEG*, Chapters 6 and 7). Similarly, neural activity produces magnetic fields, measured with *MEG* (Chapters 6 and 7). Membrane potential can also be seen with a camera after opening the scalp and applying voltage-sensitive dyes onto the surface of the cortex (voltage-sensitive dye imaging, Chapter 9, an *optical imaging* technique). Calcium enters the cell when it spikes, which can also be recorded optically with a different technique (calcium imaging, Chapter 10). More indirectly, neural activity impacts metabolism, in particular the blood vessels, which produces signals that can be recorded with intrinsic signal optical imaging (Chapter 8, also an optical imaging method) and functional magnetic resonance imaging (*fMRI*, Chapter 11).

a diffusion term which represents the current escaping through the dendrite. By analogy with an electrical cable, this is called the *cable equation*.

Thus, there are currents flowing through the membrane across all of the surface of the cell: soma, dendrites and axon. These currents create electrical and magnetic fields in the extracellular space, which derive from Maxwell's equations. These are treated in detail in Chapters 4–7. Therefore, an extracellular electrode (Figure 1.2) can record correlates of the electrical activity of a neuron, or of several neighboring neurons. These potentials tend to be small (typically less than a millivolt), because of the low resistance of the extracellular medium. The recorded extracellular potential is a complex function of electrical activity across the neuron, which is described in Chapter 4. Information about individual action potentials is typically extracted from the high-frequency band (>500 Hz) of the extracellular potentials. These signals may stem from a single neighboring neuron (single-unit recording) or more generally from an unknown number of neighboring neurons (multi-unit recording). The low-frequency part (<500 Hz) of the extracellular potentials is called the local field potential (LFP), and is described in Chapter 5.

Electrical potentials can also be recorded with electrodes at the surface of the scalp: this is the electroencephalogram (EEG), which is described in Chapters 6 and 7. Because EEG signals must propagate through various media, such as cerebrospinal fluid, dura mater, cranium, muscle and skin, they are much more filtered than LFPs. They also represent the activity of much larger neural populations. For this reason, a key issue in EEG recording is to relate the potentials measured at the scalp with their neuronal sources inside the brain. The same electrical activity also produces magnetic fields, which can be recorded with magnetoencephalographic (MEG) equipment, which raises similar issues (see Chapters 6 and 7).

Another way to measure the membrane potential is to apply a voltage-sensitive dye on the surface of the cortex. The dye molecules bind to the external surface of the membranes of all cells, and once excited with the appropriate wavelength, they emit an amount of light that depends on the membrane potential. This light can then be captured by a CCD camera (see Figure 1.2). This optical imaging technique, called voltage-sensitive dye imaging (VSDI), is described in Chapter 9. The relationship between the recorded signal and neural activity is complex, in particular because the dye penetrates several layers of the cortex and potentially because both glial and neuronal cells are dyed. Dyes are also used to image calcium concentration rather than membrane potential. Calcium is linked to many processes in all cells, in particular in neurons. For example, calcium enters the cell when an action potential is generated. The dynamics of calcium in a cell can be described by equations that are similar to the cable equation, and thus calcium signals can be related to neuronal activity, as is described in Chapter 10.

As we have seen, electrical activity in the cell relies on differences in the concentrations of various ionic species across the membrane. These differences are actively maintained by pumps, proteins which exchange Na^+ ions against K^+ ions. Because this exchange is against the diffusion flux, it consumes energy (in the form of ATP). Synaptic transmission is also a major source of energy consumption.

Thus, when a neuron fires, or when it receives synaptic currents, it consumes energy, that ultimately originates from the blood vessels (through glial cells). This results in metabolic correlates of neuronal activity that can be measured. For example, the light absorption properties of tissues change with metabolism (for example the volume of blood vessels changes with metabolism). These changes can be recorded optically with a CCD camera: this is *intrinsic signal optical recording*, and is described in Chapter 8. Finally, *functional magnetic resonance imaging* (fMRI), described in Chapter 11, measures changes in cerebral blood flow and oxygenation using the magnetic properties of hemoglobin. These two techniques measure changes in metabolism that are only indirectly related to neural activity, and the precise relationship between the measurements and neural activity is still a matter of investigation.

The ideal measurement technique would be able to resolve the activity of single neurons (high spatial resolution), simultaneously in the whole brain (large spatial scale), at a submillisecond time scale (high temporal resolution). Currently,

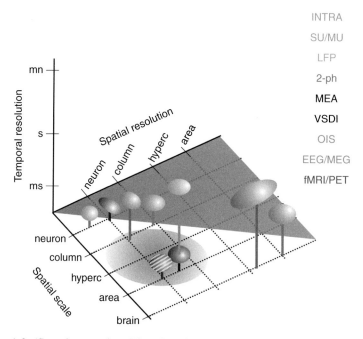

Figure 1.3 (See plate section, Plate 9.1, for color version.) Techniques classified according to their resolutions, both spatial and temporal, and their spatial scale (see figure 9.1). INTRA intracellular recording, SU/MU single-unit/multi-unit recording, LFP local field potential, 2-ph two-photon imaging, MEA multi-electrode array, VSDI voltage-sensitive dye imaging, OIS optical imaging of intrinsic signals, EEG/MEG electroencephalography/magnetoencephalography, fMRI/PET functional magnetic resonance imaging/positron emission tomography. The mesoscopic scale is represented by the oval shaded area. (Modified from Chemla and Chavane, 2010b).

no technique achieves all these goals simultaneously. Figure 1.3 summarizes the performance of the techniques described in this book along these three axes: each one has its advantages and limitations, and we hope that this book will be useful to compare and understand these techniques.

References

Dayan, P. and Abbott, L. F. (2001). *Theoretical Neuroscience*. Cambridge, MA: MIT Press.

Gerstner, W. and Kistler, W. M. (2002). *Spiking Neuron Models*. Cambridge: Cambridge University Press.

Hille, B. (2001). *Ion Channels of Excitable Membranes*. Sunderland, MA: Sinauer Associates.

Koch, C. (1999). *Biophysics of Computation: Information Processing in Single Neurons*. New York: Oxford University Press.

Tuckwell, H. (1988). *Introduction to Theoretical Neurobiology*, Vol 1: *Linear Cable Theory and Dendritic Structure*. Cambridge: Cambridge University Press.

2

Electrodes

THOMAS STIEGLITZ

2.1 Introduction

Electrodes are the first technical interface in a system for recording bioelectrical potentials. The electrochemical and biological processes at the material–tissue interface determine the signal transfer properties and are of utmost importance for the long-term behavior of a chronic implant. Here, "electrode" is used for the whole device that consists of one or multiple active recording sites, a substrate that carries these active sites, as well as interconnections, wires, insulation layers and the connectors to the next stage of a complete recording system, whether it is wire bound or wireless. The application of the electrodes in fundamental neuroscience, diagnosis, therapy, or rehabilitation determines their target specifications. The most important factors are the application site, extracorporal device or implant, acute or chronic contact, size of the electrode (device) and the recording sites, number of active sites on a device, geometrical arrangement of electrodes and type of signal to be recorded. They influence the selection process of electrodes suitable for an envisioned application and help engineers as well as neuroscientists to choose the very best materials for the active sites, substrate and insulation and the most appropriate manufacturing technique. The properties of the recorded signals are also strongly related to this selection process since the tailoring of the transfer characteristics helps to pick up the "right" signal components and to ignore, neglect and reject the "wrong" electrical potentials that might be due to the body itself or the surrounding environment or interference caused by noise from the electrode sites and the amplifier of the recording system.

Bioelectrical potentials are generated from nerves and muscles all over the body to transmit information:

- from the central nervous system to peripheral actuators, like muscles and internal organs,

Handbook of Neural Activity Measurement, ed. Romain Brette and Alain Destexhe. Published by Cambridge University Press. © Cambridge University Press 2012.

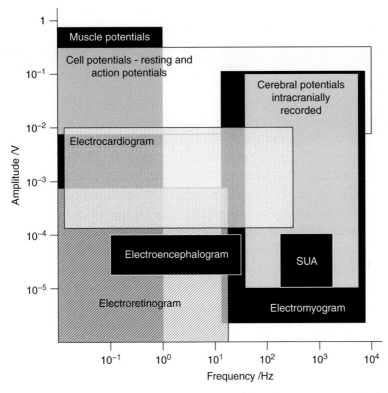

Figure 2.1 Amplitudes and frequency ranges of bioelectrical potentials in the human body. (Modified from Nagel, 2000.) SUA, single-unit activity.

- from natural sensors to the central nervous system to transfer versatile environmental modes into electrical signals as tactile sensors in the skin, the eye and the ear do, and
- within the central nervous system itself to process and compute the information in the brain.

Depending on the recording modality and the application, different amplitudes and frequency ranges will be obtained (Figure 2.1). Neural recordings in the central nervous system can be classified into (extracorporal) electroencephalograms (EEG), intracorporal electrocorticograms (ECoG) and intracortical/intracranial signals. All of these are extracellular recordings in which single-unit activity (SUA), multi-unit activity (MUA) and local field potentials (LFP) represent the spiking behavior of a single neuron, small ensembles of nerve cells and the slow potentials of local ensembles, respectively. Their amplitude is mostly in the upper microvolt range. However, the larger the electrode sites are the higher the amplitudes can be since more neurons contribute additively to the signal. In addition, strong bioelectrical signals from the heart (ECG) and the skeletal muscles might be superimposed because of their larger amplitudes and might disturb the signal. Often, these potentials and signals occur as artifacts in the signal of interest and cannot be selectively filtered out since their frequency range is identical with that of the wanted signal. Therefore, the properties of the recording electrodes and the filter–amplifier system should match the particular properties of the wanted signal and thereby reject or suppress the components of the unwanted signals.

Neurological investigations use diagnosis methods that are mostly repetitive short-term examinations where recording electrodes are placed non-invasively on the surface of the skin for a very limited time. In therapy, rehabilitation and many fundamental neuroscientific research paradigms, however, electrodes are implanted for a subchronic (up to 30 days) or chronic (longer than 30 days) time period. In these cases, electrodes must provide a reliable interface to the neural target structure and the signal quality must not degrade over time. The long-term stability and functionality of the electrode site materials as well as insulation materials is of utmost importance to obtain reliable signals and to prevent any damage of the target tissue. Especially in recordings from neurons in the brain – the cortex as well as deep brain structures – relatively stiff electrodes have to interface with the delicate brain tissue, which is still a challenge in modern electrophysiology and neurotechnology.

If one summarizes the desired target specifications of neural recording electrodes, different aspects from medicine, material sciences, biomedical engineering and ergonomics are included (Loeb et al., 1995; Stieglitz, 2004; Williams, 2008):

- the material of the electrodes and coatings must not harm the surrounding tissue and must be tolerated by the body to reduce foreign body reactions,
- tailored and stable impedance at the phase boundary delivers a functional interface,
- tailored frequency characteristics are required to record wanted signals for the envisioned application,
- low noise (thermal as well as electrochemical) allows recording of signals with small amplitudes,
- low polarization at the phase boundary and low afterpotentials if recording is combined with electrical stimulation,
- integrated pre-amplifier to minimize interfering signals from the body and the environment.

If the electrode is part of a chronic implantable system, the following specifications have to be added to obtain a useful and safe system:

- long-term stability of the material for several decades for chronic implants, i.e. no or only minimal corrosion and excellent biostability,
- wireless, transcutaneous signal transmission (and energy supply) to the acquisition system or user interface,
- at least a small number of cables if wireless transmission is not feasible,
- easy-to-use systems that do not require an additional technician during daily use.

The following sections will lead the reader step by step through these different aspects, introduce the fundamentals from electrochemical, electrical, mechanical and (partly) biological points of view and present an overview of established electrodes for different applications. Finally, the basic principles of amplifiers and filters will be summarized briefly as a guide for selection of an appropriate amplifier system for the envisioned application and the operation of such a system.

2.2 Electrochemistry at electrodes

Commonly, an electrode is described as a conducting part, mostly metallic, that mediates the transmission of an electrical current in another conducting medium, for example a fluid or gas. In biology and medicine, electrodes mediate between the ion conduction of the electrolytic medium in and between the cells and the electron conduction in the electrodes and the recording circuitry. The properties of this interface determine the performance of the electrode in any application. Therefore, the fundamental processes that influence the recording of bioelectrical signals will be introduced and discussed for selected examples of electrodes. Since electrodes need electronic amplifiers as the first stage in the process chain to visualize bioelectrical signals, the electrochemical properties of the electrodes will be translated into electrical rather than electrochemical equivalent circuit models to be matched later with the amplifiers (see Section 2.5).

2.2.1 Electrode classification

Electrodes in general can be classified with respect to the charge transfer mechanism and their voltage–current relationship, their polarization, at the phase boundary (Meyer-Waarden, 1985) between the electrode and the electrolyte. According to the desired application in analytical chemistry, biochemical monitoring or neural recording, electrode properties have to be carefully selected (Figure 2.2).

The hydrogen electrode (not included in Figure 2.2) is used as the reference for all other materials. Its potential is defined as zero in any electrolyte. This reference is called a saturated hydrogen electrode (SHE) or reversible hydrogen electrode

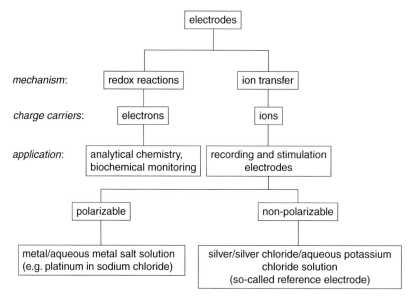

Figure 2.2 Classification of electrodes. Neural recording electrodes are usually polarizable metal electrodes. Charge carriers are transferred through phase boundaries (marked "/" in the lower boxes) from the solid state electrodes into the electrolyte of the body.

(RHE), respectively. A platinum wire is inserted into a capillary and both are placed into an electrolyte. Hydrogen gas continuously flows through the capillary and bubbles into the electrolyte. The platinum catalyzes the reaction from hydrogen gas into protons and electrons and builds a stable electrical potential $V_{RHE} = 0\,V$. This electrode with bubbling gas is not applicable for every investigation but is limited to the chemical characterization of electrodes in vitro.

Irreversible electrodes cannot be used for recording neural activity. They are intentionally designed to induce chemical reactions, for example within a Volta-element or battery to generate electrical charge (Cu and Zn electrode in diluted H_2SO_4). Reversible, i.e. stable, electrodes are preferable for reliable monitoring. They use ion transfer mechanisms at the phase boundary between the electrode and the electrolyte. Ideal polarizable electrodes cannot be used since the smallest charge transfer (i.e. current) would induce a large potential at the electrode–electrolyte phase boundary. In the real world, electrodes are always somewhere between ideal polarizable and non-polarizable electrodes. Bioelectrical signals are so small that the signal itself will not influence the transfer characteristic of the electrode. The fundamentals of polarization will be introduced and discussed below in this section. Metal electrodes in an electrolyte belong to these "real world" electrodes. If a metal is placed in an electrolyte, metal ions dissolve and

Table 2.1 *Electrochemical potentials of metal–ion interfaces at* $T = 298\ K$ *(Lide 2003).*

Material	Reaction	Potential
Al	$Al^{3+} + 3e^-$	-1.67 V
Fe	$Fe^{2+} + 2e^-$	-0.441 V
H_2	$2H^+ + 2e^-$	0.000 V (Reference)
Ag	$Ag^+ + e^-$	$+0.7996$ V
Pt	$Pt^{2+} + 2e^-$	$+1.2$ V
Au	$Au^{3+} + 3e^-$	$+1.52$ V
Au	$Au^+ + e^-$	$+1.83$ V

recombine with free electrons in the metal atoms again. These reactions form an electrochemical equilibrium that results in an electrical potential that is specific for different metals. These potentials are called the half cell potentials (Table 2.1) at the electrode–electrolyte interface. An appropriate combination of the same or similar electrode materials is responsible for stable recording systems. Noble metals are preferred since they have less tendency to dissolve and are suitable for use in recording electrodes.

The selection of the electrode material often depends not only on the chemical material properties but also on the manufacturing technology and mechanical stability. Platinum and iridium are materials of choice for thin films and coatings. Titanium nitride is sometimes chosen as a capacitive electrode. In precision mechanics approaches, in which mechanical stability is important, platinum–iridium (90/10) alloys, stainless steel and the cobalt–chromium alloys MP 35 N and Elgiloy are established materials.

Reference electrodes are metals that are covered with a scarcely soluble salt of the same metal within a solution with the same anions as the metal salt. Their electrode potential is nearly independent of their exchange current over the phase boundary, i.e. the transition of charge from the solid state electrode into the electrolyte. Their behavior will be described using the example of a silver–silver chloride electrode in the following subsection.

2.2.2 Reference electrodes

Reference electrodes are non-polarizable and reversible. They are used as substitute reversible hydrogen electrodes in cases where gas bubbling is not applicable. The common reference electrode for biomedical applications is the silver–silver chloride electrode (Figure 2.3).

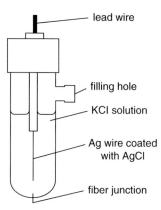

Figure 2.3 Silver–silver chloride electrode.

This consists of a silver wire that has been chlorinated and placed in 3 molar potassium chloride solution, contained in a glass tube with a fiber junction to exchange chloride ions selectively. The solution of 3 molar potassium chloride is chosen because it has a minimum voltage shift due to temperature changes. The electrode as a half cell can be operated either as a cathode (Equation (2.1)), where chloride is deposited, or as an anode where silver goes into solution (Equation (2.2)) after silver chloride has been dissociated, silver has been reduced onto the surface and chloride ions have been released into the electrolyte:

$$Ag(solid) \leftrightarrow Ag^+(solution) + e^- \text{ (metal phase)}, \quad (2.1)$$

$$Ag^+(solution) + Cl^-(solution) \leftrightarrow AgCl(solid). \quad (2.2)$$

These two equilibrium relations follow the ionic product K_s of the activities of silver and chloride (Equation (2.3)) which is constant under the conditions of reversibility. Here, the activities α, which are measures of the reactivities of species, are equal to the concentrations c because silver and chloride are monovalent ions:

$$\alpha_{Ag^+} \cdot \alpha_{Cl^-} = c_{Ag^+} \cdot c_{Cl^-} = K_s = \text{constant}. \quad (2.3)$$

Reversibility is maintained if the exchange currents at the electrode–electrolyte interface, i.e. the phase boundary between the solid state body and the fluid, are limited to small values. In this case, the equilibrium voltage U_{eq} is only dependent on constant terms (Equation (2.4)) and therefore also remains constant, since the activity which is here equal to the concentration of chloride is chosen large enough:

$$U_{eq} = \Phi_{Ag}^{0\,\prime} - \frac{RT}{zF}\ln\alpha_{Cl^-} \quad \text{with} \quad \Phi_{Ag}^{0\,\prime} = \Phi_{Ag}^{0} + \frac{RT}{zF}\ln K_s \quad (2.4)$$

where r is the gas constant, 8.3144 J mol^{-1} K^{-1}, T is the absolute temperature in degrees kelvin, z is the charge number, F is the Faraday constant, 96 500 C mol^{-1}, and Φ^0_{Ag} is the electrochemical potential of silver under equilibrium conditions.

The silver–silver chloride electrode with 3 molar KCl solution as bridge electrolyte exhibits a DC potential of +222.3 mV vs. RHE. After an initial stabilization phase where the equilibrium between silver and silver chloride settles, silver–silver chloride electrodes have a stable impedance over the electrode–electrolyte interface and exhibit low noise but produce a DC offset. These electrodes can be used for recording neural signals with high impedance amplifiers if the input current is less than 1 nA. If this value is exceeded, the current flow causes the equilibrium at the silver–silver chloride interface to deteriorate and leads to unstable recording conditions. Silver–silver chloride electrodes must not be used for chronic implantation since the antibacterial properties of molecular silver that can diffuse out of the electrode have toxic effects (McAdams et al., 1992).

2.2.3 The Helmholtz double layer

In metal electrodes, all reversible processes at the phase boundary under equilibrium conditions are mediated by water molecules. The oxygen atom is covalently bound to two hydrogen ions and forms an asymmetric, polar structure with an angle of about 105° between the hydrogen atoms. This structure results in an electrical dipole, in which the oxygen represents the negative pole and the hydrogen atoms represent the positive pole (Figure 2.4).

If a metal plate is inserted into an aqueous electrolyte, for example into interstitial or intracellular fluid, cerebrospinal fluid (CSF) or in solutions that emulate a physiological environment such as physiological saline solution or Ringer's solution, water molecules adhere on the phase boundary between the solid state body (the electrode) and the fluid (the electrolyte). A second layer of solvated ions, i.e.

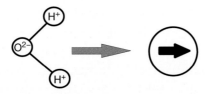

Figure 2.4 The water molecule encloses an oxygen atom between two hydrogen atoms with an angle of about 105° and thereby forms an electrical dipole.

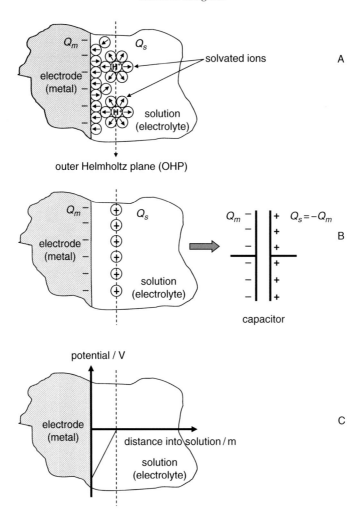

Figure 2.5 Water molecules and solvated hydrogen ions form the Helmholtz double layer that mediates electron conduction into ion conduction at the electrode–electrolyte phase boundary.

hydrated protons, covers the water dipoles and forms the so-called Helmholtz double layer (Figure 2.5A). The center of charge of the water molecules forms the outer Helmholtz plane (OHP) with the fixed diffusion that is followed by a diffuse double layer, the so-called Gouy layer. Both theories were summarized by Stern in 1924 as the Gouy–Chapman–Stern layer with the Debye length as the thickness of the Gouy–Chapman layer (Stern, 1924). In highly concentrated and therefore well conducting electrolytes, the conduction mechanisms can be approximated by the Helmholtz model. At the OHP, the same amount of charge accumulates on both sides of the double layer and acts as a capacitor (Figure 2.5B) in which all access charge is accumulated. If this capacitor is a plane, polished metal layer, the capacitance can be calculated according to the equations of a parallel plate capacitor (Equation (2.5)) with d_H as the thickness of the approximated Helmholtz double layer, A as the geometrical surface of the electrode site, the permittivity of water $\varepsilon_r = 80$ and the permittivity of vaccum $\varepsilon_0 = 8.854 \times 10^{-12}\,\mathrm{Fm^{-1}}$:

$$C_H = \varepsilon_0 \cdot \varepsilon_r \cdot \frac{A}{d_H}. \tag{2.5}$$

Its capacity can store a charge of up to $20\,\mu\mathrm{C\,cm^{-2}}$ for polished metal. Charge transfer only takes place by charging and discharging of this capacitor (Figure 2.5B). The accumulated charge on the plates causes a linear potential drop (Equation (2.6)) over the double layer (Figure 2.5C) since the relation between the charge Q and the potential U is proportional with the capacitance C:

$$Q = C \cdot U \Rightarrow U = \frac{Q}{C} = \frac{Q}{\varepsilon_0 \cdot \varepsilon_r \cdot A} \cdot d_H \Rightarrow U \sim d_H. \tag{2.6}$$

This capacitive mechanism is the desired method for charge transfer into biological tissue because it does not cause any chemical alteration of the electrode surface and does not induce any charge transfer across the phase boundary. However, only small amounts of charge can be transferred.

If these limits are exceeded due to interference or during charge injection for electrical stimulation of nerve cells, Faradaic reactions occur that alter the electrode surface as a result of chemical reactions. Charge passes the phase boundary directly. These Faradaic reactions are irreversible if diffusion of chemical reaction products and/or gas formation occurs and the reaction products are no longer available for a reverse reaction. They are modeled as a resistive component R_L in parallel to the Helmholtz capacitor C_H. The electrode is destroyed by corrosion, and corrosion by-products can change the pH value and finally damage the surrounding tissue. Reversible Faradaic reactions also alter the electrode surface but the reaction products stay stable at the interface and participate in the redox reaction again if the applied potential is reversed.

Since this section addresses only neural recording applications and not electrical stimulation, the detailed electrode reactions at elevated potentials and during excessive charge injection will not be described here. A comprehensive overview of stimulation electrodes can be found in many neural prostheses textbooks (e.g. Stieglitz, 2004).

2.2.4 *Polarization of electrodes*

Electrode polarization describes the voltage–current relationship with various, sometimes unclear, events at the phase boundary. If charge carriers pass the electrode–electrolyte phase boundary they generate a current i_F in the Faradaic resistance R_L that causes an increase in the potential at the electrode compared with its reversible equilibrium state E_{rev} (Equation (2.7)) (Fricke, 1932; McAdams et al., 1992). The degree of polarization is measured as the overpotential η at the electrode (Equation (2.8)):

$$E = E_{rev} + i_F \cdot R_L \qquad (2.7)$$
$$\eta = |E - E_{rev}| \qquad (2.8)$$
$$\eta = \eta_t + \eta_d + \eta_c + \eta_k. \qquad (2.9)$$

The overpotential can be split up into different physicochemical subparts (Equation (2.9)) (Meyer-Waarden, 1985). The charge transfer processes over the phase boundary are described by the transfer overpotential η_t. The diffusion overpotential η_d is related to diffusion of charge at higher current densities. The reaction overpotential η_c occurs during kinetic inhibition of chemical reactants and the crystallization overpotential η_k is induced by crystallization reactions or the exchange of metal ions with corresponding ions. They both play only a minor part in the polarization. The tendency of polarization at the phase boundary can be modeled as an additional polarization voltage source V_{EE} in series with the Faraday resistance.

The changes of the phase boundary properties (such as Helmholtz capacitance and Faraday resistance) vary with the applied voltage. Up to a certain limit, they can be described with a linear model (Schwan 1992; McAdams et al., 1995) and can be summarized in the polarization. In the non-linear range, because of the reaction overpotential that is related to every electrochemical reaction, phase boundaries show a diode like behavior, i.e. the Faraday resistance decreases with increasing voltage. The Helmholtz capacitance increases with increasing voltage. The thickness of the Helmholtz layer decreases with increasing strength of the electrical field.

Electrode polarization depends upon the electrode material and its surface treatment, the electrode area, the electrolyte, the temperature, and other factors, and strongly influences the electrical transfer characteristic of the electrode. These effects are in contrast to those of non-polarizable electrodes (Table 2.2).

Even though the properties of ideal non-polarizable electrodes are desirable, some important boundary conditions have to be taken into account. If electrodes are in close contact with the target tissue, the neurons, toxicity aspects of the materials have to be taken into account as well as manufacturing techniques of the electrodes and the signal amplitudes to be recorded. If the recording of neural signals is considered and not electrical stimulation, polarization due to charge transfer over the phase boundary, i.e. current flow, can be neglected. Most recording studies of bioelectrical signals assume that the signals are acquired within the linear range of the

Table 2.2 *Polarizable and non-polarizable electrodes.*

Polarizable electrode	Non-polarizable electrode
Voltage at the phase boundary depends on current flow	Voltage at the phase boundary does not depend on current flow
High resistance R_L	Low Faradaic resistance R_L
Low Helmholtz capacitance C_H	High Helmholtz capacitance C_H
High polarization voltage V_{EE}	Low polarization voltage V_{EE}
Current flows ideally as displacement current	Current flows ideally as convection current
High pass behavior $f_{cutoff} \cong 100\,\text{Hz}$	No high pass behavior, C_H not noticeable until high frequencies

electrodes. However, if the linear behavior of electrodes is limited to 50–100 mV, an a priori estimation of the voltage limits in recording studies is encouraged to prevent recording of data with non-linear signal distortion (McAdams and Jossinet, 1994a, 1994b). With respect to neural signals, this requirement is always met. Therefore, from the polarization point of view, it does not matter which contact metal is chosen. However, toxicity matters. Molecular silver is cytotoxic above a certain dose. Therefore, silver–silver chloride electrodes are not recommended for chronic implantations. Electrical potentials at liquid junctions might occur as DC potential (see Section 2.3.5 for details) in glass micropipettes but are contact potentials and not due to charge transfer.

2.2.5 *Electrical models of electrodes*

Most of the electrochemical processes at the phase boundary can be modeled with electrical equivalent circuit components. For some processes, special electrochemical elements have been derived from an analytical description of these reactions. Considering the fundamental reactions, especially under recording conditions, the electrical components will be sufficient. In a general model of an electrode in an electrolyte (Figure 2.6), the components for the linear and non-linear regimes of the electrode are included: the Helmholtz capacity C_H for reversible charging and discharging, the Faraday resistance R_L in parallel representing the charge transfer processes with the polarization voltage V_{EE} in series. R_A models the access resistance, i.e. the part of the electrolyte between the electrode and the target tissue.

The resistors with diode and Zener–diode in parallel to C_H represent the irreversible pathways of hydrogen ($R_{hydrogen}$) and oxygen (R_{oxygen}) evolution that occur selectively at negative and positive voltages, respectively, due to electrolysis of water, if the electrode specific potential limits of the so-called water window have been exceeded. This is in the range −0.6 V to 0.8 V for platinum with respect

Thomas Stieglitz

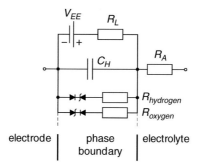

Figure 2.6 Complete equivalent circuit model of the electrode–electrolyte phase boundary.

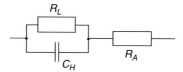

Figure 2.7 Simplified equivalent circuit model of the electrode–electrolyte phase boundary.

to a silver–silver chloride reference electrode. The water window is the safe working range of the electrode with respect to the electrical potential. In cases where the water window is exceeded, the voltage at the electrode does not increase further due to the Zener diodes and current will flow as a result of electrolysis.

Under recording conditions, the equivalent circuit of the electrode can be simplified to the direct electrode–electrolyte interface (Figure 2.7). Extracellular as well as intracellular recordings of neural signals have a signal amplitude that is far below the limits of the water window with electrolysis reactions. Amplifiers designed as voltage followers as input stages nearly block current flow at the electrodes due to their high input impedance and thereby prevent polarization since they keep the electrodes in their equilibrium state (see Equation. (2.6)).

The complex impedance Z describes the current–voltage behavior of any electronic or electrochemical component of the equivalent circuit. It is the general description of the transfer function that shows how the amplitudes of different frequency components are attenuated and how the output signal is delayed in its phase θ with respect to the input signal. It is a complex quantity that can be expressed as a vector with a magnitude $|Z|$ and a phase angle θ (Figure 2.8). The impedance as complex measure can be expressed either in Cartesian coordinates displaying the imaginary part (X) versus the real part (R) in a so-called Nyquist plot, or as the logarithm of magnitude of the impedance $|Z|$ versus the logarithm of the frequency and the phase angle θ versus the logarithm of the frequency in a so-called

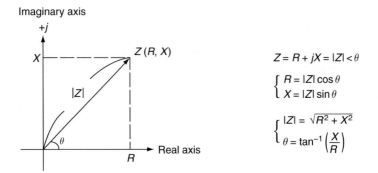

Figure 2.8 Description of impedance.

Bode plot. The two plots are equivalent and can be transferred directly without loss of information (see Figure 2.8 for transfer equations).

The impedance measurements are taken using a three-electrode setup under defined conditions (Stieglitz, 2004). The transfer properties of the simplified electrode model (Figure 2.7) act as a filter with high pass behavior (Equation (2.10)). Signal components above the cut-off frequency (Equation (2.11)) have a constant attenuation; components below f_{cutoff} are attenuated according to a passive filter of first order, i.e. one resistor with one capacitor, with 20 dB per decade:

$$Z = \frac{1}{\frac{1}{R_L} + j \cdot \omega \cdot C_H} + R_A \qquad (2.10)$$

$$f_{cutoff} = \frac{1}{2 \cdot \pi \cdot C_H \cdot R_L}. \qquad (2.11)$$

The course of the impedance is similar for most electrode materials. However, the magnitudes of the impedance and the cut-off frequencies vary with respect to the electrode size and the roughness of the surface that defines the electrochemical active surface, for example the impedance of platinum with increasing roughness (platinum, platinum gray, platinum black) of an electrode with 50 µm diameter characterized with a 10 mV sine wave (Boretius et al., 2010).

Even though the phase in the Bode diagram looks different in the three materials, the variation of the magnitude is comparable (Figure 2.9). In the equivalent circuit values, the Faraday resistance R_L and the access resistance R_A are comparable since all three materials are platinum and therefore based on the same exchange currents and energy levels at the phase boundary. The roughness of the surface only influences the Helmholtz capacitance C_H which increases the electrochemical active area at the phase boundary (Table 2.3).

If electrodes are either applied on the skin or implanted, a complex model of the electrodes, the biological components and the short circuit pathways simulates

Table 2.3 *Equivalent circuit values of different electrode materials with 60 μm diameter calculated from the impedances (Boretius et al., 2010).*

Material	C_H(nF)	R_L(MΩ)	R_A(kΩ)
Pt	6.9	1.9	10.5
Pt-gray	15.3	1.5	11.1
Pt-black	24.4	1.7	10.8

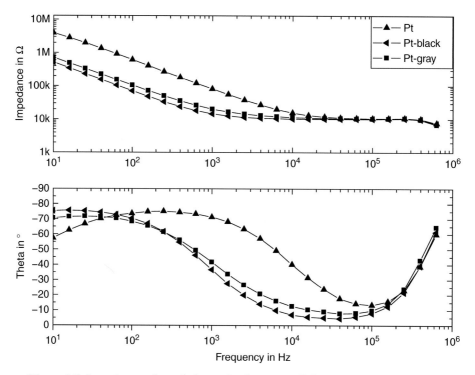

Figure 2.9 Impedance of metal electrodes for extracellular recordings with 60 μm diameter and different electrode coatings. Impedance magnitude (upper graph) and phase angle of the impedance vary with coating. Pt-black and Pt-gray have a greater surface roughness than Pt.

the behavior of the system and helps to explain the signal distortion and attenuation. In an investigation of epicortical electrode arrays, the model shown in Figure 2.10 was proposed, and parameters were identified (Henle et al., 2010) from impedance spectra that were measured over the whole implantation period (Cordeiro et al., 2008). Therefore, two arbitrary electrode sites on an electrode array are chosen as measurement probes.

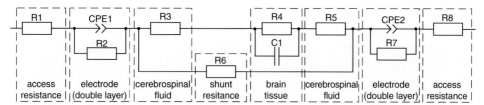

Figure 2.10 Model of two electrodes used to record neural signals from the brain. The model includes the capacitive (if present) and resistive behavior of the brain tissue, the cerebrospinal fluid and possible fluidic short cuts, i.e. the shunt resistance. If tissue reactions occur after implantation, of proteins depositing on the electrode surfaces, these reactions change the resistive and capacitive properties of the different building blocks (dashed lines) in the model. For more detailed modeling, capacitors C have been extended to constant phase elements (CPE).

The influence of tissue growth on the electrical properties of the interface can be easily identified. Protein adsorption and tissue growth as foreign body reaction on the implantation can be detected as an increase in R3 and R5 as well as in a decrease in R6, for example. During this investigation, the impedance spectra stayed stable over several weeks after the initial foreign body reaction led to an increase in impedance (Cordeiro et al., 2008). It is even possible to differentiate tissues, for example fibrotive or nervous tissue, by their electrical impedance, if the right electrode arrangement and the right frequency range are selected. In general, electrode corrosion, short circuits due to cerebrospinal fluids, and detachment of interconnection wires are examples that can be identified with this measurement method.

2.3 Electrode types

All electrodes work according to the same fundamental electrochemical principles. However, the design of the electrodes determines the exact recording specifications. After some general considerations of the selection criteria, some exemplary designs of electrodes for typical applications will be introduced and specific properties and transfer characteristics will be discussed.

2.3.1 Selection criteria for electrode types

The neuroscientific or clinical application finally determines the electrode type depending on the invasiveness of the approach, the spatial resolution of multiple electrode sites, the frequency range that is necessary to extract the desired information and the duration of the electrode–tissue contact. Often, the selected solution is a compromise between invasiveness and selectivity. Non-invasive recordings from

EEG electrodes do not harm the skin and the brain tissue but the filter properties of the skin, the skull and the dura mater attenuate high-frequency components and limit the spatial selectivity due to volume conduction effects. Therefore, only neural mass activity can be recorded. Invasive approaches are used to record epicorticograms (ECoG) epidurally or subdurally. Spatial selectivity is better and higher frequencies can be recorded. Intracortical electrodes are able to record signals from single neurons (single-unit activity, SUA), or small assemblies of neurons. Signals with high-frequency components represent multi-unit activity (MUA) while low-frequency components of the local field potential (LFP) give an insight into synaptic potentials. The electrode site size, distribution and impedance have to be tailored to match the requirements of the application and scientific objective.

In addition, adequate manufacturing technologies have to be selected or developed as prerequisites to acquire the desired signals. Often, neuroscientific target specifications do not go hand in hand with possibilities and limitations of manufacturing techniques. The design approach for recording single-unit activity in small cortical areas is completely different from that for arrays used to monitor correlation of signals over a complete hemisphere. The following examples cover a wide range of designs, technologies and applications.

2.3.2 Surface electrodes for EEG

Non-invasive measurement of neural mass activity is recorded using the electroencephalogram (EEG). Silver-silver chloride electrodes are used to record the small potentials with a stable electrode. They are often integrated in self adhesive disposable electrodes. The electrodes mediate with a gel between the silver chloride and the skin without another membrane. Therefore, electrode gels must have good conductivity and must include additional chloride ions to complement the electrochemical system. If the gel diffuses into the perspiratory glands and hair follicles it can bypass the epidermis and helps to decrease the impedance of the skin. Additional methods to decrease the skin impedance are cruder, for example mechanical abrasion of the skin with tape or sandpaper. The equivalent circuit model (Figure 2.11) describes the resistive behavior of the electrode gel and the DC potential generated between the electrode–gel system and the skin. The transfer characteristic is mainly dependent on the skin properties and the possibilities of reducing the transfer impedance.

Skin, fatty tissue and bone attenuate the neural signals from the brain. They are the main influence on the transfer behavior of the signal since the electrode sites themselves are relatively large and exhibit low impedance and cut-off frequencies. Amplitudes below 20 μV up to 150 μV have to be recorded in a frequency range from DC to about 50 Hz.

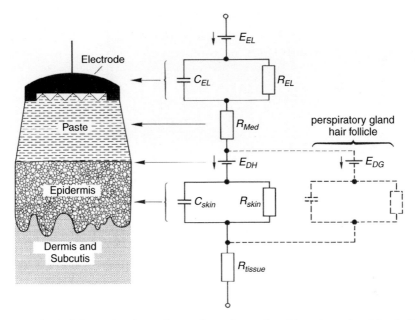

Figure 2.11 Schematic view of a surface electrode with gel on the skin (left) and the corresponding equivalent circuit (right) that represents the electrical contact potential of the electrode and its transfer characteristics (E_{EL}, C_{EL}, R_{EL}) the conductivity of the gel (R_{Med}) and electrical contact potential and transfer characteristics of the skin's epidermis (E_{skin}, C_{skin}, R_{skin}), dermis and subcutis (summarized in R_{tissue}). Perspiratory glands and hair follicles form, if present, a parallel pathway to the dermis and subcutis that can be used to bypass the high resistance of the epidermis. (From Meyer-Waarden, 1985.)

2.3.3 Electrode grids for ECoG

Electrocorticograms are recorded with electrode grid arrays that lie directly on the dura mater of the brain, i.e. epicortical. They are made by means of precision mechanics with stainless steel recording sites embedded into silicone rubber that serves as substrate and electrical insulation (Figure 2.12) (Stieglitz et al., 2009). The flexibility and elasticity of the silicone rubber ensures that the arrays adapt to the curvature of the brain and that the electrode sites stay close to the surface of the brain. The electrode site diameter typically varies between 1 mm and 10 mm and the pitch, i.e. the distance between the electrode centers, is usually fixed to 10 mm in commercially available electrodes for humans. The manufacturing technology allows variations in size and arrangement. A hexagonal instead of rectangular electrode arrangement increases the electrode density and reduces the electrode pitch. However, if the electrode pitch goes below a certain limit, the material properties of the metal dominate the material properties of the silicone

Figure 2.12 (See plate section for color version.) Epicortical grid arrays. Electrode sites (silver dots) are embedded into silicone rubber. The grid is placed on the surface of the brain. Dark lines are sulci of the brain.

rubber substrates and the devices lose flexibility and become relatively rigid owing to the stiffness of the stainless steel electrode plates.

Clinical application of epicortical electrode grid arrays is mainly limited to the presurgical diagnosis of epilepsy. Patients with severe and medical refractory epilepsy are implanted with these grid arrays for up to two weeks to determine the location of the epileptic foci before parts of the temporal lobe are removed in a neurosurgical intervention. Some of these patients have been included in clinical trials during this implantation period, to perform motor imagery and hand movement tasks, and data have been analyzed to investigate their use in future brain–computer interfaces for communication or control of robotic limbs. The epicortical recording paradigm to extract useful data for these applications has proved its feasibility but much research and development remains to be done before these systems will become commercially available products in clinical practice.

2.3.4 Metal microelectrodes

Metal microelectrodes are one of the oldest types of electrode used to interface with neurons. Two designs have been developed for neuroscientific investigations. Single wires with sharp tips are used for acute and intraoperative implantation where the insertion trajectory is determined by anatomical data and controlled by a frame that is fixed on the skull, in a so-called stereotactic implantation. Wire arrays with blunt tips are commonly chosen for chronic intracortical recordings of single-unit activity (Figure 2.13).

The small area of the electrode tip allows spatially selective recording of single cells, either extracellular or intracellular. In many cases, electrode holders are

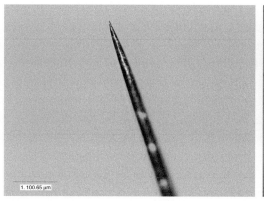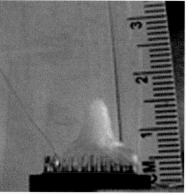

Figure 2.13 (See plate section for color version.) Photograph of a tungsten micro-electrode (left) and a wire electrode array from NB Labs (www.nblabslarry.com) (right).

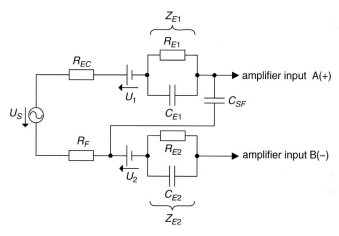

Figure 2.14 Equivalent circuit of a metal microelectrode. The electrodes (Z_{E1}, Z_{E2}) with their half cell potentials (U_1, U_2) record the neural signal (U_S). The resistances of the extracellular fluid (R_{EC}) and the electrolyte (R_F) have to be considered as well as the capacity between the microelectrode and the electrolyte (C_{SF}).

needed to insert the electrode into the cell and hold it in place. Visual control via a microscope helps to place the electrode or to penetrate the cell membrane. Metal microelectrodes consist of stainless steel, platinum–iridium alloys or tungsten. The wires can be sharpened easily with electrochemical methods (Figure 2.13, left). The detailed protocol of the electrolyte, current density and dipping speed into the electrolyte are material specific.

If one metal microelectrode E1 is combined with a large counterelectrode E2 to record neural signals U_S (Figure 2.14), two DC potentials (U_1, U_2) at the electrodes and the intracellular or extracellular fluid have to be taken into account. Metal electrodes are not reference electrodes and subtraction at a differential amplifier would not (completely) eliminate these DC electrode potentials due to different sizes and transfer functions of the electrodes. Compared to the large impedance of the microelectrode Z_1 the resistances of the extracellular fluid R_{EC} and the electrolyte resistance R_F can be neglected as well as relatively small impedance of the large counterelectrode Z_2. However, the capacity C_{SF} between the microelectrode and the extracellular fluid is of importance. It influences the transfer characteristic in combination with the input capacity of the recording system and the interconnection lines like a low pass filter. The maximum cut-off frequency of the system can be obtained if capacity compensation (see Section 2.5) is used that reduces the total capacity C_g of the recording setup to zero (Equation 2.12) but not the capacity of the electrode itself:

$$f_{cutoff} = \frac{1}{2 \cdot \pi \cdot R_{E1} \cdot (C_g + C_{E1})}. \tag{2.12}$$

The DC potentials at the electrodes strongly influence the quality of the recorded signals since they are at least of the same order as the wanted signal and result in a fluctuating curve. However, if event detection of spike activity is the only interest, they are the preferred choice because of their good mechanical properties.

Multiple wire electrodes have been established for recording single-unit activity in chronic animal experiments in several research groups (for a review see Stieglitz et al., 2009). They can be purchased (e.g. NBLabs, www.nblabslarry.com) or customized by soldering insulated wires on suitable connectors. Up to 740 wires have been implanted in one animal so far (Nicolelis et al., 2003) with stable recordings for up to 18 months.

2.3.5 Glass micropipettes

Glass microelectrodes are electrolyte-filled glass micropipettes used basically for fundamental research in electrophysiology. They have been established for many years, have proven their reproducibility for recording data and are simple to manufacture. They have been pulled out of a glass micropipette (Figure 2.15) with a shaft diameter between 1.5 mm and 2.0 mm to a minimal tip diameter of 1 μm for patch clamp electrodes and even smaller diameters for sharp electrodes. According to their diameter, glass composition and wall thickness as well as the type of electrolyte, they are used for extracellular or intracellular recording.

The pipettes are filled with an electrolyte as a conductive bridge to the recording environment. The composition of the electrolyte is chosen such that it resembles

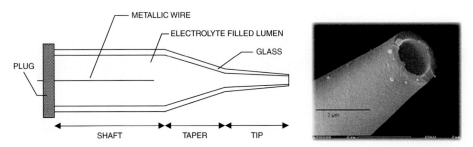

Figure 2.15 Schematic view (left) and photograph (right) of a glass micro-electrode. (N. Fertig, Wikimedia Commons, licensed under CreativeCommons-Lizenz by-sa-3.0-de, http://de.wikipedia.org/wiki/Patch-Clamp-Technik.)

the ionic concentration of the environment. Sodium chloride (0.5 M to 2 M) is chosen for extracellular recordings and potassium chloride for intracellular recording, since these ions determine the electrical potential of the extracellular and intracellular space, respectively. Having these ions in the electrolyte, an additional DC voltage is minimized. Electrical contact with the electrolyte is made best with a chlorinated silver wire that is stuck in the middle of the pipette to interface the electrode with the recording circuitry. The silver–silver chloride electrode as reference electrode allows stable recording of DC as well as AC potentials.

Recording with a glass micropipette takes place against a large indifferent electrode E2 (Figure 2.16). This metal electrode generates a contact potential U_2 in the electrolyte but its impedance can be neglected. The electrical resistance of the glass microelectrode E1, including its electrolyte, is in the range 1–200 MΩ with a tip potential U_T up to 100 mV. When recording the intracellular or extracellular potential U_S, the resistance of the intracellular fluid R_{IC} and the resistance of the fluid around the cell R_F is so small compared to the other resistances that it can be neglected. Comparable to metal microelectrodes, the capacitance between the electrode and the surrounding fluid C_{SF} together with the amplifier input capacitance and the interconnection lead capacitance (Equation (2.12)) limit the bandwidth of the system and should be compensated for (see Section 2.5).

A scenario for chronic implantation of glass microelectrodes for extracellular recordings has been developed as the "cone electrode" (Kennedy, 1989) and commercialized by Neural Systems, Inc. (Atlanta, GA, USA). In the "cone electrode" approach, a piece of a rat sciatic nerve is placed inside a glass micropipette of length 1.5 mm and diameter 100–200 μm before implantation in the cortex. The nerve graft attracts neurite ingrowth and stable recordings have been obtained over six months. After experiments in rats, implantations in a few human subjects have been performed (Brumberg et al., 2010).

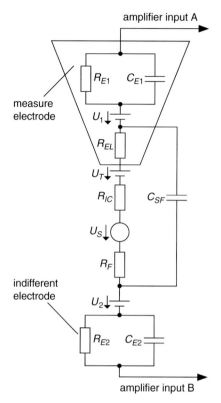

Figure 2.16 Equivalent circuit of a glass microelectrode. The measuring electrode consists of a glass cone with an inserted Ag–AgCl wire (R_{E1}, C_{E1}) that generates a DC potential (U_1) to the filling electrolyte (R_{EL}). A tip voltage (U_T) is generated if the electrode contacts the nerve cell to record its activity (U_S). The resistivity of the intracellular fluid (R_{IC}), as well as the resistivity of the extracellular fluid, or the electrolyte in the cell bath (R_F), has to be taken into account. An indifferent electrode is connected to the second input channel of the amplifier. This electrode consists of metal that is placed somewhere in the patient, animal or cell bath. The metal generates a contact or half cell voltage (U_2, see Section 2.2) and exhibits the typical high pass behavior of a metal–electrolyte interface (R_{E2}, C_{E2}). The extracellular fluid forms a capacitor (C_{SF}) with the glass micropipette.

2.3.6 Patch clamp technique

The patch clamp technique uses glass micropipettes with a very smooth surface to suck the cell membrane into the pipette and thereby generate a very high resistance to the cell, the so-called gigaohm seal. Neher and Sakman were awarded the Nobel Prize in 1991 for their investigations on ion channels with this method. The patch clamp method is suitable for investigating whole cells or isolated cell patches. The general recording principle is the same for both approaches (Figure 2.17).

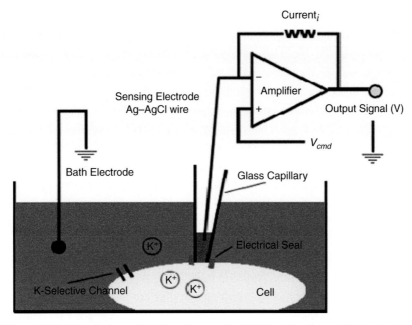

Figure 2.17 (See plate section for color version.) Principle of a patch clamp setup.

The glass micropipette is placed on the cell and suction is applied. The cell membrane gently adheres to the glass and generates a high resistance seal in the $G\Omega$ range. Now, a voltage (V_{cmd}) can be applied between the micropipette and a bath electrode. Depending on this voltage, ion channels open or close in the cell membrane. The ionic current flows into the pipette over a silver–silver chloride electrode into a recording system. The current is converted into a voltage signal and displayed or stored. Since currents have to be detected in the low picoampere range, the capacitances of the pipettes and recording systems strongly influence the system properties with respect to low pass behavior and noise. Capacitance compensation has to be applied to be able to record those low currents, even after the selection of low noise, low capacitance systems and active filters.

There are various configurations of patch clamp recordings that allow different research questions to be addressed. If a patch clamp electrode is attached to a cell and the membrane is perforated, the current–voltage behavior of the whole cell, i.e. the systemic behavior of the cell, can be investigated. In contrast, if the cell membrane is patched and the pipette is withdrawn from the cell, a membrane patch can be isolated to investigate the behavior of single ion channels in a defined environment. Variations exist to establish the gigaohm seal and to perforate the cell if this is desired, depending on the objective of the investigation.

2.3.7 Micromachined electrodes for intracortical applications

Since the first attempts to use micromachining techniques for intracortical (extra-cellular) recording 40 years ago (Wise et al., 1969), many groups have worked on the design and development of these microimplants (Stieglitz et al., 2009) to manufacture microelectrodes for extracellular recordings.

In general, two different concepts have been established (Figure 2.18). One is the so-called Utah array (Nordhausen et al., 1996), commercialized as the Brain Gate™ system from Cyberkinetics. Needles are manufactured from a block of silicon. The system consists of 100 needles with one electrode on each tip (Campbell, 1991). Recordings have been done in the central nervous system of animals (Nordhausen et al., 1996) and in a spinal cord injured human subject (lesion level C3) for 18 months (Hochberg et al., 2006) as hardware within a brain–computer interface to control a virtual keyboard.

A competitive approach has been developed at the University of Michigan. The probes are now commercially available from Neuronexus Technologies. Here, silicon is also used as substrate but the electrode sites are arranged along one surface of the shank. There is a wide range of dimensions of electrode size and arrangement as well as shank distance. This technological approach allows the monolithic integration of electronic circuitry, for example for multiplexing and amplification in the probes (Najafi and Wise, 1986). Planar comb like structures (Kim and Wise, 1996) can be assembled as three-dimensional arrays (Hoogerwerf and Wise, 1994; Bai et al., 2000) with up to 1024 electrode sites (Wise et al., 2004). Electronic circuitry for spike detection of single–unit activity reduces the amount of data to be transferred.

Both approaches are still cable bound with percutaneous connectors. However, the race is open in both groups, Michigan/Neuronexus and Utah/Cyberkinetics, to be the first with a completely integrated system with wireless energy supply and data transmission.

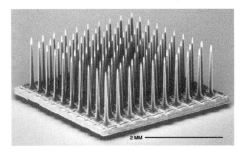 

Figure 2.18 (See plate section for color version.) Typical designs of micromachined electrode arrays: Utah electrode array-UEA (left) (Nordhausen et al., 1996, © Elsevier) and Michigan array (right) (Wise et al., 2004, © 2004 IEEE).

2.4 Reactions and processes at implanted electrodes

Cardiac pacemakers, spinal cord stimulators and cochlea implants are successful stories examples of electrodes in chronic implants (Stieglitz and Meyer, 2006a) for therapy and rehabilitation. More applications in the field of neuromodulation and neural prostheses to restore motor and sensory function are currently under development (Stieglitz and Meyer, 2006b). Even though hundreds of thousands of these electrical active medical devices have been implanted, it is still not clear how the least possible foreign body reaction can be obtained in combination with long-term functionality over decades. Even though many materials have been proven to be suitable for implantation, "[...] biocompatibility could not solely be dependent on the material characteristics but also had to be defined by the situation in which the material is used" (Williams, 2008). Therefore, this section discusses some aspects that should be taken into account when electrodes are brought in direct contact with neural tissue for chronic applications.

The foreign body reaction to the implantation of an electrode array starts immediately after contact with the neural tissue and lasts for at least some weeks (Figure 2.19) depending on the material, the surgical intervention and the insertion trauma (Delbeke, 2004). Specific and unspecific protein adsorption as well as unspecific and specific immune response triggers the various factors that lead to inflammation, healing and final scar formation.

Even though neurons are the functional cells in the central nervous system, glial cells and vascular related tissue make up 75% of the cells in the brain (Polikov et al., 2005). Oligodendrocytes are the myelin forming cells and astrocytes

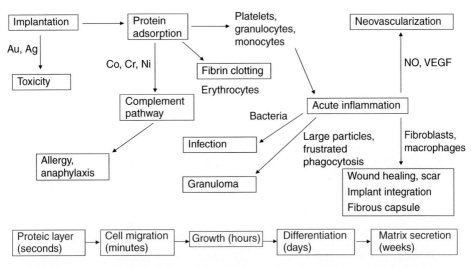

Figure 2.19 Overview of the foreign body reaction process, e.g. after implantation of an electrode. (From Delbeke, 2004.)

help to control the chemical environment of the neurons, buffer the neurotransmitters and modulate spiking activity. Microglia is the tissue involved in wound healing of the brain. Astrocytes and microglia are strongly activated after electrode implantation in the brain, migrate towards the implantation site, release factors and try to start phagocytosis.

While intracortical probes often work well in acute settings, they often fail in chronic implantations with respect to reliable single-unit recording properties. The materials themselves have been proven to be non-cytotoxic (Stensaas and Stensaas, 1978) but the brain tissue reaction is a major problem according with the structural biocompatibility of the implant, i.e. its mechanical material properties and the design itself (Polikov, 2005). The general problem for intracortical electrodes is that they can only be inserted into tissue in parallel with the shaft axis. This requires a minimal stiffness for insertion but also results in a mismatch of the mechanical material properties compared to those of the tissue. The electrodes cannot adapt to the movement of the tissue and its pulsatile changes due to blood pulsation and respiration. Electrodes with no additional fixation, so-called floating electrodes, might follow the brain movements to some extent. If the electrodes are placed in the vicinity of a neuron to record its spiking activity, any displacement will increase the distance to the signal source and result in signal attenuation or signal loss. The growth of scar tissue with low electrical conductivity will also contribute to signal attenuation and eventually all these factors result in signal loss in the recordings. Fixed electrodes that are connected to drivers for alignment can be readjusted for optimal recording but any new movement induces further tissue reaction and might lead to a vicious circle in a long time run.

A lot of research has been done and is still ongoing on the mechanical interactions at the interface between the electrode and the brain tissue that shows indications of chronic inflammation and causes foreign body reactions. Different approaches try to reduce scar tissue growth by tailoring the surface to reduce the mechanical mismatch between the stiff implant materials and the soft matter of the brain tissue (for a summary see Stieglitz et al., 2009). However, there is still no solution on the horizon that looks promising with respect to insertion, multi-channel intracortical recording of single-unit spike activity and long-term functionality at the same time.

2.5 Amplifiers and filters for extracellular and intracellular recording

The recording of neural activity requires amplifiers for mass signals from surface electrodes, single-unit spike recording from extracellular wires or detection of ion channel currents in the voltage clamp mode with intracellular electrodes. In this section, some fundamental principles and circuits will be introduced that help to measure the wanted signal with only little interference and small effect on the measurement object. This includes the rejection of unwanted signal components, the amplification of the desired components and some tips and tricks for amplifier design and amplifier "tuning" with respect to frequency range and artifact rejection (Meyer-Waarden, 1985; Nagel, 2000).

A recording system couples with the biological signal source and consists of the electrode, a preamplifier with filters to decouple DC potentials and reject unwanted components in the low-frequency range. It is connected to an isolation amplifier for patient safety and to increase the signal amplitude. Low pass filtering limits the signal to prevent non-linear aliasing in the succeeding analog to digital conversion before the data are displayed, stored and further processed.

2.5.1 Filters

Filters selectively attenuate certain frequency regions according to their circuit design. Low pass filters attenuate high frequencies, high pass filters attenuate low frequencies. The combination of both leads to band passes or notch filters that selectively attenuate certain ranges. The frequency at which the attenuation starts is called the cut-off frequency. If only resistors, capacitors and inductors are used, a filter is called passive.

Filters change the input signal (U_i) to the output signal (U_o) depending on the frequency f of the signal. The transmission behavior is described by the transfer function $H(\omega)$ with the angular frequency $\omega = 2\pi f$.

$$H(\omega) = \frac{U_o}{U_i}. \tag{2.13}$$

Passive first-order filters (Figure 2.20) consist of a simple and effective circuit of a resistor and a capacitor.

Transfer functions are calculated using the fundamental relationship of the current–voltage behavior of resistors ($U = R \cdot I$) and capacitors ($U = Z_C \cdot I$) with $Z_C = (j \cdot \omega \cdot C)^{-1}$ and $j = \sqrt{-1}$.

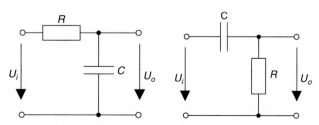

Figure 2.20 Passive filters of first order: low pass (left), high pass (right).

Describing the transfer behavior of the high pass and low pass filters by the voltage at input and output by the current flow through the components one obtains for the low pass filter

$$U_i = U_i(\omega) = I(\omega) \cdot \left(R + \frac{1}{j \cdot \omega \cdot C} \right)$$

$$U_o = U_o(\omega) = I(\omega) \cdot \left(\frac{1}{j \cdot \omega \cdot C} \right)$$

$$H(\omega) = \frac{U_o(\omega)}{U_i(\omega)} = \frac{I(\omega)}{I(\omega)} \cdot \frac{\frac{1}{j \cdot \omega \cdot C}}{R + \frac{1}{j \cdot \omega \cdot C}} = \frac{1}{j \cdot \omega \cdot R \cdot C + 1}$$

$$H(\omega \to 0) = 1$$

$$H(\omega \to \infty) = 0. \tag{2.14}$$

The same calculation has to be done for the high pass filter:

$$U_i = U_i(\omega) = I(\omega) \cdot \left(\frac{1}{j \cdot \omega \cdot C} + R \right)$$

$$U_o = U_o(\omega) = I(\omega) \cdot R$$

$$H(\omega) = \frac{U_o(\omega)}{U_i(\omega)} = \frac{I(\omega)}{I(\omega)} \cdot \frac{R}{\frac{1}{j \cdot \omega \cdot C} + R} = \frac{j \cdot \omega \cdot R \cdot C}{1 + j \cdot \omega \cdot R \cdot C}$$

$$H(\omega \to 0) = 0$$

$$H(\omega \to \infty) = 1. \tag{2.15}$$

First order filters attenuate with 20 dB per decade with a cut-off frequency of

$$f_{cutoff} = \frac{1}{2 \cdot \pi \cdot R \cdot C}. \tag{2.16}$$

A combination of x filters of first order in series leads to a filter of xth order with an attenuation of $x \cdot 20$ dB per decade. Passive filters need no additional energy source to attenuate signals.

Active filters (Tietze and Schenk 2002) use operational amplifiers that need energy to attenuate signals. Their attenuation also depends on the order of the filter, and the transfer characteristics with respect to steepness and distortion are determined by the values of the circuit. Bessel, Butterworth and Tchebychev transfer functions have different advantages and disadvantages. However, Bessel characteristics are the only ones with linear distortion of the transferred signal and are mandatory if phase relationships in different signals are investigated, for example in correlation studies. These filters need accurate design and realization to obtain stable transfer characteristics of the recorded signals and often the help of an experienced circuit design engineer.

The main objectives of different filters can be summarized as follows:

- low pass filter; reduction of effective noise bandwidth,
- high pass filter; decoupling of DC components,
- band pass; amplification of wanted signal,
- notch filter; reduction of 50/60 Hz interference from the power supply; however, this is only permitted if the wanted signal does not have (major) components around this frequency.

2.5.2 Amplifiers

Amplifiers increase the amplitude of any wanted signal without increasing signals that seem to be artifacts from either a biological or technical source. Operational amplifiers are used in the circuit design of amplifiers. They exhibit high input impedance, relatively low input capacitance, allow high amplifications and have low power consumption.

Signals at the input of an amplifier are amplified. If identical signals (in magnitude and phase) are applied at the amplifier inputs, they are named common mode signals; if the magnitude and phase are different at both amplifier inputs, they are called differential mode signals. In a physiological case, differential mode signals are the signals to be recorded while common mode signals are of physiological or technical origin but are not desired, e.g. 50 Hz power line noise, ventilation or heart beat in nerve recordings. The relation of the amplification $V = U_o/U_i$ (Figure 2.21) of common mode signals V_{GI} that are applied in parallel to the inputs and the amplification of differential signals V_D is called the common mode rejection ratio (CMRR).

The CMRR is often displayed in dB according to

$$\text{CMRR} = 20 \cdot \log \frac{V_D}{V_{GI}}. \tag{2.17}$$

Within a simple differential amplifier (Figure 2.22), minimization of the common mode amplification is obtained with a selection of the impedances according to

$$Z_1 Z_4 = Z_2 Z_3 \tag{2.18}$$

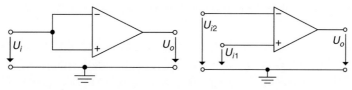

Figure 2.21 Measurement of common mode (left) and differential mode (right) amplification.

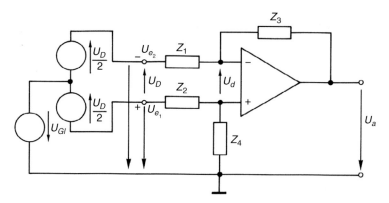

Figure 2.22 Circuit of a simple differential amplifier with common mode (U_{GI}) and differential mode (U_D) signal components.

that results in a common mode amplification of $V_{GI} = 0$, i.e. CMRR $= \infty$ and a differential mode amplification of

$$V_D = \frac{Z_3/Z_1}{Z_4/Z_2}. \qquad (2.19)$$

In most amplifiers for recording of bioelectrical signals, such as nerve or muscle signals, the impedances Z_1, Z_2, Z_3 and Z_4 are realized as resistors, i.e. $Z_i = R_i$ for $i \in \{1, 2, 3, 4\}$ delivering a constant transfer of signals over the complete frequency range of the signals.

However, tolerances in the resistors lead to asymmetries and deteriorate the CMRR. A tolerance as low as $\alpha = 0.1\%$ between $Z_1 = R_1$ and $Z_2 = R_2 = R_1 + \alpha$ reduces the CMRR from infinity to 130 dB at a differential amplification of $V_D = 1000$ according to

$$\mathrm{CMRR} = \frac{V_D}{V_{GI}} = \frac{1 + V_D + \alpha/2}{\alpha}. \qquad (2.20)$$

The low CMRR and the low impedance at the inputs of the circuit require voltage followers to deliver a high input impedance and transfer it into a low output impedance (OP1 and OP2 in Figure 2.23). The resistor network in the input stage determines the amplification of the first stage and allows an increase of the total CMRR as a product of the differential amplification of the first stage and the CMRR of the second stage.

Amplifiers for biomedical applications should have a CMRR of 80 dB and higher to ensure sufficient suppression of common mode signals. If signals with low-frequency components are being recorded, an input filter to decouple DC components might be necessary. However, the high pass reduces input impedance and CMRR. Especially in combination with metal microelectrodes, input impedances

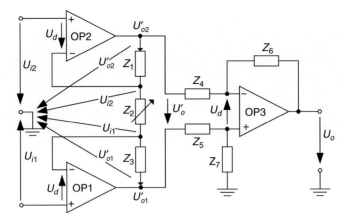

Figure 2.23 Amplifier used to record neural signals. The voltage follower input stage (OP1, OP2) transduces high input impedance into low output impedance and often has unity gain. The differential amplifier as second stage (OP3) amplifies the signals according to the ratio of impedances Z_6 to Z_4 (Equation 2.19). This arrangement increases common mode rejection, i.e. the suppression of signals that are of the same magnitude at both amplifier inputs (OP1, OP2).

might become too low and signal distortion might occur (Nelson et al., 2008). Therefore, careful adaptation of the input impedance with respect to the electrodes used is strongly encouraged.

2.5.3 Capacitance compensation of electrodes

Capacitance compensation has to be used to eliminate the influence of stray and input capacitances if slow charging and discharging processes at microelectrodes influence the amplifier behavior. The upper cut-off frequency of the complete system is affected and the discharging process is accelerated by adding a capacitance in the positive feedback loop of the amplifier. Thereby, the bandwidth of the input loop is increased. In a capacitance compensation circuit (Figure 2.24) a capacitor C_r connects the output of the second operational amplifier OP2 with the non-inverting input of the first amplifier OP1. If I_r is controlled to neutralize I_c in point A, I_{El} must be zero according to Kirchhoff's law. Thereby, the signal amplitude U_s applies directly at the input of the amplifier OP1. Ideally, every capacitance could be compensated with a circuit like this with infinite slew rate of the signals. However, with limited bandwidth the circuit tends to oscillate and a variable resistor (potentiometer) that controls the amplification of OP2 has to decrease the oscillation tendency. The adaptation of OP2 according to the actual capacitance at the input including capacitive coupling on the interconnection lines makes this circuit quite unstable and the adjustment of the capacitance is a tedious job.

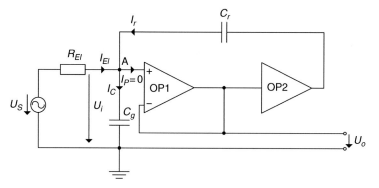

Figure 2.24 Capacitance compensation of electrodes.

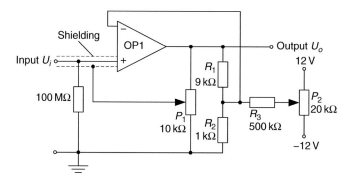

Figure 2.25 Simple preamplifier with input guarding and option to compensate for electrode polarization (P_2).

This circuit is only used if the electrode capacitance really has to be compensated to obtain the desired signal. Main applications today cover all intracellular recordings including patch clamp setups for investigation of ion channels.

2.5.4 Input guarding

Input guarding helps to compensate for the capacitance of the cables (30–200 pF m^{-1}) that limits the bandwidth at the amplifier input. These capacitances can attenuate the signal components of higher frequencies when an electrode with high impedance is used, for example a microelectrode. The bootstrapping of the potential of the shielding and the signal line results in a relief of the input network (Figure 2.25). The input network of the electrode resistance R_{EI} and the cable capacitance C_L thereby no longer works as a low pass filter.

The disadvantage of input guarding is that it only affects capacitances between the amplifier input and the shielding/guard of the cable. Electrode capacitance, shunt capacitance and common mode input capacitance of the amplifier cannot be compensated with this approach. Input guarding is used in those cases in which changes in the frequency course are mainly caused by cable capacitances, and electrode and input capacitances at the amplifier can be neglected.

2.6 Conclusions

Recording of neural activity from and in the central nervous system is a challenging task with respect to electrode selection, signal recording and processing of the obtained data. The application with its specific requirements of the origin of the signals, their spatial and temporal resolution as well as the desired frequency components drive the selection process. The volume conductivity of the brain tissue and the cerebrospinal fluid might limit the spatial arrangement of the electrode sites as well as technical design rules that ensure the absence of cross talk between adjacent recording channels on substrates and between interconnection lines due to low insulation resistance and capacitive coupling. The miniaturization of microelectronics is still under way and engineers are becoming able to develop amplifiers with smaller input capacitances, complete recording and signal processing systems on a single chip and wireless solutions to eliminate finally percutaneous plugs and cables. The neuroscientist still has to decide whether the raw signal has to be analyzed or whether event detection of spikes is sufficient. A close collaboration of neuroscientists and (biomedical) engineers often helps to understand the needs and opportunities and finally lead to better signals and systems, hopefully to new and better therapies and rehabilitation devices. For human applications, however, approved medical devices (FDA approval in the USA, CE mark in Europe) in compliance with the intended use have to be taken or clinical studies have to be launched for investigation of novel devices or applications.

References

Bai, Q., Wise, K. D. and Anderson, D. J. (2000). A high-yield microassembly structure for three-dimensional microelectrode arrays. *IEEE Trans. Biomed. Eng.*, **47**, 281–289.

Boretius, T., Badia, J., Pascual-Font, A., Schuettler, M., Navarro, X., Yoshida, K. and Stieglitz, T. (2010). A transversal intrafascicular multichannel electrode (TIME) to interface with the peripheral nerve. *Biosens. Bioelectron.*, **26**, 62–69.

Brumberg J. S., Nieto-Castanon, A., Kennedy, P. R. and Guenther, F. H. (2010). Brain–computer interfaces for speech communication. *Speech Commun.*, **52** (4), 367–379.

Campbell, P. K., Jones, K. E., Huber, R. J., Horch, K. W. and Normann, R. A. (1991). A silicon-based three-dimensional neural interface: manufacturing process for an intracortical electrode array. *IEEE Bio-Med. Eng.*, **38**, 758–768.

Cordeiro, J., Henle, C., Raab, M., Meier, W., Stieglitz, T., Schulze-Bonhage, A. and Rickert, J. (2008). Micromanufactured electrode for cortical field potentials recording: in vivo study. In: J. van der Sloten, P. Verdonck, M. Nyssen and L. Haueisen (editors), *ECIFMBE 2008, IFMBE Proceedings 22*, pp. 2375–2378.

Delbeke, J. (2004). Biocompatibility. *Workshop on Implanted Device Technology, Annual Conference of the International Functional Electrical Stimulation Society, 4 September, Bournemouth, UK.*

Fricke, H. (1932). The theory of electrolytic polarization. *Philos. Mag.*, **14**, 310–318.

Henle, C., Raab, M., Cordeiro, J., Prinz, M., Doostkam, S., Stieglitz, T., Schulze-Bonhage, A. and Rickert, J. (2010). First long term in vivo biocompatibility assessment on subdurally implanted micro-ECoG electrodes, manufactured with a novel laser technology. *Biomed. Microdev.*, DOI 10.1007/s10544-010-9471-9.

Hochberg, L. R., Serruya, M. D., Friehs, G. M., Mukand, J. A., Saleh, M., Caplan, A. H., Branner, A., Chen, D., Penn, R. D. and Donoghue, J. P. (2006). Neuronal ensemble control of prosthetic devices by a human with tetraplegia. *Nature*, **442**, 164–171.

Hoogerwerf, A. C. and Wise, K. D. (1994). A three-dimensional microelectrode array for chronic neural recording. *IEEE Trans. BioMed. Eng.*, **41**, 1136–1146.

Kennedy, P. R. (1989). The cone electrode: a long-term electrode that records from neurites grown onto its recording surface. *J. Neurosci. Methods*, **29**, 181–193.

Kim, C. and Wise, K. D. (1996). A 64-site multishank CMOS low-profile neural stimulating probe. *IEEE J. Solid-State Circuits*, **31**, 1230–1238.

Lide, D. R. (2003). *CRC Handbook on Chemistry and Physics* (83rd edition). Boca Raton, FL: CRC Press.

Loeb, G. E., Peck, R. A. and Martyniuk, J. (1995). Towards the ultimate metal microelectrode. *J. Neurosci. Methods*, **63**, 175–183.

McAdams, E. and Jossinet, J. (1994a). The detection of the onset of electrode–electrolyte interface impedance nonlinearity: a theoretical study. *IEEE Trans. Biomed. Eng.*, **41**(5), 498–499.

McAdams, E. and Jossinet, J. (1994b). Physical interpretation of Schwan's limit voltage of linearity. *Med. Biol. Eng. Comput.*, **32**, 126–130.

McAdams, E. T., McLaughlin, J. A. and Holder, D. S. (1992). Neurosensors: a review of some fundamental electrode parameters. *Satellite Symposium on Neuroscience and Technology, Ann. Int. Conf. IEEE Eng. Med. Biol. Soc.*, 226–234.

McAdams, E., Lackermeier, A., McLaughlin, J. A., Macken, D. and Jossinet, J. (1995). The linear and non-linear electrical properties of the electrode–electrolyte interface. *Biosens. Bioelectron.*, **10**, 67–74.

Meyer-Waarden, K. (1985). *Bioelektrische Signale und ihre Ableitverfahren*. Stuttgart: Schattauer-Verlag.

Nagel, J. H. (2000). Biopotential amplifiers. In: J. Bronzino (editor), *Biomedical Engineering Handbook*, Heidelberg: Springer-Verlag.

Najafi, K. and Wise, K. D. (1986). An implantable multielectrode array with on-chip signal processing. *IEEE J. Solid-State Circuits*, **21**, 1035–1044.

Nelson, M. J., Pouget, P., Nilsen, E. A., Patten, C. D. and Schall, J. D. (2008). Review of signal distortion through metal microelectrodes recording circuits and filters. *J. Neurosci. Methods*, **168**, 141–157.

Nicolelis, M. A. L., Dimitrov, D., Carmena, J. M., Crist, R., Lehew, G., Kralik, J. D. and Wise, S. P. (2003). Chronic, multisite, multielectrode recordings in macaque monkeys. *Proc. Natl. Acad. Sci. USA*, **100**, 11041–11046.

Nordhausen, C. T., Maynard, E. M. and Normann, R. A. (1996). Single unit recording capabilities of a 100 microclectrode array. *Brain Res.*, **726**, 129–140.

Polikov, V. S., Tresco, P. A. and Reichert, W. M. (2005). Response of brain tissue to chronically implanted neural electrodes. *J. Neurosci. Methods*, **148**, 1–18.

Schwan, H. P. (1992). Linear and nonlinear electrode polarization and biological materials. *Ann. Biomed. Eng.*, **20**, 269–288.

Stensaas, S. S. and Stensaas, L. J. (1978). Histopathological evaluation of materials implanted in the cerebral cortex. *Acta Neuropathol.*, **41**, 145–155.

Stern, O. (1924). Zur Theorie der Elekrolytischen Doppelschicht. *Z. Elektrochem.*, **30**, 508.

Stieglitz, T. (2004). Electrode materials for recording and stimulation. In: K. Horch and G. Dhillon (editors), *NEUROPROSTHETICS: Theory and Practice* (Series on Bioengineering and Biomedical Engineering, vol 2). Singapore: World Scientific, pp. 471–516.

Stieglitz, T. and Meyer, J.-U. (2006a). Neural implants in clinical practice. In: G. A. Urban (editor), *BIOMEMS*, Dordrecht: Springer-Verlag, pp. 41–70.

Stieglitz, T. and Meyer, J.-U. (2006b). Biomedical microdevices for neural implants. In: G. A. Urban (editor), *BIOMEMS*, Dordrecht: Springer-Verlag, pp. 71–138.

Stieglitz, T., Rubehn, B., Henle, C., Kisban, S., Herwik, S., Ruther, P. and Schuettler, M. (2009). Brain–computer interfaces: an overview of the hardware to record neural signals from the cortex. In: J. Verhaagen, E. M. Hol, I. Huitinga, J. Wijnhold, A. B. Bergen, G. J. Boer and D. F. Swaab (editors), *Neurotherapy Progress in Restorative Neuroscience and Neurology Prog. Brain Res.*, **175**, 297–315.

Tietze, U. and Schenk, Ch. (2002). *Halbleiterschaltungstechnik* (12th edition). Berlin: Springer-Verlag.

Williams, D. F. (2008). On the mechanisms of biocompatibility. *Biomater.*, **29**, 2841–2953.

Wise, K. D., Angell, J. B. and Starr, A. (1969). An integrated circuit approach to extracellular microelectrodes. *8th ICMBE, Palmer House, Chicago, IL*, 20 July 1969. *Digest of the 8th ICMBE*, vol 1, p. 14.

Wise, K. D., Anderson, D. J., Hetke, J. F., Kipke, D. R. and Najafi, K. (2004). Wireless implantable microsystems: high-density electronic interfaces to the nervous system. *Proc. IEEE*, **92**, 76–96.

3

Intracellular recording

ROMAIN BRETTE AND ALAIN DESTEXHE

3.1 Introduction

Intracellular recording is the measurement of voltage or current across the membrane of a cell. It typically involves an electrode inserted in the cell and a reference electrode outside the cell. The electrodes are connected to an amplifier to measure the membrane potential, possibly in response to a current injected through the intracellular electrode (current clamp), or the current flowing through the intracellular electrode when the membrane potential is held at a fixed value (voltage clamp). Ionic and synaptic conductances can be measured indirectly with these two basic recording modes. While spike trains can be recorded with extracellular electrodes (see Chapter 4), subthreshold events in single neurons can only be recorded with intracellular electrodes. Intracellular recordings have been used for many applications: measuring membrane potential distribution in vivo (DeWeese et al., 2003), membrane potential correlations between neurons (Lampl et al., 1999), changes in effective membrane time constant with network activity (Pare et al., 1998; Leger et al., 2005), excitatory and inhibitory synaptic conductances in response to visual stimulation (Borg-Graham et al., 1998; Anderson et al., 2000; Monier et al., 2003), current–voltage relationships during spiking activity (Badel et al., 2008), the reproducibility of neuron responses (Mainen and Sejnowski, 1995), dendritic computation mechanisms (Stuart et al., 1999), gating mechanisms in thalamocortical circuits (Bal and McCormick, 1996), oscillations of membrane potential (Engel et al., 2001; Volgushev et al., 2002), stimulus-dependent modulation of the spike threshold (Azouz and Gray, 1999; Henze and Buzsaki, 2001; Wilent and Contreras, 2005), and many others.

In this chapter, we start with a brief historical overview of intracellular recording before describing the main techniques. We explain how to interpret intracellular measurements of potential, current and conductance and we emphasize the artifacts, uncertainties and limitations of these recording techniques. We do not provide practical details about the fabrication and use of electrodes and amplifiers,

Handbook of Neural Activity Measurement, ed. Romain Brette and Alain Destexhe. Published by Cambridge University Press. © Cambridge University Press 2012.

and we invite the interested reader to refer to specialized books such as Purves (1981), Sherman-Gold (1993) and Chapter 2 of this book.

3.1.1 A brief history of intracellular recording techniques

Many discoveries in neuroscience have been triggered by the development of new tools. Figure 3.1 shows a panel of historical electrophysiological techniques developed over the last two centuries.

Animal electricity Electrophysiology started at the end of the eighteenth century when Luigi Galvani observed that the frog muscle contracted when the leg nerve and the muscle were connected through a metal conductor (Galvani, 1791). He concluded that "animal electricity" was present in the nerve and muscle and that the contraction was induced by the flow of electricity through the conductor. That discovery led to the development of the electric battery by Alessandro Volta in 1800. In the next decades (around 1840), Carlo Matteucci observed an outward current flow between the axial cut of a nerve and the undamaged surface using a galvanometer, thus showing the existence of the resting membrane potential. Inspired by Matteucci's work, Emil du Bois-Reymond later discovered the action potential by observing that the outward current was temporarily reduced during electrically induced muscle contraction. His instrument is shown in Figure 3.1A; it consisted of two electrodes applied on the muscle and connected to a galvanometer.

The first electrophysiological instrument The galvanometer could not record the time course of action potentials, but Julius Bernstein designed an ingenious device called the "differential rheotome" (Figure 3.1B): one pin on a rotating wheel closes the stimulus circuit when it touches a copper wire, while two other pins on the opposite side of the wheel close the recording circuit (a galvanometer) when passing through a mercury surface. By adjusting the position of the pins, Bernstein was able to sample the electrical response at precise times after the stimulus, and he used his instrument to produce the first recording of an action potential in 1868 (Bernstein, 1868) (Figure 3.1C). Bernstein's differential rheotome can thus be considered as the first instrument in electrophysiology. He then developed an influential theory according to which the negative resting potential is due to the membrane being permeable to potassium ions and the action potential to a nonselective increase in membrane permeability (Bernstein, 1912). For many years, the application of external electrodes was the only available technique for measuring potentials and Bernstein's hypothesis remained unchallenged.

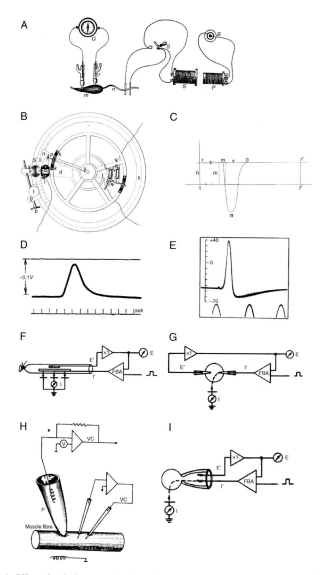

Figure 3.1 Historical electrophysiological recording techniques. A. Device used by Emil du Bois-Reymond to detect electrically triggered action potentials (APs) in a muscle (1840s). B. Bernstein's differential rheotome (1860s). The rotating wheel samples the electrical response of the muscle at a specific time following electrical stimulation. C. First (extracellular) recording of an AP, using the differential rheotome (Bernstein, 1868). D. First intracellular recording of an AP in a plant cell (Nitellia) by Umrath (1930). Each tick is a second (APs are much slower in plants than in animals). E. First intracellular recording of an AP in an animal cell, the giant squid axon, by Hodgkin and Huxley (1939). F. Voltage clamp setup in the squid axon, designed by Marmont and Cole in 1949 (illustration from Hille (2001)). G. Two-electrode voltage clamp with sharp intracellular electrodes. H. The patch clamp technique, designed by Neher and Sakmann (1976). The transmembrane current is recorded with the large patch electode while the membrane potential is held fixed with two conventional microelectrodes. I. Whole-cell patch clamp (1980, illustration from Hille (2001)). A gigaseal is formed by suction and the membrane is ruptured to give direct access to the intracellular potential.

The first intracellular recording In 1939, Cole and Curtis designed a clever experiment using extracellular electrodes on squid axons and found that the membrane resistance dropped during the action potential (Cole and Curtis, 1939), as predicted by Bernstein's theory. But around the same time, Hodgkin and Huxley managed to insert a glass microelectrode into a squid axon and made the first intracellular recording of an action potential in an animal cell (Hodgkin and Huxley, 1939) (Figure 3.1E) (the first intracellular recording of an action potential was in fact made by Umrath in 1930 in plant cells (Umrath, 1930); Figure 3.1D). They found that the intracellular membrane potential becomes significantly positive during the action potential, contradicting Bernstein's theory and leading to the finding that the action potential reflected a selective increase in sodium permeability (Hodgkin and Katz, 1949). Intracellular recordings in vertebrates were performed a few years later, in 1951 (Brock et al., 1952).

The voltage clamp Because of the explosive character of the action potential, measuring the membrane current–voltage properties that were responsible for the action potential required a new experimental device. At the end of the 1940s, Marmont and Cole designed an electronic feedback system that was able to "clamp" the membrane potential at a fixed value along the squid axon and to measure the feedback current: the voltage clamp (Cole, 1949; Marmont, 1949) (Figure 3.1F). They were shortly followed by Hodgkin and Huxley, who used that recording technique to develop their quantitative theory of the action potential based on voltage-dependent ionic currents (Hodgkin and Huxley, 1952), for which they were awarded the Nobel Prize for physiology or medicine in 1963. Mammalian cells, which are smaller than giant squid axons, became accessible to intracellular recordings with the development of pulled glass microelectrodes by Ling and Gerard in 1949 (Ling and Gerard, 1949). These electrodes have a sharp tip that can penetrate the membrane with little damage (hence the usual name "sharp electrodes") and are still used today, with minor modifications (Figure 3.1G).

The patch clamp Voltage clamping required two electrodes: one for injecting the current and another one for monitoring the voltage, which was technically difficult in small cells. In the 1970s, Brennecke and Lindemann developed a system to alternate current injection and voltage recording on the same electrode (Brennecke and Lindemann, 1971), now called the "discontinuous current clamp," and they showed that it could be used to perform a single-electrode voltage clamp (now called "discontinuous voltage clamp"). Around the same time, Neher and Sakmann developed a technique to record currents flowing through single ionic channels, by applying the tip of a glass pipette on the surface of the membrane: the patch clamp (Neher and Sakmann, 1976) (Figure 3.1H). Traditional microelectrodes were still required for voltage clamping the membrane (the two electrodes on the right in Figure 3.1H) and recording quality was limited by the background noise due to the seal between the patch and the electrode. The technique was refined in 1980 by Sigworth and Neher with the introduction of the "gigaseal" (Sigworth and Neher, 1980), which is a tight contact between patch and electrode with very high resistance (10–100 GΩ), allowing voltage clamping with the same electrode and low noise recordings. Several variations of the patch clamp method were then developed, in particular the "whole-cell" recording, in which the membrane is ruptured to make intracellular recordings in a similar way as with conventional sharp microelectrodes, but with lower access resistance and noise level (Figure 3.1I). Neher and Sakmann were awarded the Nobel Prize in 1991 for their discoveries.

3.1.2 Experimental setups

A typical setup for intracellular recording consists of a reference electrode (immersed in the bath for slice recordings or possibly in the musculature for recordings in vivo) and an intracellular microelectrode, both connected to an electronic amplifier (Figure 3.2A). The role of the amplifier is to measure the potential of the microelectrode (relative to the reference electrode) and/or to inject currents, while matching input/output impedances (since neuronal signals are typically very small). In some cases, one intracellular electrode is used to monitor the potential and another one to inject currents into the neuron (double-electrode configuration). The amplifier is connected to various electronic devices (e.g. an oscilloscope) and in general to a computer which records the measurements and possibly sends commands (e.g. current injection).

3.1.2.1 Electrodes

Intracellular electrodes are thin glass pipettes filled with an electrolyte solution (usually KCl). The tip of the pipette is in continuity with the inside of the cell, while the other end contains a metal wire (usually silver, coated with a composite of silver and silver chloride) connected to the amplifier (Figure 3.2B,C). The electrode per se is in fact the junction between the electrolyte and the wire, where electrons are exchanged for ions through the following reversible reaction:

$$Cl^- + Ag \rightleftharpoons AgCl + e^-.$$

There are two types of intracellular electrodes: sharp electrodes and patch electrodes (Figure 3.2B). *Sharp electrodes* (standard intracellular microelectrodes introduced by Ling and Gerard (1949)) are made from pulling a glass capillary tube (diameter $\approx$ 1 mm), resulting in a very fine tip (0.01 – 0.1 μm) which can

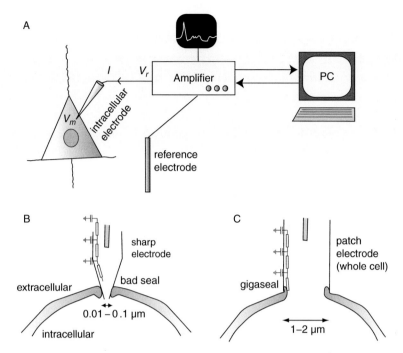

Figure 3.2 Experimental setups. A. An intracellular electrode is impaled into the cell and connected to an amplifier, which compares its potential with that of a reference electrode. The amplifier output is typically connected to an oscilloscope (top) and computer (right). B. Sharp electrodes have a small tip (equivalent electrical circuit superimposed on the left side of the electrode). C. Patch electrodes have a larger tip, with a better seal with the membrane.

penetrate the membrane of the cell (Figure 3.2B). *Patch electrodes* were initially developed by Neher and Sakmann (1976) for recording currents through small membrane patches containing few channels (hence the name). They are glass tubes with a wide round tip (1–2 μm) which are applied on the surface of the membrane (Figure 3.2C). A small suction creates a high-resistance seal (>10 GΩ) between the electrode tip and the membrane. In that original configuration, one can record the current flowing through a single ionic channel. If pressure is applied through the electrode, the membrane is ruptured and the electrode accesses the inside of the cell: this is called the *whole cell* configuration. In this chapter, we will not describe single channel recording but only intracellular recording – i.e. the whole cell configuration. We suggest the interested reader refer to Sakmann and Neher (1995) for detailed information about single channel recording.

The differences between sharp and patch recordings are summarized in Table 3.1, and result essentially from the difference in tip geometry (thin versus wide) and in seal quality (bad seal versus good seal). Electrodes have a resistance,

Table 3.1 *Properties of intracellular electrodes: sharp microelectrodes
and patch electrodes (whole-cell configuration). We have highlighted the
issues raised by each type of electrode: sharp electrodes have high
resistance, variable tip potential (hard to predict), higher noise, often
non-linear behavior and the seal with the membrane is bad (introducing
an additional leak current); patch electrodes have high resistance in
vivo and in dendrites, they replace the contents of the cell (dialysis) and
they are technically more difficult to use (especially in adult animals).*

	Sharp	Patch
Tip geometry	thin	wide
Resistance	**high** (25–125 MΩ)	low in vitro (< 20 MΩ), **higher in vivo** (up to 200 MΩ)
Tip potential	**variable**	negligible
Noise	**high**	low
Seal	**bad**	good
Non-linearity	**often non-linear**	generally linear
Dialysis	no	**yes**
Difficulty	easy	**harder** (especially adults)

which is the sum of the resistance of the electrolyte solution and of the junction of
the cell and electrode. Because sharp electrodes have a thin tip, they usually have
higher resistance than patch electrodes, which have a wider tip (although they have
a thinner tip and higher resistance when used on thin processes such as dendrites).
Junction potentials appear in both types of electrodes and produce offsets in the
potential measurement (see Section 3.2.2.1). Sharp electrodes have an additional
type of junction potential named *tip potential*, which is hard to predict. A thinner
tip also implies a higher level of noise and more non-linearities (Purves, 1981).
In addition, the seal between a sharp electrode and the membrane is bad, which
introduces an additional leak current. On the other hand, patch electrodes are technically more difficult to use, especially for adult animals in vivo. More importantly,
because the tip is wide, the electrode *dialyses* the cell, that is, the electrolyte solution diffuses into the cell and slowly replaces the soluble contents of the cell's
interior.

3.1.2.2 Amplifiers

The role of an electrophysiological amplifier is to measure currents or potentials
and to inject currents through the electrode. It includes a number of circuits to
minimize noise and various artifacts. In particular, all amplifiers include a circuit to compensate for the input capacitance (capacitance neutralization) and for

the electrode resistance (electrode compensation or series resistance compensation circuits). Electrophysiological amplifiers have two recording modes: current clamp and voltage clamp. In current clamp mode, the current flowing from the amplifier is held fixed; in voltage clamp mode, the potential at the amplifier input is held fixed (using a feedback circuit). When two intracellular electrodes are used (in addition to the reference electrode), one electrode injects a current and the other one measures the potential. When only one intracellular electrode is used, the injected current biases the measured potential, as explained below. This is compensated either by modifying the measured potential (current clamp) or modifying the voltage command (voltage clamp).

We chose to divide this chapter in sections corresponding to the quantity being measured: voltage, current or conductance. Recording the membrane potential is done in current clamp mode, currents are recorded in voltage clamp mode, and conductance recordings use various indirect techniques. Many figures in this chapter are based on numerical simulations which are explained in more detail at the end of the chapter.

3.2 Recording the membrane potential

3.2.1 The ideal current clamp

In an ideal current clamp recording, a current I is injected into the cell through an electrode with negligible resistance, while the membrane potential is recorded (Figure 3.3A). The membrane potential (voltage difference between the inside and the outside of the cell) is measured by comparing the potential at the amplifier end of the intracellular electrode with the potential of a reference electrode (outside of the cell). If the intracellular electrode has zero resistance and junction potentials are neglected, then the measured potential V_r equals the membrane potential V_m. The response of an isopotential neuron to an ideal current clamp injection $I(t)$ is described by the following differential equation:

$$C\frac{dV_m}{dt} = \sum_{ionic\ currents} I_{ionic\ current} + I(t)$$

where C is the total membrane capacitance of the neuron (Figure 3.3B). Measuring the membrane potential without injecting current (i.e. spontaneous activity) is also called a current clamp recording – referring to the fact that a null electrode current is imposed.

Real current clamp recordings differ from this idealized description in a number of ways, even when only spontaneous activity is recorded (no current injection): junction potentials develop at the interface between the electrolyte and the intracellular medium, the electrode is non-ideal and filters the signals (both the measured

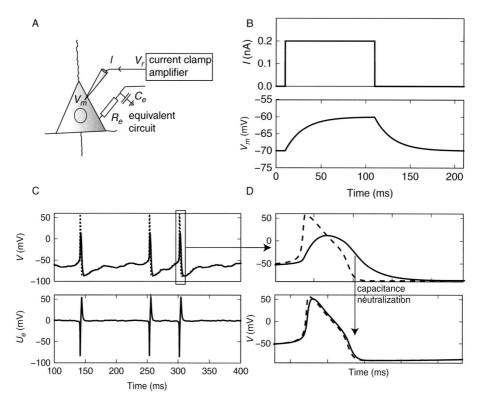

Figure 3.3 Current clamp recording (numerical simulations). A. Experimental setup: a current clamp amplifier (voltage follower) records the electrode potential (V_r) while injecting a current I through the electrode. Ideally, the recorded potential equals the membrane potential V_m but the electrode resistance (R_e) and capacitance (C_e) introduce artifacts. B. Ideal recorded response to a current pulse, when the electrode resistance is negligible (top, injected current; bottom recorded potential). C. Recording spontaneous activity with a non-ideal electrode: spikes are low pass filtered (top, dashed line, membrane potential; solid line, recording) because a voltage drop develops through the electrode during those fast events (bottom). D. Zoom on an action potential (top). The filtering is reduced with capacitance neutralization (effectively reducing C_e).

potential and the injected current), sharp electrodes damage the membrane and patch electrodes affect the ionic composition of the intracellular medium. In addition, when current is injected through an electrode with a non-zero resistance, a voltage drop appears between the two ends of the electrode $U_e = V_r - V_m$. That voltage drop must be canceled, or a second intracellular electrode must be used to measure the membrane potential. We first describe the artifacts that appear when no current is injected, i.e. when measuring spontaneous activity, then we describe the issues arising from current injection through the electrode.

3.2.2 Measuring spontaneous activity

3.2.2.1 Junction potentials

Voltage offsets of different origins arise in intracellular recordings, mostly amplifier input offsets and junction potentials, which occur wherever dissimilar conductors are in contact. The largest junction potentials occur at the liquid–metal junction formed where the wire from the amplifier input contacts the electrolyte in the micropipette and at the liquid–liquid junction formed at the tip of the micropipette, called the *liquid junction potential* (LJP). A LJP develops when two solutions of different concentrations are in contact: the more concentrated solution diffuses into the less concentrated one, and a potential develops when anions and cations diffuse at different rates. To suppress this unwanted bias, one usually starts by *zeroing* the measured potential in the bath (outside the cell, before impalement), i.e. a DC voltage offset is added so as to compensate for all voltage offsets. When the electrode accesses the interior of the cell, the LJP changes because the solution around the electrode tip changes, but all other offsets are unchanged. Thus the measured potential is $V_m + V_{LJP}^{cell} - V_{LJP}^{bath}$, where V_{LJP}^{cell} is the LJP between the cell and the electrode solution and V_{LJP}^{bath} is the LJP between the bath and the electrode solution. Because the concentrations of the bath and electrode solutions are known, V_{bath} can be calculated (using the Henderson equation, see e.g. Sakmann and Neher (1995)). With patch electrodes, the LJP between the cell and the electrode vanishes after some time and can thus be neglected. With sharp electrodes, it is very difficult to compensate for the junction potentials because, in addition to the liquid junction potential, a tip potential develops at the cell–electrode interface because the electrode tip is very thin (Purves, 1981). This tip potential is unfortunately difficult to predict with precision.

3.2.2.2 Damage induced by the electrode

Sharp microelectrodes have a very fine tip ($0.01-0.1\,\mu m$) which perforates the membrane of the cell. Thus, the membrane is damaged when the electrode impales the neuron. In particular, a leak appears because of the bad quality of the seal between the electrode and the membrane. It can be modeled as an outward current $I_{leak} = -g V_m$, where g is the conductance of that leak. The total conductance of the neuron is thus increased when the electrode perforates the membrane, so that the effective membrane time constant $\tau_m = C/g_{total}$ is decreased. This effect explains why the membrane time constant is larger and the resting potential is lower when measured with patch electrodes (whole-cell configuration) than when measured with sharp electrodes (Staley et al., 1992).

Patch electrodes do not suffer from the same problem because the electrode tip is sealed to the membrane with a "gigaseal" (resistance 10–100 GΩ). However, because the tip is wide ($1-2\,\mu$m) and the volume of the electrode is much larger than the volume of the cell, the electrolyte solution diffuses into the cell and slowly replaces the soluble contents of the cell's interior, which can alter the properties of the cell over time (>10 minutes). This phenomenon is referred to as the electrode *dialyzing* the cell. To avoid dialysis, a variant of the whole-cell configuration has been developed: the perforated patch clamp. In this configuration, instead of rupturing the membrane, the experimenter adds an antibiotic to the electrode solution, which makes small perforations in the membrane patch at the tip of the electrode. That technique prevents the dialysis but it also increases the access resistance and the recording noise.

3.2.2.3 Electrode filtering

Real electrodes have a non-null resistance, which is the sum of the resistance of the electrolyte solution and of the junction of the cell and electrode. The electrode resistance is thus more precisely referred to as the *access resistance*. If the electrode were a pure resistor, it would not affect the measurement (when no current is injected) since no current would pass through it, so that $V_r = V_m$. Unfortunately, the electrode and amplifier input have a capacitance: the *input capacitance*, on the amplifier side, and a distributed wall capacitance along the glass tube of the electrode. As a result, current can flow through the electrode and bias the potential measurement: $V_r \neq V_m$. As a first approximation, the electrode can be modeled as a resistor R_e and input capacitance C_e on the amplifier side. It follows that the electrode acts as a first-order low pass filter with cut-off frequency $f_c = 1/(2\pi R_e C_e)$. The quantity $\tau_e = R_e C_e$ is the *electrode time constant*. Electrode filtering has a very significant effect on the recording of fast phenomena such as action potentials, which appear wider and smaller than they are in reality at the recording site, as shown in Figure 3.3C,D. Thus, reliable measures of action potential width and height depend crucially on the correction of the electrode capacitance.

To reduce this problem, modern electrophysiological amplifiers include a *capacitance neutralization* circuit, which compensates for the input capacitance by an electronic feedback circuit. The current flowing through the input capacitance is $C_i dV_r/dt$; capacitance neutralization consists in inserting a "negative capacitance," that is, adding the opposite current $-C_i dV_r/dt$ to cancel the capacitive current. Since the precise value of the capacitance is unknown, it is manually adjusted by turning a knob on the amplifier, so that the actual compensating current is $-C^* dV_r/dt$. When $C^* > C_i$, the circuit becomes unstable, which can damage the cell. Tuning the capacitance neutralization circuit therefore requires careful adjustment. In reality, the capacitance can never be totally compensated because this feedback circuit can only cancel the capacitive current at the amplifier end of the electrode, and not the distributed capacitance along the glass tube of the electrode. Therefore when the input capacitance is completely canceled, further increasing the capacitance neutralization results in instability and the total capacitance is never completely suppressed at the optimal point.

This circuit reduces the effective electrode time constant and increases the cut-off frequency of the filtering, but at the same time it increases the level of noise in the recording (which appears very clearly on the oscillope as traces become thicker), for two reasons: electrode filtering masks some of the recording noise, and the capacitance neutralization circuit itself amplifies noise because it is a feedback circuit.

3.2.3 *Measuring the response to an injected current*

In many cases, the response of the membrane potential to an injected current is measured. This is obviously standard for in vitro experiments when one wants to measure neuronal properties, such as the properties of ionic channels, but also in vivo, for example to evaluate the effective membrane time constant of a neuron during spontaneous activity by observing the response to current pulses (Pare et al., 1998; Leger et al., 2005). In those cases, the main issue is that when a current is passed through an electrode with non-zero resistance, a voltage drop U_e appears between the two ends of the electrode, so that $V_r = V_m + U_e$. For a constant current I, this voltage drop is $U_e = R_e I$, where R_e is the electrode resistance (Figure 3.4). The electrode resistance is inversely correlated with the diameter of the electrode tip (Purves, 1981), so that sharp electrodes typically have high resistance (about 100 MΩ). Patch electrodes have a lower resistance because their tip is wider, although higher resistance electrodes must be used in vivo and when recording in thin processes (dendrites, axons). The electrode resistance depends partially on the interface between the electrode and the cell and thus cannot be reliably estimated before impalement. Besides, it often varies during the course of an experiment. A secondary issue, which is partially solved by capacitance neutralization, is that the injected current is filtered by the electrode.

One way of solving the electrode resistance problem is to use a second, non-injecting, intracellular electrode to measure the membrane potential (although the injected current remains filtered). However, this is technically difficult, especially in vivo, and it also increases the cell damage. The alternative solution consists in correcting the measurement bias induced by the electrode. There are essentially three methods available to suppress the electrode voltage during current injection: bridge balance, discontinuous current clamp and active electrode compensation.

3.2.3.1 *Bridge balance*

As a first approximation, the electrode can be modeled as a pure resistor with resistance R_e, so that the voltage across the electrode during current injection is

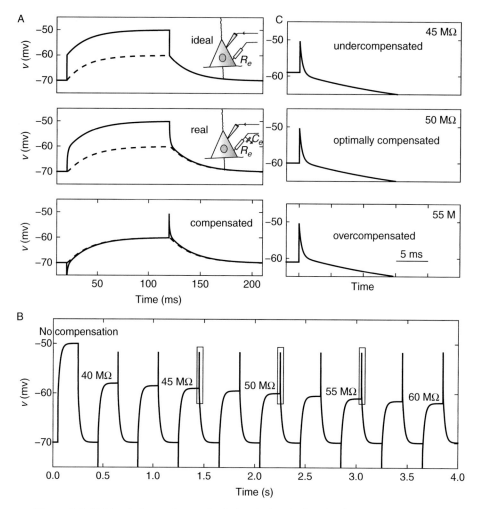

Figure 3.4 Bridge balance (numerical simulations). A. Membrane potential (V_m, dashed line) and current clamp recording (V_r, solid line) in response to a current pulse. Top: with a purely resistive electrode (resistance R_e) the recorded potential is $V_m + R_e I$, with a discontinuity at pulse onset. Middle: a real electrode has a capacitance (C_e), which smoothes the onset. Bottom: bridge balance consists in subtracting $R_e I$, which produces discontinuities at pulse onset (capacitive transients). B. Manual tuning of bridge balance. The estimated resistance is progressively increased until the trace "looks right" (real resistance $R_e = 50\,M\Omega$). The transients in boxes are magnified in C. C. The shape of capacitive transients is used to determine the optimal bridge setting.

$U_e = R_e I$ and the recorded potential is $V_r = V_m + R_e I$ (Figure 3.4A). The membrane potential can thus be recovered from the raw recording by subtracting U_e: $V_m = V_r - R_e I$. This method is named *bridge balance* or *bridge compensation*, in reference to an electrical circuit called the Wheatstone bridge, which was used in old amplifiers to perform that subtraction. Modern electrophysiological amplifiers now use operational amplifiers to perform this operation, but the name has remained. Since the electrode resistance R_e is unknown, it is estimated with an adjustable knob on the amplifier, which is tuned manually by the experimenter. The classical method to determine that resistance is to inject a current pulse into the cell and to ajust the bridge resistance until the recorded potential response "looks correct" in the eye of the experimenter (Figure 3.4B,C). That adjustment is easy if the electrode is indeed a pure resistor: in that case, the response of the electrode to a square current pulse is also a square pulse (with height $R_e I$), so that any mismatch in estimated resistance results in a discontinuity (a vertical line on the oscilloscope) at the onset of the pulse. Unfortunately, even when the capacitance neutralization circuit is used, the electrode capacitance is never completely canceled and the adjustment of the bridge resistance is more difficult. Because the electrode time constant is non null, the response of the electrode to the onset of a current pulse is approximately exponential:

$$U_e(t) = (1 - e^{-t/\tau_e})R_e I$$

where $\tau_e = R_e C_e$ (C_e is the uncompensated electrode capacitance). Bridge balance amounts to subtracting a square pulse from width τ_e, so that the compensated bridge recording is:

$$V_{bridge} = V_m + (1 - e^{-t/\tau_e})R_e I - R_e I = V_m - e^{-t/\tau_e}R_e I.$$

Thus, a negative transient appears at the onset of the pulse, with height $R_e I$ and width τ_e. Since this transient is due to the non-zero capacitance of the electrode, it is often called a "capacitive transient." Capacitive transients do not constitute a major problem if only constant currents are injected, but they can completely obscure the measured signal when a fast time-varying current is injected.

The finite capacitance of the electrode poses another problem for bridge balance, for both constant and time-varying current injection: it makes the estimation of the electrode resistance more difficult. Indeed, the adjustment of the bridge resistance relies on the discontinuity of the electrode response, which is unambiguous only when $\tau_e \ll \tau_m$ (τ_m is the membrane time constant). To our knowledge, it is not precisely known what visual cues electrophysiological experimenters implicitly use when manually balancing the bridge in face of that ambiguity. However, it seems that manually estimated resistances agree approximately with those obtained from a simple exponential fitting procedure described in Anderson et al. (2000), where the recorded response is modeled as

$$V_r(t) = V_0 + (1 - e^{-t/\tau_m})R_m I + (1 - e^{-t/\tau_e})R_e I$$

where V_0 is the resting potential and R_m is the neuron resistance. This formula is the superposition of the cell response to a direct injection of the current and of the response of the electrode alone (i.e. in the bath). Fitting the recording with this expression provides an estimated value of the electrode resistance R_e. This expression is however only an approximation, even if both the membrane and the electrode are RC circuits, because the injected current is filtered before entering the cell and current can also flow from the neuron through the electrode. If the membrane and electrode are modeled as RC circuits, then the response is indeed biexponential but with different coefficients, as described in (de Sa and MacKay, 2001):

$$V_r(t) = V_0 + (ae^{-\mu_1 t} + be^{-\mu_2 t} + c)I$$

where the coefficients are related to R_m, R_e, τ_m and τ_e by complex formulae. In general the electrode resistance R_e is not equal to the factor in front of the fastest exponential. The relationship can be inverted and gives:

$$C_e = \frac{1}{\mu_2 c - (\mu_1 - \mu_2)a}$$

$$R_e = \frac{1}{C_e(\mu_1 + \mu_2) - cC_e^2\mu_1\mu_2}$$

$$R_m = c - R_e$$

$$C_m = \frac{1}{\mu_1\mu_2 C_e R_e R_m}.$$

Thus, fitting the recorded response to a pulse to a biexponential function and using the formulae above provides a better way to estimate the electrode resistance for bridge balance. However, the method does not work so well in practice because once the input capacitance has been maximally compensated with the capacitance neutralization circuit of the amplifier, the electrode response is generally not exponential anymore (essentially because the remaining capacitance is distributed along the electrode).

Another way to estimate the electrode resistance is to take advantage of the stereotypical nature of action potentials (as in Anderson et al. (2000)). If the peak voltage of action potentials (APs) is constant, then any measured variability in AP height should be attributed to a mismatch between the bridge and the electrode resistance. Indeed, if V_{peak} is the peak value of APs, then the measured value when current is injected through the electrode should be $V_{bridge} = V_{peak} + \Delta R_e I$, where $\Delta R_e = R_e - R_{bridge}$ is the mismatch between the electrode and bridge resistances. Therefore, the slope of the linear regression between measured values of V_{bridge} and I is the difference between electrode and bridge resistances, i.e. the error in bridge balance (Figure 3.5). However, this method should be used with caution and only as

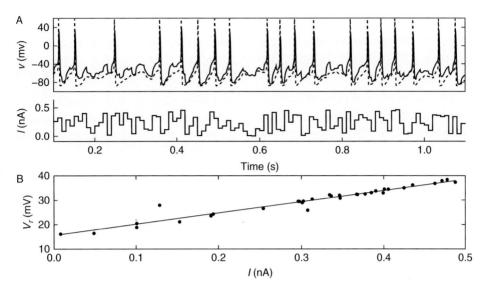

Figure 3.5 Estimating the electrode resistance (R_e) from spikes (numerical simulations). A. Membrane potential (V_m, dashed line) and (uncompensated) current clamp recording (V_r, solid line) in response to a random current injection (bottom). B. The recorded voltage at the peak of action potentials is approximately $V_r = V_m + R_e I$ (dots), where V_m is assumed constant. The slope of the I–V_r relationship is found with linear regression (line) and provides an estimate of R_e: 45 MΩ instead of 50 MΩ (real value).

a check, because the shape of APs can in fact vary as a function of the stimulation: in cortical neurons, it has been observed that AP height is inversely correlated with AP initiation threshold, which is inversely correlated with the slope of the depolarization preceding the AP (Azouz and Gray, 1999; Henze and Buzsaki, 2001; de Polavieja et al., 2005; Wilent and Contreras, 2005). This property is probably due to the inactivation of sodium channels or to the activation of potassium channels. Thus, injected current and AP height should be positively correlated, which restricts the applicability of this method.

It should thus be kept in mind that in general bridge balance is not straightforward and the resulting compensation is imperfect. In addition, the access resistance can change over time, especially in technically difficult situations such as whole-cell recordings in vivo, which can compromise the bridge balance. Finally, sharp electrodes are unfortunately not always linear. Non-linearities arise from the dissimilarity of solutions at the tip of the electrode (Purves, 1981). The amount of non-linearity is inversely correlated with the tip diameter, which is inversely correlated with resistance, so that higher resistance electrodes tend to be more non-linear. Non-linearities can be minimized by choosing an electrode solution that matches the composition of the intracellular medium.

3.2.3.2 Discontinuous current clamp

Before patch clamp recordings were developed by Neher and Sakmann (Neher and Sakmann, 1976), high resistance sharp microelectrodes were the only tool available for intracellular recording. In the early 1970s, Brennecke and Lindemann introduced a new technique (Brennecke and Lindemann, 1971) to solve the problem of the electrode resistance in current clamp mode, later adapted for voltage clamping (Brennecke and Lindemann, 1974). The technique was called *chopped current clamp* and later *discontinuous current clamp* (DCC). The idea is to alternate current passing and voltage measurement so that no current flows through the electrode when the potential is measured (Figure 3.6A). The alternation rate is determined by the electrode time constant.

In DCC mode, the current command I_{cmd} is sampled at regular intervals Δ. Current is injected through the electrode only during the initial part of each interval. The proportion of time during which current is passed is called the *duty cycle* and is usually $1/3$. During that time, the sampled current is injected through the electrode with the appropriate scaling, so as to conserve the total charge (i.e. $I = 3I_{cmd}$ if the duty cycle is $1/3$). The potential is sampled at the end of each interval, when no current is passed. Since no current is passed during the last $2/3$ of the interval, the electrode voltage $U_e(t)$ has decayed approximately as $\exp(-2\Delta/3\tau_e)$, which is small if the sampling interval Δ is large compared to the electrode time constant τ_e. In that case, the electrode voltage U_e has vanished when the potential V_r is sampled at the end the interval, so that $V_r \approx V_m$. However, the membrane potential V_m also decays when no current is passed, so that the sampling interval should be short compared to the membrane time constant τ_m. Therefore the sampling interval Δ should be such that $\tau_e \ll \Delta \ll \tau_m$, and a reasonable trade-off can be found if τ_e is at least two orders of magnitude shorter than τ_m (Finkel and Redman, 1984).

The optimal sampling frequency Suppose we want to measure the response of the membrane potential to a constant injected current I, which should ideally be $V_0 + R_m I$ in the stationary regime, where V_0 is the resting potential and R_m is the membrane resistance. If the sampling frequency is very high, then the sampled potential includes a large residual electrode component, so that the membrane potential is overestimated. As the sampling frequency is made increasingly lower, then the sampled potential tends to the resting potential, i.e. it is underestimated. Thus, there is an intermediate frequency at which the sampled stationary potential is exactly the ideal potential $V_0 + R_m I$. Note however that the real membrane potential is not constantly $V_0 + R_m I$ but it is periodically varying at the DCC sampling period.

But how can that optimal frequency be determined? The standard experimental method is empirical and based on observing the continuous electrode potential

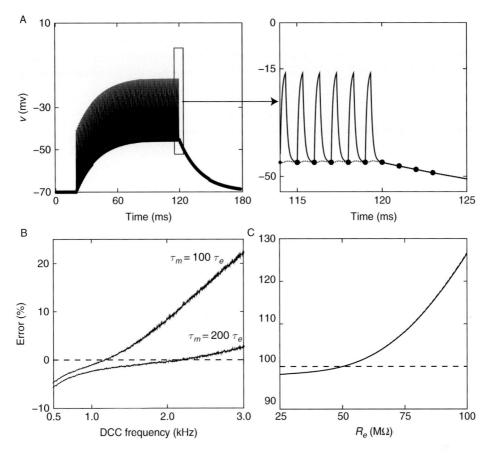

Figure 3.6 Discontinuous current clamp (DCC, numerical simulations). A. Response to a current pulse injection in DCC mode (solid, electrode potential; dashed, membrane potential; dots, sampled recording). Current injection and potential recording are alternated. B. Error in membrane potential as a function of DCC frequency, for a fast electrode ($\tau_m = 200\tau_e$) and for a slower electrode ($\tau_m = 100\tau_e$). A constant current is injected and the depolarization is measured. The measurement is less reliable for the slower electrode. C. Error in membrane potential as a function of electrode resistance (R_e) for the second electrode ($\tau_m = 100\tau_e$), with fixed DCC frequency (optimal frequency for $R_e = 50$ MΩ). This plot shows the effect of a change in electrode resistance during the course of an experiment.

on an oscilloscope synchronized to the DCC sampling clock, i.e. the electrophysiologist observes the electrode potential in response to the injected current at the time scale of one DCC period (a fraction of millisecond). The sampling frequency is the highest frequency such that the observed response at the oscilloscope looks flat, meaning that the electrode response has settled to a stationary value at the end of the duty cycle. One usually makes sure that the DCC setting matches bridge compensated recordings. In other words, the frequency tuning technique consists in adjusting the duty cycle to a few times the electrode time constant τ_e. Given that the duty cycle is $1/3$ the sampling period, the sampling period is set at about $10\tau_e$ with that standard technique. Implicitly, it is assumed that the membrane potential V_m does not change significantly between the two endpoints of the sampling interval, which might be so only if the membrane time constant τ_m is more than several tens of electrode time constants τ_e.

In fact, for any electrode time constant τ_e, there is always an optimal sampling frequency such that the measured stationary voltage is precisely $V_0 + R_m I$. Indeed, that voltage increases continuously with the frequency, is an overestimation at high frequencies and an underestimation at low frequencies. This simple fact would suggest that the use of the DCC technique is not restricted to short electrode time constants. However, there are two practical problems.

- Determining the optimal frequency is not trivial. If one plots the voltage error as a function of the DCC frequency (Figure 3.6B), the optimal frequency is near the inflexion point of that curve when the ratio τ_m/τ_e is large (>100), and there is a broad plateau so that a small error in frequency results in a small estimation error. Choosing the inflexion point as the optimal frequency is probably close to the visual procedure with the synchronized oscilloscope that we described above. However, when the ratio τ_m/τ_e is not so large, the frequency–voltage error curve does not have a broad plateau and the optimal frequency is higher than the inflexion point. Thus in that case there is no practical way to determine the optimal frequency and a small error in frequency results in a rather large voltage estimation error.
- The optimal frequency depends on both electrode properties and membrane properties. In particular, setting the optimal frequency at rest can lead to estimation errors during neuronal activity, if the membrane time constant changes.

Noise and artifacts In addition to the problem of setting the sampling frequency, DCC recordings are noisier than bridge recordings for two reasons: sampling the voltage results in aliasing noise (frequencies higher than the sampling frequency add noise at lower frequencies), and capacitance neutralization has to be used at its maximum setting in order to shorten the electrode time constant, which also increases the noise because it is a feedback circuit.

Another artifact is that the input current is distorted. Indeed, during one sampling period, the injected current is three times the command current during a third of the sampling interval. Thus, the observed neural response is the response to the command signal with additional high frequencies (harmonics of the DCC frequency). This can potentially lead to artifacts because of the non-linear nature of neurons;

for example, additional high frequencies in the input signal may trigger additional action potentials in the neuron.

Electrode non-linearities and resistance instabilities The DCC technique was introduced historically to solve the problem of resistance instability with micro-electrodes. Indeed, if the membrane and electrode time constants are well separated, then the measured voltage at the end of the sampling period contains a very small contribution from the electrode, so that changes in electrode properties have minor effects on the measured voltage. Again, this desirable property is conditional on the fact of the ratio τ_m/τ_e being very large. With reasonable ratios ($\tau_m/\tau_e = 100$), the optimal DCC frequency corresponds to a point when the electrode voltage does not completely vanish at recording times, in order to compensate for the decay in the membrane potential. Thus, an increase in electrode resistance results in overestimation of the membrane potential, but the error remains smaller than with bridge balance (Figure 3.6C).

Electrode non-linearities that arise with sharp microelectrodes are typically described as resistance changes as a function of the injected current, which suggests that DCC should reduce the impact of those non-linearities under the same assumptions (large ratio τ_m/τ_e). However, the extent of this reduction is not so clear because some non-linearities (type I non-linearities) are associated with a maximum outward current, which cannot be corrected by the DCC technique (Purves, 1981). In addition, electrode non-linearities are slow processes while DCC acts on a fast time-scale.

3.2.3.3 Active electrode compensation

Active electrode compensation is a recently introduced technique to compensate for the electrode voltage during single-electrode recordings (Brette et al., 2008). As for the classic bridge balance method, it consists in estimating the voltage drop across the electrode during current injection. The main difference is the electrode model: instead of seeing the electrode as a resistor, it is modeled as an arbitrarily complex circuit of resistances and capacitances, which can be represented by a linear time-invariant filter, i.e. the response of the electrode to a current $I_e(t)$ is expressed as a convolution:

$$U_e(t) = (K_e * I_e)(t) = \int_0^{+\infty} K_e(s)I_e(t-s)ds$$

where K_e is named the *electrode kernel*. In practice, recordings are digitized and the formula reads:

$$U_e(n) = \sum_0^{+\infty} K_e(p)I_e(n-p).$$

The technique consists of (1) identifying the electrode kernel by observing the response to a known noisy current and (2) estimating the electrode voltage during current clamp injection and subtracting it from the measured potential (Figure 3.7). The main difficulty is that the electrode kernel can only be estimated when the electrode impales the neuron (because electrode properties change after impalement). Thus the estimation algorithm consists in (1) finding the kernel of the full system neuron + electrode (+ amplifier) from the voltage response to a known input current and (2) extracting the electrode kernel from the full kernel.

By using small white noise currents, the voltage response of the system is approximately linear and reads in the digital domain:

$$V_n = V_0 + \sum_{0}^{+\infty} K_p I_{n-p}$$

where V_0 is the resting potential and K is the unknown kernel of the full system (neuron + electrode). Assuming Gaussian noise, the best estimation of K and V_0 is found by solving the linear least-squares problem, as explained in Brette et al. (2008, 2009). The difficult part, which involves more assumptions, is to extract the electrode kernel K_e from the full kernel K. It is useful to observe that in a linear system, the kernel or impulse response K completely characterizes its responses, so that K is all the information that one can ever obtain about the system using a single electrode. Therefore, without further assumption, there is no way to separate the membrane and the electrode contributions. The full kernel K can be approximated as $K = K_m + K_e$, where K_m is the membrane kernel, but this is a poor approximation because the injected current is filtered by the electrode before entering the neuron, so that a better approximation is:

$$K = K_m * \frac{K_e}{R_e} + K_e \qquad (3.1)$$

where $R_e = \int K_e$ is the electrode resistance. If K_m is known, that convolution equation can be solved by various methods, for example by using the Z-transform or by expressing it as a linear system where the unknowns are the vector components of K_e. Unfortunately K_m is unknown so that further assumptions are required. In the AEC technique, two assumptions are then made.

- The electrode is faster than the membrane, i.e. its electrode kernel vanishes before the membrane kernel.
- The membrane kernel is that of a first-order low pass filter (an exponential function), so that it can be parameterized by the (unknown) membrane resistance R_m and the membrane time constant τ_m.

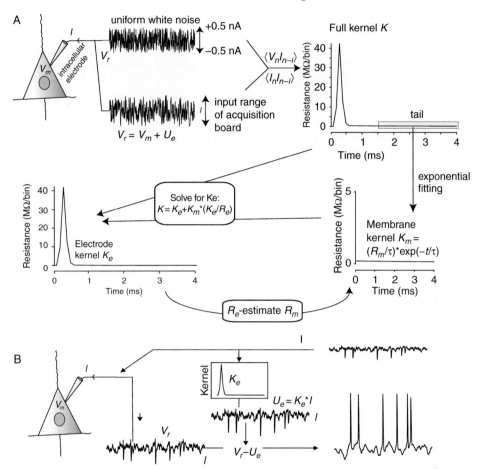

Figure 3.7 Active electrode compensation (AEC). A. A noisy current is injected into the neuron and the total response $V_r = V_m + U_e$ (U_e is the voltage drop through the electrode) is recorded. The cross-correlation between the input current and the output voltage and the autocorrelation of the current give the kernel K (or impulse response) of the neuronal membrane + electrode system (full kernel K, right). The tail of the kernel is fit to an exponential function, which gives a first estimation of the membrane kernel K_m (note, the resistance of each bin is very small since the kernel is distributed over a long duration). The electrode kernel K_e is recovered from K and K_m by solving the equation $K = K_e + K_m * (K_e/R_e)$ (convolution). The process is iterated several times to obtain a better estimation of the membrane kernel. B. Once the electrode kernel has been calibrated, it is then used in real time for electrode compensation: the injected current is convolved with the electrode kernel to provide the electrode response U_e to this current. U_e is then subtracted from the total recorded voltage V_r to yield the estimated V_m.

For every value of (R_m, τ_m) there is a solution to the convolution equation (3.1). The first assumption means that we are looking for the solution K_e with the smallest tail (i.e. minimizing $\int_T^{+\infty} K_e^2$, where T is the expected duration of the electrode kernel), which involves an optimization algorithm.

There are three main difficulties and limitations with the AEC technique.

Ratio of time constants Equation (3.1) is a good approximation when the electrode time constant is significantly shorter than the membrane time constant. The quality of electrode kernel estimation degrades with larger ratios τ_e/τ_m: as a rule of thumb, the error in signal reconstruction grows as τ_e/τ_m. Empirically, the method is useful when the electrode time constant is about one order of magnitude shorter than the membrane time constant, which is better than with DCC (two orders of magnitude). Bridge balance also requires a good separation of time constants in order to estimate the bridge resistance. As we mentioned previously, because the full kernel K is the only information available in single-electrode recordings, there is no way to distinguish between electrode and membrane kernels if they act on the same time-scale.

Dendrites To extract the electrode kernel from the full kernel, an assumption (i.e. a model) has to be made about the membrane kernel K_m. In the AEC technique, K_m is modeled as a single exponential function, which amounts to seeing the neuron as a sphere with no dendrites. When the dendritic tree is taken into account, the kernel includes additional faster exponential functions, some of which can have similar time constants to the electrode kernel. In this case, these additional functions due to the dendrites are mistakenly included in the estimated electrode kernel, leading to an overestimation of the electrode resistance R_e. The magnitude of the resulting error depends on the geometry of the cell and the recording point (soma or dendrite). In somatic recordings of cortical pyramidal cells, that error was found to be small (Brette et al., 2008) (using numerical simulations of morphologically reconstructed cells). It could be larger when recording in thin processes such as dendrites or axons. In that case, a different model for K_m could be used (Brette and Destexhe, work in progress).

Electrode non-linearities The central assumption of the AEC technique is that the electrode is linear. This is not always the case for sharp microelectrodes, which can show current-dependent resistance changes. Physical modeling of non-linearities (Purves, 1981) indicates that these are slow processes due to redistribution of ions near the electrode tip. Non-linearities are stronger for electrode tips with a small radius (which is inversely correlated with the resistance) and when the concentrations of the two solutions (intracellular and inside the electrode) differ. However, in practice the amount of electrode non-linearity is highly variable and unfortunately cannot be assessed before the electrode is impaled into the cell – although electrodes with an unusually high resistance in the slice can be discarded from the start. This non-linearity problem is not different with AEC than with standard bridge balance, however AEC provides a simple way to measure it, and possibly discard the recordings if the non-linearity is too significant. Electrode nonlinearities are usually measured before impalement from the *I–V* curve of the electrode, but it is not possible to use the same approach intracellularly because the *I–V* curve of the electrode could be confused with the *I–V* curve of the neuron. AEC can be used to measure the electrode resistance by running the kernel estimation procedure intracellularly with different levels of constant injected current, corresponding to the typical (average) levels that will be used subsequently, and checking that the amount of non-linearity is acceptable (in the experiments in Brette et al. (2008), about half the electrodes were not significantly non-linear).

3.3 Recording currents

3.3.1 The ideal voltage clamp

In an ideal voltage clamp recording, the membrane potential of the cell is held at fixed value V_{clamp} while one measures the current flowing through the electrode. Assuming an isopotential neuron with an ideal voltage clamp setup, the membrane potential is constant ($V_m = V_{clamp}$) and its derivative is null, so that:

$$0 = \sum_{ionic\ currents} I_{ionic\ current} + I(t) \tag{3.2}$$

where $I(t)$ is the current flowing through the electrode. Thus, the voltage clamp configuration is used to measure ionic currents flowing when the membrane is held at a given potential. For this reason, the reported voltage clamp current is generally $-I(t)$, the opposite of the current through the electrode. In a typical voltage clamp experiment, the clamp potential is switched instantaneously from a resting value to a target value (step change) and transient currents from voltage-dependent channels are measured. In that case, the speed of clamping is an important issue (i.e. how fast the membrane potential follows the command potential). In other experiments (essentially in vivo), the voltage is held fixed and time-dependent changes in currents, typically resulting from synaptic activity, are measured. In both types of experiments, the two main issues are the quality of membrane potential clamping (the difference between V_m and V_{clamp}) and the quality of current recording (Figure 3.8).

The voltage clamp is implemented as a negative feedback circuit (either analog or digital): the clamp error $V_m - V_{clamp}$ is measured and a feedback current is

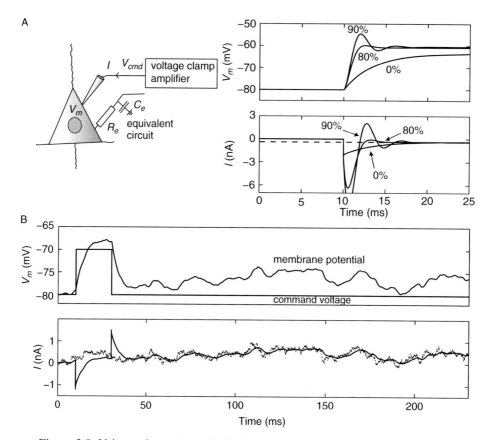

Figure 3.8 Voltage clamp (numerical simulations). A. In a passive neuron (only leak current), the voltage command is set at −60 mV at time $t = 10$ ms and the actual membrane potential response is shown (top) together with the measured current (bottom, dashed line, ideal measured current). When the electrode resistance is not compensated (0%), the response is slow and does not reach the command potential. The settling time and the clamp error are reduced with compensation (80%) but the resistance cannot be completely compensated because of capacitive effects, inducing oscillatory instability (90%). B. The neuron receives a noisy excitatory synaptic current (bottom, dashed) and is measured in voltage clamp with offline compensation (bottom, solid). A square voltage pulse (top) is used to estimate R_e, then the command voltage is −80 mV. The estimated resistance is $\Delta V/\Delta I$ (ΔI is the current discontinuity, bottom). Offline compensation corrects the error in the mean current but not the high-frequency components.

injected, such as $I = g(V_{clamp} - V_m)$, where g is a large (ideally infinite) gain. One can see that when the system is stabilized (implying $dV_m/dt = 0$), the injected current I necessarily satisfies Equation (3.2). There are a number of difficulties with this technique:

- The neuron potential can be clamped at only one point: the soma may be clamped at a given potential while remote dendritic locations are not. This is a problem when recording currents originating from dendrites and is called the *space clamp* problem.
- The membrane potential needs to be measured, which requires an electrode compensation technique if there is a single electrode. Because compensation errors can destabilize the system, it is common that only partial electrode compensation be applied.
- Because of various capacitive currents and imperfections, the feedback gain cannot be made arbitrarily large without destabilizing the system. Lower feedback gains result in an imperfect clamp ($V_m \neq V_{clamp}$).

In modern amplifiers, an additional control feedback is inserted to ensure that the membrane potential is clamped at the correct value: $I = g(V_{clamp} - V_m) + I_c$, where the control current I_c is proportional to the integral of the clamp error, i.e.:

$$\frac{dI_c}{dt} = g_c(V_{clamp} - V_m)$$

where g_c is another gain parameter (in units of conductance per time). This is called a *proportional-integral controller* (PI) in control engineering. When the system is stationary, the equality $dI_c/dt = 0$ implies that the membrane potential is clamped at the correct value $V_{clamp} - V_m$ (assuming that there is no error on measuring the membrane potential V_m).

3.3.1.1 Space clamp issues

In principle, the membrane potential can only be imposed at one point of the neuron morphology. If the neuron is not electrotonically compact, then the membrane is imperfectly clamped far from the voltage clamp electrode. For example, if the membrane is clamped at a potential V_{clamp} and the neuron is passive (no voltage-dependent ionic channels), then the potential on a dendrite at electrotonic distance d from the soma is $V_0 + (V_{clamp} - V_0)e^{-d}$, where V_0 is the resting potential (Koch, 1999). Thus, when recording currents (whether synaptic or intrinsic) with somatic voltage clamp, it should be kept in mind that the clamp is imperfect if those currents originate from a distal location. It is difficult to compensate for a poor space clamp (for example by changing the voltage command at the soma), first because the electrotonic distance is generally unknown and second because the potential at the distal location is time dependent, even with an ideal voltage clamp (the expression above is the stationary value, with passive membrane properties only).

What is the spatial extent of voltage clamping in a neuron? From the expression above, attenuation of the potential is within 10% up to a distance of 5% the electrotonic length of the dendrite, which is given by the following formula:

$$\lambda = \sqrt{\frac{a r_m}{2 r_L}}$$

where a is the radius of the dendrite, r_m is the specific membrane resistance and r_L is the intracellular resistivity. Unfortunately, this analysis only holds when active ionic channels are neglected. When ionic channels open, their conductance increases so that the effective membrane resistance decreases. As a result, the effective electrotonic length decreases, which decreases the spatial extent of voltage clamping. For example, the effective time constant of cortical neurons is about five times smaller in vivo than in vitro (as assessed by somatic injection of current pulses), presumably because of intense synaptic activity (Destexhe et al., 2003), which increases the total conductance (hence the name *high-conductance state*). If the increase is homogeneous, this conductance increase results in a decrease of electrotonic length by a factor greater than two – or, in other words, the effective size of the neuron doubles. Similarly, intrinsic conductances such as voltage-gated K^+ channels can open with the voltage clamp command, resulting in serious space clamp problems even in small neurons (Bar-Yehuda and Korngreen, 2008).

3.3.2 Double-electrode voltage clamp

In double-electrode setups, one electrode is used to measure the membrane potential while the other one is used to inject the feedback current $I = g(V_{clamp} - V_m)$. Using two different electrodes ensures that the measure of the membrane potential V_m is not distorted by the injection of the feedback current. This ensures that the membrane potential matches the command clamp potential when the clamp is established (provided junction potentials are properly compensated), because no current passes through the measuring electrode in the stationary regime. Similarly, the measured current in the stationary regime is also correct. However, several factors make the double-electrode voltage clamp non-ideal. The most serious problem is capacitive coupling between the two electrodes, which is destabilizing. That coupling limits the gain of the feedback circuit, which results in poorer clamp, longer settling time and distortions in the measured current. Experimentally, capacitive coupling can be reduced by inserting the two electrodes at a wide angle. Unfortunately, this does not suppress all capacitive currents in the recording circuit, in particular the electrode capacitance (through the electrode capillary tube) and the input capacitance (at the amplifier input), which cause similar instability problems.

To reduce the problems due to capacitances in the recording circuit, voltage clamp amplifiers either introduce a delay in the feedback current or reduce the gain of the feedback. In voltage clamp experiments with a step command potential, the initial transient in the measured current is generally suppressed offline, which makes the measurement of fast activating currents such as sodium channel currents difficult.

3.3.3 Single-electrode voltage clamp

In many cases, it is not possible to insert two electrodes in the neuron and one must use a single electrode to clamp the cell, either a sharp microelectrode or a patch electrode (whole-cell configuration). This introduces an additional problem: the measurement of the membrane potential is contaminated by the injection of the feedback current through the same electrode. There are two kinds of methods to deal with this issue: compensating for the electrode bias (series resistance compensation and AEC) and alternating voltage measurement and current injection (discontinuous voltage clamp).

3.3.3.1 Series resistance compensation

The nature of the problem is similar to current clamp single-electrode recordings, but the strong feedback makes it more serious. The electrode resistance acts as a voltage divider. In the stationary regime, the command potential and the membrane potential are related by the following relationship:

$$V_m = \frac{R_m}{R_m + R_e} V_{clamp}$$

where R_m is the membrane resistance and R_e is the electrode resistance (the potentials are relative to the resting potential). Thus, the clamp error increases with R_e. If electrode and membrane resistances have the same magnitude then the error is dramatic, since the membrane potential is only half the command potential. If the electrode resistance is small, the problem might seem minor at first sight, but the electrode resistance results not only in an error on the stationary potential but also in a non-zero settling time and errors on the measured current. Indeed, consider a simple model where the membrane has only passive properties (resistance R_m and capacitance C_m) and the electrode is a resistor (resistance R_e). Applying Kirchhoff's law gives the following differential equation:

$$C_m \frac{dV_m}{dt} + \frac{V_m}{R_m} = \frac{V_{clamp} - V_m}{R_e}.$$

It appears that the membrane potential approaches the stationary voltage exponentially with the following time constant:

$$\tau_{settle} = \frac{C_m R_m R_e}{R_e + R_m} \approx C_m R_e = \tau_m \frac{R_e}{R_m}$$

where the approximation is valid when $R_e \ll R_m$. That settling time can be long: for example, if $R_e = R_m/10$ and $\tau_m = 20$ ms, then the stationary clamp error is about 10%, which might be acceptable, but $\tau_{settle} = 2$ ms, which is long for fast activating channels. This settling time results in a transient in the measured current. Since the measured current is the opposite of the current injected through the electrode, it equals

$$I_{clamp} = (V_m - V_{clamp})/R_e.$$

Ideally, in our passive neuron model, that measured current should equal the leak current at the command potential, i.e. $-V_{clamp}/R_m$ (except for an infinite current at onset). With a non zero electrode resistance, the measured current starts at $-V_{clamp}/R_e$ and relaxes exponentially to $-V_{clamp}/(R_m + R_e)$ with time constant τ_{settle} (Figure 3.8A, 0% compensation). Thus, even if the electrode resistance is small, the measured current is completely wrong during the time of the transient (about τ_{settle}). This issue arises even without taking voltage-dependent channels into account, which make the problem much worse. The settling time of the voltage clamp can be shortened by a technique named *supercharging*, which consists in adding a brief pulse at the onset of a voltage step. Although the membrane potential reaches the target potential quicker, it does not enhance the resolution of the measured currents after the onset. It is important to keep this issue in mind when applying offline series resistance compensation. In many situations, the series resistance cannot be compensated during the recording because of instability problems (see below), or can only be partially compensated. The electrode resistance R_e can be estimated from the peak of the transient current ($-V_{clamp}/R_e$, all potentials are relative to the resting potential). The imperfect clamp can then be corrected by applying the following correction to the measured current:

$$I_{corrected} = I_{measured} \frac{R_m + R_e}{R_m}$$

if the membrane resistance R_m can be measured. However, this correction works in the stationary regime, when the clamp is established, while the transient current is barely modified.

When recording time-varying currents, the electrode resistance reduces the bandwidth of the measured current, with an approximate cutoff frequency $f_c = 1/(2\pi \tau_{settle})$, which cannot be corrected by offline compensation (Figure 3.8B). This filtering property is best understood by considering a simple model of a neuron with a synaptic current $I_s = g_s(t)(E_s - V_m)$. To measure the synaptic conductance, we clamp the neuron at the resting potential (which we choose as the reference potential). The system is governed by the following differential equation:

$$C_m \frac{dV_m}{dt} + g_s(t)(V_m - E_s) + \frac{V_m}{R_m} = -\frac{V_m}{R_e}$$

and the measured current is $-V_m/R_e$; it is proportional to the membrane potential. That equation can be written more clearly as

$$C_m \frac{dV_m}{dt} + g_{tot}(t)(V_m - E_{eff}(t)) = 0$$

where $g_{tot}(t) = g_s(t) + R_m^{-1} + R_e^{-1}$ is the total conductance and

$$E_{eff}(t) = \frac{g(t)E_s}{g_{tot}}$$

is the effective reversal potential. The membrane potential follows $E_{eff}(t)$ with a time constant $C_m/g_{tot}(t)$, which is close to τ_{settle} if the synaptic conductance is small compared to the electrode conductance R_e^{-1} (if it is not small, then the membrane potential is far from the clamp potential and the recording is probably not useful). In summary, the resolution of current recordings is about R_e/R_m in units of the membrane time constant, which can be a severe restriction.

It is therefore important to reduce the electrode resistance as much as possible. Correcting the clamp potential by multiplying the clamp potential by $(R_m+R_e)/R_m$ provides similar benefits to offline compensation, i.e. the stationary value is corrected but neither the transient current nor the current filtering are affected. *Series resistance compensation* consists in adding an offset to the command potential that depends on the current injected through the electrode. If I is the current flowing through the electrode (to the neuron) and the electrode is a simple resistor with resistance R_e, then the voltage across the electrode is $R_e I$. Thus, compensating for the electrode consists in applying a command potential $V_{clamp} + R_e I$ instead of V_{clamp}. This is in fact the same as bridge balance for current clamp, if one looks at how the feedback current I is implemented: $I = g(V_{clamp} - U)$, where U is the measured potential and g is the gain of the feedback. Correcting U by bridge balance means changing the feedback current into $I = g(V_{clamp} - (U - R_e I)) = g(V_{clamp} + R_e I - U)$, which corresponds to changing the clamp command into $V_{clamp} + R_e I$. Unfortunately, because I is a feedback current which depends on the measured potential, series resistance compensation is destabilizing. Indeed, with the pure resistor electrode model and an estimated electrode resistance R_e^*, the electrode current reads:

$$I = \frac{V_{clamp} + R_e^* I - V_m}{R_e}$$

which simplifies to

$$I = \frac{V_{clamp} - V_m}{1 - R_e^*/R_e}$$

and that current goes to infinity and changes sign near the ideal setting $R_e^* = R_e$. In fact, the instability point is reached well before that point when considering other capacitances in the circuit such as the electrode capacitance or the amplifier input capacitance (Figure 3.8A). Thus, series resistance compensation cannot be applied directly in this way. Most amplifiers address this problem by delaying the command offset $R_e^* I$, which enhances the stability of the system. Even with this strategy, in many cases the electrode resistance can only be partially compensated, especially with high-resistance electrodes. In those cases, an alternative strategy consists of alternating current injection and voltage measurement, in the same way as for discontinuous current clamp.

3.3.3.2 Discontinuous voltage clamp

The principle of the discontinuous voltage clamp is identical to the discontinuous current clamp, and works in current clamp mode (the current is imposed, not the voltage). Current injection and potential measurement are alternated so as to minimize the effect of the electrode on the measured potential. Thus, it is subject to the same limitations as DCC: the electrode time constant must be two orders of magnitude smaller than the membrane time constant, and the optimal sampling frequency cannot be determined unambiguously, which results in measurement errors. The principle of the feedback is the same as for continuous voltage clamp, except the current is discretized (Figure 3.9). During one time step $[t_n, t_{n+1}]$, the injected current is $I_n = g(V_{clamp} - U(t_n))$, where g is the gain and U is the estimated membrane potential at the end of the previous time step. More precisely, a current $I_n = (g/D)(V_{clamp} - U(t_n))$ is injected during $[t_n, t_n + D\Delta]$ and current is injected in $[t_n + D\Delta, t_{n+1}]$, where D is the duty cycle (typically about $1/3$) and $\Delta = 1/f$ is the sampling step (f is the sampling frequency). It is expected that $V_m(t_n) \approx U(t_n)$, that is, the electrode voltage vanishes at the end of a time step. Under that assumption, the statibility of this feedback depends on the size of the sampling step $\Delta = 1/f$ and on the gain g. Consider that the electrode resistance has indeed been canceled and that the effective recording circuit consists of a membrane modeled as a resistor and capacitor. Then the membrane potential $V_n = V_m(t_n)$ is governed by the following difference equation:

$$V_{n+1} = e^{-1/\tau_m f} V_n + e^{-(1-D)/\tau_m f}(1 - e^{-D/\tau_m f})\frac{R_m}{D}g(V_{clamp} - V_n)$$

$$\approx \left(1 - \frac{R_m g}{\tau_m f}\right)V_n + \frac{R_m g}{\tau_m f}V_{clamp} = \left(1 - \frac{g}{C_m f}\right)V_n + \frac{g}{C_m f}V_{clamp}$$

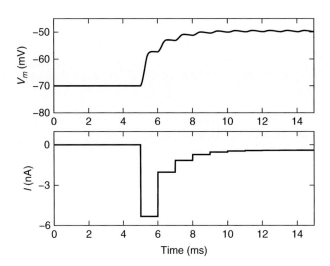

Figure 3.9 Discontinuous voltage clamp (numerical simulations). A. The membrane potential of a passive neuron is clamped at -50 mV (from $t = 5$ ms) with discontinuous voltage clamp. The real membrane potential is shown (top) together with the measured current (bottom), which is ideally the constant leak current at -50 mV. The sampling frequency is 1 kHz.

where we used the fact that $\Delta \ll \tau_m$ ($\tau_m = R_m C_m$). The gain is optimal when $g = C_m f$ and stable if $g < 2C_m f$. The stationary membrane potential at maximum gain is then:

$$V_m = \left(1 - \frac{1}{2\tau_m f}\right) V_{clamp}$$

(after Taylor expansion in $(\tau_m f)^{-1}$). Typically when the frequency is properly adjusted, $\tau_m f \approx 10$, so that the clamp error is about 5% according to the formula above. However, a number of factors contribute to increasing that error: errors in setting the optimal sampling frequency result in measurement errors, which are a source of instability; non-idealities, in particular the electrode capacitance and other capacitances in the circuit, also reduce the maximum gain. The stationary clamp error can be reduced by inserting an additional control current as mentioned in Section 3.3.1, but it affects neither the settling time nor the resolution of the measured current. Noise is also higher with discontinous voltage clamp, in particular because of aliasing noise in the potential measurement (Finkel and Redman, 1984).

3.3.3.3 *Voltage clamp with AEC*

Active electrode compensation can be used in exactly the same way as discontinous voltage clamp, i.e. the amplifier is in current clamp mode and a feedback current is injected at every time step: $I_n = g(V_{clamp} - U(t_n))$, where U is the AEC estimation of the membrane potential. The main differences are that (1) the membrane potential is estimated with AEC and (2) the sampling frequency is not limited by the electrode time constant. Therefore the technique is perhaps closer to a continuous voltage clamp with series resistance compensation. An integral control can also be added to the current to improve the quality of the clamp. The AEC based voltage clamp is still under investigation at this time.

3.4 Recording conductances

The earliest recording of the conductance of a neuron is probably the recording of the increase in total conductance during action potential performed by Cole and Curtis in 1939 with an ingenious electrical circuit. This proved that the initiation of the action potential was indeed due to an increase in membrane permeability, as was hypothesized by Bernstein. When recording with a current clamp or voltage clamp amplifier, conductances can only be inferred indirectly, using a model for the recorded currents or voltages. Conductances can be intrinsic (e.g. conductances of sodium channels) or synaptic, but since we focus on recording neural activity in this chapter, we will mainly discuss the measurement of synaptic conductances.

3.4.1 Models for conductance measurements

3.4.1.1 Current clamp model

Let us start with a simple case where there is only one non-constant conductance in an isopotential neuron. In a current clamp experiment, the membrane potential of that neuron is governed by the following differential equation:

$$C\frac{dV_m}{dt} = g_l(E_l - V_m) + g(t)(E - V_m) + I(t)$$

where $I(t)$ is the injected current, $g(t)$ is the conductance to be measured and E is the corresponding reversal potential. We assume that E is known. The first term is the leak current, which is assumed to be constant. Such a situation with only one additional current may be obtained by suppressing the expression of other ionic channels using pharmacological methods. In that case the conductance $g(t)$ can be directly derived from the equation:

$$g(t) = \left(C\frac{dV_m}{dt} - g_l(E_l - V_m) - I(t)\right)/(E - V_m)$$

provided that the parameters C, g_l and E_l are known. These values can be obtained for example from the response of the neuron to a current pulse. It is often easier to suppress the capacitive current by measuring in voltage clamp mode (see below).

Difficulties arise when several time-varying conductances are present:

$$C\frac{dV_m}{dt} = \sum_i g_i(t)(E_i - V_m) + I(t)$$

where $g_i(t)$ is the ith conductance and E_i is the corresponding reversal potential. Ambiguities in the measurement come from the fact that several unknown quantities (the conductances) contribute to the single physical quantity being measured ($V_m(t)$), so that most existing techniques rely on multiple measurements with different injected currents $I(t)$. Since the right hand side is linear with respect to V_m, the equation can be written equivalently as

$$C\frac{dV_m}{dt} = g(t)(E(t) - V_m) + I(t)$$

where

$$g(t) = \sum_i g_i(t)$$

is the total conductance, and

$$E(t) = \frac{\sum_i g_i(t)E_i}{\sum_i g_i(t)}$$

is the effective reversal potential. We observe that the conductances are mapped to the quantities g and gE through a linear mapping $(g_1, g_2, \ldots, g_n) \mapsto (g, gE)$, which is defined by the reversal potentials (which are assumed to be distinct). That mapping is invertible only if there are no more than two unknown conductances. Otherwise, since the mapping has rank 2, there are an infinite number of linearly related possibilities for the conductances that give the same measurements for the membrane potential and there is no principled way to distinguish between them (except that they must be positive). Besides, even when there are only two time-varying conductances, their values are determined by the choice of the reversal potentials E_i. Thus, for any measurement technique, one can only hope to recover two independent variables at most and their values depend on the choice of reversal potentials, which cannot be inferred from the data. In general, the constant leak current is estimated independently and excitatory and inhibitory conductances are measured. The leak current can be estimated for example from the response to pulses during periods of low activity.

There are several issues with the model we have described. First, a neuron is not isopotential, which we discuss briefly in Section 3.4.1.3. Second, the response of the neuron is non-linear because of intrinsic voltage-dependent currents (e.g. sodium and potassium currents). That issue can be addressed with more complex models, including polynomial models of the $I-V$ curve or more complex models that can be obtained with white noise injection (Badel et al., 2008). Spike-related conductances (such as those responsible for spike frequency adaptation) can also produce artifacts.

3.4.1.2 Voltage clamp model

In voltage clamp mode, the measured current for an isopotential neuron model reads:

$$I(t) = \sum_i g_i(t)(E_i - V_{clamp})$$

where V_{clamp} is the holding potential. When there is a single time-varying conductance, it is obtained directly from this formula. When there are several time-varying conductances, the same issues arise as in the current clamp mode. The equation can be rewritten as

$$I(t) = g(t)(E(t) - V_{clamp})$$

where $g(t)$ is the total conductance and $E(t)$ is the effective reversal potential. Here V_{clamp} is imposed and $I(t)$ is measured. Again, even with many measurements with different holding potentials, only two independent variables can be measured unambiguously, and recovering the values of the conductances depends on the choice of the reversal potentials.

3.4.1.3 Visibility of dendritic synaptic inputs

Two issues arise if the neuron is not electrotonically compact: first, the membrane equation includes a current flowing to the dendrites, second, the currents may be generated distally in the dendrites. In the latter case, there is no direct access to the synaptic conductances if the distance to the dendritic site is large (in units of the space length of the neuron) and one can only talk of "effective" synaptic conductances seen at the soma.

The effect of distal location of synaptic inputs on the measurement is twofold: conductances measured at the soma seem smaller and reversal potentials seem further away from the resting potential. This effect can be understood in a simplified neuron model consisting of an isopotential soma connected to a semi-infinite cylindrical dendrite (a "ball-and-stick" model). A synaptic current $g_s(E_s - v)$ is inserted on the dendrite at distance x_s from the soma. We assume for simplicity that the synaptic conductance g_s is constant. A more detailed study can be found in Koch et al. (1990). We consider a voltage clamp experiment in which the voltage is held fixed at the soma at a value v_0 and the injected current I is measured. According to passive cable theory (Tuckwell, 1988; Dayan and Abbott, 2001), the stationary membrane potential $v(x)$ satisfies the following second-order differential equation on the two segments $[0, x_s]$ and $[x_s, +\infty[$:

$$\lambda^2 \frac{d^2 v}{dt} = v$$

where λ is the electrotonic length (we chose the resting potential as the reference potential). The solution of this equation is $v(x) = ae^{x/\lambda} + be^{-x/\lambda}$, where the coefficients a and b must be determined by boundary conditions. At the soma ($x = 0$), the injected current I is the sum of the leak current leaving the membrane and the current flowing through the dendrite:

$$I = g_l v_0 - \frac{1}{R_a} \frac{dv}{dx}(0)$$

where R_a is the axial resistance of the dendrite. At the synaptic site ($x = x_s$), there is a discontinuity in the axial current that equals the synaptic current:

$$-\frac{1}{R_a} \frac{dv}{dx}(x_s^+) = -\frac{1}{R_a} \frac{dv}{dx}(x_s^-) + g_s(E_s - v(x_s)).$$

Finally, the membrane potential must vanish at infinity. With these boundary conditions and the continuity at $x = x_s$, one can calculate the potential $v(x)$ over the two segments $[0, x_s]$ and $[x_s, +\infty[$ and ultimately obtain the current I as a function of v_0 and g_s. After some algebra, we obtain:

$$I(v_0, g_s) = I(v_0, 0) + g^*(v_0 - E^*)$$

where $I(v_0, 0)$ is the effective leak current (current in the absence of synaptic input), g^* is the effective conductance as measured at the soma and E^* the effective reversal potential, which are given by the following expressions:

$$E^* = e^{x_s/\lambda} E_s$$

$$g^* = \frac{2g_s}{1 + e^{2x_s/\lambda}}.$$

We observe that $E^* > E_s$ and $g^* < g_s$. The effective reversal potential E^* does not depend on the value of the synaptic conductance g_s. For distal dendrites, the effective reversal potential is further away from the resting potential than the actual reversal potential, and the effective conductance is reduced.

3.4.1.4 Sharp electrodes and patch electrodes

Another important point should be kept in mind: measurements with a sharp micro-electrode and with a patch electrode (whole-cell configuration) are not equivalent. The main effect of the sharp electrode is to damage the membrane of the cell, which inserts a non-selective leak current. In particular, the leak conductance is larger with a sharp electrode than with a patch electrode. On the other hand, patch electrodes have a large tip which lets the contents of the electrode diffuse in the cell (except with the perforated patch clamp technique, in which antibiotics are used to perforate the membrane). This phenomenon is called *dialysis* and has important consequences. For the measurement of conductances, the main effects are firstly that the resistivity of the intracellular medium is changed (which changes the electrotonic dimension of the neuron) and secondly that synaptic reversal potentials can change over time as the cell is dialyzed (because of changes in ionic concentrations).

3.4.2 *Multi-trial conductance measurements*

As we noted earlier, when there are several conductances to be measured, ambiguities arise from the fact that only one quantity is measured (the membrane potential in current clamp or the current in voltage clamp). To solve that problem, most current techniques combine measurements on several trials with the same stimulus and different experimental conditions: different injected currents (current clamp, Figure 3.10B) or different holding potentials (voltage clamp, Figure 3.10A). One

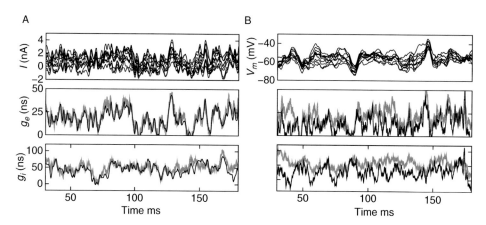

Figure 3.10 Recording synaptic conductances (numerical simulations with passive neuron model). The neuron model includes a leak current and excitatory and inhibitory noisy synaptic conductances, which are partly reproducible over trials: $g(t) = g_{same}(t) + g_{different}(t)$. The reproducible and the variable parts have the same magnitude (i.e. the SNR ratio is 1). A. Measurement of synaptic conductances with voltage clamp (10 different holding potentials; the electrode is 10 MΩ with 90% compensation). Top: measured current (10 trials). Middle: reconstructed excitatory conductance (black) and real one (gray). Bottom: reconstructed inhibitory conductance (black) and real one (gray). B. Current clamp (10 different injected currents; the electrode has negligible resistance). Top: measured membrane potential (10 trials). Middle, bottom: as in A.

obvious limitation of this type of technique, which is reviewed in Monier et al. (2008), is that only stimulus-locked activity can be recorded in this way.

3.4.2.1 Voltage clamp

We start with voltage clamp measurements (Figure 3.10A). Consider n measurements of the response to the same stimulus, with different holding potentials V_k. Assuming that the synaptic conductances are identical on all trials, the measured current $I_k(t)$ is

$$I_k(t) = g(t)(E(t) - V_k)$$

where $g(t)$ is the total conductance and $E(t)$ is the effective reversal potential. For a given time t, the measure $I_k(t)$ is an affine function of V_k whose slope is the total conductance and whose intercept is the reversal potential multiplied by the slope. In principle, two trials are sufficient to recover those values but in practice more trials are used to make the measurement more reliable. In that case the conductances are obtained with a linear regression.

Methods based on voltage clamp are mathematically simpler than those based on current clamp because the capacitive current vanishes. However, they raise experimental issues because for practical reasons most intracellular recordings in vivo use a single electrode. In many cases, the access resistance cannot be fully compensated, which results in imperfect clamping. If the electrode resistance R_e is known, then the measured current is related to the holding potential V_k according to the following equation:

$$I_k(t) = g(t)(E(t) - V_k + R_e I_k(t))$$

which simplifies to

$$I_k(t) = \frac{g(t)}{1 - g(t)R_e}(E(t) - V_k)$$

and the same linear regression can be applied to recover $g(t)$ and $E(t)$. However, this is only an approximation because the membrane equation should now include a capacitive current, since the membrane potential is no longer fixed. Other non-idealities such as the input capacitance also make this formula less accurate.

3.4.2.2 Current clamp

Conductance measurements in current clamp mode consist in repeating the same voltage measurements in response to a given stimulus with different injected currents I_k (Figure 3.10B). The membrane potential $V_k(t)$ satisfies the following differential equation:

$$I_k - C\frac{dV_k}{dt} = g(t)(V_k(t) - E(t)).$$

As in voltage clamp mode, for any given time t, the current $I_k - CdV_k/dt(t)$ is an affine function of $V_k(t)$, whose slope is the total conductance and whose intercept is the reversal potential multiplied by the slope. Provided that the membrane capacitance C is known (estimated for example from the response to a current pulse), both $g(t)$ and $E(t)$ can be recovered.

This method is experimentally easier than in voltage clamp mode but many other issues arise:

- Differentiating the membrane potential adds noise to the measurements, which may be reduced by filtering.
- Voltage-dependent conductances may be activated. That issue also arises in voltage clamp experiments, but it results in constant biases in the $I-V$ curve, which are easier to compensate for.
- The neuron may fire action potentials. Synaptic conductances cannot be estimated during action potentials because they are masked by the spike-related increase in total conductance (Guillamon et al., 2006). Unfortunately, part of this increase may last for a very long time. For example, pyramidal cortical cells exhibit spike frequency adaptation, related to a slow spike-triggered adaptation conductance whose stationary value increases with the firing rate. Since the firing rate is most likely related to the injected current I_k, the effect on conductance estimation is potentially significant. To avoid this problem, one can block the action potentials pharmacologically or use hyperpolarizing currents.
- Because the membrane potential is not controlled, conductance measurements are less robust to noise in current clamp than in voltage clamp mode (by noise, we mean any activity that is not locked to the repeated stimulus), as is illustrated by Figure 3.10.

3.4.3 Statistical measurements

Measuring the time course of synaptic conductances is difficult, either because of technical difficulties (voltage clamp) or because the measurements are not robust to noise (current clamp). A different approach consists in looking for statistical information about the conductances, such as their mean and variance, by using a stochastic model for the neuron and its synaptic inputs. One such model, the "point-conductance" model, consists in a single-compartment model with time-varying excitatory and inhibitory conductances $g_e(t)$ and $g_i(t)$ described by Ornstein–Uhlenbeck processes (Destexhe et al., 2001), i.e. Gaussian Markov processes with mean g_{e0} (respectively g_{i0}) and standard deviations σ_e (respectively

σ_i). That stochastic description derives from a diffusion approximation of the total conductance as a sum of random postsynaptic conductances modeled as exponential functions with time constants τ_e and τ_i. The complete model is described by the following equations:

$$C \frac{dV_m}{dt} = -g_l (V_m - E_l) - g_e (V_m - E_e) - g_i (V_m - E_i) + I$$

$$\frac{dg_e}{dt} = -\frac{1}{\tau_e} [g_e - g_{e0}] + \sqrt{\frac{2\sigma_e^2}{\tau_e}} \xi_e(t)$$

$$\frac{dg_i}{dt} = -\frac{1}{\tau_i} [g_i - g_{i0}] + \sqrt{\frac{2\sigma_i^2}{\tau_i}} \xi_i(t)$$

where C denotes the membrane capacitance, I a stimulation current, g_l the leak conductance, E_l the leak reversal potential, E_e the excitatory reversal potential and E_i the excitatory reversal potential. That model has been used to estimate the distribution of synaptic conductances, synaptic time constants, spike-triggered averages of conductances, and the time course of synaptic conductances.

3.4.3.1 Estimating synaptic conductance distributions

The point-conductance model has been thoroughly studied theoretically and numerically. Different analytic approximations have been proposed to describe the steady-state distribution of the V_m activity of the point-conductance model (Rudolph and Destexhe, 2003; Richardson, 2004; Rudolph et al., 2005; Lindner and Longtin, 2011; for a comparative study, see Rudolph and Destexhe, 2011). One of these expressions can be inverted (Rudolph and Destexhe, 2003; Rudolph et al., 2005), which enables one to estimate the synaptic conductance parameters (g_{e0}, g_{i0}, σ_e, σ_i) directly from experimentally obtained V_m distributions. This constitutes the basis of the VmD method (Rudolph et al., 2004), which we outline below.

The VmD method consists of estimating the statistical properties of the conductances (mean and variance) from the statistics of the intracellularly recorded activity (mean and variance of the V_m). The following analytic expression provides a good approximation to the steady-state probability distribution $\rho(V_m)$ of the membrane potential (Rudolph and Destexhe, 2003; Rudolph et al., 2005):

$$\rho(V_m) = N \exp \left[A_1 \ln \left[\frac{u_e(V_m - E_e)^2}{C^2} + \frac{u_i(V_m - E_i)^2}{C^2} \right] \right.$$
$$\left. + A_2 \arctan \left[\frac{u_e(V_m - E_e) + u_i(V_m - E_i)}{(E_e - E_i)\sqrt{u_e u_i}} \right] \right], \quad (3.3)$$

where $u_e = \sigma_e^2 \tilde{\tau}_e$, $u_i = \sigma_i^2 \tilde{\tau}_i$, A_1 and A_2 are voltage-independent terms depending on the parameters of the membrane equation (see details in Rudolph et al., 2004). Here, N denotes a normalization constant such that $\int_{-\infty}^{\infty} dV \, \rho(V) = 1$ and $\tilde{\tau}_{\{e,i\}}$ are effective synaptic time constants, given by Rudolph et al. (2005) (see also Richardson, 2004):

$$\tilde{\tau}_{\{e,i\}} = \frac{2\tau_{\{e,i\}}\tilde{\tau}_m}{\tau_{\{e,i\}} + \tilde{\tau}_m} , \tag{3.4}$$

where $\tilde{\tau}_m = C/(g_l + g_{e0} + g_{i0})$ is the effective membrane time constant. Due to the multiplicative coupling of the stochastic conductances to the membrane potential, the V_m probability distribution (Equation (3.3)) in general takes an asymmetric form. However, it is well approximated by a Gaussian distribution, which can be obtained by Taylor expansion around the maximum $\bar{V}_m$ of the probability distribution $\rho(V_m)$. The mean and variance of that approximation can be expressed as a function of the parameters (Rudolph et al., 2004). This Gaussian approximation provides an excellent fit to V_m distributions obtained from models and experiments (Rudolph et al., 2004), because the V_m distributions obtained experimentally show little asymmetry (for up-states and activated states; for specific examples, see Rudolph et al., 2004, 2005, 2007; Piwkowska et al., 2008).

The main advantage of this Gaussian approximation is that it can be easily inverted, which leads to expressions for the synaptic noise parameters as a function of the measured V_m distribution, specifically $\bar{V}_m$ and σ_V. By fixing the values of τ_e and τ_i, which are related to the decay time of synaptic currents and can be estimated from voltage clamp data and/or current clamp data by using power spectral analysis (see below), four parameters remain to be estimated: the means (g_{e0}, g_{i0}) and standard deviations (σ_e, σ_i) of excitatory and inhibitory synaptic conductances. Since the Gaussian distribution is only characterized by two values ($\bar{V}_m$ and σ_V), at least two recordings with different constant levels of injected current I are required, as for multi-trial conductance measurements (Rudolph et al., 2004). The quality of the estimation can then be assessed by comparing the full expression (Equation (3.3)) with the experimental data.

These relations enable us to estimate global characteristics of network activity, such as mean excitatory (g_{e0}) and inhibitory (g_{i0}) synaptic conductances, as well as their respective variances (σ_e^2, σ_i^2), solely from knowledge of the V_m distributions computed from intracellular measurements. This VmD method has been tested using computational models (Figure 3.11A) and dynamic clamp experiments (Figure 3.11B,C; Rudolph et al., 2004; Piwkowska et al., 2008) and has also been used to extract conductances from different experimental conditions in vivo (Rudolph et al., 2005, 2007; Zou et al., 2005). In particular, it was applied to analyze intracellular recordings in anesthetized (Rudolph et al., 2005), as well as naturally sleeping and awake cats (Rudolph et al., 2007).

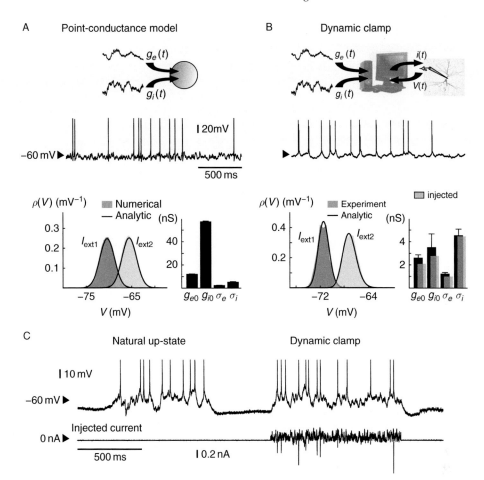

Figure 3.11 Numerical and dynamic clamp test of the VmD method to extract conductances. A. Simulation of the point-conductance model (top trace) and comparison between numerically computed V_m distributions (bottom, left) and the analytic expression (black, conductance values shown in the bar graph). B. Dynamic clamp injection of the point-conductance model in a real neuron. Right: conductance parameters are re-estimated (black bars, error bars are standard deviations obtained when the same injected conductance parameters are re-estimated in different cells) from the V_m distributions and compared to the known parameters of the injected conductances (grey bars). Left: The experimental V_m distributions are compared to the analytic distributions calculated using the re-estimated conductance parameters. C. Comparison of a spontaneous up-state (natural up-state) with an artificial up-state recreated using conductance injection (dynamic clamp). (Modified from Rudolph et al., 2004.)

3.4.3.2 Estimating synaptic time constants from the power spectrum

Synaptic time constants (τ_e and τ_i) can be estimated from the power spectral density (PSD) of the membrane potential which, for the point-conductance model, can be well approximated by the following expression (Destexhe and Rudolph, 2004):

$$S_V(\omega) = \frac{4}{G_T^2} \frac{1}{1 + \omega^2 \, \tilde{\tau}_m^2} \left[\frac{\sigma_e^2 \tau_e \, (E_e - \bar{V})^2}{1 + \omega^2 \, \tau_e^2} + \frac{\sigma_i^2 \tau_i \, (E_i - \bar{V})^2}{1 + \omega^2 \, \tau_i^2} \right], \qquad (3.5)$$

where $\omega = 2\pi f$, f is the frequency, $G_T = g_L + g_{e0} + g_{i0}$ is the total membrane conductance, $\tilde{\tau}_m = C/G_T$ is the effective time constant, and $\bar{V} = (g_L E_L + g_{e0} E_e + g_{i0} E_i)/G_T$ is the average membrane potential. The "effective leak" approximation used to derive this equation consisted in incorporating the average synaptic conductances into the total leak conductance, and then considering that fluctuations around the obtained mean voltage are subjected to a constant driving force (Destexhe and Rudolph, 2004).

This expression is very accurate for single-compartment models and provides an excellent fit for neurons stimulated with dynamic clamp in vitro up to frequencies of about 500 Hz, above which the mismatch was presumably due to instrumental noise (Piwkowska et al., 2008). However, the fit with in vivo recordings is more approximate for frequencies above 100 Hz (Rudolph et al., 2005), where the PSD scales as $1/f^{2.5}$ instead of $1/f^4$. This different scaling may be due to the attenuation of synaptic inputs occurring on dendrites, as well as to the non-ideal aspect of the membrane capacitance (Bédard and Destexhe, 2008). Nevertheless, the matching of the expression above to the low-frequency end (<100 Hz) of the PSD yielded values of time constants of $\tau_e = 3$ ms and $\tau_i = 10$ ms, with a precision of the order of 30% (Rudolph et al., 2005).

3.4.3.3 Estimating spike-triggered average conductances

The VmD method can be used to extract the spike-triggered averages (STAs) of conductances from recordings of V_m (Pospischil et al., 2007). The basis of the STA method is first to calculate the STA of the V_m activity, and then to search for the "most likely" spike-related conductance time courses ($g_e(t)$, $g_i(t)$) that are compatible with the observed voltage STA. Assuming that both conductances are realizations of Ornstein–Uhlenbeck processes whose means (g_{e0}, g_{i0}) and variances (σ_e^2, σ_i^2) are known (estimated with the VmD method), the probability of a given conductance time course ($g_e(t)$, $g_i(t)$) can be calculated from the definition of the stochastic processes. Because Ornstein–Uhlenbeck processes are Gaussian Markov processes, the probability of ($g_e(t + dt)$, $g_i(t + dt)$) only depends on the value ($g_e(t)$, $g_i(t)$) and is normally distributed. It follows that, if time is discretized, the log of the probability of a given conductance time course is a sum of quadratic terms, so that the maximum likelihood solution can be found with linear algebra (Pospischil et al., 2007).

The STA method predicted the correct results in numerical simulations and in vitro using dynamic clamp injection of known patterns in real neurons (Pospischil et al., 2007). It was also applied to intracellular recordings in awake and naturally sleeping cats (Rudolph et al., 2007), where it was found that for the majority of neurons, spikes are correlated with a decrease in inhibitory conductance, suggesting that inhibition is most effective in determining spiking activity. This observation matches the dominance of inhibition observed using the VmD method in the same neurons (see above).

3.4.3.4 Estimating the time course of synaptic conductances

The two different strategies outlined above, the VmD and STA methods, can be merged into a new method. This method, called "VmT," extracts synaptic conductance parameters, similar to the VmD method, but using a maximum-likelihood estimation similar to the STA method, thus applicable to single V_m traces (Pospischil et al., 2009). By following a similar procedure as for the STA method, one obtains estimates of the "most likely" values for g_{e0}, g_{i0}, σ_e and σ_i from single V_m traces. Similar to above, the method was tested using computational models and dynamic clamp experiments; the VmT method performs remarkably well for high-conductance states (see details in Pospischil et al., 2009).

3.5 Conclusion

Intracellular electrophysiology is one of the oldest techniques for measuring neural activity. There are essentially two recording modes: current clamp, in which the membrane potential is measured, and voltage clamp, in which currents are measured while the membrane potential is held fixed. Conductance measurements are based on these two recording modes. Most of the experimental difficulties come from two unavoidable aspects: firstly, the non-ideality of the electrode biases the measurements and causes stability problems; secondly, current can only be injected at a single point of the cell, which makes it difficult to control the potential at distal sites in the neuron.

Although electrodes and amplifiers are well established experimental devices, we might expect new developments in measuring techniques in the future, either in the way recordings are analyzed or in the way the experimental devices are controlled. We list below a few areas where new techniques might emerge in the future:

- recording techniques using numerical models of neurons and/or of the experimental apparatus (e.g. electrodes), as were introduced recently for current clamp and dynamic clamp recordings (Brette et al., 2008),
- dynamic clamp techniques–dynamic clamp recordings consist in injecting a current that depends in real time on the measured potential (Destexhe and Bal, 2009), which poses specific technical problems (Brette et al., 2009),
- single-trial conductance measurements–model based and/or statistical techniques could be used to estimate the time course of synaptic conductances in single trials.

Numerical simulations

All numerical simulations were done using the Brian simulator (Goodman and Brette, 2008), which is freely available at www.briansimulator.org. The scripts for the figures can be downloaded at www.briansimulator.org/ electrophysiology. All neuron models were single-compartment models, with either passive properties (Figures 3.4, 3.6, 3.8, 3.9, 3.10) or ionic channels with Hodgkin–Huxley type dynamics (Figures 3.3, 3.5), adapted from Mainen et al. (1995). Synaptic activity (Figures 3.5, 3.8, 3.10) was modeled as fluctuating excitatory and inhibitory conductances represented by halfwave rectified Ornstein–Uhlenbeck processes. Electrodes were modeled as RC circuits or two RC circuits in series. Amplifier models include bridge balance, capacitance neutralization, discontinuous current clamp and voltage clamp.

References

Anderson, J., Carandini, M. and Ferster, D. (2000). Orientation tuning of input conductance, excitation, and inhibition in cat primary visual cortex. *J. Neurophysiol.*, **84** (2), 909.

Azouz, R. and Gray, C. M. (1999). Cellular mechanisms contributing to response variability of cortical neurons in vivo. *J. Neurosci.*, **19** (6), 2209–2223.

Badel, L., Lefort, S., Brette, R., Petersen, C. C. H., Gerstner, W. and Richardson, M. J. E. (2008). Dynamic *I–V* curves are reliable predictors of naturalistic pyramidal-neuron voltage traces. *J. Neurophysiol.*, **99** (2), 656–666.

Bal, T. and McCormick, D. A. (1996). What stops synchronized thalamocortical oscillations? *Neuron*, **17** (2), 297.

Bar-Yehuda, D. and Korngreen, A. (2008). Space-clamp problems when voltage clamping Neurons expressing voltage-gated conductances. *J. Neurophysiol.*, **99** (3), 1127–1136.

Bédard, C. and Destexhe, A. (2008). A modified cable formalism for modeling neuronal membranes at high frequencies. *Biophys. J.*, **94** (4), 1133–1143.

Bernstein, J. (1868). Ueber den zeitlichen Verlauf der negativen Schwankung des nervenstroms. *Pflügers Archiv Eur. J. of Physiol.*, **1** (1), 173–207.

Bernstein, J. (1912). *Elektrobiologie*. Braunschweig: F. Vieweg.

Borg-Graham, L. J., Monier, C. and Fregnac, Y. (1998). Visual input evokes transient and strong shunting inhibition in visual cortical neurons. *Nature*, **393** (6683), 369.

Brennecke, R. and Lindemann, B. (1971). A chopped-current clamp for current injection and recording of membrane polarization with single electrodes of changing resistance. *T-I-T-J Life Sci.*, **1**, 53–58.

Brennecke, R. and Lindemann, B. (1974). Design of a fast voltage clamp for biological membranes, using discontinuous feedback. *Rev. Sci. Instrum.*, **45** (5), 656–661.

Brette, R., Piwkowska, Z., Monier, C., Rudolph-Lilith, M., Fournier, J., Levy, M., Frgnac, Y., Bal, T. and Destexhe, A. (2008). High-resolution intracellular recordings using a real-time computational model of the electrode. *Neuron*, **59** (3), 379–391.

Brette, R., Piwkowska, Z., Monier, C., Gomez Gonzales, J. F., Frégnac, Y., Bal, T. and Destexhe, A. (2009). Dynamic clamp with high-resistance electrodes using active electrode compensation in vitro and in vivo. In: A. Destexhe and T. Bal (editors), *Dynamic-Clamp: From Principles to Applications*, pp. 347–382. New York: Springer.

Brock, L., Coombs, J. and Eccles, J. (1952). The recording of potentials from motoneurones with an intracellular electrode. *J. Physiol.*, **117** (4), 431–460.

Cole, K. S. (1949). Dynamic electrical characteristics of the squid axon membrane. *Arch. Sci. Physiol.*, **3** (25), 3–25.

Cole, K. S. and Curtis, H. J. (1939). Electric impedance of the squid giant axon during activity. *J. Gen. Physiol.*, **22** (5), 649–670.

Dayan, P. and Abbott, L. F. (2001). *Theoretical Neuroscience*. Cambridge, MA: MIT Press.

de Polavieja, G. G., Harsch, A., Kleppe, I., Robinson, H. P. C. and Juusola, M. (2005). Stimulus history reliably shapes action potential waveforms of cortical neurons. *J. Neurosc.*, **25** (23), 5657–5665.

de Sa, V. R. and MacKay, D. J. (2001). Model fitting as an aid to bridge balancing in neuronal recording. *Neurocomputing*, **38–40**, 1651–1656.

Destexhe, A. and Bal, T. (editors), (2009). *Dynamic-Clamp: From Principles to Applications*. New York: Springer.

Destexhe, A. and Rudolph, M. (2004). Extracting information from the power spectrum of synaptic noise. *J. Comput. Neurosci.*, **17** (3), 327–345.

Destexhe, A., Rudolph, M., Fellous, J. M. and Sejnowski, T. J. (2001). Fluctuating synaptic conductances recreate in vivo-like activity in neocortical neurons. *Neuroscience*, **107** (1), 13.

Destexhe, A., Rudolph, M. and Pare, D. (2003). The high-conductance state of neocortical neurons in vivo. *Nat. Rev. Neurosci.*, **4** (9), 739.

DeWeese, M. R., Wehr, M. and Zador, A. M. (2003). Binary spiking in auditory cortex. *J. Neurosci.*, **23** (21), 7940–7949.

Engel, A. K., Fries, P. and Singer, W. (2001). Dynamic predictions: oscillations and synchrony in top-down processing. *Nat. Rev. Neurosci.*, **2** (10), 704.

Finkel, A. S. and Redman, S. (1984). Theory and operation of a single microelectrode voltage clamp. *J. Neurosci. Methods*, **11** (2), 101–127.

Galvani, L. (1791). *De viribus electricitatis in motu musculari: Commentarius*. Bologna: Tip. Istituto delle Scienze.

Goodman, D. and Brette, R. (2008). Brian: a simulator for spiking neural networks in python. *Frontiers Neuroinformatics*, **2**, 5.

Guillamon, A., McLaughlin, D. W. and Rinzel, J. (2006). Estimation of synaptic conductances. *J. Physiol. Paris*, **100** (1–3), 31–42.

Henze, D. A. and Buzsaki, G. (2001). Action potential threshold of hippocampal pyramidal cells in vivo is increased by recent spiking activity. *Neuroscience*, **105** (1), 121–30.

Hille, B. (2001). *Ion Channels of Excitable Membranes.* Sinauer Associates.

Hodgkin, A. and Huxley, A. (1939). Action potentials recorded from inside a nerve fibre. *Nature*, **144** (3651), 710.

Hodgkin, A. and Huxley, A. (1952). A quantitative description of membrane current and its application to conduction and excitation in nerve. *J. Physiol.* (London), **117**, 500.

Hodgkin, A. L. and Katz, B. (1949). The effect of sodium ions on the electrical activity of the giant axon of the squid. *J. of Physiol.*, **108** (1), 37.

Koch, C. (1999). *Biophysics of Computation: Information Processing in Single Neurons.* Oxford University Press.

Koch, C., Douglas, R. and Wehmeier, U. (1990). Visibility of synaptically induced conductance changes: theory and simulations of anatomically characterized cortical pyramidal cells. *J. Neurosci.*, **10** (6), 1728–1744.

Lampl, I., Reichova, I. and Ferster, D. (1999). Synchronous membrane potential fluctuations in neurons of the cat visual cortex. *Neuron*, **22** (2), 361.

Leger, J. F., Stern, E. A., Aertsen, A. and Heck, D. (2005). Synaptic integration in rat frontal cortex shaped by network activity. *J. Neurophysiol.*, **93** (1), 281–293.

Lindner, B. and Longtin, A. (2011). Comment on "Characterization of subthreshold voltage fluctuations in neuronal membranes," by M. Rudolph and A. Destexhe. *Neural Comput.*, **18** (8), 1896–1931.

Ling, G. and Gerard, R. (1949). The normal membrane potential of frog sartorius fibers. *J. Cell. Physiol.*, **34** (3), 383–96.

Mainen, Z. and Sejnowski, T. (1995). Reliability of spike timing in neocortical neurons. *Science*, **268**, 1503.

Mainen, Z. F., Joerges, J., Huguenard, J. R. and Sejnowski, T. J. (1995). A model of spike initiation in neocortical pyramidal neurons. *Neuron*, **15** (6), 1427–1439.

Marmont, G. (1949). Studies on the axon membrane: a new method. *J. Cell. Physiol.*, **34** (3), 351–382.

Monier, C., Chavane, F., Baudot, P., Graham, L. J. and Fregnac, Y. (2003). Orientation and direction selectivity of synaptic inputs in visual cortical neurons: a diversity of combinations produces spike tuning. *Neuron*, **37** (4), 663.

Monier, C., Fournier, J. and Fregnac, Y. (2008). In vitro and in vivo measures of evoked excitatory and inhibitory conductance dynamics in sensory cortices. *J. Neurosci. Methods*, **169** (2), 323–365.

Neher, E. and Sakmann, B. (1976). Single-channel currents recorded from membrane of denervated frog muscle fibres. *Nature*, **260** (5554), 799–802.

Pare, D., Shink, E., Gaudreau, H., Destexhe, A. and Lang, E. J. (1998). Impact of spontaneous synaptic activity on the resting properties of cat neocortical pyramidal neurons in vivo. *J. Neurophysiol.*, **79** (3), 1450.

Pospischil, M., Piwkowska, Z., Rudolph, M., Bal, T. and Destexhe, A. (2007). Calculating event-triggered average synaptic conductances from the membrane potential. *J. Neurophysiol.*, **97** (3), 2544.

Piwkowska, Z., Pospischil, M., Brette, R., Sliwa, J., Rudolph-Lilith, M., Bal, T. and Destexhe, A. (2008). Characterizing synaptic conductance fluctuations in cortical neurons and their influence on spike generation. *J. Neurosci. Methods*, **169** (2), 302–322.

Pospischil, M., Piwkowska, Z., Bal, T. and Destexhe, A. (2009). Extracting synaptic conductances from single membrane potential traces. *Neuroscience*, **158** (2), 545–552.

Purves, R. D. (1981). *Microelectrode Methods for Intracellular Recording and Ionophoresis.* New York: Academic Press.

Richardson, M. J. E. (2004). Effects of synaptic conductance on the voltage distribution and firing rate of spiking neurons. *Phys. Rev. E*, **69** (5), 051918.

Rudolph, M. and Destexhe, A. (2003). Characterization of subthreshold voltage fluctuations in neuronal membranes. *Neural Comput.*, **15** (11), 2577.

Rudolph, M. and Destexhe, A. (2011). On the use of analytical expressions for the voltage distribution to analyze intracellular recordings. *Neural Comput.*, **18** (12), 2917–2922.

Rudolph, M., Piwkowska, Z., Badoual, M., Bal, T. and Destexhe, A. (2004). A method to estimate synaptic conductances from membrane potential fluctuations. *J. Neurophysiol.*, **91** (6), 2884–2896.

Rudolph, M., Pelletier, J. G., Paré, D. and Destexhe, A. (2005). Characterization of synaptic conductances and integrative properties during electrically induced EEG-activated states in neocortical neurons in vivo. *J. Neurophysiol.*, **94** (4), 2805–2821.

Rudolph, M., Pospischil, M., Timofeev, I. and Destexhe, A. (2007). Inhibition determines membrane potential dynamics and controls action potential generation in awake and sleeping cat cortex. *J. Neurosci.*, **27** (20), 5280–5290.

Sakmann, B. and Neher, E. (1995). *Single-Channel Recording*. New York: Plenum Press.

Sherman-Gold, R. (1993). *The Axon Guide for Electrophysiology and Biophysics: Laboratory Techniques*. Foster City, CA: Axon Instruments.

Sigworth, F. J. and Neher, E. (1980). Single Na channel currents observed in cultured rat muscle cells. *Nature*, **287** (2), 447.

Staley, K. J., Otis, T. S. and Mody, I. (1992). Membrane properties of dentate gyrus granule cells: comparison of sharp microelectrode and whole-cell recordings. *J. Neurophysiol.*, **67** (5), 1346–1358.

Stuart, G., Spruston, N. and Hausser, M. (1999). *Dendrites*. Oxford University Press.

Tuckwell, H. (1988). *Introduction to Theoretical Neurobiology*, Vol 1: *Linear Cable Theory and Dendritic Structure*. Cambridge: Cambridge University Press.

Umrath, K. (1930). Untersuchungen über Plasma und Plasmaströmung an Characeen. *Protoplasma*, **9** (1), 576–597.

Volgushev, M., Pernberg, J. and Eysel, U. (2002). A novel mechanism of response selectivity of neurons in cat visual cortex. *J. Physiol.*, **540** (1), 307.

Wilent, W. B. and Contreras, D. (2005). Stimulus-dependent changes in spike threshold enhance feature selectivity in rat barrel cortex neurons. *J. Neurosci.*, **25** (11), 2983–2991.

Zou, Q., Rudolph, M., Roy, N., Sanchez-Vives, M., Contreras, D. and Destexhe, A. (2005). Reconstructing synaptic background activity from conductance measurements in vivo. *Neurocomputing*, **65–66**, 673–678.

4

Extracellular spikes and CSD

KLAS H. PETTERSEN, HENRIK LINDÉN, ANDERS M. DALE AND GAUTE T. EINEVOLL

4.1 Introduction

Extracellular recordings have been, and still are, the main workhorse when measuring neural activity in vivo. In single-unit recordings sharp electrodes are positioned close to a neuronal soma, and the firing rate of this particular neuron is measured by counting *spikes*, that is, the standardized extracellular signatures of action potentials (Gold et al., 2006). For such recordings the interpretation of the measurements is straightforward, but complications arise when more than one neuron contributes to the recorded extracellular potential. For example, if two firing neurons of the same type are at about the same distance from their somas to the tip of the recording electrode, it may be very difficult to sort the spikes according to from which neuron they originate.

The use of two (*stereotrode* (McNaughton et al., 1983)), four (*tetrode* (Recce and O'Keefe, 1989; Wilson and McNaughton, 1993; Gray et al., 1995; Jog et al., 2002)) or more (Buzsáki, 2004) close-neighbored recording sites allows for improved spike sorting, since the different distances from the electrode tips or contacts allow for triangulation. With present recording techniques and clustering methods one can sort out spike trains from tens of neurons from single tetrodes and from hundreds of neurons with multi-shank electrodes (Buzsáki, 2004).

Information about spiking is typically extracted from the high-frequency band ($\gtrsim 500$ Hz) of extracellular potentials. Since these high-frequency signals generally stem from an unknown number of spiking neurons in the immediate vicinity of the electrode contact, this is called *multi-unit activity (MUA)*. The low-frequency part ($\lesssim 500$ Hz) of extracellular potentials is called the *local field potential (LFP)*. In in vivo recordings the LFP is typically due to dendritic processing of synaptic inputs, not firing of action potentials (Mitzdorf, 1985; Einevoll et al., 2007; Pettersen et al., 2008; Lindén et al., 2010). The interpretation of the LFP is difficult as it is a less local measure of neural activity than MUA; the LFP measured at any point will

Handbook of Neural Activity Measurement, ed. Romain Brette and Alain Destexhe. Published by Cambridge University Press. © Cambridge University Press 2012.

typically have sizable contributions from neurons located several hundred micrometers away (Kreiman et al., 2006; Liu and Newsome, 2006; Leski et al., 2007; Berens et al., 2008; Pettersen et al., 2008; Lindén et al., 2009a, 2009b, 2010; Xing et al., 2009). The analysis of LFP data has thus generally been restricted to the estimation of *current source density (CSD)*, the volume density of net transmembrane currents through the neuronal membranes (Nicholson and Freeman, 1975; Mitzdorf, 1985; Pettersen et al., 2006), based on linear (laminar) multi-electrode recordings (Rappelsberger et al., 1981; Di et al., 1990; Ulbert et al., 2001; Einevoll et al., 2007; Pettersen and Einevoll, 2008). While CSD analysis cannot separate contributions from different spatially intermingled neuronal populations (unlike the newly developed *laminar population analysis (LPA)* (Einevoll et al., 2007)), the CSD is still easier to interpret than the less localized LFP signal. New silicon-based multi-contact probes in various other geometrical arrangements, such as "multi-shank" (Buzsáki, 2004) or "needlepad" (Normann et al., 1999), are rapidly being developed, and the *inverse current source density (iCSD)* method has been introduced to estimate CSDs in such situations (Pettersen et al., 2006; Leski et al., 2007, 2011).

The estimation of CSD from the measured LFP is a so-called "inverse problem" which cannot be solved without imposing additional constraints on the form of the CSD (Nicholson and Freeman, 1975; Pettersen et al., 2006; Leski et al., 2007, 2011). However, the corresponding "forward problem", i.e. calculation of the LFP from a known CSD distribution, is well posed (Nicholson and Freeman, 1975; Pettersen et al., 2006; Einevoll et al., 2007; Leski et al., 2007). Likewise, the extracellular potential generated by neurons, both the LFP and the MUA, can be calculated if one knows all the transmembrane currents and their respective spatial positions, as well as the extracellular conductivity in the surrounding medium (Holt and Koch, 1999; Gold et al., 2006; Einevoll et al., 2007; Pettersen and Einevoll, 2008; Pettersen et al., 2008; Lindén et al., 2010).

In the 1960s Rall used such a neuronal forward-modeling scheme to calculate extracellular potentials related to action potential firing and synaptic interaction using simplified equivalent-cylinder geometries (Rall, 1962; Rall and Shepherd, 1968). Thirty years later Holt and Koch combined this scheme with compartmental modeling based on morphologically reconstructed pyramidal neurons, to calculate the extracellular signature of an action potential (Holt and Koch, 1999). This modeling scheme was later used to calculate other extracellular spike signatures of single neurons (Gold et al., 2006, 2007; Milstein and Koch, 2008; Pettersen and Einevoll, 2008), MUA from populations of firing neurons (Pettersen et al., 2008), and LFP from synaptically activated neurons and neuronal populations (Einevoll et al., 2007; Pettersen et al., 2008; Lindén et al., 2010). A convenient feature of the forward-modeling scheme is that due to the linearity of Maxwell's equations, the contributions to the extracellular potential from the various neuronal sources add up linearly, and the calculation of extracellular potentials from joint activity in populations with thousands of morphologically reconstructed neurons may even be done on desktop computers (Pettersen et al., 2008).

In the next section we describe the biophysical origin of the extracellular potentials and the mathematical formalism connecting it to the underlying neural activity. In Section 4.3 we illustrate the biophysical forward-modeling scheme by investigating the LFP generated by a single pyramidal neuron activated by apical synapses. This example also illustrates some general salient features of the LFP, in particular an unavoidable low-pass filtering effect due to the dendritic distribution of transmembrane return currents (Lindén et al., 2010) (also in the absence of inherent frequency-dampening in the extracellular medium (Bédard et al., 2006a; Logothetis et al., 2007)). In Section 4.4 we describe results from a forward-modeling study of the influence of the dendritic morphology on the size and shape of the extracellular spike (Pettersen and Einevoll, 2008), and in Section 4.5 we correspondingly investigate the LFP and MUA generated by a synaptically activated model population of about 1000 morphologically reconstructed pyramidal neurons, mimicking the sensory-evoked response in a population of layer-5 neurons in rat whisker (barrel) cortex (Pettersen et al., 2008). In Section 4.6 we discuss the problem of CSD estimation, and in particular outline the principles behind the iCSD method (Pettersen et al., 2006; Leski et al., 2007, 2011). Some concluding remarks are given in the final section.

4.2 Biophysical origin of extracellular potentials

From an electrical point of view cortical tissue consists of a tightly packed collection of neurons and other cells embedded in a low-resistance extracellular medium filling less than a fifth of the total volume (Nunez and Srinivasan, 2006). The low resistance of the extracellular medium implies that neighboring cells typically are electrically decoupled and that the difference between the extracellular potential recorded at different positions will be small, typically less than a millivolt. In contrast, the potential difference across the highly resistant cell membranes, that is, the membrane potential, is typically between 50 and 100 mV.

4.2.1 Biophysical forward-modeling formula

The extracellular potentials are generated by transmembrane currents, and in the commonly used *volume conductor theory* the system can be envisioned as a three-dimensional smooth extracellular continuum with the transmembrane currents represented as *volume current sources* (Nunez and Srinivasan, 2006). In this theoretical framework the fundamental relationship describing the extracellular potential $\phi(t)$ at position $\mathbf{r}$ due to a transmembrane current $I_0(t)$ at position $\mathbf{r}_0$ is given by (Hämäläinen et al., 1993; Nunez and Srinivasan, 2006)

$$\phi(\mathbf{r}, t) = \frac{1}{4\pi\sigma} \frac{I_0(t)}{|\mathbf{r} - \mathbf{r}_0|}. \tag{4.1}$$

Here the extracellular potential ϕ is set to be zero infinitely far away from the transmembrane current, and σ is the *extracellular conductivity*, assumed to be *real*, *scalar* (the same in all directions) and *homogeneous* (the same at all positions).

The validity of Equation (4.1) relies on several assumptions.

1. *Quasistatic approximation of Maxwell's equations.* This amounts to neglecting the terms with the time derivatives of the electric field $\mathbf{E}$ and the magnetic field $\mathbf{B}$ from the original Maxwell equation, i.e.

$$\nabla \times \mathbf{E} = -\frac{\partial \mathbf{B}}{\partial t} \approx 0, \tag{4.2}$$

$$\nabla \times \mathbf{B} = \mu_0 \mathbf{j} + \mu_0 \epsilon_0 \frac{\partial \mathbf{E}}{\partial t} \approx \mu_0 \mathbf{j}, \tag{4.3}$$

so that the electric (Equation (4.2)) and magnetic (Equation (4.3)) field equations effectively decouple (Hämäläinen et al., 1993). With $\nabla \times \mathbf{E} = 0$ it follows that the electric field $\mathbf{E}$ in the extracellular medium is related to an extracellular potential ϕ via

$$\mathbf{E} = -\nabla \phi. \tag{4.4}$$

For the frequencies inherent in neural activity, i.e. less than a few thousand hertz, the quasistatic approximation seems to be well justified (see, e.g., argument on p. 426 of Hämäläinen et al. (1993)).

2. *Linear extracellular medium.* A linear relationship is assumed between the current density $\mathbf{j}$ and the electrical field $\mathbf{E}$,

$$\mathbf{j} = \sigma \mathbf{E}. \tag{4.5}$$

This constitutive relation is quite general, and σ in Equation (4.5) may in principle be (i) a *tensor*, accounting for different conductivities in different directions (Nicholson and Freeman, 1975), (ii) *complex*, accounting also for capacitive effects (Nunez and Srinivasan, 2006), and/or (iii) *position dependent*, that is, varying with spatial position. (Note that Equation (4.5) is valid only in the frequency domain. In the time domain $\mathbf{j}$ is generally given as a temporal convolution of σ and $\mathbf{E}$ (Bédard and Destexhe, 2009). However, in the case of a frequency independent σ, cf. point 5 below, Equation (4.5) will also be valid in the time domain.)

3. *Ohmic (resistive) medium.* The imaginary part of the conductivity σ is assumed to be zero, that is, the capacitive effects of the neural tissue are assumed to be negligible compared to resistive effects. This appears to be well fulfilled for the relevant frequencies in extracellular recordings (Nunez and Srinivasan, 2006; Logothetis et al., 2007).

4. *Isotropic (scalar) extracellular conductivity.* Conductivity σ is assumed to be the same in all directions, i.e. $\sigma_x = \sigma_y = \sigma_z = \sigma$. Measurements from visual cortex in monkeys indeed found the conductivities to be comparable across different directions in cortical gray matter; in white matter, however, the conductivity was found to be anisotropic (Logothetis et al., 2007). However, recent measurements from somatosensory barrel cortex in rats found the conductivity in the vertical direction to be up to a factor of two larger than in the horizontal directions (Goto et al., 2010). Early measurements on frog and toad cerebella also revealed anisotropy in the conductivity (Nicholson and Freeman, 1975).

5. *Frequency-independent extracellular conductivity.* Conductivity σ is assumed to be the same for all relevant frequencies, i.e. $\sigma(\omega)$ is constant. The validity of this assumption is still debated: while some studies have measured negligible frequency dependence (Nicholson and Freeman, 1975; Logothetis et al., 2007), other investigations have suggested otherwise (Gabriel et al., 1996; Bédard et al., 2004, 2006a, 2006b); cf. Chapter 5.

6. *Homogeneous extracellular conductivity.* Extracellular medium is assumed to have the same conductivity everywhere. This appears to be roughly fulfilled within cortical gray matter (Logothetis et al., 2007; Goto et al., 2010) and frog and toad cerebella (Nicholson and Freeman, 1975), but perhaps not in the hippocampus (López-Aguado et al., 2001). Further, white matter has a lower conductivity than cortical gray matter which in turn has a lower conductivity than the cell-free cerebral spinal fluid (CSF) (Nunez and Srinivasan, 2006).

While Equation (4.1) requires all assumptions 1–6 to be fulfilled, the expression can be generalized to apply also for other situations for example the following.

- If assumption 5 is violated and σ varies with frequency, Equation (4.1) can still be used separately for each Fourier component $\hat{I}_0(\omega)$ of the transmembrane current $I_0(t)$ with $\sigma(\omega)$ inserted in the denominator of the equation. Since the extracellular potential ϕ is linear in the transmembrane current I_0, a simple Fourier sum over the contributions from all Fourier components will provide the total extracelluar potential (Pettersen and Einevoll, 2008); see also Chapter 5.
- For the case where the conductivity is anisotropic, i.e. assumption 4 is violated, the equations still apply if the denominator $4\pi\sigma|\mathbf{r} - \mathbf{r}_0|$ is

replaced by $4\pi\sqrt{\sigma_y\sigma_z(x-x_0)^2+\sigma_z\sigma_x(y-y_0)^2+\sigma_x\sigma_y(z-z_0)^2}$ (Nicholson and Freeman, 1975).

- In situations with piecewise constant conductivities, for example with disconti-nuities in σ at the interfaces between gray and white matter or between the gray matter and the cortical surface, assumption 6 is violated. However, a general-ized version of Equation (4.1) can be derived based on the "method of images" (Nicholson and Llinas, 1971; Gold et al., 2006; Pettersen et al., 2006; Einevoll et al., 2007).

Equation (4.1) applies to the situation with a single transmembrane current I_0, but since contributions from several transmembrane current sources add linearly, the equation generalizes straightforwardly to a situation with many transmembrane current sources. With N current point sources, Equation (4.1) generalizes to

$$\phi(\mathbf{r},t)=\frac{1}{4\pi\sigma}\sum_{n=1}^{N}\frac{I_n(t)}{|\mathbf{r}-\mathbf{r}_n|}. \tag{4.6}$$

In Figure 4.1 we illustrate this formula for the situation where all transmembrane currents comes from a single compartmentalized "ball-and-stick" neuron; it is clear that the measured extracellular potential will depend not only on the position of the electrode, but also on the distribution of transmembrane currents.

Figure 4.1 further illustrates an important "conservation" law when calculating extracellular potentials due to neural activity: Kirchhoff's current law implies that

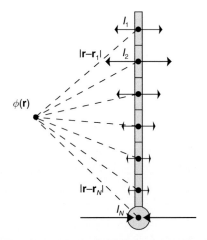

Figure 4.1 Illustration of the mathematical formula Equation (4.6) providing the extracellular potential from transmembrane currents in a single neuron. The size and direction of the arrows illustrate the amplitudes and directions of the transmembrane currents.

the net transmembrane current (including the capacitive current) coming out of a neuron at all times must equal zero. Thus with the neuron depicted in Figure 4.1 divided into N compartments, one must at all times have $\sum_{n=1}^{N} I_n(t) = 0$. Therefore a one-compartment model cannot generate any extracellular potential since the net transmembrane current necessarily will be zero. The simplest model producing an extracellular potential is a two-compartment model where transmembrane current entering the neuron at one compartment leaves at the other compartment. The simplest possible multipole configuration is thus the current *dipole*.

4.2.2 Numerical forward-modeling scheme

The numerical evaluation of extracellular potentials can be separated into two steps (Holt and Koch, 1999; Pettersen and Einevoll, 2008; Pettersen et al., 2008; Lindén et al., 2010):

1. calculation of transmembrane currents for all neuronal membrane segments using multi-compartment neuron models (Segev and Burke, 1998), typically using neural simulation tools such as NEURON (Carnevale and Hines, 2006) or Genesis (Bower and Beeman, 1998),
2. calculation of the extracellular potential on the basis of the modeled transmembrane currents and their spatial position using a forward-modeling formula similar to Equation (4.6).

When a neuron is split into N compartments, the formula in Equation (4.6) should be used with $\mathbf{r}_n$ corresponding to a characteristic "mean" position for compartment n, for example the center of a spherical soma compartment or the mid-point of a cylindrical dendritic compartment. This scheme corresponds to the so-called *point-source* approximation (Holt and Koch, 1999; Pettersen and Einevoll, 2008) since all transmembrane currents into the extracellular medium from a particular compartment are assumed to go through a single point. Another scheme, the *line-source* approximation, assumes the transmembrane currents from each cylindrical compartment to be evenly distributed along a line corresponding to the cylinder axis (Holt and Koch, 1999; Pettersen and Einevoll, 2008). A line-source formula, analogous to the point-source formula in Equation (4.6), can be found in Pettersen and Einevoll (2008) (Eq. 2). Unless otherwise noted, all forward-modeling calculations with morphologically reconstructed neurons presented in this chapter use the line-source approximation. Further, a frequency-independent, scalar and homogeneous extracellular conductivity with a numerical value of $\sigma = 0.3$ S m^{-1} (Hämäläinen et al., 1993) is assumed.

4.2.3 *Current source density (CSD)*

The forward-modeling formula in Equation (4.6) can be reformulated mathematically as

$$\phi(\mathbf{r}, t) = \frac{1}{4\pi\sigma} \iiint_V \frac{C(\mathbf{r}', t)}{|\mathbf{r} - \mathbf{r}'|} d^3 r', \tag{4.7}$$

when we introduce the quantity $C(\mathbf{r}, t) \equiv \sum_{n=1}^{N} I_n(t) \delta^3(\mathbf{r} - \mathbf{r}_n)$. Here $\delta^3(\mathbf{r})$ is the three-dimensional Dirac δ-function, and the volume integral goes over all transmembrane currents. The quantity $C(\mathbf{r}, t)$ is called the *current source density (CSD)*, has dimension A m^{-3}, and is in general interpreted as the volume density of current entering or leaving the extracellular medium at position $\mathbf{r}$ (Nicholson and Freeman, 1975; Mitzdorf, 1985; Nunez and Srinivasan, 2006). A negative $C(\mathbf{r}, t)$ corresponds to current leaving the extracellular medium and is thus conventionally called a *sink*. Likewise, current entering the extracellular medium is called a *source*. The CSD is easier to relate to the underlying neural activity than the extracellular potential itself, and current source density analysis has thus become a standard tool for analysis of the low-frequency part (LFP) of such potentials recorded with linear (laminar) multi-electrodes (Nicholson and Freeman, 1975; Pettersen et al., 2006).

While Equation (4.7) gives the numerical recipe for calculating the extracellular potential given the CSD, a formula providing the opposite relationship can also be derived. Following Nicholson and Llinas, (1971), Nicholson and Freeman, (1975) and Nunez and Srinivasan (2006), we have for the situation with an Ohmic extracellular medium that current conservation requires

$$\nabla \cdot \mathbf{j}_{tot} = \nabla \cdot (\sigma \mathbf{E} + \mathbf{j}_s) = 0, \tag{4.8}$$

where $\mathbf{j}_s$ is the so-called *impressed* transmembrane currents entering the extracellular medium (Nicholson and Llinas, 1971; Nunez and Srinivasan, 2006). With the additional use of Equation (4.4) one obtains

$$\nabla \cdot \left(\sigma(\mathbf{r})\nabla\phi(\mathbf{r}, t) \right) = -C(\mathbf{r}, t), \tag{4.9}$$

where $C(\mathbf{r}, t) \equiv -\nabla \cdot \mathbf{j}_s(\mathbf{r}, t)$. This equation is not only valid for the case with position-dependent σ, but also when it depends on direction, i.e. is a tensor (Nicholson and Freeman, 1975). In the special case where σ is isotropic and homogeneous, the equation simplifies to

$$\sigma\nabla^2\phi(\mathbf{r}, t) = -C(\mathbf{r}, t). \tag{4.10}$$

This equation, called Poisson's equation, is well known from standard electrostatics where it describes how potentials are generated by electrical charges (with the conductivity σ replaced by the dielectric constant ε) (Jackson, 1998). However, emphasized by Nunez and Srinivasan (2006), these two versions of Poisson's equation represent different physical processes.

4.3 Local field potential (LFP) from a single neuron

4.3.1 Characteristic features of the LFP

To illustrate the forward-modeling scheme and highlight some salient features of the LFP we here calculate the extracellular potential around a reconstructed layer-5 model pyramidal neuron from cat visual cortex (Mainen and Sejnowski, 1996) receiving a single excitatory synaptic input in the apical dendrite. For simplicity the neuron is considered to have purely passive neuronal membranes and to be excited by a synaptic input current $I_s(t)$ modeled as an α-function, that is,

$$I_s(t) = I_0\, t/\tau_s\, e^{1-t/\tau_s}\, \theta(t), \tag{4.11}$$

where $\theta(t)$ is the Heaviside unit step function. A time constant $\tau_s = 1$ ms is chosen, and I_0 is set to give a peak EPSP amplitude in the soma of about 0.5 mV. The model is linear, that is, all calculated extracellular and intracellular potentials are proportional to I_0, making the model somewhat easier to analyze than when non-linear currents are involved. However, most qualitative features are expected to be unchanged if we, for example, consider excitation by a set of conductance-based synapses instead.

In Figure 4.2A we show the calculated extracellular potential traces at a set of positions outside the neuron. An important feature which is immediately apparent is that the shape and amplitude of the extracellular potentials depend on position. Near the apical synaptic input the extracellular signature is always negative, reflecting that the excitatory current-synapse providing a current sink dominates the sum in the forward-model formula, cf. Equation (4.6). At positions close to the soma the extracellular potential is always positive, reflecting that return currents in the soma area dominate the sum. At other positions, for example above the synapse, a biphasic extracellular potential is observed. Interestingly, there is not a monotonous decay of the amplitude with distance from the synaptic input: large extracellular responses are observed close to the soma, almost a millimeter away.

Another important feature is the observed increased half-width of the extracellular potentials recorded close to the soma compared to those in the vicinity of the synaptic input. This is illustrated by the two insets showing magnified extracellular potential traces in Figure 4.2A. In the upper inset close to the synapse the width is 4.2 ms, while the width at the lower inset close to soma is 7.1 ms, both widths measured at 50% of the trace's peak amplitudes. Thus the extracellular potential close to the synaptic input contains higher frequencies than the extracellular potential far away from the synaptic current generator.

This feature can be understood on the basis of passive cable properties of the neuron. The transmembrane currents dominating the extracellular potentials close to the soma have been low-pass filtered and have a wider temporal profile compared to the transmembrane currents close to the synaptic input. This is illustrated in Figure 4.2C where the transmembrane current profile is seen to have a much larger half-width at the soma ($\sim$6.5 ms) compared to at the dendritic segment containing the synapse ($\sim$2.5 ms).

An analogous low-pass filtering is seen from the temporal shapes of the apical and somatic membrane potentials, respectively, in Figure 4.2E. Here the apical EPSP peaks already a couple of milliseconds after synaptic onset and has a half-width of about 4 ms. In contrast the somatic EPSP peaks about 15 ms after synaptic onset and has a half-width of more than 30 ms. The low-pass filtering effect is thus stronger for the membrane potential than for the transmembrane current, and thus also compared to the extracellular potentials.

In Figure 4.2B we also show calculated extracellular potential traces for an analogous two-compartment model, the simplest neuron model that produces an extracellular potential. The spatial extension corresponds to the distance between the single synapse and the soma for the reconstructed neuron in Figure 4.2A. This model has only five parameters, the resistances (r_a, r_s) and capacitances (c_a, c_s) of the apical and soma compartments, respectively, and the intercompartment resistance (r_{as}).

The pattern of extracellular responses in the two-compartment model is seen to resemble the pattern for the reconstructed pyramidal neuron in that large negative responses are observed close to the apical compartment while large positive responses are observed close to the soma compartment. However, in the two-compartment model the net transmembrane current in the soma compartment is forced by Kirchhoff's current law to be identical in size, but with opposite sign, compared to the apical compartment. What goes in at one compartment, must leave at the other. Since only these two compartments contribute to the sum in the forward-modeling formula for the extracellular potential (that is, $N = 2$ in Equation (4.6)), the temporal form of the extracellular potential will be the same everywhere; only the sign and size of an overall amplitude will vary. This is illustrated by the two insets showing magnified extracellular traces in Figure 4.2B which both have half-widths of 11.3 ms. There is thus no *position-dependent* filtering of frequency components in the two-compartment model. At least three neuron compartments are needed to capture such an effect.

There is, however, low-pass filtering also inherent in the two-compartment model as illustrated by the larger half-width of the extracellular potential (11.3 ms)

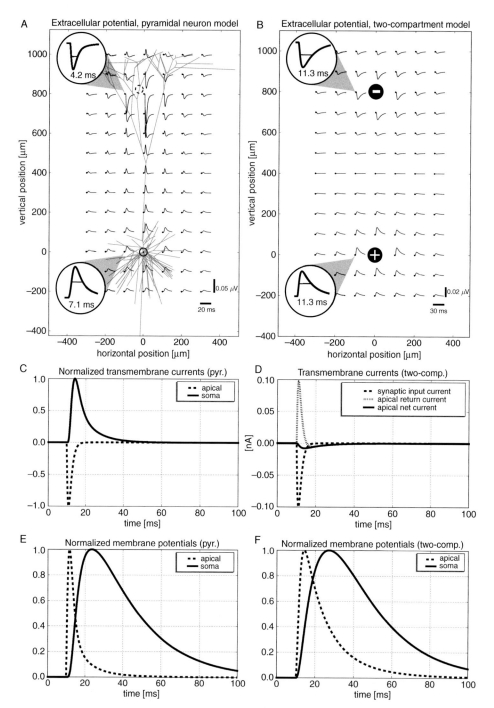

Figure 4.2 Calculated extracellular potentials following an excitatory synaptic input into purely passive neuron models. The synapse is current-based and modeled as an α-function $I_s(t - t_{on})$ (Equation (4.11)) with $\tau_s = 1$ ms, $I_0 = 0.1$ nA and the onset time t_{on} set to 10 ms. A. Results for reconstructed L5 pyramidal

observed in Figure 4.2B compared to the half-width of the synaptic input current (2.5 ms) in Figure 4.2D. This reflects that in a two-compartmental model like this, where both compartments have a resistive and a capacitive component, the axial current going between the compartments is not equal to the imposed synaptic current in the apical compartment. Instead it is the difference between the synaptic current and the return current of the apical compartment. This axial current corresponds in magnitude to the net transmembrane currents at the two compartments, and as illustrated in Figure 4.2D these net transmembrane currents are both smaller in amplitude and temporally wider than the synaptic current. In Figure 4.2F we in fact observe an even larger low-pass filtering effect for the membrane potential compared to results for the reconstructed model neuron in Figure 4.2E.

In Lindén et al., (2010) we discuss in detail how the LFP patterns depend on neuronal morphologies, spatial positions of the driving synapse, as well as electrode recording positions.

4.3.2 Low-pass filtering of the LFP

The frequency content of LFP and EEG signals has attracted significant interest, in particular since power laws, i.e. power spectra scaling as $1/f^\beta$, have commonly been observed (Pritchard, 1992; Linkenkaer-Hansen et al., 2001; Freeman et al., 2003; Bédard et al., 2006a; Buszáki, 2006; Monto et al., 2008; Bédard and

---

Caption for Figure 4.2 (cont.)

neuron from Mainen and Sejnowski (1996) with active channels removed. Passive parameters: membrane resistivity R_m = 30000 $\Omega\,\mathrm{cm}^2$, axial resistivity R_i = 150 $\Omega\,\mathrm{cm}$, membrane capacitance C_m = 0.75 $\mu\mathrm{F\,cm}^{-2}$. Potentials are shown in a 20 ms window starting 2 ms prior to synaptic onset. Dashed circle denotes position of the synapse. B. Results for an analogous two-compartment neuron model. The apical (top) and soma (bottom) compartments have resistive (r_a, r_s) and capacitive (c_a, c_s) membrane elements, and are connected to each other via the resistance r_{as}. The same synaptic current as in A is inserted into the apical compartment. Model parameters: r_a = 318 MΩ, r_s = 95 MΩ, r_{as} = 358 MΩ, c_a = 71 pF, c_s = 236 pF. The point-source approximation is used, cf. Equation (4.6). C. Normalized transmembrane currents at the synaptic input segment and at the soma for the pyramidal neuron in A. Half-widths are 2.5 ms and 6.5 ms, respectively. D. Synaptic input current, return current, and net transmembrane current for the apical compartment in the two-compartment model. Half-widths are 2.5 ms, 2.3 ms and 5.2 ms, respectively. E. Normalized membrane potential for synaptic input segment and soma segment for the pyramidal neuron model. Half-widths are 4.1 ms and 33 ms, respectively. F. Normalized membrane potential of apical and soma compartments of the two-compartment model. Half-widths are 13 ms and 38 ms, respectively. Extracellular potentials in insets in A and B are scaled arbitrarily.

Destexhe, 2009; Miller et al., 2009; Milstein et al., 2009; He et al., 2010). Suggested explanations for these observed power laws have invoked a variety of neural network mechanisms (Levina et al., 2007; Freeman and Zhai, 2009; Miller et al., 2009; Milstein et al., 2009), as well as frequency filtering inherent in the extracellular medium (Bédard and Destexhe, 2006a, 2009; Freeman and Zhai, 2009). The results above, elaborated in Lindén et al. (2010), point to an additional source of frequency filtering of the LFP and EEG: extended dendritic morphologies, because of their passive cable properties, will unavoidably give a separate frequency-filtering effect for the extracellular potentials. In fact there are two dendrite-based filtering mechanisms: (i) a higher fraction of the apical synaptic input current will propagate to the soma for low frequencies than for high frequencies, and (ii) extracellular potentials recorded far away from the synaptic input current will have more low frequencies than those recorded close to the input current due to the low-pass filtering of the return current by the dendritic tree. The simple two-compartment model only displayed the first type of filtering, while the reconstructed pyramidal neuron model displayed both types.

A comprehensive investigation of these filtering effects is beyond the scope of this chapter; we refer to Lindén et al. (2010). However, some example results illustrating the important principles are shown in Figure 4.3. The same pyramidal neuron as in Figure 4.2A is considered, now with sinusoidal currents $I_s(t) = I_0 \cos(2\pi f t)$ inserted at ten apical synapses. The extracellular potential is simulated along an axis oriented perpendicular to the primary apical dendrite at the level of the soma.

The amplitude of the extracellular potential is plotted in the main (middle) panel. The most obvious feature is the difference in amplitude in the extracellular potential for the different frequencies: the amplitude is much larger for the lowest frequency ($f = 10$ Hz) than for the highest frequency ($f = 100$ Hz), even with the same input current amplitude I_0.

The somatic transmembrane current is usually the most important source for the extracellular potential for proximal recordings at the level of soma. As the frequency increases, the current profile of the return currents tend to become more localized around the synaptic inputs, i.e. a larger fraction of the current returns through the dendrites near the synapses. This is clearly seen in the current profile to the left in Figure 4.3. The 100 Hz sinusoid has a much larger current apically, and a much smaller current basally, than the 10 Hz sinusoid.

In the right part of Figure 4.3 we illustrate how the low-pass filtering effect of the extracellular potential depends on the distance from soma. Here all curves are normalized to unity for the lowest frequency considered, $f = 10$ Hz. When the frequency is increased, more of the current returns apically, further away from any recording position at the depth level of the soma. This implies that the extracellular potential becomes smaller, since the difference in distance between the

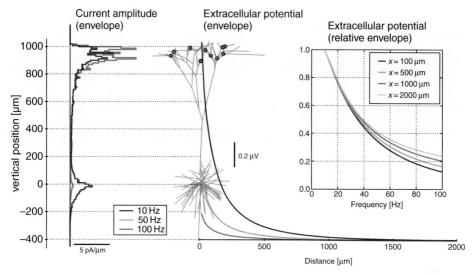

Figure 4.3 (See plate section for color version.) Illustration of frequency filtering of LFP for the passive layer-5 pyramidal model neuron in Figure 4.2A receiving simultaneous sinusoidal input currents $I_s(t) = I_0 \cos(2\pi f t)$ at 10 apical synapses (red dots in middle panel). The middle panel shows the envelope (amplitude) of the sinusoidally varying extracellular potential plotted at different lateral positions at the level of the soma (x-direction). The left panel shows the envelope of the linear current source density of the *return current* along the depth direction (z-direction) for $f = 10$ Hz and $f = 100$ Hz. The right panel shows the relative magnitude of envelopes of the extracellular potential as a function of frequency for different lateral distances from the soma. Here curves are normalized to unity for the lowest frequency considered, $f = 10$ Hz.

contributions to the potential from the synaptic input current and the return current will be smaller. Since the distance between the synaptic current generator and the return currents is relatively larger for recordings near the soma than for recordings further away in the lateral direction, the frequency decay of the extracellular potential will be steeper near the soma (small x) than for the distal recordings (large x).

The decay in extracellular amplitude as a function of frequency is not only seen in recordings at the level of the soma, but is also prominent for recordings at the level of the synaptic input (results not shown). The reason is the same: the potential is the sum of the transmembrane currents weighted inversely with distance to the sources, and when the typical distance between the synaptic current generator and the return currents becomes smaller, the extracellular potential will also become smaller.

The low-pass filtering effect described here is a general feature always present for spatially extended neuronal-membrane structures (Lindén et al., 2010), and in the next section we will show its impact on the extracellularly recorded signature from an action potential.

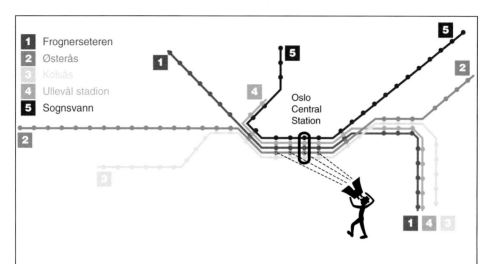

1 Frognerseteren
2 Østerås
3 Kolsås
4 Ullevål stadion
5 Sognsvann

Oslo Central Station

Extracellular versus intracellular potentials Intracellular and extracellular potentials are often confused: modelers sometimes compare their predictions of *intracellular* potentials (which are easier to model) with recorded *extracellular* potentials (which are easier to measure). As seen in Figures. 4.2 and 4.4 the connection between intracellular and extracellular is not trivial, however. A light-hearted metaphor is illustrated by the map of the Oslo subway system (see plate section for color version). With its branchy structure of different lines ("dendrites") stretching out from the hub at Oslo Central Station ("soma"), the subway system resembles a neuron. If we pursue this analogy, the subway stations (marked with dots) may correspond to "neuronal compartments" and the net number of passengers entering or leaving the subway system at each station to the net "transmembrane current" at this compartment. If more passengers enter than leave the subway system at a point in time, it means that the number of people in the subway system, i.e. the "intracellular membrane potential," increases. The intracellular soma membrane potential, crucial for predicting the generation of neuronal action potentials (which luckily have no clear analogy in normal subway traffic), would then correspond to the number of passengers within the subway station at Oslo Central Station. The extracellular potential on the other hand would be more similar to what could be measured by an eccentric observer counting passengers flowing in and out of a few neighboring subway stations (with binoculars on the top of a large building maybe). While the analogy is not perfect, it should illustrate that intracellular and extracellular potentials are correlated, but are really two different things. (Adapted from Pettersen and Einevoll, 2009. See plate section for color version.)

4.4 Extracellular signatures of action potentials

4.4.1 Example forward-modeling result

In a typical single- or multi-electrode recording, spikes from tens of neurons may be intermingled (Buzsáki, 2004). When developing automated algorithms for detecting and sorting these spikes according to their true neural source (Fee et al., 1996; Lewicki, 1998; Wehr et al., 1999; Harris et al., 2000; Qian Quiroga et al., 2004; Shoham and Nagarajan, 2004; Rutishauser et al., 2006; Kim and McNames, 2007; Quian Quiroga, 2007; Smith and Mtetwa, 2007; Pouzat and Chaffiol, 2009; Franke et al., 2010; Takekawa et al., 2010), several issues arise. For example, which types of neurons are most likely to be seen in the recordings, which neuronal parameters are important for the spike amplitude and shape, and which parameters determine the decay of the spike amplitude with increasing distance from the neuron? These are important questions also for the interpretation of multi-unit activity (MUA), the high-frequency content of the extracellular potential (Schroeder et al., 2001; Ulbert et al., 2001; Einevoll et al., 2007; Pettersen and Einevoll, 2008, 2009), and for the question of why the firing of neurons in the brain appears to be so sparse (Shoham et al., 2006).

An example forward-modeling result for the extracellular potential related to an action potential is shown in Figure 4.4A. Again the layer-5 pyramidal model neuron of Mainen and Sejnowski, (1996) is used, this time including active conductances. A combined pattern of apical excitation and basal inhibition is used to excite the action potential, similar to what is labeled "stimulus input pattern 1" (SIP1) in Pettersen et al. (2008). The largest extracellular responses are seen closest to the soma (thick lines in Figure 4.4A). As the shortest distance considered is as large as 100 μm, the spike amplitudes depicted in the figure are nevertheless all smaller than 20 μV (see Fig. 3 in Pettersen et al. (2008) for a close-up picture of spike shapes closer to the soma). The lowest inset in the figure, showing a magnified extracellular potential, illustrates the typical shape of recorded extracellular spikes: a sharp, deep dip (sodium phase) followed by a shallower, but longer-lasting, positive bump (potassium phase).

As for the spatial LFP patterns in Figure 4.2, the extracellular spike is also seen to have an inverted sign apically compared to basally. Further, a position-dependent low-pass filtering effect is also observed: the magnified extracellular potential in the top inset in Figure 4.4A is seen to be wider than at the lower inset closer to the soma. With the extracellular spike-width defined as the width of the sodium phase at 25% of its maximum, a widening from 0.625 to 0.75 ms is observed. This implies that the higher frequencies attenuate faster than the lower frequencies when moving away from the soma. An increase in spike-width with increasing distance from the soma has been seen experimentally, and explanations for this

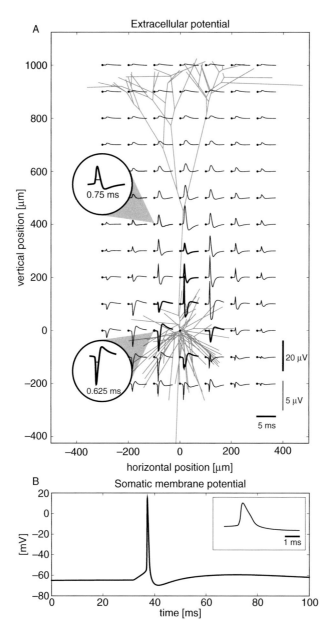

Figure 4.4 A. Calculated extracellular signature of an action potential in layer-5 pyramidal model neuron taken from Mainen and Sejnowski (1996). The neuron is stimulated with apical excitation and basal inhibition similar to "stimulus input pattern 1" (SIP1) in Pettersen et al. (2008). Traces show extracellular potential in a 5 ms window around the time of spiking. Thick lines corresponds to 20 μV scaling, thin lines to 5 μV scaling. Extracellular potentials in the two insets are scaled arbitrarily. B. Somatic membrane potential during simulation. Inset shows soma potential for same 5 ms time window as for the extracellular potentials in A.

in terms of extracellular-medium effects has been suggested (Bédard et al., 2004, 2006b). However, it is still debated whether such effects are present in cortical tissue: while some investigators have measured low-pass filtering effects in the extracellular medium (Gabriel et al., 1996), other investigators found no such effect (Nicholson and Freeman, 1975; Logothetis et al., 2007).

Below we outline how the neuron *morphology*, combined with its *cable properties*, can provide an alternative, or supplementary, explanation for the distance-dependent low-pass filtering effect of extracellular spikes (Pettersen and Einevoll, 2008). Section 4.2 explained why a neuron model has to contain at least two compartments to produce an extracellular potential at all, and in Section 4.3 it was shown that a two-compartment model could not produce any *distance-dependent* low-pass filtering effect. Pettersen and Einevoll (2008) investigated the effect of the neuronal morphology and the passive dendritic parameters on the extracellular spike signature in detail, in particular the distance dependence of the spike amplitude and low-pass filtering. A variety of neuronal morphologies was considered, both morphologically reconstructed pyramidal (Figure 4.5A) and stellate cells (Figure 4.5C) and simplified models built up of dendritic sticks ("ball-and-stick," "ball-and-star," "ball-and-bush," cf. Fig. 1 in Pettersen and Einevoll (2008)). While the shape of the intracellular action potential varies from neuron to neuron, we wanted to focus on how the dendritic structure affects the relationship between the intracellular and extracellular potentials (Wehr et al., 1999; Henze et al., 2000). We thus imposed a standardized intracellular action potential (cf. Figure 4.5B) in the somas of the neurons in the numerical evaluation of the extracellular spike signatures. In accordance with the qualitative observation in Figure 4.4A, all neuron models were found to exhibit a distance-dependent low-pass filtering effect, that is, larger spike widths further away from soma, cf. Figs. 6 and 7 in Pettersen and Einevoll (2008). However, the amplitudes of the spikes were found to be quite different, both their size and their dependence on distance from soma. For example, with identical intracellular action potentials, the spike amplitude 60 μm away from soma was found to be about 40 μV for the pyramidal neuron, but only about 10 μV for the stellate neuron.

To obtain a better understanding of the phenomenon we also developed a conceptually simpler and more intuitive theory accounting for the observed variation in spike shape and amplitude (Pettersen and Einevoll, 2008). This theory also produced analytical predictions of the dependence of the spike amplitude on the dendritic parameters, predictions that later were confirmed by numerical calculations. The essential idea behind the theory is that during an action potential, the soma can be viewed as a voltage source driving current into the soma-attached dendrites, and that the size and shape of the extracellular signature will depend qualitatively on (i) the magnitudes of the axial currents entering the dendrites from

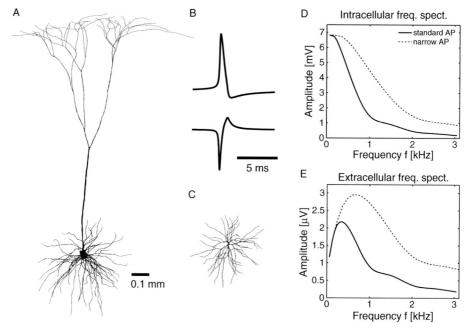

Figure 4.5 A. Pyramidal layer-5 neuron used in investigation of extracellular spikes (Pettersen and Einevoll, 2008). B. Upper: "standard" action potential (AP) used in model study, half-amplitude spike width is 0.55 ms. Lower: typical shape of corresponding extracellular spike near soma for "standard" AP. Extracellular spike width is 0.44 ms (see text for definition). C. Stellate layer-4 neuron used in investigation (Pettersen and Einevoll, 2008). D. Frequency spectrum of intracellular voltage for "standard" AP in B, and a corresponding "narrow" AP with identical form but exactly half the spike width. E. Frequency spectrum of extracellular voltage traces of "standard" spike in B (solid), and corresponding extracellular voltage trace for "narrow" spike (dashed). Reconstructed neuron morphologies are taken from Mainen and Sejnowski (1996). See Pettersen and Einevoll (2008) for further information.

the soma, (ii) what distances from the soma the imposed axial currents on average return through the dendritic membranes and (iii) the number and geometrical arrangement of dendrites. In fact it was found that many of the salient features of the extracellular spike could be understood by considering the simple ball-and-stick neuron model where the soma is modeled as a single compartment and the dendrite as a simple cable stick (Johnston and Wu, 1994; Pettersen and Einevoll, 2008). With the soma considered as a voltage source, the various soma-attached dendrites are effectively decoupled from each other. Consequently the total extracellular potential generated by a more complex neuron can be approximated as a superposition of contributions from a collection of soma-attached dendritic sticks pointing in different directions (Pettersen and Einevoll, 2008).

4.4.2 Dendritic sticks and AC length constant

A concept we found essential to get both an intuitive and quantitative handle on the crucial spatial distribution of the return current along the dendritic stick, is the so-called *alternating current (AC) length constant* (Pettersen and Einevoll, 2008).

Imagine a ball-and-stick neuron model (cf. Figure 4.6) where the dendritic stick is infinitely long. This *infinite* ball-and-stick neuron is assumed to receive a constant (DC) somatic transmembrane current. Since the membrane currents at all times have to sum to zero, the same amount of current has to return to the extracellular medium through the dendritic stick. The density function of the dendritic return current has the functional form of an exponential decay with the *length constant* (or *space* constant) λ describing the steepness of the decay (Dayan and Abbott, 2001; Pettersen and Einevoll, 2008). More precisely, the length constant is the dendritic position where the steady-state transmembrane return current has decreased to $1/e$ of its value at the soma end, or equivalently, λ is the position where the dendritic return current has its center of gravity. The center of gravity is then defined as the mean of the normalized transmembrane current density weighted by dendritic position.

The length constant is not only useful for describing the neuron's intrinsic qualities (for example electrotonic compactness), it is also useful for understanding the extracellular potentials generated by the neuron. For example, when computing the extracellular potential far away from the neuron (far-field limit), the ball-and-stick neuron model can be approximated by a dipole model (Pettersen and Einevoll, 2008). The parameters of the dipole model will then be the dipole current, which equals the somatic current, and the dipole size, which essentially is given by the dendritic length constant, cf. Figure 4.7A. For infinite dendritic sticks under the DC condition this length constant is given by $\lambda = \sqrt{dR_m/4R_i}$, where d is the stick diameter, R_m is the specific membrane resistance [$\Omega\,cm^2$] and R_i is the axial resistivity [$\Omega\,cm$] (Dayan and Abbott, 2001).

As the length constant is important for understanding several aspects of a neuron (electrotonic compactness, extracellular far-field potential), it is useful to define a general length constant which is not restricted to infinite sticks and DC conditions. In analogy to the definition of the standard DC space constant λ, Pettersen and Einevoll (2008) define the AC length constant, $\lambda_{AC}(\omega)$, to be the mean of the absolute value of the current density amplitude weighted with distance, when the dendritic stick is driven by a sinusoidal voltage in the soma end of the stick. This length constant will be frequency dependent through the angular frequency $\omega = 2\pi f$. For a finite stick this corresponds to (in complex notation)

$$\lambda_{AC}(\omega) = \frac{\int_0^l z |\hat{\mathbf{i}}_m(z)| dz}{\int_0^l |\hat{\mathbf{i}}_m(z)| dz}, \tag{4.12}$$

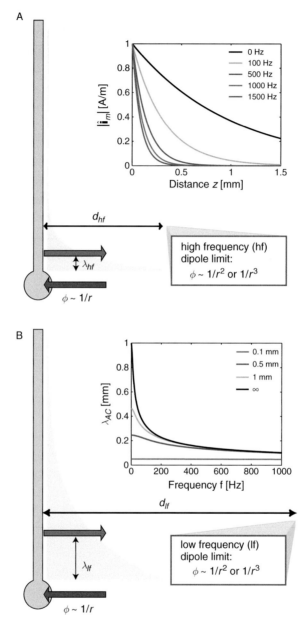

Figure 4.6 (See plate section for color version.) Illustration of ball-and-stick neuron and its frequency-dependent dipole sizes and corresponding far-field limits. A. For high frequencies (hf), the center of gravity (blue arrow) of the dendritic return current is close to soma. Therefore, the AC length constant λ_{hf} is small and transition to the far-field limit occurs around a distance d_{hf}, relatively close to the neuron. Inset: transmembrane return-current profile along an infinite dendritic stick for different frequencies (Pettersen and Einevoll, 2008). Parameters: stick diameter 2 mm, membrane and axial resistivities $R_m = 30000\ \Omega\,cm^2$,

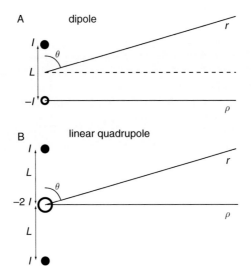

Figure 4.7 Illustration of a current dipole (A) and linear current quadrupole (B).

where the stick is assumed to be extended along the positive z-axis from $z = 0$ (soma end) to $z = l$. Here $\hat{\mathbf{i}}_m(z)$ denotes the complex transmembrane current density at position z along the stick (where the real part corresponds to the physical transmembrane current). In the inset in Figure 4.6 A we show normalized values for $|\hat{\mathbf{i}}_m(z)|$ as a function of distance from the soma for an infinite ball-and-stick neuron for different frequencies. The higher the frequency, the closer to the soma the return current is seen to be. This is reflected in the frequency dependence of the AC length constant $\lambda_{AC}(\omega)$ as seen in the inset of Figure 4.6B: the highest frequencies have the shortest AC length constants. This latter panel also shows that shorter dendritic sticks have shorter $\lambda_{AC}(\omega)$, as expected since the closed ends will force the return currents out closer to the soma. This effect will be most pronounced for the lower frequencies.

For an infinite stick an analytical formula can be found for the AC length constant. In this special case Equation (4.12) reduces to (Johnston and Wu, 1994; Pettersen and Einevoll, 2008)

Caption for Figure 4.6 (cont.)
$R_i = 150\ \Omega\,\mathrm{cm}^2$, membrane capacitance $C_m = 1\ \mu\mathrm{F}/\mathrm{cm}^2$. λ^{∞}_{AC} is 317 mm, 145 mm, 103 mm, and 84 mm for 100 Hz, 500 Hz, 1000 Hz, and 1500 Hz, respectively. B. For low frequencies (lf) the AC length constant λ_{lf} is relatively large and the far-field limit is reached for a larger distance d_{lf} than for the higher frequency in A. Inset: AC length constant $\lambda_{AC}(\omega)$ as a function of frequency for ball-and-stick models of different length; parameter values for diameter, resistivity and capacitance are the same as in A.

$$\lambda_{AC}^{\infty}(\omega) = \lambda\sqrt{2/[1 + \sqrt{1 + (\omega\tau)^2}]}\,, \tag{4.13}$$

where τ denotes the membrane time constant, $\tau = R_m C_m$.

4.4.3 Low-pass filtering for the ball-and-stick neuron

In Pettersen and Einevoll (2008) numerical investigations of the extracellular signature of action potentials in ball-and-stick neurons also revealed a characteristic spike-width increase when moving away from soma, similar to what is seen for the pyramidal neuron in Figure 4.4A. Here we will outline how a reduced model, a dipole model with a soma compartment attached to a conflated dendritic stick, can explain the phenomenon. In Section 4.3 we showed that a two-compartment neuron model, i.e. a dipole model with a *fixed* dipole length, cannot express such position-dependent low-pass filtering. The crucial element introduced here is that the dipole model must have a *frequency-dependent dipole length* based on λ_{AC}.

Far away (i.e. far-field limit) from a current dipole with current strength I and length L the extracellular potential is given by (Jackson, 1998; Pettersen and Einevoll, 2008)

$$\phi_{far,d}(r, \theta) = \frac{1}{4\pi\sigma}\frac{IL}{r^2}\cos\theta, \tag{4.14}$$

when polar coordinates are used, cf. Figure 4.7A. This model shows a $1/r^2$ decay when moving in any direction where θ is fixed. However, when moving perpendicular to the dipole (e.g. along the ρ-axis in Figure 4.7A) the extracellular potential decays as a quadrupole, i.e. as $1/r^3$ (Pettersen and Einevoll, 2008).

The distance dependence is more complicated for proximal extracellular potentials than for far-field potentials. Close to the soma compartment, the soma current will dominate the potentials, and in this region the distance dependence will be given by the monopole expression

$$\phi_m(r, \theta) = \frac{1}{4\pi\sigma}\frac{I}{r}, \tag{4.15}$$

that is, the amplitude decays as $1/r$. This dipole neuron model therefore predicts a transition in the power of the distance dependence of the extracellular potential from -1 close to the soma to -2 (or -3) in the far-field limit.

If the soma membrane potential oscillates at an angular frequency ω, current will flow from the extracellular medium through the soma and up into the dendritic stick with the same frequency with an amplitude we denote $I(\omega)$. A simple model for the extracellular potential generated around the ball-and-stick neuron can now be made: Near the soma the amplitude of oscillating extracellular potential can be descibed by Equation (4.15) with I replaced by $I(\omega)$, and in the far-field limit the extracellular potential amplitude can be described by Equation (4.14) with I

replaced by $I(\omega)$ and L replaced with $\lambda_{AC}(\omega)$ from Equation (4.12) (Pettersen and Einevoll, 2008). From the dendrite's point of view the soma action potential can be seen as a voltage source enforcing the characteristic intracellular voltage waveform. Since the dendritic stick itself has linear response properties, this waveform can be Fourier decomposed, and each frequency can be treated separately. The extracellular signature of the action potential can thus be found by a simple linear superposition (Pettersen and Einevoll, 2008).

The extracellular signature at a particular position will depend crucially on whether the frequency components are in the "close-to-soma-regime" (Equation (4.15)), in the "far-field limit" (Equation (4.14)), or somewhere in between. The following question thus arises: what decides the distance for which the far-field limit is reached? Clearly, the transition to the far-field limit must depend on the dipole length, that is, $\lambda_{AC}(\omega)$. Thus the transition to the far-field limit for each component will depend on frequency. Further, since the highest frequencies will have the smallest $\lambda_{AC}(\omega)$, these components will reach the far-field limit (where the distance-decay is sharper and the signal rapidly diminishes) closer to soma. Figure 4.6 illustrates this low-pass filtering effect for the dipole model approximation of the ball-and-stick neuron.

4.4.4 Parameter dependence of spike amplitude

In addition to explaining the position-dependent low-pass filtering of the extracellular spike, the dipole model approximation of the ball-and-stick neuron can also explain essential features of the size and distance dependence of the spike amplitude (Pettersen and Einevoll, 2008). For the infinite ball-and-stick neuron it is possible to derive an analytical expression for the frequency-dependent *transfer function* $\mathbf{T}$ describing how a soma membrane potential "transfers" to an extracellular potential. With a complex notation (boldface) the soma membrane potential for a given angular frequency ω can be represented as $\mathbf{V}_0(t; \omega) = \hat{\mathbf{V}}_0(\omega)e^{j\omega t}$, where $\hat{\mathbf{V}}_0$ contains both the amplitude and phase of the sinusoidal potential and $j = \sqrt{-1}$. The physical soma membrane potential will then be the real part of this complex quantity, $V_0(t; \omega) = \text{Re}\{\mathbf{V}_0(t; \omega)\}$. The complex Fourier amplitude of the extracellular potential $\hat{\Phi}(\mathbf{r}, \omega)$ for a ball-and-stick model can thus be related to the complex soma potential $\hat{\mathbf{V}}_0(\omega)$ through the transfer function $\mathbf{T}(\mathbf{r}, \omega)$, i.e. $\hat{\Phi}(\mathbf{r}, \omega) = \mathbf{T}(\mathbf{r}, \omega)\hat{\mathbf{V}}_0(\omega)$ (Pettersen and Einevoll, 2008). Since the DC-subtracted intracellular somatic action potential $V_0(t)$ can be expressed by a Fourier series, $V_0(t) = \sum_{k=1}^{\infty} \text{Re}\{\hat{\mathbf{V}}_0(\omega_k)e^{j\omega_k t}\}$, the measured extracellular response to any such DC-subtracted somatic action potential can be expressed as

$$\phi(\rho, z, t) = \sum_{k=1}^{\infty} \text{Re}\{\mathbf{T}(\mathbf{r}, \omega_k)\hat{\mathbf{V}}_0(\omega_k)e^{j\omega_k t}\} . \qquad (4.16)$$

The transfer function for the ball-and-stick neuron has a rather complex analytical form (Pettersen and Einevoll, 2008). To investigate the parameter dependence of the spike amplitude we instead use the much simpler dipole model with a frequency-dependent dipole length given by the length constant of the infinite ball-and-stick model in Equation (4.13).

Near the soma the monopole contribution from the soma membrane current will dominate, and the extracellular potential will decay as $|\hat{\Phi}(\omega)| \sim |\hat{I}(\omega)|/4\pi\sigma r$, where the somatic membrane current I is related to the somatic membrane potential through the dendrite's admittance, $\hat{I} = \hat{Y}\hat{V}_0$ (see Pettersen and Einevoll, 2008). The transfer function $\mathbf{T}$ will therefore be given by $\hat{Y}/4\pi\sigma r$ in the near-field approximation, and for high frequencies ($\omega\tau \gg 1$), the transfer function can be shown to be (Pettersen and Einevoll, 2008)

$$|\mathbf{T}_{near}| \sim \frac{d^{3/2}}{\sigma r} \left(\frac{f C_m}{R_i} \right)^{1/2} . \tag{4.17}$$

In the far-field approximation the potential is given by the dipole or quadrupole expressions (when moving laterally, see above), $\hat{\Phi}(\omega) \sim \hat{I}(\omega)L/4\pi\sigma r^2$ or $\hat{\Phi}(\omega) \sim \hat{I}(\omega)L^2/4\pi\sigma r^3$, respectively (Pettersen and Einevoll, 2008). We then assume $L \approx \lambda_A C^{\infty}(\omega)$ (Equation (4.13)), and for $\omega\tau \gg 1$ we have $\lambda_{AC}^{\infty}(\omega) \sim \lambda/\sqrt{\omega\tau} \sim \sqrt{d/f R_i C_m}$. With a reasonable time constant such as $\tau = 20$ ms, this high-frequency approximation holds for frequencies $f \gg 8$ Hz, i.e. all the dominant frequencies of the action potential. With these assumptions the far-field transfer functions can be shown to be (Pettersen and Einevoll, 2008)

$$|\mathbf{T}_{far,d}| \sim \frac{d^2}{\sigma r^2 R_i} , \quad |\mathbf{T}_{far,q}| \sim \frac{d^{5/2}}{\sigma r^3 f^{1/2} R_i^{3/2} C_m^{1/2}} , \tag{4.18}$$

where "*far, d*" means far-field dipole expression and "*far, q*" means far-field quadrupole expression (applicable when moving perpendicular to the ball-and-stick neuron). In Pettersen and Einevoll (2008) a host of numerical simulations were done to investigate to what extent these analytical predictions are accurate, not only for individual frequency components but also for the full action potential. The numerical calculations indeed confirmed their validity, cf. Fig. 9 in Pettersen and Einevoll (2008).

The transfer-function expressions in Equations (4.17) and (4.18) give some interesting qualitative insights.

1. Close to the soma the higher frequencies are amplified compared to the low frequencies, $|T| \sim \sqrt{f}$. Thus close to the soma the extracellular action potential will typically appear sharper than the intracellular action potential.
2. Far away this high-frequency amplification either vanishes ("*far,d*," $|T| \sim f^0$) or is reversed ("*far,q*," $|T| \sim 1/\sqrt{f}$).

3. $|T|$ is independent of the membrane resistivity R_m.
4. $|T|$ decreases with increasing intracellular resistivity R_i.
5. $|T|$ may, depending on distance from soma, increase or decrease with increasing capacitance C_m.
6. $|T|$ increases with increasing dendritic diameter d, that is, $T \sim d^k$ where k is in the range 1.5–2.5.

While Equations (4.17) and (4.18) were derived for a simple ball-and-stick model, similar expressions can easily be derived for more complicated neuron models, see Pettersen and Einevoll (2008) for details.

The last entry in the above list (point 6) suggests an important connection between the extracellular spike amplitude and the dendritic diameters. Since the contributions from different soma-attached dendrites add up, this point suggests a rule of thumb: *a neuron's extracellular spike amplitude is approximately proportional to the sum of the dendritic cross-sectional areas of all dendritic branches connected to the soma* (Pettersen and Einevoll, 2008). Thus, neurons with many, thick dendrites connected to soma will produce large-amplitude spikes, and will therefore have the largest radius of visibility.

Pettersen and Einevoll (2008) confirmed this rule of thumb for the two morphologically reconstructed cells shown in Figure 4.5A and C. The pyramidal neuron had more soma-attached dendrites than the stellate cell (11 versus 6) and they were thicker as well (average diameter 3.0 μm versus 2.1 μm). In the numerical simulations the ratio between the peak-to-peak extracellular spike amplitudes of the pyramidal and stellate neurons were found to be 3.3, 3.7, and 4.0 at 20 μm, 60 μm, and 100 μm distances, respectively (see Table 1 in Pettersen and Einevoll (2008) for details). The above rule of thumb ($|T| \sim d^2$) predicts this ratio to be 4.5, in reasonable agreement with the numerical results. The agreement is even better if one considers that the d^2-rule is expected to be best far away from the soma. Strictly speaking, a $d^{3/2}$-rule is predicted close to the soma (Equation (4.17)). The latter rule predicts the ratio to be 3.3, exactly what is calculated for the smallest distance (20 μm).

4.4.5 Active dendritic conductances

So far, we have only considered action potentials from neurons with electrically passive dendrites. This assumption makes the problem of translating intracellular potentials in the soma to extracellular potentials recorded outside the neuron *linear* and, importantly, independent of the detailed form of the intracellular action potential, i.e. independent of the detailed properties of the active soma conductances responsible for generating the action potential. Thus the analytical insights reviewed in the previous subsection apply in principle to all different intracellular action potential waveforms.

However, real neurons also have active conductances in the dendrites (Stuart et al., 2007). In general, this makes the problem non-linear, and the trick of considering each frequency component of the action potential separately is no longer applicable. Instead one has to use comprehensive compartmental models including all active conductances explicitly. Gold et al. (2006, 2007) have done thorough investigations of the extracellular signatures of spikes from pyramidal neurons in hippocampus CA1 and fitted compartmental models to reproduce simultaneously recorded intracellular and extracellular waveforms. An important result from their studies was that extracellular waveforms provide tighter constraints on the model parameters than the intracellularly recorded somatic action potentials. This suggested that extracellular action potentials could be a good source of data for constraining compartmental models (Gold et al., 2007).

Their results are also in qualitative agreement with many of the observations seen above for the purely passive dendrites: (i) the spike width was seen to increase with distance from the soma (cf. Fig. 5A in Gold et al. (2006)), (ii) the amplitude was seen to decay with soma distance with a power between 1 and 2 for distances less than 50 μm (cf. Fig. 14 in Gold et al. (2006)), and (iii) the amplitude was seen to change significantly to varying intracellular resistivity R_i and capacitance C_m, but not so much to varying membrane resistivity (Gold et al., 2007).

4.5 Extracellular potentials from columnar population activity

In Sections 4.3 and 4.4 we considered extracellular potentials generated by activity in single neurons. Extracellularly recorded signals like LFP and MUA do not stem from single neurons, however, but rather from *populations* of neurons. The forward-modeling scheme applied above for single neurons applies equally well to populations of neurons, and here we outline results from our modeling study of the generated LFP and MUA by a synaptically activated, spatially confined population of layer-5 neurons (Pettersen et al., 2008), mimicking a population of large pyramidal cells in a sensory neocortical column (Mountcastle, 1997).

In this pilot MUA and LFP forward-modeling study we sought to answer questions like: is the MUA really a more local measure of neural activity than LFP? how sharply does the MUA and LFP decay outside the active population? to what extent is the MUA a measure of the *population firing rate*? do existing CSD analysis methods estimate the true CSD accurately?

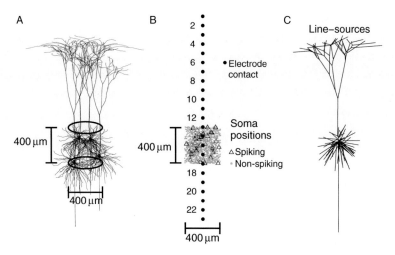

Figure 4.8 A. Schematic illustration of the population of reconstructed layer-5 pyramidal neurons considered in a forward-modeling study (Pettersen et al., 2008). B. Somas of pyramidal neurons placed randomly inside a cylindrical annulus with height 0.4 mm, outer diameter 0.4 mm and inner diameter 0.1 mm. One thousand neurons were non-spiking (dots), while 40 neurons (triangles) produced a single spike following synaptic stimulation. Extracellular potential was simulated at assumed electrode contact positions along the center axis of the population (filled circles). C. Illustration of line-source method: the transmembrane current from each neural segment is modeled as a linear current source of uniform current density. Note that each neural branch (section) depicted in the panel may consist of several segments.

4.5.1 Columnar population model

The simulated population in Pettersen et al. (2008) consisted of 1040 layer-5 pyramidal neurons of the type shown in Figure 4.5A. Their somas were placed stochastically in a cylinder with both diameter and height of 0.4 mm, see Figure 4.8. The population was constructed based on a single neuron model template, but with two different synaptic input patterns. To obtain MUA responses in reasonable agreement with experimental data (Schroeder et al., 2001; Einevoll et al., 2007), only 40 of the neurons received a net synaptic input sufficiently strong to generate a single action potential within a time window of about 20 ms. The net synaptic input to the remaining 1000 neurons was tuned such that no action potential was generated. To introduce temporal jitter in the synaptic activation of the neurons in the population, the neuronal templates were stochastically shifted in the time domain assuming a Gaussian distribution with a standard deviation of 5 ms. The extracellular potential was computed at 23 positions, every 0.1 mm along the center axis of the population, see Figure 4.8B. For the present modeling example we found that a balanced combination of apical excitation and basal inhibition was needed to obtain realistic LFP amplitudes compared to experimental LFP data, since apical excitation alone did not give large enough LFP amplitudes for a population of about 1000 cells. Both single-trial and trial-averaged population responses were calculated. A set of 40 trials was considered in the trial-averaging procedure, and each trial differed in the stochastic distribution of both the position and time-shifting for the individual neurons.

4.5.2 *Population response*

With more than 1000 synaptically activated pyramidal neurons, the LFP response was found to be very robust, different trials giving virtually identical results. For the MUA, the detailed temporal structure of single-trial signals was found to vary considerably on a millisecond scale, reflecting the stochastic firing of specific neurons located close to the electrode contacts (cf. Fig. 4 in Pettersen et al. (2008)). The stochastic placement of the soma positions in the model thus makes single-trial MUA a much more noisy measure of neural activity than LFP. However, the trial-averaged MUA over 40 trials was seen to be quite reproducible and, importantly, independent of the form of synaptic input pattern providing the excitation.

Trial-averaged responses for the LFP and MUA data obtained in our forward modeling scheme are shown in Figure 4.9A and B. Note that while synaptically evoked LFP can be seen at most electrode contacts, MUA can essentially only be seen at the contacts inside the vertical distribution of somas in the population, i.e. between contacts 13 and 17.

4.5.3 *Spatial spread of LFP and MUA signals*

If one measures extracellular potentials in the cortex with two adjacent electrodes, and finds that their LFP signals are correlated, two possible interpretations come to mind. Either (i) the two electrodes may measure neural activity from two separate neural populations which happen to have correlated synaptic input activity, or (ii) the LFP generated by a single population may spread to the two electrodes by *volume conduction* (Nunez and Srinivasan, 2006). Likewise, if the MUA signals of the two electrodes are observed to be correlated, this can also be due to correlated firing in two spatially separated populations or volume conduction of the MUA signal from a single population.

From the discussion in Section 4.4 one would expect the LFP generated by a population to have a larger spatial spread, i.e. larger volume conduction, than the corresponding MUA: the LFP contains lower frequencies than the MUA and will thus have longer characteristic AC length constants and consequently decrease less steeply with distance. These expectations were confirmed by numerical simulations

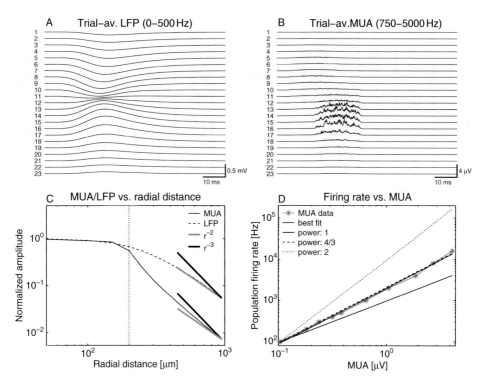

Figure 4.9 Trial-averaged local field potential (A) and multi-unit activity (B) recorded at the center axis of a cylindrical population of 1040 layer-5 pyramidal neurons receiving apical excitation and basal inhibition. The depicted LFP is obtained by (i) low-pass filtering of the calculated extracellular potential (<500 Hz) and (ii) trial-averaging ($n = 40$). The MUA is obtained by (i) band-pass filtering between 750–5000 Hz, (ii) rectification, and (iii) trial-averaging ($n = 40$). 40 of 1040 neurons fire an action potential stochastically within a time window of 20 ms. For details of numerical simulation, see Pettersen et al. (2008). C. Decay of amplitude of "area under graph" MUA and LFP calculated at electrode 15 (middle of population) as a function of lateral distance from the population center, cf. Fig. 15 in Pettersen et al. (2008). The vertical dotted line illustrates the lateral edge of the soma distribution corresponding to a radial distance of 200 µm. D. Relationship between "true" firing rate and estimates based on MUA signal, cf. Fig. 13 in Pettersen et al. (2008). The depicted MUA is the average MUA for the five electrode contacts running through the center of the population (electrode contacts 13–17, see Figure 4.8). The population size was varied and all neurons within the population were spiking. The power law giving the best fit to the data has a coefficient of 1.346.

(Pettersen et al., 2008): Figure 4.9C illustrates the distance-decay of the MUA and LFP signals from our model population in a direction perpendicular to its center axis. The MUA is seen to decay sharply outside the population ($r > 200$ µm), whereas the LFP is seen to spread much further. For example, at a position 0.3

mm outside the population cylinder, i.e. 0.5 mm from the population center, the magnitude of the LFP signal is seen to be reduced with about a factor five compared to the value at the population center, whereas the MUA is reduced by a factor 30. Thus compared to the LFP, observed correlations of the MUA signal between adjacently placed electrodes are more likely to be due to correlated firing in two different populations. However, these are just example results, and more systematic studies are needed to elucidate, for example, the neural origin of LFPs recorded in the cortex (Lindén et al., 2009a, 2009b, 2010).

4.5.4 MUA as a measure of population firing rate

The MUA, obtained by high-pass filtering ($\gtrsim 500$ Hz) with subsequent rectification of the extracellular signal, has been assumed to measure the population firing rate for a group of neurons around the electrode contacts (Schroeder et al., 2001; Ulbert et al., 2001; Einevoll et al., 2007; Blomquist et al., 2009). Pettersen et al. (2008) used the present population forward-modeling study to test this assertion: since we can set the population firing rate ourselves in this model world, we have a gold standard against which the calculated MUA signal can be compared.

Two different regimes can be expected (Pettersen et al., 2008). The first regime corresponds to very low firing rates. Here the various extracellular signatures from firing in the nearby neurons contributing to the MUA will not overlap significantly in time. Thus even with biphasic extracellular signatures (cf. Figure 4.4A), there will be little cancelation between positive and negative phases of the extracellular potential. A linear relationship between the MUA and the population firing rate is thus expected.

The other regime corresponds to very high population firing rates. In this high-firing limit the MUA was found to grow roughly as the square root of the population firing rate (Pettersen et al., 2008). Here there will be strong temporal overlap in the sum over extracellular signatures from all contributing neurons, and the summation is better viewed as a sum over randomly drawn positive and negative contributions to the extracellular potential. If the positive and negative contributions are similar in size, the rectified summed signal is expected to grow as the square root of the number of contributions, i.e. as the square root of the population firing rate.

It is, however, unclear a priori what ranges of population firing rates correspond to the different regimes; this will depend on neuronal morphologies and densities (as well as the physical characteristics of the electrode). Pettersen et al. (2008) found that for realistic population firing rates and trial averages over 40 trials, there is a large regime where the relationship between the MUA and population firing rate is well approximated by raising the population firing rate to a power of 3/4, i.e. intermediate between the linear and the square-root regimes. This implied that a good estimate of the population firing rate can be obtained by raising the MUA to the power 4/3, see double-logarithmic plot in Figure 4.9D. Indeed, this rule clearly improved the population firing rate estimates for the examples considered by Pettersen et al. (2008) compared to results using the standard linear rule. However, it is unclear to what extent this rule extends to other situations.

4.6 Estimation of current source density (CSD) from LFP

The previous Sections 4.3, 4.4, and 4.5 all considered forward modeling of extra-cellular potentials, that is, the calculation of extracellular potentials from known activity in neurons. The present section deals with the opposite problem, namely how the underlying neural activity can be estimated based on measurements of the extracellular potential or more specifically the LFP. An estimation of transmem-brane current through a particular segment of a particular neuron is in practice out of the question; in principle, one can only extract one unknown current source per electrode contact and an infinite number of different current source constellations can produce the extracellular potential recorded on a finite number of electrode contacts.

A common strategy has been to use multi-contact LFP recordings to estimate the current source density (CSD), that is, the volume density of net current entering or leaving the extracellular medium, see Section 4.2.3. A microscopic view inside the cortical tissue reveals an inhomogeneous, densely packed collection of neural segments acting as current sources. The CSD is a more mesoscopic concept and can be interpreted as the net transmembrane current density found by averaging over a volume element a few tens of micrometers across, limiting the possible spatial resolution of the CSD to lengths of this order.

4.6.1 Standard CSD method

The traditional CSD estimation is based on LFP recordings with laminar (linear) multi-electrode arrays with a constant inter-contact distance h inserted perpendic-ularly to the cortical surface (Rappelsberger et al., 1981; Mitzdorf, 1985; Di et al., 1990; Schroeder et al., 2001; Ulbert et al., 2001; Einevoll et al., 2007). Motivated by the prominent laminar structure of cortical tissue where the changes in the lat-eral directions are much smaller than in the vertical direction, it has been common to assume an infinite activity diameter in the lateral (xy) plane, i.e. perpendicular to the laminar electrode oriented in the z-direction. Variation of the extracellular potential in the x- and y-directions can then be neglected, so that Equation (4.10) simplifies to its one-dimensional version:

$$\sigma \frac{d^2\phi(z, t)}{dz^2} = -C(z, t). \tag{4.19}$$

A natural estimator for the CSD at electrode position z_j has thus been (Nicholson and Freeman, 1975)

$$C(z_j) = -\sigma \frac{\phi(z_j + h) - 2\phi(z_j) + \phi(z_j - h)}{h^2} \tag{4.20}$$

or variations thereof, including additional spatial smoothing filters (Nicholson and Freeman, 1975; Ulbert et al., 2001). With N electrode contacts the above estimator can predict the CSD only at the $N - 2$ interior contact positions. However, a trick allowing for the estimation of the CSDs also at the top and bottom electrodes has been suggested (Vaknin et al., 1988).

If we define the domain of electrode contact j located at position z_j as the domain from $z_j - h/2$ to $z_j + h/2$, it is natural to assume that the estimate $C(z_j)$ should correspond to the average CSD within this domain. Pettersen et al., (2006) instead showed from electrostatic theory that the process of discretizing the one-dimensional Poisson equation into Equation (4.20) corresponds to assuming all CSD within each electrode's domain to be located in an infinitely thin (and infinitely wide) sheet at the height of the electrode contact. However, a possibly larger source of estimation error stems from the assumption of an infinite activity diameter perpendicular to the laminar electrode. This was noted already by Nicholson and Freeman (1975) who showed that small "columnar" activity diameters ($\sim$ 1 mm or less) may give large errors in the estimated CSD. The numerical example from Nicholson and Freeman (1975) is reproduced here in Figure 4.10, and for the small source diameters the estimated CSD is clearly seen to be erroneous, predicting, for example, spurious sinks and sources. Indeed Nicholson and Freeman (1975) recommended and later pursued a full three-dimensional CSD analysis based on the full Poisson equation (4.10) which required technically demanding measurements of extracellular potentials in all three spatial directions (Nicholson and Llinás, 1975). With the advent of the present silicon-based multi-electrodes such a CSD estimation scheme can now become more practically feasible (Buzsáki, 2004).

4.6.2 Inverse CSD methods

Pettersen et al. (2006) introduced a new method for estimation of the CSD, the *inverse CSD (iCSD)* method. The core idea behind this method is to exploit the well-known forward-modeling scheme for calculation of the LFP from given a CSD distribution. With an assumed form of the CSD distribution parameterized by N unknown parameters, the forward solution can be calculated and inverted to give estimates of these N parameters based on N recorded potentials. This iCSD approach has several inherent advantages:

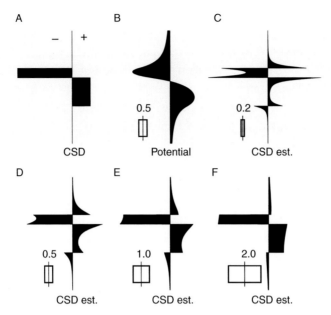

Figure 4.10 Illustration of errors inherent in the standard CSD estimation method for one-dimensional recordings for simplified CSD profiles, similar to (Nicholson and Freeman, 1975). A. Example CSD depth profile. B. Corresponding LFP at center axis when the CSD distribution has a diameter-to-height ratio of 0.5. C–F. Estimated CSDs for increasing diameter-to-height ratios, as indicated by the number and inset in the lower left of each panel. All estimates are based on the double spatial-derivative formula of the standard CSD method, i.e. Equation (4.20). Arbitrary units, negative values to the left and positive to the right.

- The method does not rely on a particular geometrical arrangement of the N electrode contacts recording the LFP signals. It is thus not only applicable to linear multi-electrodes (Pettersen et al., 2006), but can also be generalized straightforwardly to other multi-electrode geometries (Leski et al., 2007, 2011),
- A priori constraints, such as knowledge of the lateral size of columnar activity, can be built directly into the iCSD estimator (Pettersen et al., 2006, 2008; Einevoll et al., 2007; Leski et al., 2011),
- Unlike the standard CSD method, the iCSD method can also predict CSD at the positions of the boundary electrode contacts (Pettersen et al., 2006; Leski et al., 2007, 2011),
- Discontinuities and direction dependence of the extracellular conductivity can be incorporated (Pettersen et al., 2006; Einevoll et al., 2007).

To present the iCSD idea more explicitly we now consider a situation where one has recordings from N electrode contacts. Further, the CSD has been parameterized by N parameters describing the weights of the different contributions to the CSD (Pettersen et al., 2006). Regardless of the choice of parametrization, the CSD is now uniquely determined by the N weight parameters $\{C_1, C_2, \ldots, C_N\}$. The LFP due to this CSD distribution can then be calculated at the N electrode contact positions using electrostatic theory (e.g. using Equation (4.7) if σ is scalar and homogeneous). Due to the linearity of electrostatic theory the LFP grows linearly with the CSD weight parameters, and their relationship can thus be formulated in matrix form as (Pettersen et al., 2006)

$$\Phi = \mathbf{F}\mathbf{C}. \tag{4.21}$$

Here $\Phi = [\phi_1 \ \phi_2 \ \ldots \ \phi_N]^T$ is a vector containing the extracellular potential, and $\mathbf{C} = [C_1 \ C_2 \ \ldots \ C_N]^T$ is a corresponding vector containing the CSD parameters. $\mathbf{F}$ is an $N \times N$ matrix containing the mapping from CSDs to extracellular potentials found from electrostatic theory. If $\mathbf{F}$ is constructed properly, it will be invertible, and the N unknown CSD parameters $\hat{\mathbf{C}}$ can then be estimated from the N recorded potentials by a simple matrix multiplication with the inverse matrix $\mathbf{F}^{-1}$:

$$\hat{\mathbf{C}} = \mathbf{F}^{-1}\Phi. \tag{4.22}$$

To illustrate the calculation of the matrix $\mathbf{F}$ we can consider the most common situation where a laminar electrode array with equidistant electrode contacts is inserted perpendicularly into, say, sensory cortex (Rappelsberger et al., 1981; Mitzdorf, 1985; Di et al., 1990; Schroeder et al., 2001; Ulbert et al., 2001; Einevoll et al., 2007). For simplicity we further assume the stimulus-evoked CSD to be located in infinitely thin, circular disks centered on the N electrode contacts. Each disk is further assumed to have the same CSD throughout the disk and to be positioned in the horizontal plane perpendicular to an inserted laminar electrode array (Pettersen et al., 2006). For this "δ-source" method a simple formula is obtained for the matrix elements, and the method also has some additional interest since it turns out to correspond to the standard CSD method in the limit of infinitely large discks (Pettersen et al., 2006). From electrostatic theory we have that the extracellular potential at a position z at the center axis due to an infinitely thin current source disk placed in z' is given by $\phi = (\sqrt{(z - z')^2 + R^2} - |z - z'|)C^p/2\sigma$, where C^p now is the *planar* CSD, R is the radius of the disks, and σ is the extracellular conductivity (Nicholson and Llinas, 1971; Pettersen et al., 2006). This implies that the matrix elements f_{jk} of the matrix $\mathbf{F}$ are given by

$$f_{jk} = \left(\sqrt{(z_j - z_k)^2 + R^2} - |z_j - z_k| \right) h/2\sigma, \tag{4.23}$$

where $C_j = C_j^p/h$ and $z_j - z_k = h(j - k)$. This δ-source iCSD method is now completely specified by Equations (4.22) and (4.23), and as shown, for example in

Fig. 7 of Pettersen et al. (2008), even this simple δ-source CSD method completely outperforms the standard CSD method when the population activity is spatially confined.

Pettersen et al. (2008) also investigated two other variations of the iCSD method: the *step* iCSD method, where the CSD is assumed to be stepwise constant in the z-direction, and the *spline* iCSD method based on cubic-splines interpolation. A GUI-based MATLAB toolbox for estimating the CSD from laminar multi-electrode recordings has been developed based on these three iCSD methods and can be downloaded from http://software.incf.org/.

These two latter methods were generalized and further developed by Leski et al. (2007) who developed the iCSD method for estimation of CSD based on three-dimensional recordings. In this situation the advantage of the iCSD method is even larger compared to the standard CSD method in that the fraction of electrode contacts at the boundary is much higher. For example, their electrode grid consisted of $4 \times 5 \times 7$ contacts, for which 110 of the 140 electrode contacts are the boundary and thus outside the scope of the standard CSD method. From their studies Leski et al. concluded that a spline iCSD method was a good choice for this three-dimensional situation (Leski et al., 2007).

The iCSD method has now also been implemented for the case with two-dimensional recordings (Leski et al., 2011), for example recordings done with multi-shank laminar electrodes (Buzsáki, 2004). As for the one-dimensional method, a GUI-based MATLAB toolbox has been developed to facilitate easy use of the method (Leski et al., 2011) and can be downloaded from http://software.incf.org/.

4.6.3 Validation of iCSD with population forward modeling

The results from forward modeling of synaptically evoked activity in a population of morphologically reconstructed pyramidal neurons in Section 4.5 are well suited for testing the iCSD approach, and for comparing the accuracy of this method with the standard CSD method. Here we consider the situation where the LFP is recorded by a laminar electrode array oriented perpendicular to the cortical layers and penetrating the population through its center (Pettersen et al., 2008), but these forward modeling population results have also been used to test the iCSD approach for recordings with multi-shank laminar electrodes (Leski et al., 2011).

As discussed above, the CSD should be considered as the average net transmembrane current within a particular volume element. In Figure 4.11A we show the actual CSD at the center of the columnar model population shown in Figure 4.8, i.e. the average transmembrane current of a centered cylindrical volume element of height 0.1 mm and radius 0.2 mm, plotted as a function of time. The spatial

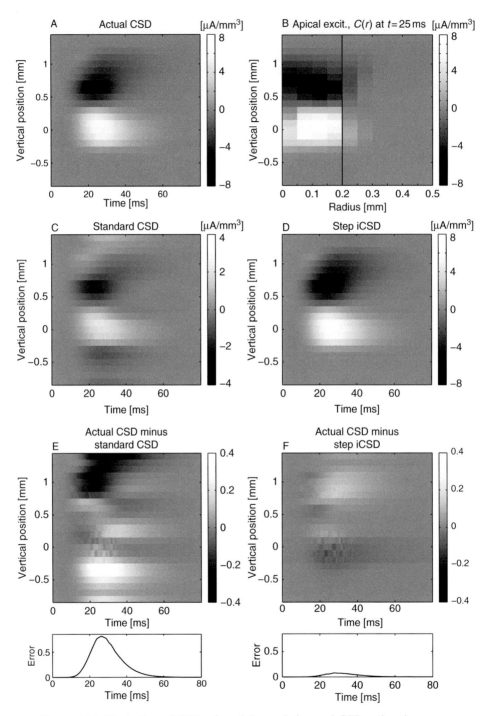

Figure 4.11 Illustration of CSD of model population and CSD estimation errors for different CSD methods. A. Actual CSD within the column, i.e. average of CSD in a centered cylindrical volume of height 0.1 mm and diameter 0.4 mm.

spread of this CSD is illustrated in Figure 4.11B. Here, the CSD at a particular time ($t = 25$ ms, cf. Figure 4.9A) was computed for cylindrical annuli with rectangular cross-sectional areas of 0.05 mm × 0.1 mm. It is seen that the CSD varies only moderately as a function of radial distance when inside the column. One further sees that even though the somas are restricted to radial distances less than 0.2 mm, the dendrites give non-negligible CSD outside this boundary.

The next four panels in Figure 4.11 illustrate the accuracy of the CSD estimation methods in this model situation. The estimated CSD along the (virtual) laminar electrode from the standard CSD method is shown in Figure 4.11C. This CSD is estimated by using Equation (4.20) on model LFPs at the (virtual) electrode contacts at the center axis of the column, cf. Figure 4.8B, except at the top and bottom contacts where the method of (Vaknin et al., 1988) is used. Comparison with Figure 4.11A shows that the standard CSD method predicts spurious sinks and sources below and above the actual CSD. Further, the sizes (amplitudes) of the actual sources and sinks are underestimated by about a factor of two.

Figure 4.11D shows the CSD estimated from the step iCSD method (Pettersen et al., 2008). This method assumes piecewise constant CSD distribution in the vertical direction. A columnar diameter of 0.4 mm is assumed in the iCSD method, the diameter of the cylindrical box to which the soma positions are restricted. This is clearly less than the spatial extension of the actual CSD seen in Figure 4.11B, but anyhow a natural parameter choice. Despite the somewhat unnatural assumptions regarding the form of the CSD, the step iCSD estimates are seen to be very similar to the actual CSD seen in Figure 4.11A. The much improved CSD estimates from the step iCSD method compared to the standard CSD method are further illustrated in the panels E and F showing the relative mean-square differences between the actual and estimated CSD for the two CSD estimation methods (see Pettersen et al., (2008) for detailed specification of the error estimate); the error of the standard CSD method estimates is much larger than for the iCSD step method. For further discussion see Pettersen et al. (2008).

Caption for Figure 4.11 (cont.)
B. Radial CSD distribution at particular point in time ($t = 25$ ms) as a function of depth and radial distance from the population center, computed by averaging over volume elements consisting of annuli with rectangular cross-sectional area of 0.05 mm × 0.1 mm. C. CSD estimate using the standard CSD method on the modeled LFPs along the center axis of the population. Top and bottom estimates are found by the using the method of Vaknin et al. (1988). D. CSD estimate from the step iCSD method. E–F. Difference between the actual CSD and estimates from the standard (E) and step iCSD methods (F), respectively. Both the actual CSD and the estimated CSDs were normalized to have a maximum amplitude of unity prior to error estimation and plotting (Pettersen et al., 2008).

4.7 Concluding remarks

The main topic of this chapter has been the forward-modeling of extracellular potentials, i.e. calculation of the extracellular potentials from activity in neurons or populations of neurons. So far there have been relatively few modeling studies pursuing such calculations; in fact, the first full-fledged study of this type using morphologically reconstructed neurons was done less than fifteen years ago (Holt and Koch, 1999). With the advent of new public databases of reconstructed neurons such as www.neuromorpho.org/ and ever more powerful computers, we expect that the relatively straightforward forward-modeling scheme for calculating extracellular potentials will be more frequently used in the years to come. A Python package, LFPy, has been developed in our group to facilitate such calculations (http://software.incf.org/software/lfpy/).

An important set of applications of this forward modeling scheme will be the validation of methods for analysis of LFP and MUA data, as exemplified by the test of the iCSD method in Figure 4.11 or the test of the MUA as a measure of population firing rate, cf. Figure 4.9D and Pettersen et al. (2008). The forward-modeling scheme will likewise be useful for testing and development of new methods for analysis of multi-electrode data such as the so-called *laminar population analysis (LPA)* (Einevoll et al., 2007) or spike-sorting algorithms (Lewicki, 1998; Buzsáki, 2004).

To improve the accuracy and reliability of the forward-modeling scheme it is important to establish good, experimentally validated models for the impedance properties of the extracellular media in all relevant types of neural tissue. Ideally one could envision experimental setups and protocols allowing for in situ measurement of the extracellular conductivity in conjunction with each multi-electrode recording. Further, a more detailed understanding and accurate model representation of the electrical properties of the various types of multielectrodes are needed (Moffit and McIntyre, 2005; Grimnes and Martinsen, 2008; Nelson et al., 2008; Nelson and Pouget, 2010).

Acknowledgments

Trygve Solstad and Øystein Sørensen are acknowledged for a thorough reading of the chapter. This work was supported by the Research Council of Norway (eVita, NOTUR, NevroNor).

References

Bédard, C. and Destexhe, A. (2009). Macroscopic models of local field potentials and the apparent 1/*f* noise in brain activity. *Biophys. J.*, **96**, 2589–2603.
Bédard, C., Kröger, H. and Destexhe, A. (2004). Modeling extracellular field potentials and the frequency-filtering properties of extracellular space. *Biophys. J.*, **86**, 1829–1842.

Bédard, C., Kröger, H. and Destexhe, A. (2006a). Does the 1/*f* frequency scaling of brain signals reflect self-organized critical states? *Phys. Rev. Lett.*, **97**, 118102.

Bédard, C., Kröger, H. and Destexhe, A. (2006b). Model of low-pass filtering of local field potentials in brain tissue. *Phys. Rev. E*, **73**, 051911.

Berens, P., Keliris, G. A., Ecker, A. S., Logothetis, N. and Tolias, A. S. (2008). Comparing the feature selectivity of the gamma-band of the local field potential and the underlying spiking activity in primate visual cortex. *Front. Syst. Neurosci.*, 10.3389/neuro.06/002.2008

Blomquist, P., Devor, A., Indahl, U. G., Ulbert, I., Einevoll, G. T. and Dale, A. M. (2009). Estimation of thalamocortical and intracortical network models from joint thalamic single-electrode and cortical laminar-electrode recordings in the rat barrel system. *PLOS Comput. Biol.*, **5** (3), e1000328.

Bower, J. M. and Beeman, D. (1998). *The Book of GENESIS: Exploring Realistic Neural Models with the GEneral NEural SImulation System* (2nd edition). New York: Springer.

Buzsáki, G. (2004). Large-scale recording of neuronal ensembles. *Nature Neurosci.*, **7**, 446–451.

Buzsáki, G. (2006). *Rhythms of the Brain.* New York: Oxford University Press.

Carnevale, N. T. and Hines, M. L. (2006). *The NEURON Book.* Cambridge: Cambridge University Press.

Csicsvari, J., Hirase, H., Czurkó, A., Mamiya, A. and Buzsáki, G. (1999). Oscillatory coupling of hippocampal pyramidal cells and interneurons in the behaving rat. *J. Neurosci.*, **19**, 274–287.

Dayan, P. and Abbott, L. F. (2001). *Theoretical Neuroscience.* Cambridge, MA: MIT Press.

Di, S., Baumgartner, C. and Barth, D. S. (1990). Laminar analysis of extracellular field potentials in rat vibrissa/barrel cortex. *J. Neurophysiol.*, **63**, 832–840.

Einevoll, G. T., Pettersen, K. H., Devor, A., Ulbert, I., Halgren, E. and Dale, A. M. (2007). Laminar population analysis: estimating firing rates and evoked synaptic activity from multielectrode recordings in rat barrel cortex. *J. Neurophysiol.*, **97**, 2174–2190.

Fee, M. S., Mitra, P. M. and Kleinfeld, D. (1996). Automatic sorting of multiple unit neuronal signals in the presence of anisotropic and non-Gaussian variability. *J. Neurosci. Methods*, **69**, 175–188.

Franke, F., Natora, M., Boucsein, C., Munk, M. H. J. and Obermayer K. (2010). An online spike detection and spike classification algorithm capable of instantaneous resolution of overlapping spikes. *J. Comput. Neurosci.*, **29**, 127–148.

Freeman, J. A. and Nicholson, C. (1975). Experimental optimization of current source-density technique for anuran cerebellum. *J. Neurophysiol.*, **38**, 369–382.

Freeman, W. J. and Zhai, J. (2009). Simulated power spectral density (PSD) of background electrocorticogram (ECoG). *Cogn. Neurodyn.*, **3**, 97–103.

Freeman, W. J., Holmes, M. D., Burke, B. C. and Vanthalo, S. (2003). Spatial spectra of scalp EEG and EMB from awake humans. *Clin. Neurophysiol.*, **114**, 1053–1068.

Gabriel, S., Lau, R. W. and Gabriel, C. (1996). The dielectric properties of biological tissues: III. Parametric models for the dielectric spectrum of tissues. *Phys. Med. Biol.*, **41**, 2271–2293.

Gold, C., Henze, D. A., Koch, C. and Buzsáki, G. (2006). On the origin of the extracellular action potential waveform: a modeling study. *J. Neurophysiol.*, **95**, 3113–3128.

Gold, C., Henze, D. A. and Koch, C. (2007). Using extracellular action potential recordings to constrain compartmental models. *J. Comput. Neurosci.*, **23**, 39–58.

Goto, T., Hatanaka, R., Ogawa, T., Sumiyoshi A., Riera, J. and Kawashima, R. (2010). An evaluation of the conductivity profile in the somatosensory barrel cortex of Wistar rats. *J. Neurophysiol.*, **104**, 3388–3412.

Gray, C. M., Maldonado, P. E., Wilson, M. and McNaughton, B. (1995). Tetrodes markedly improve the reliability and yield of multiple single-unit isolation from multi-unit recordings in cat striate cortex. *J. Neurosci. Methods*, **63**, 43–54.

Grimnes, S. and Martinsen, Ø. G. (2008). *Bioimpedance and Bioelectricity Basics* (2nd edition). Academic Press.

Hämäläinen, M., Hari, R., Ilmoniemi, R., Knuutila, J. and Lounasmaa, O. (1993). Magnetoencephalography theory, instrumentation, and applications to noninvasive studies of the working human brain. *Rev. Mod. Phys.*, **65**, 413–497.

Harris, K. D., Henze, D. A., Csicsvari, J., Hirase, H. and Buzsaki, G. (2000). Accuracy of tetrode spike separation as determined by simultaneous intracellular and extracellular measurements. *J. Neurophysiol.*, **84**, 401–414.

He, B. J., Zempel, J. M., Snyder, A. Z. and Raichle, M. E. (2010). The temporal structures and functional significance of scale-free brain activity. *Neuron*, **66**, 353–369.

Henze, D. A., Borhegyi, Z., Csicsvari, J., Mamiya, A., Harris, K. D. and Buzsaki, G. (2000). Intracellular features predicted by extracellular recordings in the hippocampus in vivo. *J. Neurophysiol.*, **84**, 390–400.

Holt, G. R. and Koch, C. (1999). Electrical interactions via the extracellular potential near cell bodies. *J. Comput. Neurosci.*, **6**, 169–184.

Hubel, D. H. (1957). Tungsten microelectrode for recording from single units. *Science*, **125**, 549–550.

Jackson, J. (1998). *Classical Electrodynamics*. Hoboken, NJ: Wiley.

Jog, M. S., Connolly, C. I., Kubota, Y., Iyengar, D. R., Garrido, L., Harlan, R. and Graybiel, A. M. (2002). Tetrode technology: advances in implantable hardware, neuroimaging, and data analysis techniques. *J. Neurosci. Methods*, **117**, 141–152.

Johnston, D. and Wu, S. M.-S. (1994). *Foundation of Cellular Neurophysiology*. Cambridge, MA: MIT Press.

Katzner, S., Nauhaus, I., Benucci, A., Bonin, V., Ringach, D. L. and Carandini, M. (2009). Local origin of field potentials in visual cortex. *Neuron*, **61**, 35–41.

Kim, S. and McNames, J. (2007). Automatic spike detection based on adaptive template matching for extracellular neural recordings. *J. Neurosci. Methods*, **165**, 165–174.

Kreiman, G., Hung, C. P, Kraskov, A., Quiroga, R. Q., Poggio, T. and DiCarlo, J. J. (2006). Object selectivity of local field potentials and spikes in the macaque inferior temporal cortex. *Neuron*, **49**, 433–445.

Leski, S., Wojcik, D. K., Tereszczuk, J., Swiejkowski, D. A., Kublik, E. and Wrobel, A. (2007). Inverse current-source density method in 3D: reconstruction fidelity, boundary effects, and influence of distant sources. *Neuroinformatics*, **5**, 207–222.

Leski, S., Pettersen, K. H., Tunstall, B., Einevoll, G. T., Gigg, J. and Wojcik, D. K. (2011). Inverse current source density method in two dimensions: inferring neural activation from multielectrode recordings. *Neuroinformatics*, **9** (4), 401–425.

Levina, A., Herrmann, J. M. and Geisel, T. (2007). Dynamical synapses causing self-organized criticality in neural networks. *Nature Physics*, **3**, 857–860.

Lewicki, M. S. (1998). A review of methods for spike sorting: the detection and classification on neural action potentials. *Network: Comput. Neural Syst.*, **9**, R53–R78.

Lindén, H., Pettersen, K. H., Tetzlaff, T., Potjans, T., Denker, M., Diesmann, M., Grün, S. and Einevoll, G. T. (2009a). Estimating the spatial range of local field potentials in a cortical population model. *BMC Neurosci.*, **10** (Supplement 1), 224.

Lindén, H., Potjans, T., Einevoll, G. T., Denker, M., Grün, S., and Diesmann, M. (2009b). Modeling the local field potential by a large-scale layered cortical network model. *Fronti. Neuroinf. Conference Abstract: Neuroinformatics.* doi: 10.3389/conf.neuro.11.2009.08.046.

Lindén, H., Pettersen K. H. and Einevoll, G. T. (2010). Intrinsic dendritic filtering gives low-pass power spectra of local field potentials. *J. Comput. Neurosci.*, **29**, 423–444.

Linkenkaer-Hansen, K., Nikouline, V. V., Palva, J. M. and Ilmoniemi, R. J. (2001). Long-range temporal correlations and scaling behavior in human brain oscillations. *J. Neurosci.*, **21**, 1370–1377.

Liu, J. and Newsome, W. T. (2006). Local field potential in cortical area MT. Stimulus tuning and behavioral correlations. *J. Neurosci.*, **26**, 7779–7790.

Logothetis, N. K., Kayser, C. and Oeltermann, A. (2007). In vivo measurement of cortical impedance spectrum in monkeys: implications for signal propagation. *Neuron*, **55**, 809–823.

López-Aguado, L., Ibarz, J. M. and Herreras, O. (2001). Acitivity-dependent changes of tissue resisitivity in the CA1 region in vivo are layer-specific: modulation of evoked potentials. *Neuron*, **108**, 249–262.

Mainen, Z. F. and Sejnowski, T. J. (1996). Influence of dendritic structure on firing pattern in model neocortical neurons. *Nature*, **382**, 363–366.

McNaughton, B. L., O'Keefe, J. and Barnes, C. A. (1983). The stereotrode: a new technique for simultaneous isolation of several single units in the central nervous system from multiple unit records. *J. Neurosci. Methods*, **8**, 391–397.

Miller, K. J., Sorensen, L. B., Ojemann, J. G. and den Nijs, M. (2009). Power-law scaling in the brain surface electric potential. *PLoS Comp. Biol.*, **5**, e1000609.

Milstein, J. N. and Koch, C. (2008). Dynamic moment analysis of the extracellular electric field of a biologically realistic spiking neuron. *Neural Comput.*, **20**, 2070–2084.

Milstein, J., Mormann, F., Fried, I. and Koch, C. (2009). Neuronal shot noise and Brownian $1/f^2$ behavior in the local field potential. *PLoS ONE*, **4**, e4338.

Mitzdorf, U. (1985). Current source-density method and application in cat cerebral cortex: investigation of evoked potentials and EEG phenomena. *Physiol. Rev.*, **65**, 37–100.

Moffitt, M. A. and McIntyre C. C. (2005). Model-based analysis of cortical recording with silicon microelectrodes. *Clin. Neurophysiol.*, **116**, 2240–2250.

Monto, S., Palva, S., Voipio J. and Palva, J. M. (2008). Very slow EEG fluctuations predict the dynamics of stimulus detection and oscillation amplitudes in humans. *J. Neurosci.*, **28**, 8268–8272.

Mountcastle, V. B. (1997). The columnar organization of the neocortex. *Brain*, **120**, 701–722.

Nadasdy, Z., Csicsvari, J., Penttonen, M., Hetke, J., Wise, K. and Buzsaki, G. (1998). Extracellular recording and analysis of neuronal activity: from single cells to ensembles. In: H. Eichenbaum and J. L. Davis (editors), *Neuronal Ensembles*. New York: Wiley.

Nelson, M. J. and Pouget, P. (2010). Do electrode properties create a problem in interpreting local field potential recordings? *J. Neurophysiol.*, **103**, 2325–2317.

Nelson, M. J., Pouget, P., Nilsen, E. A., Patten, C. D. and Schall, J. D. (2008). Review of signal distortion through metal microelectrode recording circuits and filters. *J. Neurosci. Methods*, **169**, 141–157.

Nicholson, C. and Freeman, J. A. (1975). Theory of current source-density analysis and determination of conductivity tensor for anuran cerebellum. *J. Neurophysiol.*, **38**, 356–368.

Nicholson, C. and Llinas, R. (1971). Field potentials in the alligator cerebellum and theory of their relationship to Purkinje cell dendritic spikes. *J. Neurophysiol.*, **34**, 509–531.

Nicholson, C. and Llinás, R. (1975). Real time current source-density analysis using multi-electrode array in cat cerebellum. *Brain Res.*, **100**, 418–424.

Normann, R. A., Maynard, E. M., Rousche, P. J. and Warren, D. J. (1999). A neural interface for a cortical vision prosthesis. *Vision Res.*, **39**, 2577–2587.

Nunez, P. L. and Srinivasan, R. (2006). *Electric Fields of the Brain: The Neurophysics of EEG*. Oxford: Oxford University Press.

Pettersen, K. H. and Einevoll, G. T. (2008). Amplitude variability and extracellular low-pass filtering of neuronal spikes. *Biophys. J.*, **94**, 784–802.

Pettersen, K. H. and Einevoll, G. T. (2009). Neurophysics: what the telegrapher's equation has taught us about the brain. In: Ø. Martinsen and Ø. Jensen (editors), *An Anthology of Developments in Clinical Engineering and Bioimpedance: Festschrift for Sverre Grimnes*. Unpublished, Oslo.

Pettersen, K. H., Devor, A., Ulbert, I., Dale, A. M. and Einevoll, G. T. (2006). Current-source density estimation based on inversion of electrostatic forward solution: effects of finite extent of neuronal activity and conductivity discontinuities. *J. Neurosci. Methods*, **154**, 116–133.

Pettersen, K. H., Hagen, E. and Einevoll, G. T. (2008). Estimation of population firing rates and current source densities from laminar electrode recordings. *J. Comput. Neurosci.*, **24**, 291–313.

Pouzat C. and Chaffiol, A. (2009). Automatic spike train analysis and report generation. An implementation with R, R2HTML and STAR. *J. Neurosci. Methods*, **181**, 119–144.

Pritchard, W. S. (1992). The brain in fractal time: $1/f$-like power spectrum scaling of the human electroencephalogram. *Int. J. Neurosci.*, **66**, 119–129.

Quian Quiroga, R. (2007). Spike sorting. *Scholarpedia*, **2**, 3583.

Quian Quiroga, R., Nadasdy, R. and Ben-Shaul, Y. (2004). Unsupervised spike detection and sorting with wavelets and superparamagnetic clustering. *Neural Comput.*, **16**, 1661–1687.

Rall, W. (1962). Electrophysiology of a dendritic neuron model. *Biophys. J.*, **2**, 145–167.

Rall, W. and Shepherd, G. M. (1968). Theoretical reconstruction of field potentials and dendrodendritic synaptic interactions in olfactory bulb. *J. Neurophysiol.*, **31**, 884–915.

Rappelsberger, P., Pockberger, H. and Petsche, H. (1981). Current source density analysis: methods and application to simultaneously recorded field potentials of the rabbit's visual cortex. *Pflügers Arch.*, **389**, 159–170.

Recce, M. and O'Keefe, J. (1989). The tetrode: a new technique for multi-unit extracellular recording. *Soc. Neurosci. Abstr.*, **15**: 1250.

Rutishauser, U., Schuman, E. M. and Mamelak, A. N. (2006). Online detection and sorting of extracellularly recorded action potentials in human medial temporal lobe recordings in vivo. *J. Neurosci. Methods*, **154**, 204–224.

Schroeder, C. E., Lindsley, R. W., Specht, C., Marcovici, A., Smiley, J. F. and Javitt, D. C. (2001). Somatosensory input to auditory association cortex in the macaque monkey. *J. Neurophysiol.*, **85**, 1322–1327.

Segev, I. and Burke, R. (1998). Compartmental model of complex neurons. In: C. Koch and I. Segev (editors), *Methods in Neuronal Modeling: From Ions To Network*. Cambridge, MA: MIT Press.

Shoham, S. and Nagarajan, S. S. (2004). The theory of CNS recording. In: K. W. Horch and G. S. Dhillon (editors), *Neuroprosthetics: Theory and Applications*. Singapore: World Scientific, pp. 448–471.

Shoham, S., O'Connor, D. H. and Segev, R. (2006). How silent is the brain: is there a "dark matter" problem in neuroscience? *J. Comp. Physiol. A*, **192**, 777–784.

Smith, L. S. and Mtetwa, N. (2007). A tool for synthesizing spike trains with realistic interference. *J. Neurosci. Methods*, **159**, 170–180.

Stuart, G., Spruston, N. and Häusser, M. (editors) (2007). *Dendrites* (2nd edition). Oxford: Oxford University Press.

Sukov, W. and Barth, D. S. (1998). Three-dimensional analysis of spontaneous and thalamically evoked gamma oscillations in auditory cortex. *J. Neurophysiol.*, **79**, 2875–2884.

Takekawa, T., Isomura, Y. and Fukai, T. (2010). Accurate spike sorting for multiunit recordings. *Eur. J. Neurosci.*, **31**, 263–272.

Ulbert, I., Halgren, E., Heit, G. and Karmos, G. (2001). Multiple microelectrode-recording system for human intracortical applications. *J. Neurosci. Methods*, **106**, 69–79.

Vaknin, G., DiScenna, P. G. and Teyler, T. J. (1988). A method for calculating current source density (CSD) analysis without resorting to recording sites outside the sampling volume. *J. Neurosci. Methods*, **24**, 131–135.

Wehr, M., Pezaris, J. S. and Sahani, M. (1999). Simultaneous paired intracellular and tetrode recordings for evaluation the performance of spike sorting algorithms. *Neurocomputing*, **26–27**, 1061–1068.

Wilson, M. A. and McNaughton, B. L. (1993). Dynamics of the hippocampal ensemble code for space. *Science*, **261**, 1055–1058.

Xing, D., Yeh, C.-I. and Shapley, R. M. (2009). Spatial spread of the local field potential and its laminar variation in visual cortex. *J. Neurosci.*, **29**, 11540–11549.

5

Local field potentials

CLAUDE BÉDARD AND ALAIN DESTEXHE

5.1 Introduction

Extracellular electric potentials, such as local field potentials (LFPs) or the electroencephalogram (EEG), are routinely measured in electrophysiological experiments. LFPs are recorded using micrometer-size electrodes, and sample relatively localized populations of neurons, as these signals can be very different for electrodes separated by 1 mm (Destexhe et al., 1999a) or by a few hundred micrometers (Katzner et al., 2009). In contrast, the EEG is recorded from the surface of the scalp using millimeter-scale electrodes and samples much larger populations of neurons (Niedermeyer and Lopes da Silva, 1998). LFPs are subject to much less filtering compared to EEG, because EEG signals must propagate through various media, such as cerebrospinal fluid, dura mater, cranium, muscle and skin. LFP signals are also filtered, because the recording electrode is separated from the neuronal sources by portions of cortical tissue. Besides these differences, EEG and LFP signals display the same characteristics during wake and sleep states (Steriade, 2003).

The observation that action potentials have a limited participation in the genesis of the EEG or LFPs dates from early studies. Bremer (1938, 1949) was the first to propose that the EEG is not generated by action potentials, based on the mismatch of the time course of EEG waves with action potentials. Eccles (1951) proposed that LFP and EEG activities are generated by summated postsynaptic potentials arising from the synchronized excitation of cortical neurons. Intracellular recordings from cortical neurons later demonstrated a close correspondence between EEG/LFP activity and synaptic potentials (Klee et al., 1965; Creutzfeldt et al., 1966a, 1966b). The current view is that EEG and LFPs are generated by synchronized synaptic currents arising on cortical neurons, possibly through the formation of dipoles (Nunez, 1981; Niedermeyer and Lopes da Silva, 1998).

The fact that action potentials have little participation in EEG-related activities indicates strong frequency-filtering properties of cortical tissue. High frequencies

Handbook of Neural Activity Measurement, ed. Romain Brette and Alain Destexhe. Published by Cambridge University Press. © Cambridge University Press 2012.

(greater than $\approx$100 Hz), such as that produced by action potentials, are subject to a severe attenuation, and therefore are visible only for electrodes immediately adjacent to the recorded cell. On the other hand, low-frequency events, such as synaptic potentials, attenuate less with distance. These events can therefore propagate over large distances in extracellular space and be recordable as far as on the surface of the scalp, where they can participate in the genesis of the EEG. This frequency-dependent behavior is also seen routinely in extracellular unit recordings: the amplitude of extracellularly recorded spikes is very sensitive to the position of the electrode, but slow events show much less sensitivity to the position. In other words, an extracellular electrode records slow events that originate from a large number of neighboring neurons, and are more stable to small changes of electrode position. In contrast, action potentials are recorded only for the cell(s) immediately adjacent to the electrode, and are therefore very sensitive to changes in electrode position. This property is fundamental because it allows the resolution of single units from extracellular recordings.

EEG and LFP measurements also display approximately $1/f$ frequency scaling in their power spectra (Pritchard, 1992; Novikov et al., 1997; Bhattacharya and Petsche, 2001; Bédard et al., 2006a) (see Figure 5.1). The origin of such $1/f$ "noise" is at present unclear. $1/f$ spectra can result from self-organized critical phenomena (Jensen, 1998), suggesting that neuronal activity may be working according to such states (Beggs and Plenz, 2003). Alternatively, the $1/f$ scaling may be due to filtering properties of the currents through extracellular media (Bédard et al., 2006a). This conclusion was reached by noting that the global activity reconstructed from multi-site unit recordings scales identically as the LFP if a "$1/f$ filter" is assumed, and without the need to assume self-organized critical states in neural activity.

Despite its potential importance for resolving single cells or explaining the power spectral structure of LFPs, little is known about the physical basis of the frequency dependence of extracellular potentials in the cortex. By contrast to intracellular events, for which biophysical mechanisms have been remarkably well characterized during the last 60 years (reviewed in Koch, 1999), comparatively little has been done to investigate the biophysical mechanisms underlying the genesis of extracellular field potentials (see review by Nunez, 1981). The reason is that LFPs result from complex interactions involving many factors, such as the spatial distribution of current sources, the spatial distribution of positive and negative electric charges (forming dipoles), their time evolution (dynamics), as well as the conductive and permittivity properties of the extracellular medium. One of the simplest and widely used models of LFP activity considers current sources embedded in a homogeneous extracellular medium (Nunez, 1981; Koch and Segev, 1998). Although this formalism has been successful in many instances (Rall and

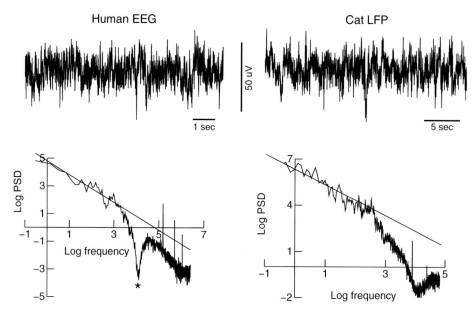

Figure 5.1 $1/f$ frequency scaling of electroencephalogram and local field potentials. The top traces show examples of human EEG recordings (left, vertex EEG) and cat LFP recordings (right, parietal cortex) during awake and attentive states. The corresponding power spectral densities (lower panels) display approximate $1/f$ scaling at low frequencies. The straight lines indicate a slope of -1 in this log–log representation. The signals were not filtered, except for a notch filter at 60 Hz (*) for the EEG, and the data acquisition filters at high frequencies (not visible at this scale).

Shepherd, 1968; Klee and Rall, 1977; Destexhe, 1998; Protopapas et al., 1998), it does not account for the frequency-dependent attenuation and therefore is not precise enough to model extracellular field potentials containing both spikes and synaptic potential activity.

In this chapter, we review recent work investigating possible physical bases for the frequency-filtering properties of LFPs. We start from first principles (Maxwell equations) and consider different conditions of current sources and extracellular media. We delineate the cases leading to frequency-filtering properties consistent with physiological data. We show that the assumption of a resistive homogeneous extracellular medium cannot account for the frequency-dependent attenuation. It is necessary to take into account the inhomogeneous structure of the extracellular medium (in both permittivity and conductivity), the "reaction" of the medium (such as polarization) and ionic diffusion, in order to account for frequency-dependent attenuation.

The chapter is organized as follows: we start from the simplest model of LFPs, which consists of current sources in a homogeneous and resistive extracellular fluid. We next consider different causes of frequency dependence, such as inhomogeneities of conductivity, polarization phenomena and ionic diffusion. We also introduce a macroscopic formalism in which all these effects can be merged together into a unified model, and finish by discussing conflicting experiments and a proposition on how to reconcile them, as well as how to reproduce the $1/f$ frequency-scaling of LFPs from plausible physical causes.

5.2 Modeling LFPs in resistive media

5.2.1 Extracellular potential in homogeneous resistive media

The simplest model for the extracellular potential assumes that the extracellular medium is purely *resistive*, with no capacitive component. In this case, one considers a set of punctual current sources embedded in a homogeneous conductive medium of conductivity σ. The extracellular potential due to a single punctual current source can be deduced simply as follows. Starting from Ohm's law ($\vec{j} = \sigma \vec{E}$), combined with the law of current conservation ($\vec{j} = \frac{i}{4\pi r^2}\hat{r}$) in spherical symmetry around the source, and by integrating along a straight line from the source to a given point in extracellular space, we obtain:

$$\Delta V(r) = V(\infty) - V(r) = -\int_{\infty}^{r} \vec{E} \cdot d\vec{s} = -\frac{i}{4\pi\sigma} \int_{\infty}^{r} \frac{\hat{r} \cdot d\vec{s}}{r^2}$$
$$= \frac{i}{4\pi\sigma} \int_{\infty}^{r} \frac{dr}{r^2} = -\frac{i}{4\pi\sigma} \frac{1}{r}. \tag{5.1}$$

The extracellular potential at some distance r from the source i is then given by:

$$V(r) = \frac{1}{4\pi\sigma} \frac{i}{r}, \tag{5.2}$$

if we assume that the reference $V(\infty) = 0$. This relation can also be deduced directly from Coulomb's law and the first Maxwell equation in integrated form.

In the case of a set of n current sources i_j, the superposition principle applies and one can write:

$$V(r) = \frac{1}{4\pi\sigma} \sum_{j=1}^{n} \frac{i_j}{r_j}, \tag{5.3}$$

where r_j is the distance from the source i_j to the position r in extracellular space. This expression captures many effects such as dipoles or multipolar configurations. Equation (5.3) has been used to model extracellular potentials, from the early models (Rall and Shepherd, 1968) to today's models of extracellular activity (Protopapas et al., 1998; Nunez and Srinivasan, 2005; see also Chapter 4).

5.2.2 *Example of modeling LFPs in a homogeneous resistive medium*

To illustrate an example of such simple models of LFPs, we consider a model of absence seizure activity in the thalamocortical system (Destexhe, 1998). Absence seizures, like many other types of epileptic phenomena, are characterized by the genesis of typical oscillatory EEG patterns which consist of the alternation of negative sharp deflections ("spike") and slow positive waves ("wave"), which repeats at slow frequencies (typically around 3 Hz). This "spike-and-wave" pattern is an important signature of epileptic activity (Niedermeyer and Lopes da Silva, 1998) and is seen in humans and diverse animal models, in the EEG and in the LFPs.

The thalamocortical mechanisms leading to absence seizures are complex and involve recurrent inhibition in the cortex, as well as the rebound burst properties of thalamic neurons (Destexhe et al., 1999b; Steriade, 2003). This type of activity was reproduced by biophysical models of thalamocortical networks which included the intrinsic properties of cortical and thalamic neurons, as well as the different synaptic receptors present in these circuits (Destexhe, 1998; Destexhe et al., 1999b). This model reproduced the hypersynchronized neuronal discharges in epileptic seizures, and how such activity generates the typical spike-and-wave patterns in the LFPs (Figure 5.2). In this case, the LFP was calculated using Equation (5.3), using a linear arrangement of pyramidal (PY) neurons (20 μm intercellular distance), where i_j was the transmembrane current of each pyramidal neuron, excluding the currents responsible for action potentials (see details in Destexhe, 1998). The simulation showed that spike-and-wave patterns can be reproduced by this simple model, where synchronized postsynaptic potentials (EPSP/IPSP sequences) generate the negative "spike," while slow positive currents – also synchronized – generate the slow positive "wave" (Figure 5.2).

It is important to note that the exclusion of action potentials to generate the LFP was done here artificially. In reality, fast events (action potentials) have a much steeper attenuation compared to slow events (synaptic currents), such that only the latter contribute to the genesis of LFPs. Models with resistive media cannot account for such frequency filtering properties, and one must use more sophisticated models with non-resistive media, as we review in the next sections.

5.2.3 *Multipolar configurations*

Models with resistive media can also be made using multipolar source distributions. In this case, the electric potential at a point P exterior to the sources can be written as[1]

---

[1] This expression is obtained by Taylor expansion of V around the center of charge, as a function of the inverse of the distance between point P and the center of charge.

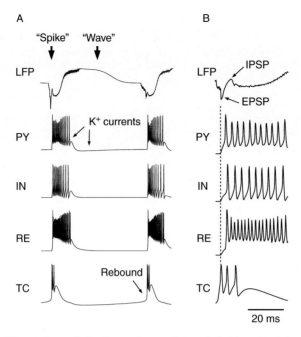

Figure 5.2 Simulation of "spike-and-wave" local field potentials in a model of absence epileptic seizures. The model consisted of a thalamocortical network where neurons were single-compartment Hodgkin–Huxley type models, and synaptic interactions were modeled by conductance-based kinetic models for glutamate (AMPA, NMDA) and γ-amino-butyric acid (GABA$_A$ and GABA$_B$) synaptic receptors. Reducing the fast (GABA$_A$-mediated) inhibition in the cortex resulted in the emergence of hypersynchronized oscillations at a frequency of around 3 Hz. The LFP calculated from monopolar sources (pyramidal neurons only) generated the typical spike-and-wave patterns seen experimentally during seizures. When all cells fired in synchrony, the LFP generated a negative "spike," while the neuronal silences were associated with a slow positive "wave" (A). The "spike" was generated by hypersynchronized EPSP/IPSP sequences (see magnification in B), while the "wave" was generated by slow K$^+$ currents in pyramidal neurons. (PY cortical pyramidal neurons, IN cortical interneurons, RE thalamic reticular neurons, TC thalamic relay cells). (Modified from Destexhe, 1998.)

$$V(\vec{r}) = \frac{1}{4\pi\epsilon} \left[\frac{Q}{r} + \frac{\hat{r} \cdot \vec{P}}{r^2} + \frac{k_0}{r^3} + \frac{k_1}{r^4} + \cdots \right] \qquad (5.4)$$

where $Q = \int \rho(\vec{r}')dv'$ is the total amount of charge involved in the different sources, $\vec{P} = \int \rho(\vec{r}')\vec{r}'dv'$ is the total dipole moment of the source distribution, ϵ is the electric permittivity of the medium, and r is the distance between the center of charge and point P. When the total charge (first term) is zero and the second term is non-zero, then we have a dipole by definition. We have a quadrupole when the two first terms are zero (null charge and null dipolar moment) while the third term is non-zero. In general, we have a 2^n-pole when the n first terms are zero, while the $(n+1)$th term is non-zero. In the case of a dipole, for a distance r which is large compared to the spatial scale of charge distributions, the electric potential is given by:

$$V(\vec{r}) = \frac{\hat{r} \cdot \vec{P}}{4\pi\epsilon r^2}. \tag{5.5}$$

Note that $\vec{P} = q\|\vec{r}_1 - \vec{r}_2\| = qd$ when the distribution of charge is given by $q(\delta(\vec{r} - \vec{r}_1) - \delta(\vec{r} - \vec{r}_2))$. The decision to express the electrical potential as a function of charge distribution in space allows us to stress the necessity of having a variation of charge distribution in a given region of space in order to have a variation of electric field outside that region. Thus, it avoids making hypotheses on the nature of the biological sources that produce the electric field.

Although Equation (5.3) is adequate for many purposes, such as modeling extracellular spikes (see Chapter 4) or spike-and-wave patterns (Figure 5.2), this expression may be too simple for other purposes. One example is the modeling of the correct distance dependence of extracellular field potentials, or their frequency content. As we will see in the next sections, the inhomogeneities of extracellular medium have a strong impact on the propagation and frequency filtering of field potentials.

5.2.4 Is extracellular space electrically uniform?

Extracellular space is highly non-uniform, and is made from the alternation of fluids and membranes (Peters et al., 1991; see Figure 5.3). Only about 6% of the extracellular space in the cortex is devoted to interstitial space (extracellular fluid), while the core of the volume is made up of axons (34%), neuronal dendrites (35%), spines (14%) and glial cells (11%); see details in Braitenberg and Schüz (1998). Because these media have very different conductivity and permittivity, one can expect the extracellular space to be electrically highly non-homogeneous, and therefore to contradict directly Equation (5.3) which assumes homogeneity.

To integrate such inhomogeneities of extracellular space, one cannot simply use the above equations, because this would violate the assumption that σ and ϵ are constant in the Maxwell equations. Therefore, to build a model of extracellular potentials that allows variations in these parameters, one needs to start from first principles. This is the object of the next section.

5.3 Modeling LFPs in non-resistive media: general theory

In this section, we start from the Maxwell equations of electromagnetism and apply these equations to neural tissue. Because Maxwell's theory is a mean-field theory

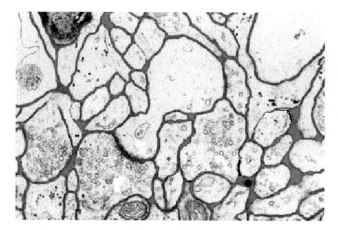

Figure 5.3 Geometry of extracellular space. Electron micrograph of a small region $4 \times 6 \ \mu m^2$ of rat cerebral cortex. The extracellular space (ECS) is outlined in gray. Note the simple convex cell surface and the presence of regions where the ECS widens. Note that the ECS is probably slightly reduced in width due to the fixation procedure. (Figure modified from Nicholson and Sykova, 1998.)

of electromagnetism, it is necessary to consider mean values, as we will describe below. We will model neural tissue using the three standard parameters, electric conductivity (σ), electric permittivity (ϵ) and magnetic permeability (μ).[2] Because neural tissue is not ferromagnetic, the magnetic permeability μ can be considered constant and equal to that of vacuum.

We derive the equations governing the time evolution of the extracellular potential for a locally neutral medium.[3] We will consider two different scales: a first scale, which will be called "microscopic," will consider averages of electromagnetic parameters (σ, ϵ, μ) over volumes of the order of cubic micrometers. These volumes are small compared to neuronal and glial processes in extracellular space. At larger scales, which will be called "macroscopic," we will consider averages over volumes of the order of cubic millimeters, which are large compared to cellular processes.

The electromagnetic parameters have very different properties at these two scales. At microscopic scales, the conductivity is a fixed number that depends on whether a fluid or a membrane is considered. At macroscopic scales, the volumes considered contain fluids and membranes, and have capacitive effects, such as the

---

[2] It is important to note that the notions of conductivity and permittivity of a given medium do not have a physical sense at a subatomic level, and thus these notions only apply to a mean-field theory. This is analogous to the notion of pressure and temperature in classical thermodynamics.

[3] We define as *locally neutral* a medium in which the charge density is zero when there is no electrical field. For the considered scales, we assume that the neural tissue is locally neutral because these scales are large compared to the membrane thickness (around 5–7 nm), which is the value at which charge inhomogeneities will be apparent.

conductivity is a complex number in the frequency domain of Fourier analysis. However, we can treat both scales using the same formalism, as we will see below.

In the next sections, we follow a formalism similar to that developed previously (Bédard et al., 2004), except that we reformulate the model macroscopically, with frequency-dependent electrical parameters (conductivity σ and permittivity ϵ). This will allow us to integrate directly macroscopic measurements of these electric parameters, which were found to be strongly frequency dependent in some cases (Gabriel et al., 1996a, 1996b, 1996c).

5.3.1 Microscopic model

We begin by deriving a general equation for the electrical potential when the electrical parameters are scalar numbers and are frequency dependent. We start from the Maxwell equations, taking the $\nabla \cdot \vec{D} = \rho^{free}$ (Gauss' law), and $\nabla \times (\vec{B}/\mu) = \vec{j} + (\partial \vec{D}/\partial t)$ (Ampère–Maxwell law) in a medium with constant magnetic permeability, which gives:

$$\nabla \cdot \vec{D} = \rho^{free}$$

$$\nabla \cdot \vec{j} + \frac{\partial \rho^{free}}{\partial t} = 0 \tag{5.6}$$

where $\vec{D}$, $\vec{j}$ and ρ^{free} are respectively the electric displacement, current density and charge density in the medium surrounding the sources.

In a linear medium, the equations linking the electric field $\vec{E}$ with electric displacement $\vec{D}$ and with current density $\vec{j}$, give:

$$\vec{D}(\vec{x}, t) = \int_{-\infty}^{\infty} \epsilon(\vec{x}, \tau)\vec{E}(\vec{x}, t - \tau)d\tau \tag{5.7}$$

and

$$\vec{j}(\vec{x}, t) = \int_{-\infty}^{\infty} \sigma(\vec{x}, \tau)\vec{E}(\vec{x}, t - \tau)d\tau. \tag{5.8}$$

This model is microscopic, in the sense that the electric parameters σ and ϵ take their microscopic values and are assumed to be independent of frequency. Indeed, the electric parameters of the extracellular fluid can be considered constant for frequencies lower than 1000 Hz (Gabriel et al., 1996c). In this case, the Fourier transforms of the above equations are respectively $\vec{D}_\omega = \epsilon_\omega \vec{E}_\omega = \epsilon \vec{E}_\omega$ and $\vec{j}_\omega = \sigma_\omega \vec{E}_\omega = \sigma \vec{E}_\omega = \sigma_z \vec{E}_\omega$.[4] It is important to note that σ_z is a real number in this microscopic model, which will not be the case in the macroscopic model of the next section.

----

[4] Note that $\omega = 2\pi f$ where f is the frequency.

Given the limited precision of measurements, we can consider $\nabla \times \vec{E} \approx 0$ for frequencies smaller than 1000 Hz. Thus, we can assume that $\vec{E} = -\nabla V$ such that the complex Fourier transform of Equations (5.6) can be written as:

$$\nabla \cdot (\epsilon(\vec{x})\nabla V_\omega) = -\rho_\omega^{free}$$

$$\nabla \cdot (\sigma(\vec{x})\nabla V_\omega) = i\omega\rho_\omega^{free}.$$

Consequently, we have

$$\nabla \cdot ((\sigma + i\omega\epsilon)\nabla V_\omega) = \nabla(\sigma + i\omega\epsilon) \cdot \nabla V_\omega + (\sigma + i\omega\epsilon)\nabla^2 V_\omega = 0. \qquad (5.9)$$

This equation was derived previously (Bédard et al., 2004) and is general enough to calculate the propagation of the extracellular potential in extracellular media which can have a complex or inhomogeneous structure, as well as frequency-dependent electric parameters.

Equation (5.9) reduces to the Laplace equation ($\nabla^2 V_\omega = 0$) when the medium is homogeneous with respect to σ and ϵ. Thus, Equation (5.9) is a generalization of the Laplace equation for a medium where σ and ϵ are non-homogeneous. Except for particular cases where the ratio ϵ/σ is independent of position, Equation (5.9) shows that, in non-homogeneous media, the extracellular potential will necessarily have a different power spectral structure compared to that of the current sources, because the extracellular potential must be solution of a differential equation with frequency-dependent coefficients.

5.3.2 Macroscopic model

In principle, it is sufficient to solve Equation (5.9) in the extracellular medium to obtain the frequency dependence of LFPs. However, in practice, this equation cannot be solved because the structure of the medium is too complex to define the limit conditions properly. The associated values of electric parameters must be specified for every point of space and for each frequency, which represents a considerable difficulty. One way to solve this problem is to consider a macroscopic or mean-field approach at a larger scale, noting that the Maxwell equations are invariant under change of scale.[5] This approach is justified here by the fact that the values measured experimentally are averaged values, with the precision depending on the measurement technique. Because our goal is to simulate these macroscopic measurements, we will use a macroscopic model, in which we take spatial averages of Equation (5.9), and make a continuous approximation for the spatial variations of these average values (see Bédard and Destexhe, 2009).

---

[5] Note that a change of scale can necessitate renormalizing electromagnetic parameters.

To this end, we define macroscopic electric parameters, ϵ^M and σ^M, as follows:

$$\epsilon_\omega^M(\vec{x}) = \langle \epsilon_\omega(\vec{x}) \rangle_{|V} = f(\vec{x}, \omega)$$

and

$$\sigma_\omega^M(\vec{x}) = \langle \sigma_\omega(\vec{x}) \rangle_{|V} = g(\vec{x}, \omega)$$

where V is the volume over which the spatial average is taken. We assume that V is of the order of mm³, and is thus much smaller than the cortical volume, so that the mean values will be dependent of the position in the cortex.

A priori, it may seem surprising here that macroscopic parameters depend on frequency. However, as we have seen in the preceding section, the independence of frequency of microscopic parameters does not imply that there is no frequency dependence macroscopically. As shown by Equation (5.9), the impedance inequalities can generate a dependence on frequency. Thus, if we replace the electric parameters of a non-homogeneous region by their mean values, we have to include parameters which depend on frequency, even if the microscopic parameters are frequency independent.

Because the average values of electric parameters are statistically independent of the mean value of the electric field, we have:

$$\langle \vec{j}^{total} \rangle_{|V}(\vec{x}, t) = \int_{-\infty}^{\infty} \sigma^M(\tau) \langle \vec{E} \rangle_{|V}(\vec{x}, t - \tau) d\tau$$
$$+ \int_{-\infty}^{\infty} \epsilon^M(\tau) \frac{\partial \langle \vec{E} \rangle_{|V}}{\partial t}(\vec{x}, t - \tau) d\tau,$$

where the first term on the right hand side represents the "dissipative" contribution, and the second term represents the "reactive" contribution (reaction from the medium, such as polarization). Here, all physical effects, such as diffusion, resistive and capacitive phenomena, are integrated into the frequency dependence of σ^M and ϵ^M. We will examine this frequency dependence more quantitatively in Section 5.6.

The complex Fourier transform of $\langle \vec{j}^{total} \rangle_{|V}(\vec{x}, t)$ then becomes:

$$\langle \vec{j}_\omega^{total} \rangle_{|V} = (\sigma_\omega^M + i\omega\epsilon_\omega^M)\langle \vec{E}_\omega \rangle_{|V}) = \sigma_z^M \langle \vec{E}_\omega \rangle_{|V}, \tag{5.10}$$

where σ_z^M is the complex macroscopic conductivity. We can also assume

$$\sigma_z^M = i\omega\epsilon_z^M \tag{5.11}$$

such that

$$\nabla \cdot \langle \vec{j}_\omega^{total} \rangle_{|V} = \nabla \cdot (\sigma_z^M \langle \vec{E}_\omega \rangle_{|V}) = \nabla \cdot (i\omega\epsilon_z^M \langle \vec{E}_\omega \rangle_{|V}) = 0. \tag{5.12}$$

Because $\sigma_z^M = (\sigma_\omega^M + i\omega\epsilon_\omega^M)$ and $\langle \vec{E}_\omega \rangle = -\nabla\langle V_\omega \rangle$, the expressions above (Equations (5.12)) can also be written in the form:

$$\nabla \cdot ((\sigma_\omega^M + i\omega\epsilon_\omega^M)\nabla\langle V_\omega \rangle|_V) = 0. \tag{5.13}$$

We note that starting from the continuum model (Bédard et al., 2004), where only spatial variations were considered, and generalizing this model by including frequency-dependent electric parameters, gives the same mathematical form as the original model (compare with Equation (5.9)). This form invariance will allow us to introduce surface polarization phenomena as well as the physical effects of ionic diffusion by including an ad hoc frequency dependence in σ_ω^M and ϵ_ω^M (see Section 5.7). The physical cause of this macroscopic frequency dependence is that the cortical medium is microscopically non-neutral (although the cortical tissue is macroscopically neutral). Such a local non-neutrality was already postulated in a previous model of surface polarization (Bédard et al., 2006a). This situation cannot be accounted for by Equation (5.9) if σ_ω^M and ϵ_ω^M are frequency independent (in which case $\rho_\omega = 0$ when $\nabla V_\omega = 0$). Thus, including the frequency dependence of these parameters enables the model to capture a much broader range of physical phenomena.

Finally, a fundamental point is that the frequency dependences of the electrical parameters σ_ω^M and ϵ_ω^M cannot take arbitrary values, but are related to each other by the Kramers–Kronig relations (Kronig, 1926; Landau and Lifshitz, 1984; Foster and Schwan, 1989):

$$\Delta\epsilon^M(\omega) = \epsilon^M(\omega) - \epsilon^M(\infty) = \frac{2}{\pi} \int_0^\infty \frac{\sigma^M(\omega')}{\omega'^2 - \omega^2} d\omega' \tag{5.14}$$

and

$$\sigma^M(\omega) = \sigma^M(0) - \frac{2\omega^2}{\pi} \int_0^\infty \frac{\Delta\epsilon^M(\omega')}{\omega'^2 - \omega^2} d\omega' \tag{5.15}$$

where principal value integrals are used. These equations are valid for any linear medium (i.e. when Equations (5.7) and (5.8) are linear). These relations will turn out to be critical for relating the model to experiments, as we will see in Section 5.8 below.

Note that, in contrast to the frequency dependence, the spatial dependences of σ_ω^M and ϵ_ω^M are independent of each other, because these dependences are related to the spatial distribution of elements within the extracellular medium.

5.3.3 Simplified geometry for macroscopic parameters

To obtain an expression for the extracellular potential in the macroscopic model, we need to solve Equation (5.13), which is possible analytically only if we consider a simplified geometry of the source and surrounding medium. The first simplification is to consider the source as monopolar. The choice of a monopolar source does not intrinsically reduce the validity of the results because multipolar configurations can be composed from the arrangement of a finite number of monopoles (Purcell, 1984). In particular, if the physical nature of the extracellular medium determines a frequency dependence for a monopolar source, it will also do so for multipolar configurations. A second simplification is to consider that the current source is spherical and that the potential is uniform on its surface. This simplification will enable us to calculate exact expressions for the extracellular potential and should not affect the results on frequency dependence. A third simplification is to consider the extracellular medium as isotropic. This assumption is certainly valid within a macroscopic approach, and is justified by the fact that the neuropil of the cerebral cortex is made of a quasi-random arrangement of cellular processes of very diverse size (Braitenberg and Schüz, 1998). This simplified geometry will allow us to determine how the physical nature of the extracellular medium can determine a frequency dependence of the LFPs, independently of other factors (such as more realistic geometry, propagating potentials along dendrites, etc.).

Thus, considering a spherical source embedded in an isotropic medium with frequency-dependent electrical parameters, combining with Equation (5.13), we have:

$$\frac{d^2 \langle V_\omega \rangle_{|V}}{dr^2} + \frac{2}{r} \frac{d \langle V_\omega \rangle_{|V}}{dr} + \frac{1}{(\sigma_\omega + i\omega\epsilon_\omega)} \frac{d(\sigma_\omega + i\omega\epsilon_\omega)}{dr} \frac{d \langle V_\omega \rangle_{|V}}{dr} = 0. \quad (5.16)$$

Integrating this equation gives the following relation between two points r_1 and r_2 in the extracellular space,

$$r_1^2 \frac{d \langle V_\omega \rangle_{|V}}{dr}(r_1) \left[\sigma_\omega(r_1) + i\omega\epsilon_\omega(r_1) \right] = r_2^2 \frac{d \langle V_\omega \rangle_{|V}}{dr}(r_2) \left[\sigma_\omega(r_2) + i\omega\epsilon_\omega(r_2) \right]. \quad (5.17)$$

Assuming that the extracellular potential vanishes at large distances ($\langle V_\omega \rangle = 0$), we find

$$\langle V_\omega \rangle_{|V}(r_1) = \frac{I_\omega(R)}{4\pi\sigma_z(R)} \int_{r_1}^{\infty} dr' \frac{1}{r'^2} \frac{\sigma_\omega(R) + i\omega\,\epsilon_\omega(R)}{\sigma_\omega(r') + i\omega\,\epsilon_\omega(r')}, \quad (5.18)$$

where $I_\omega(R)$ is the current produced by the source and $\sigma_z(R)$ is the extracellular conductivity at the border of the spherical source. This expression is valid for both microscopic and macroscopic models for spherical symmetry. The difference between the two types of models lies in the expression for σ_z: in the microscopic model σ_z is real, while it is complex in the macroscopic model.

In the following, we will use the simplified notations $\vec{j}_\omega$, $\vec{E}_\omega$ and V_ω instead of $\langle \vec{j}_\omega \rangle_{|V}, \langle \vec{E}_\omega \rangle_{|V}$ and $\langle V_\omega \rangle_{|V}$, respectively in order to keep the same formalism for both microscopic and macroscopic models.

Finally, using the relation $V_\omega = Z_\omega I_\omega$, the impedance Z_ω is given by:

$$Z_\omega(r) = \frac{1}{4\pi\sigma_z(R)} \int_r^\infty dr' \frac{1}{r'^2} \frac{\sigma_\omega(R) + i\omega \, \epsilon_\omega(R)}{\sigma_\omega(r') + i\omega \, \epsilon_\omega(r')} \qquad (5.19)$$

where r is the distance between the center of the source and the position defined by $\vec{r}$. This expression will be used in the next sections, every time the impedance needs to be evaluated.

Note that this expression is valid only for a source and surrounding medium in spherical symmetry. This of course constitutes an approximation compared to the complex structure of biological media. This approximation enables us to treat the system analytically, and evaluate the influence of inhomogeneities of conductivity on local field potentials in a simple configuration. More complex configurations will be considered in Sections 5.5 and 5.6.

5.3.4 Different models of non-resistive media

In the rest of the chapter, we will consider different physical mechanisms giving rise to non-resistive extracellular media (still within the macroscopic model). These mechanisms involve the complex structure of extracellular media, with alternating fluids and membranes. In such media, we will successively investigate the effect of abrupt variations of impedance (fluids and membranes) in a continuum model (Bédard et al., 2004), the effect of polarization (Bédard et al., 2006a) and the effect of ionic diffusion (Bédard and Destexhe, 2009), as illustrated in Figure 5.4.

5.4 Modeling LFPs in non-resistive media: the continuum model

The simplest model of non-resistive media considers a continuum approximation of the inhomogeneities of electric parameters in extracellular space (Bédard et al., 2004). The inhomogeneous composition of extracellular space is essentially due to the alternation of fluids and membranes. This structure is represented here by *smooth* spatial profiles of variation of the electric parameters, conductivity and permittivity (σ_ω and ϵ_ω), respectively. This model is a direct application of Equation (5.19) above. To enable one to treat the problem analytically, a further assumption is that the current source is spherical, and that the spatial variations of σ_ω and ϵ_ω follow a spherical symmetry around the source or, in other words, they vary only according to the radial distance from the source. We will examine how the spatial pattern of σ and ϵ affects the frequency filtering of the LFP signal.

It is important to note that this type of model is *microscopic*, in the sense that the values of σ_ω and ϵ_ω are supposed to represent the values of conductive fluids or

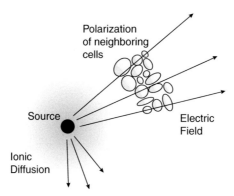

Figure 5.4 Illustration of several physical phenomena involved in the genesis of local field potentials. A given current source produces an electric field, which propagates across media consisting of fluids and membranes. The electric field will also tend to polarize the charged membranes around the source, as schematized at the top. The flow of ions across the membrane of the source will also involve ionic diffusion to re-equilibrate the concentrations. This diffusion of ions will also be responsible for inducing currents in extracellular space. These phenomena influence the frequency filtering and the genesis of LFP signals. (Modified from Bédard and Destexhe, 2009.)

membranes, at scales of the order of micrometers. The particular choices of values were those derived in Bédard et al. (2004).

We begin by considering the impedance of homogeneous media, and show that it is equivalent to a macroscopic resistance. We then discuss the order of magnitude of the electric parameters from known experimental results. Next, we present a few particular spatial profiles of $\sigma_\omega = \sigma(r)$ and $\epsilon_\omega = \epsilon(r)$ in order to illustrate how these (frequency-independent) variations can lead to frequency filtering effects. Finally, we will illustrate this frequency filtering by a numerical simulation of LFPs in a neuron model.

5.4.1 Frequency independence in homogeneous media

In an extracellular medium where the ratio ϵ/σ is constant in space, Equation (5.13) becomes (dividing by $1 + i\omega\frac{\epsilon}{\sigma}$):

$$\nabla \cdot \left[\sigma\left(1 + i\omega\frac{\epsilon}{\sigma}\right)\nabla V_\omega\right] = \nabla \cdot [\sigma\nabla V_\omega] = 0. \tag{5.20}$$

To determine the influence the frequency filtering properties of the medium, we consider a white noise source. Because the above second-order differential equation has a unique solution when the boundary conditions are specified and independent of frequency we can write that the extracellular potential is necessarily frequency independent because for boundary conditions independent of frequency we must have the same solution for every frequency. Thus, the spatial homogeneity of ϵ/σ and the independence of σ of frequency are sufficient conditions for a frequency-independent impedance of the medium. The same applies if the ratio ϵ/σ is independent of frequency and if the medium is homogeneous.

Now we show that the above condition is also a necessary condition for frequency independence of the impedance. If we suppose that ϵ/σ depends on space, and that the extracellular potential does not depend on frequency, then Equation (5.13) is completely different for $f = 0$ and $f \neq 0$. The solution of Equation (5.13) would then depend on frequency, which is contradictory to one of the assumptions.

In the rest of the chapter, we will designate the quantity $\tau_{MW} = \epsilon_\omega/\sigma_\omega$ as the *Maxwell–Wagner time*. The physical meaning of this time, also called "dielectric time," is the mean time taken by a given non-stationary electric charge distribution to reach equilibrium in the dielectric in the absence of an electric field (see Chapter 10 and pp. 374–375 in Maxwell, 1873; see also Raju, 2003).

We can conclude at this point that:

> if one of the electric parameters is independent of frequency, then the spatial homogeneity of τ_{MW} is a necessary and sufficient condition to have a perfectly resistive medium.

Thus, in these conditions, the impedance of the extracellular medium is equivalent to a macroscopic resistance. We will show that this is not necessarily the case in neural tissue.[6] It is interesting to note that if the medium is homogeneous with respect to τ_{MW} and σ (or equivalently with respect to σ and ϵ), then we recover the simple LFP model of Equation (5.2). In this case, Equations (5.9) and (5.19) become:

$$\nabla^2 V_\omega = 0 \qquad (5.21)$$

and

$$V_\omega = Z_\omega(r)i_\omega = \frac{i_\omega}{4\pi} \int_r^\infty dr' \frac{1}{r'^2} \frac{1}{\sigma} = \frac{1}{4\pi\sigma} \frac{i_\omega}{r}. \qquad (5.22)$$

In general, σ and ϵ will take very different values according to whether they represent fluids or membranes, and these values correspond to very different values of τ_{MW} (see next section). Thus, we can say that the extracellular potentials will be frequency dependent, but there is no information on whether this frequency

---

[6] Note that a medium homogeneous with respect to τ_{MW} is not necessarily homogeneous relative to σ and ϵ, because these parameters could display identical spatial variations such that their ratio, τ_{MW}, remains invariant.

dependence is strong or negligible. In the next section, we evaluate this as a function of the order of magnitude of the electric parameters.

5.4.2 Conductivity and permittivity of neural tissue

Precise experimental data on the variations of permittivity ϵ_ω and conductivity σ_ω at microscopic scales in the extracellular medium have not been measured so far. However, averaged values of these parameters are available from macroscopic measurements. A value for σ_ω, averaged over large extracellular distances, σ_{av}, was measured by Ranck (1963) and was between 0.28 Sm^{-1} and 0.43 Sm^{-1}, for 5 Hz and 5 kHz, respectively. The macroscopic frequency dependence of conductivity therefore seems relatively weak. However, the situation is different microscopically. As reviewed in Nunez and Srinivasan (2005), the conductivity of the CSF fluid is 1.56 Sm^{-1} while the typical conductivity of membranes is 10^{-9} to 3.5 $\times 10^{-9}$ Sm^{-1}. This value was obtained from the resting (leak) membrane conductance of cortical neurons, typically around 4.5×10^{-5} S cm^{-2}, multiplied by the thickness of the membrane (2–8 nm; Peters et al., 1991). Other types of membranes, such as myelinated or unmyelinated fibers, and glial cells, have different membrane conductances, which are in the range of 0.1 to 10^{-6} S cm^{-2} (Hille, 2001). At microscopic scales, there is therefore approximately nine orders of magnitude variation of conductivity.

The value of ϵ of a membrane is between 10^{-11} and 8×10^{-11} Fm^{-1}. This value is derived from the specific capacitance of membranes, $C_\omega = C = 1\,\mu\text{F}\,\text{cm}^{-2}$ (Johnston and Wu, 1999), assuming a membrane thickness of 2 to 8 nm (Peters et al., 1991). The value ϵ_ω of extracellular fluid is not known, but was roughly estimated from conductivity measurements of salted water (Gabriel et al., 1996b), which was reported to be of the order of 10^{-2} to 10^{-3} Fm^{-1} for frequencies between 10 and 100 Hz. We will see in the next section that these variations can be responsible for significant frequency filtering effects.

5.4.3 Non-homogeneous extracellular media

We have shown above that, in a homogeneous medium relative to the Maxwell–Wagner time, and if the electric parameters do not depend on frequency, the extracellular electric potential does not display frequency-dependent properties. We now turn to a possible source of frequency-dependent attenuation, namely the presence of inhomogeneities in the Maxwell–Wagner time, by varying the conductivity of the extracellular medium, while keeping a constant permittivity. We will also consider the case where both parameters vary in space.

In the following, we assume $\sigma_\omega = \sigma$ and $\epsilon_\omega = \epsilon$ and we will use normalized values for conductivity $\sigma(r)/\sigma(R)$ and permittivity $\epsilon(r)/\sigma(R)$. Because the membrane is always surrounded by extracellular fluid, the conductivity at the source is $\sigma(R) = 1.56 \text{ Sm}^{-1}$. Based on the values estimated above, in simulations we use the values for the normalized conductivity $\sigma(r)/\sigma(R)$ included between large values (equal to unity) and a low value of 2×10^{-9}. Similarly, the normalized (constant) value of permittivity will be $\epsilon(r)/\sigma(R) \approx 0.01$ s. We have verified that no qualitative change results from variations of these parameters.

In a first example, we considered the case of a localized reduction of conductivity (Figure 5.5A) while the permittivity was kept constant $\epsilon(r)/\sigma(R) \approx 0.01$ s. The resulting impedance measured at different distances from the source is shown in Figure 5.5A as a function of frequency f. In this case, for distances around the conductivity drop, there is a moderate frequency dependence with low-pass characteristics (Figure 5.5A, dotted and dashed lines).

Because the extracellular space is composed of alternating fluids and membranes (Peters et al., 1991), which have high and low conductivity, respectively, we next considered the situation where the conductivity fluctuates periodically with distance (Figure 5.5B). Considering a cosine function of conductivity with constant permittivity leads to a rather strong frequency-dependent attenuation with low-pass characteristics. There was a strong attenuation with distance for all frequencies (Figure 5.5B). Very similar results were obtained with other periodic functions (for example by replacing cos by sin in the function used in Figure 5.5B), different oscillation periods, or even for damped oscillations of conductivity (not shown).

It could be argued that although fluids and membranes alternate in extracellular space, there is efficient diffusion of ions only in the extracellular fluid around the membrane. For larger distances, diffusion becomes increasingly difficult because of the increased probability of meeting obstacles. In this case, the conductivity would be highest around the source and would decrease progressively to an "average" conductivity level for larger distances. This situation is illustrated in Figure 5.5C. We have considered, as a third example, that the conductivity is highest at the source, then decreases exponentially with distance with a space constant λ (Figure 5.5C; note that in this case, real distances were used). The resulting impedance displayed pronounced frequency filtering properties with low-pass characteristics. In particular, the attenuation with distance revealed strong differences between low and high frequencies of the spectrum (Figure 5.5C). Similar results can be obtained with other decreasing functions of connectivity (not shown).

Another example that could be considered is the case where the electric parameters σ and ϵ fall abruptly at a short distance from the source. This situation would account for the fact that the conductivity is high only in the immediate

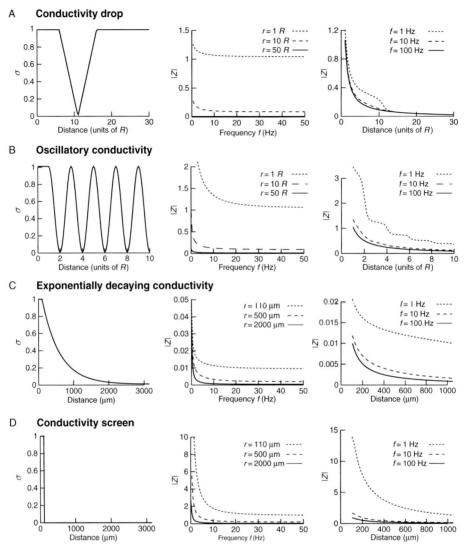

Figure 5.5 Frequency filtering properties obtained for four different profiles of spatial variation of σ. Left: spatial profile of σ shown as a function of relative distance d/R, where R is the radius of the source. Middle: impedance $|Z|$ as a function of frequency for three values of d/R (1, 10 and 100). Right: impedance as a function of distance d/R for three different frequencies (1, 10 and 100 Hz). A. Drop in conductivity. $\sigma(r)/\sigma(R) = 1 - 0.2(r - 6R)/R$ for $6R < r < 11R$, $\sigma(r)/\sigma(R) = -1 + 0.2(r - 6R)/R$ for $11R < r < 16R$, and $\sigma(r)/\sigma(R) = 1$ otherwise. B. Oscillatory conductivity. $\sigma(r)/\sigma(R) = 0.501 + 0.5 * \cos[2\pi(r - R)/2R]$. C. Exponentially decaying conductivity. $\sigma(r)/\sigma(R) = \sigma_0 + (1 - \sigma_0)\exp[-(r - R)/\lambda]$, with a space constant $\lambda = 500$ μm and $\sigma_0 = 10^{-9}$Sm^{-1}. D. Abrupt variation of conductivity. $\sigma(r)/\sigma(R) = 1$ and $\epsilon(r)/\sigma(R) = 0.01$ for $r < 100$ μm, while $\sigma(r) = 10^{-9}$ and $\epsilon(r)/\sigma(R) = 10^{-11}$ s for $r > 100$ μm. Permittivity is constant in A–C, with $\epsilon(r)/\sigma(R) = 0.01$ s.

neighborhood of the source, in the extracellular fluid, and drops abruptly at a distance corresponding to the position of the membrane of neighboring cells. This configuration also gave frequency dependence (not shown).

5.4.4 Comparison of different conductivity profiles

The above examples show that there can be a strong frequency filtering behavior, with low-pass characteristics as observed in experiments. However, although these examples show a more effective filtering for high frequencies, it still remains to be shown that the high frequencies attenuate more steeply with distance compared to low frequencies. To this end, we define the quantity:

$$Q_{100} = Z_{100}(r)/Z_1(r), \qquad (5.23)$$

where Z_1 and Z_{100} are the impedances computed at 1 Hz and 100 Hz, respectively. This ratio quantifies the differential filtering of fast and slow frequencies as a function of distance r. Figure 5.6A displays the Q_{100} values obtained for some of the examples considered above. In the case of a localized drop in conductivity (Figure 5.5A, *Drop*), there was an effect of distance for $r < 16R$, then the Q_{100} remained equal to unity for further distances. This behavior is in agreement with the impedance shown in Figure 5.5A, in which case there was no frequency filtering for $r > 16R$. For oscillatory conductivities (Figure 5.6, *Osc*), Q_{100} was always less than unity, consistent with the low-pass frequency filtering behavior observed

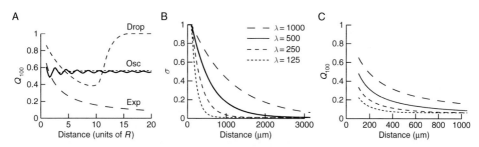

Figure 5.6 Distance dependence of frequency filtering properties. A. Ratio of impedance at fast and slow frequencies (Q_{100}) represented as a function of distance r (units of R). The Q_{100} ratios are represented for different profiles of conductivity. *Drop*: localized drop in conductivity (short dash; same parameters as in Figure 5.5A). *Osc*: oscillatory profile of conductivity (solid line; same parameters as in Figure 5.5B; the dotted line indicates a damped cosine oscillation). *Exp*: exponential decrease in conductivity (long dash; same parameters as in Figure 5.5C except $R = 1$, $\lambda = 10R$). B. Profiles of conductivity with exponential decay (same parameters as in Figure 5.5C; space constants λ indicated in μm). C. Q_{100} ratios obtained for the conductivity profiles shown in B.

in Figure 5.5B. However, Q_{100} oscillated around a value of 0.6 and did not decrease further with distance. Thus, in this case, although there was a clear low-pass filtering behavior, all frequencies still contribute by the same relative amount to the extracellular potential, regardless of distance. On the other hand, with exponential decay of conductivity, Q_{100} decreased monotonically with distance (Figure 5.6, *Exp*). Thus, this case shows both low-pass filtering behavior (Figure 5.5) and a stronger attenuation of high frequencies compared to low frequencies (Figure 5.5C, *Exp*), which is in qualitative agreement with experiments. Analyzing exponentially decaying conductivities of different space constants (Figure 5.6B) revealed that the various patterns of distance dependence approximately followed the pattern of conductivity (Figure 5.6C). This type of conductivity profile is relatively simple and plausible, and will be the one considered in the biophysical model investigated in Section 5.4.5.

If the conductivity and permittivity fall abruptly at a short distance λ from the source (Figure 5.5D), the Q_{100} value is independent of distance for $r > \lambda$, while at large distances it is approximately equal to the Q_{100} value of the exponentially decaying conductivity (not shown).

5.4.5 *Biophysical model of the frequency-filtering properties of local field potentials*

We now apply the above formalism to model the frequency dependence of the extracellular field potentials stemming from a conductance-based spiking neuron model (Figure 5.7). We consider a simple biophysical model of a spiking neuron containing voltage-dependent and synaptic conductances. The single-compartment model neuron includes conductance-based models of voltage-dependent conductances and synaptic conductances. This model is described by the following membrane equation:

$$C_m \frac{dV}{dt} = -g_L(V - E_L) - g_{Na}(V - E_{Na}) - g_{Kd}(V - E_K)$$
$$- g_M(V - E_K) - g_e(V - E_e), \qquad (5.24)$$

where $C_m = 1 \ \mu\text{F cm}^{-2}$ is the specific membrane capacitance, $g_L = 4.52 \times 10^{-5} \ \text{S cm}^{-2}$ and $E_L = -70$ mV are the leak conductance and reversal potential. $g_{Na} = 0.05 \ \text{S cm}^{-2}$ and $g_{Kd} = 0.01 \ \text{S cm}^{-2}$ are the voltage-dependent Na$^+$ and K$^+$ conductances responsible for action potentials and were described by a modified version of the Hodgkin and Huxley (1952) model. $g_M = 5 \times 10^{-4} \ \text{S cm}^{-2}$ is a slow voltage-dependent K$^+$ conductance responsible for spike-frequency adaptation. $g_e = 0.4 \ \mu\text{S}$ is a fast glutamatergic (excitatory) synaptic conductance. The

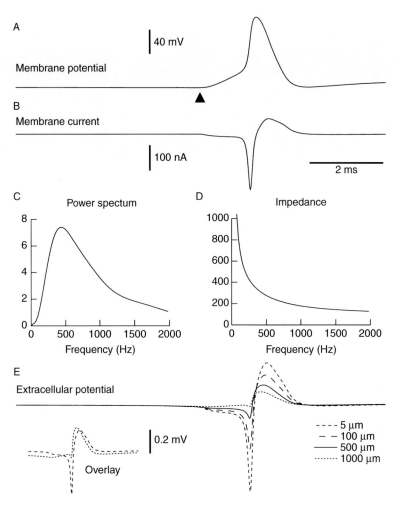

Figure 5.7 Frequency-filtered extracellular field potentials in a conductance-based model. A. Membrane potential of a single-compartment model containing voltage-dependent Na^+ and K^+ conductances and a glutamatergic synaptic conductance. The glutamatergic synapse was stimulated at $t = 5$ ms (arrow) and evoked an action potential. B. Total membrane current generated by this model. Negative currents correspond to Na^+ and glutamatergic conductances (inward currents), while positive currents correspond to K^+ conductances (outward currents). C. Power spectrum of the total current shown in B. D. Impedance at 500 μm from the current source assuming a radial profile of conductivity and permittivity. E. Extracellular potential calculated at various distances from the source (5, 100, 500 and 1000 μm). The frequency filtering properties can be seen by comparing the negative and positive deflections of the extracellular potential. The fast negative deflection almost disappeared at 1000 μm whereas the slow positive deflection was still present. The inset in E (*Overlay*) shows the traces at 5 and 1000 μm overlayed (only the amplitude was scaled).

voltage-dependent conductances are described by conventional Hodgkin–Huxley type models adapted for modeling neocortical neurons, and the synaptic conductance is described by a first-order kinetic model of neurotransmitter binding to postsynaptic receptors. These models and their kinetic parameters were described in detail in a previous publication (Destexhe and Paré, 1999), and all numerical simulations were performed using the NEURON simulation environment (Hines and Carnevale, 1997).

The profile of conductivity and permittivity used in the model is shown in Figure 5.5C. We calculate the total membrane current generated by a single-compartment model of an adapting cortical neuron, containing voltage-dependent Na^+ and K^+ conductances for generating action potentials and a slow voltage-dependent K^+ conductance responsible for spike-frequency adaptation. The model contained a fast glutamatergic excitatory synaptic conductance, which was adjusted to evoke a postsynaptic potential just above threshold, in order to evoke a single action potential (Figure 5.7A). The total membrane current (Figure 5.7B) was calculated and stored in order to calculate its Fourier transform (power spectral density shown in Figure 5.7C). The impedance of the extracellular medium (Figure 5.7D) was calculated using absolute values of the parameters (Equation (5.19)).

We now use this model to calculate the field potentials at different radial distances assuming the neuron is a spherical source (radius of 105 μm). The extracellular potential is indicated for 5, 100, 500 and 1000 μm away from the source (see Figure 5.7E) and strong frequency filtering properties are apparent: the fast negative deflection of extracellular voltage shows a steep attenuation and almost disappeared at 1000 μm (although it has the highest amplitude at 5 μm). In contrast, the slow positive deflection of the extracellular potential shows less attenuation with distance and becomes dominant at large distances (500 and 1000 μm in the example of Figure 5.7E). This is best seen from the overlayed traces (see Figure 5.7E, inset), showing the marked difference in fast and slow components in the field potentials recorded at 5 and 1000 μm.

Thus, this example illustrates that the approach provided here can lead to a relatively simple model for calculating local field potentials with frequency filtering properties. The exact profiles of filtering and attenuation depend on the exact shape of the gradients of conductivity/permittivity as well as on the spherical symmetry inherent to this model. This is illustrated in Figure 5.8 for the other radial profiles of conductivity considered earlier (drop, periodic and damped conductivity). This figure shows that for these particular profiles, the attenuation of fast and slow deflections is similar (the global shape of the LFP remains similar although attenuated in amplitude; see the almost perfect overlay in the inset of Figure 5.8C). There is therefore a negligible frequency-dependent attenuation in these cases. This is in agreement with the quasi-absence of frequency-dependent attenuation evidenced in

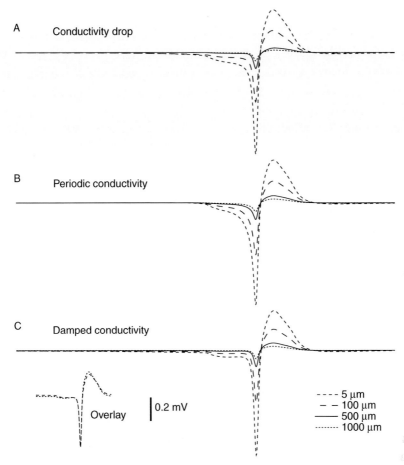

Figure 5.8 Frequency-filtered extracellular field potentials for different radial profiles of conductivity. The extracellular potential was calculated from a conductance-based spiking neuron model (identical to that of Figure 5.7) and was shown at various distances from the source (5, 100, 500 and 1000 μm) for different profile of conductivity. A. Localized drop in conductivity (constant conductivity, dropping between $r = 120$ μm and $r = 280$ μm, as described in Bedard et al., 2004). B. Same simulation using a periodic conductivity profile (periodic function with same extremal values as in Figure 5.7 and a period of 2 μm). C. Same simulation using damped oscillations of conductivity (same parameters as in B, with a space constant of $\lambda = 500$ μm). In all cases, the attenuation of the fast negative deflection was similar to the slow positive deflection, in contrast with Figure 5.7E. The inset in C (*Overlay*) shows the traces at 5 and 1000 μm overlayed, which are almost superimposable (compare with inset in Figure 5.7E).

the quantitative analysis of Figure 5.6 (see above). In contrast, Figure 5.7E shows that a radial model with exponentially decaying conductivity replicates the experimental observation that only slow frequencies propagate through large extracellular distances.

5.5 Modeling LFPs in non-resistive media: the polarization model

A second type of approach to modeling the non-resistive properties of extracellular media is to include another physical phenomenon, electric polarization (Bédard et al., 2006a). In this approach, we also consider that the extracellular medium is non-homogeneous, but we assume an explicit structure consisting of a current source, surrounded by a number of other cellular processes,[7] all surrounded by a conductive fluid. The effect of the electric field will be to polarize these cells, as illustrated in Figure 5.4.

It is important to note that in this section we consider the movement of charges in volumes of the order of cubic nanometers. This scale corresponds to the surface polarization phenomena on membranes of cells in the neighborhood of the sources. At this scale, the medium cannot be considered as locally neutral (typical membrane thickness is around 7 to 8 nm), because local excesses of charge will appear. This would not be the case for example at larger scales of μm^3. At nanometer scales, the general equations (Equations (5.13) and (5.19)), which were introduced for a locally neutral medium, cannot apply directly. However, we will see in Section 5.7 that it is still possible to use these general equations to simulate the polarization effects of neighboring cells and its consequences on the frequency dependence of current propagation in extracellular media. The "trick" will be to consider the system as locally neutral, but incorporate an appropriate frequency dependence in the electric parameters (Bédard and Destexhe, 2009).

Indeed, a consequence of the polarization mechanisms is that the medium is considered as "reactive" in the sense that it will react to the electric field. Neighboring cells maintain charge distributions around their membrane, and these charges will be affected by the electric field and they will polarize. In other words, a given current source will necessarily polarize the medium around it, and in particular will shift the charges on the membrane surface of its neighbors. Thus, when the smallest considered scale is of the order of μm^3, the medium can be considered as approximatively neutral, except for some small spatial fluctuations.

Not only will the medium react by polarizing, but this polarization is not instantaneous and will take a characteristic time to operate. This time, known as the *Maxwell–Wagner time*, represents the inertia of charge movement (Pethig, 1979; Raju, 2003). Thus, the frequency dependence of the electric parameters σ and ϵ, as caused by the inertial time τ_{MW} (Maxwell–Wagner time) of polarization phenomena does not vanish, even at a scale (approximately μm^3) where the medium can appear as locally neutral ($\rho \approx 0$) at all times.

---

[7] We will consider these neighboring cellular processes as "passive" cells, in the sense that they maintain a membrane potential. Such processes include axons, dendrites, somas, spines, glial cells and astrocytes, all of which are part of the neuropil.

Thus, the values of the apparent (macroscopic) electric parameters at μm^3 scales necessarily have a frequency dependence which comes from this inertia time. This frequency dependence is a function of the magnitude of frequencies with respect to $1/\tau_{MW}$. We will examine the phenomenon of polarization of cell surfaces, which is not at all taken into account by models with resistive media. We will show that it does affect the propagation of field potentials.

In a first step, we will illustrate this with a simple model of surface polarization taken from a previous study (Bédard et al., 2006a). This simple model will be used to determine the frequency dependence of polarization. In a second step, we will present an application of this model in the case of a spherical current source, surrounded by different layers of adjacent cells. In a third step, we will show that the non-instantaneous character of surface polarization is important and is responsible for a form of frequency dependence. It also affects the propagation of field potentials over large distances.

5.5.1 A simple model of cell surface polarization

Neurons are characterized by various voltage-dependent and synaptic ion channels, and they will be considered here as the sole source of the electric field in extracellular space. On the other hand, glial cells are very densely packed in interneuronal space, sometimes surrounding the soma or the dendrites of neurons (Cajal, 1909; Peters et al., 1991). Glial cells normally do not have dominant voltage-dependent channel activity, rather they play a role in maintaining extracellular ionic concentrations. Like neurons, they have an excess of negative charges inside the cell, which is responsible for a negative resting potential (for most central neurons, this resting membrane potential is around –60 to –80 mV). They will be considered here as "passive" and representative of all non-neuronal cell types characterized by a resting membrane potential. We will show that such passive cells can be polarized by the electric field produced by neurons. This polarization has an inertia and a characteristic relaxation time which may have important consequences for the properties of propagation of local field potentials. These different cell types are separated by extracellular fluid, which plays the role of a conducting medium, i.e. allows for the flow of electric currents. In the remainder of this text, we will use the term "passive cell" to represent the various cell types around neurons, but bearing in mind that they may represent other neurons as well.

Another simplification is that we will consider these passive cells to be of elementary shape (spherical or cubic). Under such a simplification, it will be possible to treat the propagation of field potentials analytically and design simulations using standard numerical tools. Our primary objective here is to explore one essential physical principle underlying the frequency filtering properties of extracellular space, based on the polarization of passive membranes surrounding neuronal sources. We assume that such a principle will be valid regardless of the morphological complexity and spatial arrangement of neurons and other cell types in extracellular space. As a consequence of these simplifications, the present work does not attempt to provide a quantitative description but rather an exploration of first principles that could be applied in later work to the actual complexity of biological tissue.

The arrangement of charges in our model is schematized in Figure 5.9A, where we delimited five regions. The membrane of the passive cell (region 3) separates the

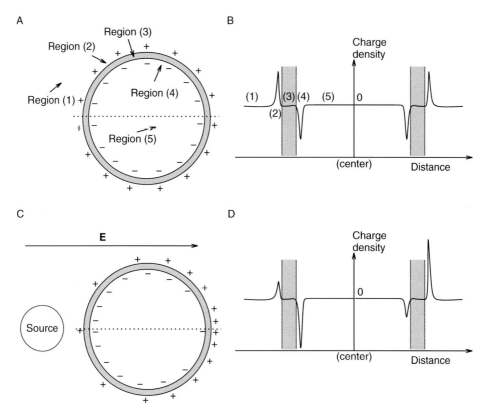

Figure 5.9 Scheme of charge distribution around the membrane of a passive cell. A. Charge distribution at rest. The following regions are defined: the extracellular fluid (region 1), the region immediately adjacent to the exterior of the membrane where positive charges are concentrated (region 2), the membrane (region 3, in gray), the region immediately adjacent to the interior of the membrane where negative charges are concentrated (region 4), and the intracellular (cytoplasmic) fluid (region 5). B. Schematic representation of the charge density as a function of distance (along the horizontal dotted line in A). C. Redistribution of charges in the presence of an electric field. The ions move away from or towards the source, according to their charge, resulting in polarization of the cell. D. Schematic representation of the charge density predicted from C.

intracellular fluid (region 5) from the extracellular fluid (region 1), both of which are electrically neutral. The negative charges in excess in the intracellular medium agglutinate in the region immediately adjacent to the membrane (region 4), while the analogous region at the exterior surface of the membrane (region 2) contains the positive ions in excess in the extracellular space. This arrangement results in a charge distribution (schematized in Figure 5.9B) which creates a strong electric field inside the membrane and a membrane potential.

The behavior of such a system depends on the values of conductivity and permittivity in these different regions (they are considered constant within each region). The extracellular fluid (region 1) has good electric conductance properties. We have taken as conductivity $\sigma_1 = 4 \, \Omega^{-1} \, m^{-1}$, consistent with biological data $\sigma = 3.3 - 5 \, \Omega^{-1} \, m^{-1}$, taken from measurements of specific impedance of rabbit cerebral cortex (Ranck, 1963). This value is comparable to the conductivity of salt water ($\sigma_{sw} = 2.5 \, \Omega^{-1} \, m^{-1}$). The permittivity is given by $\epsilon_1 = 70 \, \epsilon_0$, corresponding to salt water. Here $\epsilon_0 = 8.854 \times 10^{-12} \, F \, m^{-1}$ denotes the permittivity of the vacuum. In region 2, to our best knowledge, there are no experimental data on conductivity close to the membrane. We have chosen the values of $\sigma_2 = 0.7 \times 10^{-7} \, \Omega^{-1} \, m^{-1}$ and $\epsilon_2 = 1.1 \times 10^{-10} \, F \, m^{-1} \approx 12\epsilon_0$ for region 2. Such a choice is not inconsistent with biological observations. First, electron microscopic photographs taken from the region near the membrane reflect very little light, which hints at quite low conductivity compared to the conductivity of region 1. We consider it plausible that the permittivity in region 2 should be smaller than that in region 1. Our choice of σ_2 and ϵ_2 corresponds to a Maxwell time τ_M yielding a cut-off frequency $f_c \approx 100$ Hz, which was also the choice given in a previous study investigating signal propagation along cable structures (Bédard and Destexhe, 2008).[8]

For passive cells, we neglect ion channels and pumps located in the membrane, which is equivalent to assuming the absence of any electric current across the membrane. Therefore, region 3 has zero conductivity perpendicular to the membrane surface. The capacity of a cellular membrane has been measured and is about $C = 10^{-2} \, F \, m^{-2}$ (Johnston and Wu, 1999). Approximating the membrane by a parallel plate capacitor (with surface area S and distance d, obeying $C = \epsilon S/d$), one estimates the electric permittivity of the membrane to be $\epsilon_3 = 10^{-10} \, F \, m^{-1}$. Hence we used the parameters $\sigma_3 = 0$, and $\epsilon_3 = 12\epsilon_0$.

Thus, the basic idea behind the model is as follows. As represented in Figure 5.9, we consider a single spherical passive cell under the influence of an electric field. The electric field will induce a polarization of the cell by reorganizing its charge distribution (Figure 5.9C, D). This polarization will create a secondary electric field, with field lines connecting these opposite charges. It is customary to call the

---

[8] Note that the value of the cut-off frequency could possibly vary according to cell type. To estimate such a cut-off, one would need to estimate the tangential resistivity, which has never been done experimentally.

original electric field the *source field*, or the *primary field*, while the field due to polarization is called the *induced field*, or the *secondary field*. The physical electric field is the sum (in the sense of vectors) of both the source and the induced fields. This induced field will be highly dependent on frequency, for high frequencies, the "inertia" of charge movement in regions 2 and 4 will limit such a polarization, and will reduce the effect of the induced field. This phenomenon is the basis of the model of frequency-dependent local field potentials presented in this section.

5.5.2 Frequency dependence of the polarization model

We now examine the frequency dependence of the secondary field produced by surface polarization. We still consider the simple case of one layer of passive cells surrounding the source. It can be shown that a very good approximation of the transfer function[9] is given by the following expression (see Bédard et al., 2006a):

$$F_{TM}(\vec{x}, \omega) = \frac{1}{1 + i\omega\tau_{MW}} \tag{5.25}$$

where $\tau_{MW} = \epsilon^S/\sigma^S$ is the characteristic time for the inertia of surface polarization. It is important to note that ϵ^S and σ^S are the *tangential* conductivity and permittivity, respectively, and that they are therefore related to the membrane and not the extracellular medium. The physical interpretation of τ_m^S is the characteristic time to set charges into movement during the polarization; this charge movement is not instantaneous ($\tau_{MW} = 0$ would mean an infinite conductivity or zero permittivity). It is important to note that we consider here that all membranes have a unique τ_{MW}, but in reality this value may vary as a function of the molecular composition of the membrane.

The simulations (Figure 5.10) show that the induced potential vanishes for very high frequencies of the source field, a fact that can also be deduced from Equation (5.25). In other words, for very high frequencies ($\gg \tau_{MW}^{-1}$), the extracellular field will be equal to the source field, since the induced field will vanish. The space dependence is easy to deduce in such a case, and the extracellular potential attenuates with distance according to a $1/r$ law (for a spherical source), as if the source was surrounded only by conducting fluid.

However, for very low frequencies ($\ll \tau_{MW}^{-1}$), the space dependence of the extracellular potential will be a complex function depending on both the $1/r$ attenuation

---

[9] The transfer function is given by:

$$F_{TM}(\vec{x}, \omega) = \lim_{t \to \infty} \frac{V_{ind}^{\omega}(\vec{x}, t) \exp(-i\omega t)}{V_{ind}^{\omega=0}(\vec{x}, t)}.$$

This function gives the ratio of the amplitude of the secondary field at a given frequency, with the amplitude of the null frequency in Fourier analysis.

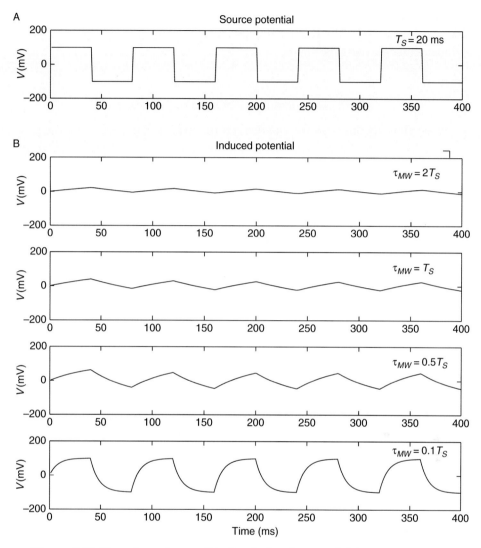

Figure 5.10 Time dependence of induced electric potential in response to an external source given by a periodic function. A. Source potential given by a periodic step function $H(\sin(\omega t))$, with period $T_S = 2\pi/\omega = 20$ ms. B. Induced electric potential for various values of T_S/τ_{MW}. In the case $T_S/\tau_{MW} \ll 1$ (top), the induced potential fluctuates closely around the mean value of the source potential. In the case $T_S/\tau_{MW} \gg 1$ (bottom) the induced potential relaxes and fluctuates between $V_{min} = 0$ mV and $V_{max} = 100$ mV of the source potential. As a result, the amplitude of oscillation becomes more attenuated with increased τ_{MW} with respect to T_S.

of the source field, and the contribution from the induced field. Such a space dependence is not easy to deduce, since it depends on the spatial arrangement of fluids and membranes around the source.

5.5.3 Attenuation as a function of distance

In this section, to illustrate the importance of surface polarization on the propagation of local field potentials, we derive the low-frequency space dependence alluded to above for a simplified arrangement of source and surrounding cells. We consider a system of densely packed and regularly arranged cells, as illustrated in Figure 5.11A.

To constrain the behavior to low frequencies, we only consider the zero-frequency limit by using a constant source field. We proceeded in two steps. First, we calculate the electric potential at the surface of a passive cell (Section 5.5.3.1). Second, we calculate the spatial profile of LFPs in a system of densely packed spheres of identical shape (Section 5.5.3.2).

5.5.3.1 Electric potential at the surface of passive membranes at equilibrium

Let us assume a spherical passive cell embedded in a perfect dielectric medium, and exposed to a constant electric field. At equilibrium, we have seen above that the effect of the electric field is to polarize the charge distribution at the surface of the cell, such as to create a secondary electric field (see Figure 5.9B), but the induced electric field is zero inside the cell. In this case, the conservation of charges on the surface implies:

$$\int_{Surf} \rho_{Surf}\, dS = 0, \tag{5.26}$$

where ρ_{Surf} is the charge density on the surface of the cell. The resulting electric potential is given by:

$$V_{tot}(x, y, z) = V_{source}(x, y, z) + \int_{Surf} \frac{\rho_{Surf}}{4\pi\epsilon\, r}\, dS \tag{5.27}$$

where V_{source} is the electric potential due to the source field, V_{tot} is the total resulting electric potential due to the source field and the induced field, and r is the distance from point (x, y, z) to the center of the cell. Because at the center of the cell, (a, b, c), we necessarily have $r = R$ (where R is the cell's radius), the value of the resulting electric potential at the center is given by:

$$V_{tot}(a, b, c) = V_{source}(a, b, c) + \int_{Surf} \frac{\rho_{Surf}}{4\pi\epsilon\, R}\, dS = V_{source}(a, b, c). \tag{5.28}$$

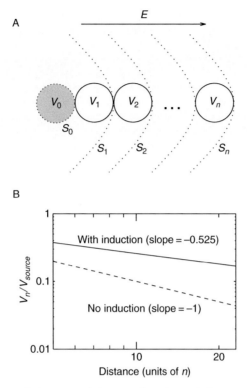

Figure 5.11 Extracellular potentials as a function of distance for a system of densely packed spherical cells. A. Scheme of the arrangement of successive layers of identical passive cells packed around a source (in gray). The potential of the source is indicated by V_0. V_1, V_2, ..., V_n indicate the potential at the surface of passive cells in layers 1, 2, ..., n, respectively. The dotted lines indicate isopotential surfaces, which are concentric spheres centered around the source, and which are indicated here by S_1, S_2, ..., S_n. B. Extracellular potential as a function of distance (in units of cell radius), comparing two cases: with induction (solid line, corresponding to the arrangement schematized in A), and without induction (dashed line, source surrounded by conductive fluid only). The two cases predict a different scaling of the electric potential with distance (see text for details).

Thus, the electric potential at the surface of a spherical passive cell at equilibrium equals the potential due to the primary field at the center of the cell. In other words, the effect of the secondary field in this case perfectly compensates for the distance dependence of the primary field, such that the surface of the cell becomes isopotential, as discussed above.

5.5.3.2 Attenuation of electric potential in a system of packed spheres

Keeping the assumption of a constant electric field, we now calculate how the extracellular potential varies as a function of distance in a simplified geometry.

We consider a system of packed spheres as indicated in Figure 5.11A. From the solution of Laplace's equation, the extracellular potential at a distance r from the source center is given by:

$$V(r) = \frac{k}{r},$$ (5.29)

where k is a constant, which is evaluated from the potential at the surface of the source (S_0 in Figure 5.11A),

$$V(R) = V_0 = \frac{k}{R},$$ (5.30)

where R is the radius of the source. Thus, the potential due to the source field at a given point r in extracellular space is given by:

$$V(r) = \frac{R V_0}{r}.$$ (5.31)

Considering the arrangement of Figure 5.11A, if all cells of a given layer are equidistant from the source, their surface will be at the same potential (see Section 5.5.3.1), which we approximate as a series of isopotential concentric surfaces ($S_1, S_2, \ldots, S_n$ in Figure 5.11A). A given layer (n) of isopotential cells is therefore approximated by a new spherical source of radius r_n, which will polarize cells in the following layer $(n + 1)$. According to such a scheme, the potential in layer $n + 1$ is given by:

$$V_{n+1} = \frac{r_n V_n}{d_{n+1}},$$ (5.32)

where d_{n+1} is the distance from the center of cells in layer $(n + 1)$ to the center of the source. According to the scheme of Figure 5.11A, we have $r_n = (2n + 1)R$ and $d_n = 2nR$. Thus, we can write the following recurrence relation:

$$V_{n+1} = \frac{2n + 1}{2n + 2} V_n.$$ (5.33)

Consequently:

$$V_{n+1} = \left(\prod_{j=1}^{n} \frac{2j + 1}{2j + 2} \right) V_0,$$ (5.34)

which can be written, for large n

$$V_n = \frac{(2n + 1)!}{2^{2n}(n + 1)!n!} V_0 \approx \frac{2(2n)!}{2^{2n}(n!)^2} V_0.$$ (5.35)

Using Stirling's approximation, $n! \simeq (n/e)^n \sqrt{2\pi n}$ for large n, leads to:

$$V_n \simeq \frac{2}{\sqrt{\pi n}} V_0. \tag{5.36}$$

Thus, in a system of densely packed spherical cells, the extracellular potential falls-off like $1/\sqrt{r}$ (Figure 5.11B, continuous line).

In contrast, in the absence of passive cells in extracellular space, the electric potential is given by the source field only (Equation (5.31)), which, using the same distance notation as above, is given by:

$$V_{n+1} = \frac{V_0}{2(n+1)}. \tag{5.37}$$

Such $1/r$ behavior is illustrated in Figure 5.11B (dashed line). Note that other theories predict a steeper decay. For instance the Debye–Hückel theory of ionic solutions (Debye and Hückel, 1923) predicts a fall off as $\exp(-kr)/r$.

Thus, for this particular configuration, there is an important difference in the attenuation of extracellular potential with distance. The extracellular potential in a system of densely packed spheres falls off approximately like $1/\sqrt{r}$, in contrast to a $1/r$ behavior in a homogeneous extracellular fluid. Note that a $1/\sqrt{r}$ behavior can also be found in a system in which the source is defined as a current. For a constant current source I_0, with variations of conductivity following a spherical symmetry around the source, the extracellular potential is given by (Bédard et al., 2004):

$$V(r) = \frac{I_0}{4\pi} \int_r^\infty \frac{1}{r^2 \sigma(r)} dr, \tag{5.38}$$

where $\sigma(r)$ is the radial profile of conductivity around the source. Assuming that $\sigma(r) = \sigma_0/\sqrt{r}$, gives

$$V(r) = \frac{\sigma_0 I_0}{4\pi \sqrt{r}} = \sqrt{\frac{R}{r}} V_0. \tag{5.39}$$

Consequently, the $1/\sqrt{r}$ behavior found above is functionally equivalent to a medium with conductivity varying like $1/\sqrt{r}$. This "effective conductivity" is similar to that introduced in Section 5.3.

5.5.4 Polarization of isotropic disorganized media

In the preceding section, we considered a very "organized" medium consisting of regularly arranged cells of the same size. Such an arrangement is of course very different from the "disorganized" arrangement of cells and fluids seen in actual neural tissue (Peters et al., 1991). In this section, we show that the phenomenon of surface polarization implies that the first layer of cells around a given source forms

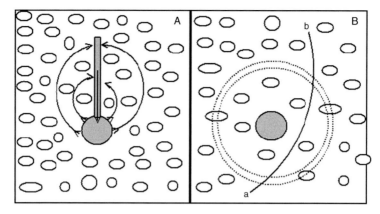

Figure 5.12 Monopole and dipole arrangements of current sources. A. Scheme
of the extracellular medium containing a quasi-dipole (gray) representing a pyramidal neuron, with soma and apical dendrite arranged vertically. B. Illustration
of one of the monopoles of the dipole. The extracellular space is represented by
cellular processes of various size (circles) embedded in a conductive fluid. The
dashed lines represent equipotential surfaces. The line $\widehat{ab}$ illustrates the fact that
the extracellular fluid is linearly connex.

a perfect screen to the electric field produced by the source, but this is only the
case for a "disorganized" medium. This conclusion is very different from that of
the preceding section.

Let us now determine for zero frequency the amplitude of the secondary field $\vec{E}_0^S$
produced between two cells embedded in a given electric field. First, we assume
that it is always possible to trace a continuous path which links two arbitrary points
in the extracellular fluid (see Figure 5.12B). Consequently, the domain defined by
extracellular fluid is said to be *linearly connex*. In this case, the electric potential arising from a current source is necessarily continuous in the extracellular
fluid. Second, in a first approximation, we can consider that the cellular processes
surrounding sources are arranged randomly (as opposed to being regularly structured) and their distribution is therefore approximately isotropic. Consequently,
the field produced by a given source in such a medium will also be approximately
isotropic. Also consequent to this quasi-random arrangement, the equipotential surfaces around a spherical source will necessarily cut the cellular processes around
the source (Figure 5.12B).

Because in complex Fourier analysis, the partials of the spectrum correspond to
asymptotic states ($t \rightarrow \infty$), the zero-frequency component implies that the value
of the charge is independent of time inside the cell (for a passive membrane). Thus,
for zero frequency, there is a constant charge on cell surfaces such that the surface
polarization is at steady state. This implies that the electric field inside the cells
is zero (otherwise one would have a charge migration towards cell surface, which

is in contradiction with the constant charge). Therefore, there cannot be a voltage difference between two equipotential surfaces (illustrated in Figure 5.12B). One can therefore conclude that *a passive inhomogeneous medium with randomly distributed passive cells is a perfect dielectric at zero frequency.* In this case, the conductivity must tend to zero when the frequency tends to zero. Thus, in the applications considered in Section 5.7, we assume that $\vec{E}_0^S = -\vec{E}_0^P$ where $\vec{E}_0^P$ is the field produced by the source.

5.6 Modeling LFPs in non-resistive media: the diffusion model

In this section, we consider another possible source for frequency dependence of extracellular potentials: ionic diffusion. The opening of ion channels will create ion fluxes, which will induce ionic diffusion to re-establish the local concentrations. This movement of charged particles will create an ionic diffusion current. We will show below that this contribution is significant for local field potentials, and in particular for their frequency dependence. Ionic diffusion can account for $1/f$ filtering effects for spherical current sources.

5.6.1 Is ionic diffusion important for local field potentials?

To estimate whether ionic diffusion has a potential influence on LFPs, we estimate the ratio between ionic diffusion currents (perpendicular to the membrane) and the electric field currents in the extracellular space directly adjacent to the source. We will designate this ratio by the term r_{ie}.

We have in general

$$\vec{j}_{Total} = eD\nabla[C] + \sigma\nabla V, \tag{5.40}$$

where the first term on the right hand side is the electric current density produced by ionic diffusion, while the second term is that produced by differential Ohm's law.

For a displacement $\Delta r = 10\text{nm}$ in the direction across the membrane (from inside to outside), we have approximately:

$$\vec{j}_{Total} \cdot \vec{dr} \simeq \vec{j}_{Total} \cdot \vec{\Delta r} = eD\Delta[C]_{\Delta r=10nm} + \sigma\Delta V_{\Delta r=10nm}. \tag{5.41}$$

Suppose that we have a spherical cell of 10 μm radius, at resting potential and embedded in sea water. The resting membrane potential is a dynamic equilibrium between inflow and outflow of charges, in which these two fluxes are equal on (temporal) average. Fluctuations of current around this average in the extracellular medium around the membrane have all the characteristics of thermal noise (Nyquist, 1928) because the shot noise (see Vasilyev, 1983; Buckingham, 1985) is zero when the current is zero on average, such that the net charge on the external side of the membrane varies around a mean value with the same characteristics as white noise (thermal noise). These fluctuations will therefore also be present at the level of the membrane potential. In this section, we evaluate the order of magnitude of the electric current caused by ionic diffusion, relative to the electric field for this situation of dynamic equilibrium.

First, the ratio between the membrane voltage noise and the variation of total charge concentration is given by

$$\Delta Q_{tot} = C \Delta V_{membrane} = k_1 \Delta V_{membrane} = 1.25 \times 10^{-11} \Delta V_{membrane}, \quad (5.42)$$

because the cell's capacitance is given by $C = 4\pi R^2 C_m = 0.04\pi R^2$ where $C_m = 10^{-2}\,\mathrm{F\,m^{-2}}$ is the specific capacitance of cellular membranes.

Second, mass conservation imposes

$$\Delta Q_{tot} = e \cdot v_{eff} \cdot \Delta[C]_{tot}, \quad (5.43)$$

where $e = 1.69 \times 10^{-19}$ C and v_{eff} is the volume of the spherical shell containing the charges. Because the charges are not distributed uniformly inside the cell, but rather distributed within a thin spherical shell adjacent to the membrane, because the electric field developed across the membrane is very intense (of the order of $\frac{70 \times 10^{-3}}{7 \times 10^{-9}}\,\mathrm{V\,m^{-1}} = 10^7\,\mathrm{V\,m^{-1}}$), Thus, the width of the shell is of the order of $dR \simeq 10^{-4}R < 1$ nm, where the volume of the spherical shell is approximately equal to $4\pi R^2 dR$. In this case, we have for monovalent ions ($|z| = 1$)

$$\Delta Q_{tot} = k_2 \Delta[C]_{tot} \simeq 2.2 \times 10^{-38} \Delta[C]_{tot} = 2.2 \times 10^{-38} \Delta[C]_{\Delta r=10nm} \quad (5.44)$$

if we assume that the variation of concentration on the adjacent border of the exterior surface of the cell is over a width of 10 nm.

In this case, we have

$$\frac{\Delta[C]_{\Delta r=10nm}}{\Delta V_{membrane}} = \frac{k_1}{k_2} \simeq 10^{27}\,\mathrm{C\,m^{-3}V^{-1}}. \quad (5.45)$$

Third, the potential difference between the cell surface and 10 nm away from it, is given by

$$\Delta V_{\Delta r=10nm} = \Delta V_{membrane} - \frac{R \Delta V_{membrane}}{R + \Delta r} \simeq 10^{-3} \Delta V_{membrane}. \quad (5.46)$$

Consequently, and assuming that the electric conductivity of the extracellular fluid is close to that of sea water (2 S m^{-1}), the ratio between the ionic diffusion current and electric diffusion current caused by thermal noise is given by:

$$r_{ie} \approx \frac{e D_{sea} \Delta[C]_{\Delta r=10nm} \Delta t}{\sigma_\omega^{sea} \Delta V_{\Delta r=10nm}} = \frac{e D_{sea} k_1}{\sigma_\omega^{sea} k_2} \simeq \frac{10^2}{\sigma_\omega^{sea}}, \quad (5.47)$$

where the diffusion constant of K^+ or Na^+ in sea water is of the order of 10^{-9} m^{-2} s^{-1}. This implies that the ratio r_{ie} is much larger than 1 for frequencies less than 1000 Hz because σ_ω^{sea} of sea water is necessarily less than 2 S m^{-1}.

Because tortuosity is given by $\lambda = \sqrt{D_{sea}/D_{cortex}}$, and is between 1.6 and 2.2 (for small and large molecules, respectively) in the cerebral cortex (Nicholson and Sykova, 1998; Rusakov and Kullmann, 1998; Nicholson, 2005), the macroscopic diffusion constant in the cortex is certainly larger than $D_{sea}/10$. Thus, we have

$$r_{ie}^{cortex} > \frac{10}{\sigma_\omega^{cortex}} \tag{5.48}$$

where $\sigma_{f=100\ Hz}^{cortex} \simeq 0.1$ S m^{-1} (see Gabriel et al., 1996c).

This evaluation shows that the phenomenon of ionic diffusion has a determinant effect and must be taken into account to calculate the current field in the cortex.

Finally, we note that we did not need to evaluate the absolute magnitude of ΔV. This evaluation is valid for a physical situation where we have a permanent white noise over a distance of 10 nm, independently of the intensity of this noise (which in practice will depend on many factors, such as the size of the cell, the number of ion channels, etc.).

5.6.2 Frequency scaling of ionic diffusion

We now calculate the frequency dependence of an ionic diffusion current outside a spherical current source. We consider a constant variation of ionic concentration, ΔC_i, on the surface of the source, and null variation at an infinite distance (Warburg conditions).

The diffusion equation for a given ionic species is

$$\frac{\partial \Delta C_i}{\partial t} = D_i \nabla^2 \Delta C_i, \tag{5.49}$$

where ΔC_i is the perturbation of the concentration C_i of ion i around the steady-state value, and D_i is the associated diffusion coefficient. This diffusion coefficient depends on the ionic species considered and the structure of the medium.

Because the geometry of the problem and the boundary conditions respect spherical symmetry, we use spherical coordinates. In this coordinate system, we have

$$\frac{\partial \Delta C_i}{\partial t} = D_i \left[\frac{\partial^2 \Delta C_i}{\partial r^2} + \frac{2}{r} \frac{\partial \Delta C_i}{\partial r} \right] \tag{5.50}$$

because ΔC_i does not depend on θ and Φ (spherical symmetry).

The Fourier transform of ΔC_i with respect to time gives:

$$\frac{\partial^2 C_{i_\omega}}{\partial r^2} + \frac{2}{r}\frac{\partial \Delta C_{i_\omega}}{\partial r} = \frac{d^2 C_{i_\omega}}{dr^2} + \frac{2}{r}\frac{d \Delta C_{i_\omega}}{dr} = \frac{i\omega}{D_i}\Delta C_{i_\omega} \qquad (5.51)$$

with general solution given by

$$\Delta C_{i_\omega} = A(\omega)\frac{e^{\sqrt{\frac{i\omega}{D_i}}\,r}}{r} + B(\omega)\frac{e^{-\sqrt{\frac{i\omega}{D_i}}\,r}}{r}. \qquad (5.52)$$

For a variation of concentration at the source border which is independent of frequency and which satisfies the Warburg hypothesis (the variation of concentration tends to zero at an infinite distance (Taylor and Gileadi, 1995; Diard et al., 1999)), we have:

$$\Delta C_{i_\omega}(r) = \Delta C_{i_\omega}(R) \cdot \frac{R\,e^{-\sqrt{\frac{i\omega}{D_i}}\,(r-R)}}{r} \qquad (5.53)$$

where r is the distance to the center of the source and R is the radius of the source.

Thus, the electric current density produced by ionic diffusion is given by:

$$\vec{j}_{i_\omega}(r) = ZeD_i\frac{\partial \Delta C_{i_\omega}}{\partial r}\hat{r} = -ZeD_i\left(\frac{1}{r} + \sqrt{\frac{i\omega}{D_i}}\right)\Delta C_{i_\omega}(r)\,\hat{r} \qquad (5.54)$$

where Ze is the charge of ions i, and $\hat{r}$ is a unit vector in the direction of the current. This current is in the direction of $\hat{r}$ by spherical symmetry.

Because we can consider that the source and extracellular medium form a spherical capacitor, the voltage difference between the surface of the source and infinite distance is given by $ZeC_p\Delta C_{i_\omega}(R)$ where C_p is the capacitance value. Thus, the electric impedance of the medium is given by:

$$Z_\omega = \frac{C_p}{D_i(1/R + \sqrt{i\omega/D_i})}. \qquad (5.55)$$

For a source of radius $R = 10\ \mu m$ and a macroscopic ionic diffusion coefficient of the order of $10^{-11}\ m^2\ s^{-1}$, and for frequencies greater than 1 Hz, we can approximate the impedance by:

$$Z_\omega \approx \frac{C_p}{\sqrt{i\omega D_i}}. \qquad (5.56)$$

The same expression for the impedance is also obtained in cylindrical coordinates or planar Cartesian coordinates (not shown).

Note that if several ionic species are present, then the superposition principle applies (Fick equations are linear) and therefore the contribution of each ion will add up. The diffusion constants for different ions are of the same order of magnitude (for Na^+, K^+, Cl^-, Ca^{2+}), so no particular ion would be expected to dominate.

5.7 Synthesis of the different models

In this section, we integrate the two phenomena, polarization and ionic diffusion, considered in previous sections, into a general macroscopic model. The objective of this synthesis is to analyze experimental results at macroscopic scales, and estimate the magnitude of the frequency dependence of the extracellular potential, as well as the respective influence of polarization and ionic diffusion. We will also mention other possible causes for frequency dependence in the discussion.

We consider below successively more complex models and how these models account for experimental data such as macroscopic measurements of conductivity.

5.7.1 Non-reactive media with ionic diffusion (model D)

Because current sources are ionic currents, there is flow of ions inside or outside the membrane, and another physical phenomenon underlying current flow is ionic diffusion. Let us consider a resistive medium such as a homogeneous extracellular conductive fluid in which the ionic diffusion coefficient is D. When the extracellular current is exclusively due to ionic diffusion, the current density depends on frequency as $\sqrt{\omega}$ (see Section 5.6.2). A resistive medium behaves as if it had a resistivity equal to $(1/\sigma^m)(1 + (k/\sqrt{\omega}))$, where b is complex. The parameter σ^m is the conductivity for very high frequencies, and reflects the fact that the effect of ionic diffusion becomes negligible compared to calorific dissipation (Ohm's law) for very high frequencies. When ionic diffusion is dominant compared to electric field effects, the real part of b is much larger than a.

The frequency dependence of conductivity will be given by:

$$\sigma_\omega^M = \frac{\sigma^m \sqrt{\omega}}{\sqrt{\omega} + k}. \tag{5.57}$$

Applying Equation (5.19) to this configuration gives the following expression for the electric potential as a function of distance:

$$V_\omega(\vec{r}) = \frac{1}{4\pi\sigma_z^M} \cdot \frac{I_\omega}{r} = \frac{\sqrt{\omega} + k}{\sqrt{\omega}} \cdot \frac{1}{4\pi\sigma^m} \cdot \frac{I_\omega}{r}. \tag{5.58}$$

This expression shows that, in a non-reactive medium, when the extracellular current is dominated by ionic diffusion ($k > \sqrt{\omega}$), then the impedance of the medium V_ω/I_ω will be frequency dependent and will scale as $1/\sqrt{\omega}$. This model will be referred to as "model D" in the following. Note that if the electric field dominates over ionic diffusion, then we have the opposite situation, as described in Section 5.2.

5.7.2 Reactive media with electric fields (model P)

In reality, extracellular media contain different charge densities, for example because cells have a non-zero membrane potential by maintaining differences in ionic concentrations between the inside and outside of the cell. Such charge densities will necessarily be influenced by the electric field or by ionic diffusion. As above, we first consider the case with only electric field effects and will consider next the influence of diffusion and the two phenomena taken together.

Electric polarization is a prominent type of "reaction" of the extracellular medium to the electric field. In particular, the ionic charges accumulated over the surface of cells will migrate and polarize the cell under the action of the electric field. It was shown previously in a theoretical study that this "surface polarization" phenomenon can have important effects on the propagation of local field potentials (Bédard et al., 2006a). If a charged membrane is placed inside an electric field $\vec{E}_0^S$, there is production of a secondary electric field $\vec{E}_\omega^S$ given by (Bédard et al., 2006a):

$$\vec{E}_\omega^S = \frac{\vec{E}_0^S}{1 + i\omega\tau_{MW}}. \tag{5.59}$$

This expression is the frequency-domain representation of the effect of the inertia of charge movement associated with surface polarization, reflecting the fact that the polarization does not occur instantaneously but requires a certain time to set up. This frequency dependence of the secondary electric field was derived in Bédard et al. (2006a) for a situation where the current was exclusively produced by electric field. The parameter $\tau_M W$ is the characteristic time for charge movement (Maxwell–Wagner time) and equals $\epsilon^{memb}/\sigma^{memb}$, where ϵ^{memb} and σ^{memb} are respectively the absolute (tangential) permittivity and conductivity of the membrane surface, and are in general very different from the permittivity and the conductivity of the extracellular fluid.

Let us now determine for zero frequency the amplitude of the secondary field $\vec{E}_0^S$ produced between two cells embedded in a given electric field. First, we assume that it is always possible to trace a continuous path which links two arbitrary points in the extracellular fluid (see Figure 5.12B). Consequently, the domain defined by extracellular fluid is said to be *linearly connex*. In this case, the electric potential arising from a current source is necessarily continuous in the extracellular fluid. Second, in a first approximation, we can consider that the cellular processes

surrounding sources are arranged randomly (as opposed to being regularly structured) and their distribution is therefore approximately isotropic. Consequently,
the field produced by a given source in such a medium will also be approximately
isotropic. Also consequent to this quasi-random arrangement, the equipotential surfaces around a spherical source will necessarily cut the cellular processes around
the source (Figure 5.12B).

Because in complex Fourier analysis, the partials of the spectrum correspond
to asymptotic states ($t \to \infty$), the zero-frequency component necessarily implies
that the value of the charge is independent of time inside the cell (for a passive
membrane). Thus, for zero frequency, there is a constant charge on cell surfaces
such that the surface polarization is at steady state. This implies that the electric
field inside the cells is zero (otherwise one would have a charge migration towards
cell surface, which is in contradiction with the constant charge). Therefore, there
cannot be a voltage difference between two equipotential surfaces (illustrated in
Figure 5.12B). One can therefore conclude that *a passive inhomogeneous medium
with randomly distributed passive cells is a perfect dielectric at zero frequency.*
In this case, the conductivity must tend to 0 when frequency tends to 0. Thus, in
the following, we assume that $\vec{E}_0^S = -\vec{E}_0^P$ where $\vec{E}_0^P$ is the field produced by the
source.

It follows that the expression for the current density in extracellular space as a
function of the electric field is given by:

$$\vec{j}_\omega = \sigma_z^M \cdot \vec{E}_\omega^P = \sigma^m \cdot \vec{E}_\omega^{resul} = \sigma^m \cdot (\vec{E}^P + \vec{E}^S) = \sigma^m \cdot \frac{i\omega\tau_{MW}}{1 + i\omega\tau_{MW}} \cdot \vec{E}_\omega^P.$$

Thus, the conductivity can be written as:

$$\sigma_z^M = \sigma^m \cdot \frac{i\omega\tau_{MW}}{1 + i\omega\tau_{MW}}. \tag{5.60}$$

Applying Equation (5.19) gives:

$$V_\omega(\vec{r}) = \frac{1}{4\pi\sigma_z^M} \cdot \frac{I_\omega}{r} = \frac{i\omega\tau_{MW}}{1 + i\omega\tau_{MW}} \cdot \frac{1}{4\pi\sigma^m} \cdot \frac{I_\omega}{r}. \tag{5.61}$$

This model describes the effect of polarization in reaction to the source electric
field, and will be referred to as "model P" in the following.

5.7.3 Reactive media with electric field and ionic diffusion (model DP)

The propagation of current in the medium is dominated by ionic diffusion currents
or by currents produced by the electric field, according to the values of k and k_1
with respect to $\sqrt{\omega}$. The values of k and k_1 are respectively inversely proportional

to the square root of the global ionic diffusion coefficient in the extracellular fluid, and of the membrane surface (see Section 5.6.2).

We apply the reasoning based on the connex topology of the cortical medium (see above) to deduce the order of magnitude of the induced field for zero frequency $\vec{E}_0^S$

$$\vec{E}_\omega^S = -\frac{\vec{E}_0^P}{1 + i\sqrt{\omega}\,\tau}. \tag{5.62}$$

where

$$\tau = (\sqrt{\omega} + k_1)\tau_{MW} = \sqrt{\omega}\,\frac{\epsilon^{memb}}{\sigma^{memb}}.$$

Because the "tangential" conductivity on membrane surface is given by

$$\sigma_\omega^{memb} = \frac{\sigma^{Memb}\sqrt{\omega}}{\sqrt{\omega} + k_1}$$

when the current is dominated by either electric field or ionic diffusion (see Equation (5.57)).

It follows that the expression for the current density in extracellular space as a function of the electric field is given by

$$\vec{j}_\omega = \sigma_z^M \vec{E}_\omega^P = \sigma_\omega^M \cdot \left(1 + i\frac{\omega\epsilon_\omega^M}{\sigma_\omega^M}\right) \cdot \vec{E}_\omega^{resul} = \frac{\sigma^m\sqrt{\omega}}{\sqrt{\omega} + k} \cdot \left(1 + i\frac{\omega\epsilon_\omega^M}{\sigma_\omega^M}\right) \cdot (\vec{E}^P + \vec{E}^S).$$

We then obtain

$$\vec{j}_\omega \approx \frac{\sigma^m\sqrt{\omega}}{\sqrt{\omega} + k} \cdot \frac{i\sqrt{\omega}\,\tau}{1 + i\sqrt{\omega}\,\tau} \cdot \vec{E}_\omega^P$$

because $1 + i(\omega\epsilon_\omega^M/\sigma_\omega^M) \approx 1$ in cortical tissue for frequencies greater than 10 Hz and less than 1000 Hz (see Gabriel et al., 1996b).

Thus, we have the following expression for the complex conductivity of the extracellular medium:

$$\sigma_z^M \approx \frac{\sigma^m\sqrt{\omega}}{\sqrt{\omega} + k} \cdot \frac{i\sqrt{\omega}\,\tau}{1 + i\sqrt{\omega}\,\tau}. \tag{5.63}$$

where $\tau = (\sqrt{\omega} + k_1)\tau_{MW}$.

Thus, we have obtained a unique expression (Equation (5.63)) for the apparent conductivity in extracellular space outside of the source, and its frequency dependence due to differential Ohm's law, electric polarization phenomena and ionic diffusion. These phenomena are responsible for an apparent frequency dependence of the electric parameters, which will be compared to the frequency dependence observed in macroscopic measurements of conductivity (Section 5.8).

Finally, Equations (5.19) and (5.63) imply that the macroscopic impedance of a homogeneous spherical shell of width $R_2 - R_1$ is given by:

$$Z_\omega \approx \frac{1}{4\pi} \int_{R_1}^{R_2} \frac{1}{r'^2} \frac{dr'}{\sigma_\omega^M + i\omega\epsilon_\omega^M} = \frac{R_2 - R_1}{4\pi R_1 R_2} \cdot \frac{1}{\sigma_\omega^M}. \qquad (5.64)$$

In the following, this model will be referred as the "diffusion-polarization" model, or "DP" model, and we will use the above expressions (Equations (5.63) and (5.64)) to simulate experimental measurements.

5.8 Application of non-resistive LFP models to experimental data

One of the great advantages of the macroscopic LFP model is not only that it allows integration of different physical characteristics such as non-uniform media, electric polarization and ionic diffusion, but also it can directly integrate experimental measurements of conductivity. In this section, we consider two applications to illustrate the use of this model to interpret experimental data and suggest physical phenomena which are consistent with the experimental measurements.

5.8.1 Macroscopic measurements of brain conductivity

A first application of the formalism is to reproduce the macroscopic measurements of conductivity in brain tissue. The experiments of Gabriel et al., published in three companion papers (Gabriel et al., 1996a, 1996b, 1996c), provided a comprehensive set of measurements of conductivity in different biological media. In these experiments, the biological tissue was placed between two capacitor plates, and a sinusoidal current of frequency ω was applied. This setup was used to measure the capacitance and leak current using the relation $I_\omega = YV_\omega$, imposing the same current amplitude at all frequencies. Because the admittance value is proportional to $\sigma_\omega^M + i\omega\epsilon_\omega^M$, measuring the admittance provides direct information about σ_ω^M and ϵ_ω^M. In the case of brain tissue, these experiments revealed that the electric parameters were strongly dependent on frequency (Figure 5.13, curve G).

To stay consistent with the formalism developed above, we will assume that the capacitor has a spherical geometry. The exact geometry of the capacitor should in principle have no influence on the frequency dependence of the admittance, because the geometry will only affect the proportionality constant between σ_z and Y_ω. In the case of a spherical capacitor, by applying Equation (5.64), we obtain:

$$Y_\omega = \frac{1}{R} + i\omega C = 4\pi \frac{R_1 R_2}{R_2 - R_1} [\sigma_\omega^M + i\omega\epsilon_\omega^M] = 4\pi \frac{R_1 R_2}{R_2 - R_1} \sigma_z^M. \qquad (5.65)$$

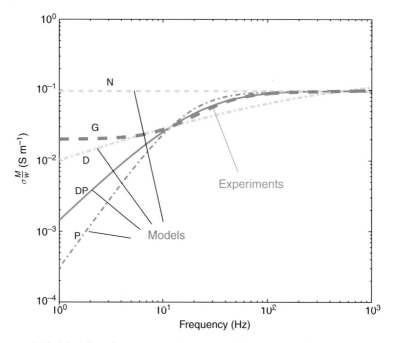

Figure 5.13 Models of macroscopic extracellular conductivity compared to experimental measurements in the cerebral cortex. The experimental data (labeled "G") show the real part of the conductivity measured in cortical tissue by the experiments of Gabriel et al. (1996b). The curve labeled "N" represents the macroscopic conductivity calculated according to the effects of electric field in a non-reactive medium. The curve labeled "D" is the macroscopic conductivity due to ionic diffusion in a non-reactive medium. The curve labeled "P" shows the macroscopic conductivity calculated from a reactive medium with electric-field effects (polarization phenomena). The curve labeled "DP" shows the macroscopic conductivity in the full model, combining the effects of electric polarization and ionic diffusion. Each model was fit to the experimental data using a least-square procedure, and the best fit is shown. The DP model's conductivity is given by $\frac{1}{\sigma_\omega^M} = K_0 + \frac{K_1}{f^{1/2}} + \frac{K_2}{f} + \frac{K_3}{f^{3/2}}$, with $K_0 = 10.84$, $K_1 = -19.29$, $K_2 = 180.35$ and $K_3 = 52.56$. The experimental data (G) is the parametric Cole–Cole model (Cole and Cole, 1941) which was fit to the experimental measurements of Gabriel et al. (1996b). This fit is in agreement with experimental measurements for frequencies larger than 10 Hz. No experimental measurements exist for frequencies lower than 10 Hz, and for those frequencies, the different curves show different predictions from the phenomenological model of Cole–Cole and the present models.

We also take into account the fact that the resistive part is always greater than the reactive part for low frequencies (less than 1000 Hz), which is expressed by

$$\omega \epsilon_\omega^M / \sigma_\omega^M \ll 1.$$

This relation can be verified for example from the Gabriel et al. (1996b) measurements, where it is valid for the whole frequency band investigated experimentally (between 10 and 10^{10} Hz).

The real part of $\sigma_\omega^M = \sigma_z$ then takes the form

$$\sigma_\omega^M \approx \frac{\sigma^M \sqrt{\omega}}{\sqrt{\omega} + k} \cdot \frac{\omega \tau^2}{\omega \tau^2 + 1} \tag{5.66}$$

where $\tau = (\sqrt{\omega} + k_1)\tau_{MW}$.

By substituting this value of τ, the inverse of the conductivity (the resistivity) is given by:

$$\frac{1}{\sigma_\omega^M} \approx \frac{1}{\sigma^M} \cdot \left(1 + \frac{k}{\sqrt{\omega}}\right) \cdot \left(1 + \frac{1}{\omega \tau_M^2 \left(\sqrt{\omega} + k_1\right)^2}\right). \tag{5.67}$$

Rearranging these terms, we can write:

$$\frac{1}{\sigma_\omega^M} \approx \frac{1}{\sigma^M} \cdot \left[1 + \frac{k}{\sqrt{\omega}} + \left(\frac{1}{\omega \tau_M^2} + \frac{k}{\omega^{3/2} \tau_M^2}\right)\left(\frac{1}{\omega + 2k_1\sqrt{\omega} + k_1^2}\right)\right].$$

Finally, by assuming that the current produced by the electric field on membrane surfaces is negligible compared to the current produced by ionic diffusion, we have $k_1 \gg \sqrt{\omega}$ and by developing in series, the last term (in parentheses) of the last equation, we have:

$$\frac{1}{\sigma_\omega^M} \approx \overline{K}_0 + \frac{\overline{K}_1}{\omega^{1/2}} + \frac{\overline{K}_2}{\omega} + \frac{\overline{K}_3}{\omega^{3/2}} = K_0 + \frac{K_1}{f^{1/2}} + \frac{K_2}{f} + \frac{K_3}{f^{3/2}}. \tag{5.68}$$

Equation (5.68) corresponds to the conductivity σ^M, as measured under the experimental conditions of Gabriel et al. (the permittivity ϵ^M is obtained by applying Kramers–Kronig relations). Figure 5.13 shows that this expression for the conductivity can explain the measurements in the frequency range of 10 to 1000 Hz, which are relevant for LFPs. To obtain this agreement, we had to assume in Equation (5.63) a relatively low Maxwell–Wagner time of the order of 0.15 s ($f_c = 1/(2\pi\tau_M)$) between 1 Hz and 10 Hz), $k_1 > \sqrt{\omega} > k$ (for frequencies smaller than 100 Hz).

Thus, the model predicts that in the Gabriel et al. experiments, the transformation of electric current carried by electrons to ionic current in the biological medium necessarily implies an accumulation of ions at the plates of the capacitor. This ion accumulation will in general depend on frequency, because the conductivity and permittivity of the biological medium are frequency dependent. This will create a concentration gradient across the biological medium, which will cause an ionic diffusion current opposite to the electric current. This ionic current will allow a greater resulting current because surface polarization is opposite to the electric field. Figure 5.13 shows that such conditions give frequency-dependent macroscopic parameters consistent with the measurements of Gabriel et al.

The choice of parameter needed to obtain this agreement can be justified qualitatively because the ionic diffusion constant on cellular surfaces is probably much smaller than in the extracellular fluid, such that $k_1 \gg k$. This implies the existence of a frequency band B_f for which $\sqrt{\omega}$ is negligible with respect to k_1, but not with respect to k because these constants are inversely proportional to the square root of their respective diffusion coefficients. Thus, the approximation that we suggest here is that this band B_f finishes around 100 Hz in the Gabriel et al. experimental conditions. It is important to note that this parameter choice is entirely dependent on the ratio between the ionic diffusion current and the current produced by the electric field, and thus will depend on the particular experimental conditions.

It is interesting to note that the present model and the phenomenological Cole–Cole model (Cole and Cole, 1941) predict different behaviors of the conductivity for low frequencies (less than 10 Hz). In the present model, the conductivity tends to zero when the frequency tends to zero, while in the Cole–Cole extrapolation, it tends to a constant value (Gabriel et al., 1996a). The main difference between these models is that the Cole–Cole model is phenomenological and has never been deduced from physical principles for low frequencies, unlike the present model which is entirely deduced from well-defined physical phenomena.

5.8.2 Frequency dependence of the power spectral density of local field potentials

A second type of application of the formalism is to model the frequency dependence of LFPs. As outlined in the introduction (see Figure 5.1), the power spectral density (PSD) of LFPs or EEG signals displays $1/f$ frequency scaling (Bhattacharya and Petshe, 2001; Bédard et al., 2006a; Novikov et al., 1997; Pritchard, 1992). To examine whether this $1/f$ scaling can be accounted for by the present formalism, we consider a spherical current source embedded in a continuous macroscopic medium. We also assume that the PSD of the current source is a Lorentzian, which could derive for example from randomly occurring exponentially decaying postsynaptic currents (Bédard et al., 2006b) (see Figure 5.14).

To simulate this situation we used the "diffusion-polarization" (DP) model with ionic diffusion and electric field effects in a reactive medium. We have estimated above that surface polarization phenomena have a cut-off frequency of the order of 1 Hz, and will not play a role above that frequency. So, if we focus on the PSD of extracellular potentials in the frequency range larger than 1 Hz, we can consider only the effect of ionic diffusion (in agreement with the Gabriel et al. experiments, see above).

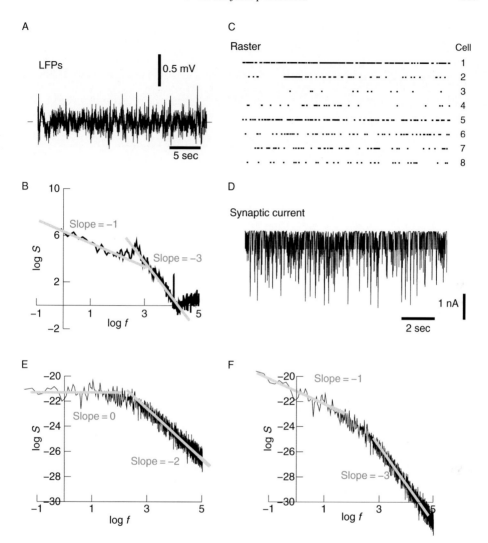

Figure 5.14 Simulation of $1/f$ frequency scaling of LFPs. A. LFP recording in the parietal cortex of an awake cat. B. Power spectral density (PSD) of the LFP on a log scale, showing two different scaling regions with a slope of -1 and -3, respectively. C. Raster of eight simultaneously recorded neurons in the same experiment as in A. D. Synaptic current calculated by convolving the spike trains in C with exponentials (decay time constant of 10 ms). E. PSD calculated from the synaptic current, showing two scaling regions of slope 0 and -2, respectively. F. PSD calculated using a model including ionic diffusion (see text for details). The scaling regions are of slope -1 and -3, respectively, as in the experiments in B. Experimental data taken from Destexhe et al. (1999a); see also Bédard et al. (2006b) for details of the analysis in B–D.

Thus, we can approximate the conductivity as (see Equation (5.63)):

$$\sigma_\omega^M = a\sqrt{\omega} \qquad (5.69)$$

where a is a constant.

It follows that the extracellular voltage around a spherical current source is given by (see Equation (5.19)):

$$V(r, \omega) = \frac{I_\omega}{4\pi a\sqrt{\omega}\,r} = \frac{V(r, 1)}{\sqrt{\omega}} = \frac{V(R, 1)R}{r\sqrt{\omega}} \qquad (5.70)$$

where R is the radius of the source.

In other words, we can say that the extracellular potential is given by the current source I_ω convolved with a filter in $1/\sqrt{\omega}$, which is essentially due to ionic diffusion (Warburg impedance; see Hooge, 1962; Hooge and Bobbert, 1997; Diard et al., 1999). A white noise current source will thus result in a PSD scaling as $1/f$, and can explain the experimental observations, as shown in Figure 5.14. Experimentally recorded LFPs in cat parietal cortex display LFPs with frequency scaling as $1/f$ for low frequencies, and $1/f^3$ for high frequencies (Figure 5.14A, B). Following the same procedure as in Bédard et al. (2006b), we reconstructed the synaptic current source from experimentally recorded spike trains (Figure 5.14C, D). The PSD of the current source scales as a Lorentzian (Figure 5.14E) as expected from the exponential nature of synaptic currents. Calculating the LFP around the source and taking into account ionic diffusion, gives a PSD with two frequency bands, scaling as $1/f$ for low frequencies, and $1/f^3$ for high frequencies (Figure 5.14F). This is the frequency scaling observed experimentally for LFPs in awake cat cortex (Bédard et al., 2006b). We conclude that ionic diffusion is a plausible physical cause of the $1/f$ structure of LFPs for low frequencies.

5.9 Discussion

In the present chapter, we have proposed a framework to model local field potentials, which synthesizes previous measurements and models. This framework integrates microscopic measurements of electric parameters (conductivity σ and permittivity ϵ) of extracellular fluids, with macroscopic measurements of those parameters (σ_ω^M, ϵ_ω^M) in cortical tissue (Gabriel et al., 1996b; Logothetis et al., 2007). It also integrates previous models of LFPs, such as the *continuum model* (Bédard et al., 2004), which was based on a continuum hypothesis of variations of electric parameters in extracellular space, the *polarization model* (Bédard et al., 2006a) and models including the phenomenon of ionic diffusion (Bédard and Destexhe, 2009). The latter models explicitly considered different media (fluid and membranes), their polarization by current sources, and the mode of propagation of current field in the medium. This "diffusion-polarization" model also accounts for observations of $1/f$ frequency scaling of LFP power spectra, which is due here to ionic diffusion, and is therefore predicted to be a consequence of the genesis of the LFP signal, rather than being solely due to neuronal activity (see bedard et al., 2006a). Finally, this work suggests that ephaptic interactions between neurons can occur not only through electric fields and that ionic diffusion should also be considered in such interactions.

An additional source of frequency dependence, not considered here, is the complex neuronal morphology (Pettersen and Einevoll, 2008). Even in a homogeneous and resistive medium, the neuronal morphology can influence the frequency dependence, as occurs for example for dipoles. Each monopole is responsible for a $1/r$ frequency dependence, which will be seen in the immediate vicinity of the monopole. For larger distances, at first approximation, the neuron acts as a linear combination of a dipole and a quadrupole, which gives a distance dependence as $a/r^2 + b/r^4$ (see Section 5.2).

Pettersen and Einevoll (2008) have pointed out that there is necessarily a frequency dependence of extracellular potential even in a homogeneous medium when the morphology is taken into account. In this case, the attenuation law produced by the soma is $1/r$, while that produced by currents distributed in dendrites will contribute as $a/r^2 + b/r^3$ (with respect to the "center" of the dendritic arbor). There is also an inherent frequency dependence of the signal propagation along dendrites (low-pass filter of the membrane RC circuit), such that the potential at a given position is approximately given by

$$\frac{A(f)}{r_{soma}} + B(f)\left[\frac{a}{r^2_{center}} + \frac{b}{r^3_{center}}\right]$$

where $A(f)$ and $B(f)$ are two different functions. Thus, the frequency dependence will be different according to position; near the soma, the $1/r$ term will be dominant, while closer to the "center" of dendrites, the term $a/r^2_{center} + b/r^3_{center}$ will be the dominant one. Note that, as shown in Chapter 4, if the neuron is described by an ideal dipole, the two functions $A(f)$ and $B(f)$ will be equal, such that there will be no frequency dependence due to position.

It is important to note that such effects will only be prominent at fast frequencies. At frequencies less than 10Hz, the difference between functions A and B will become negligible because the effects due to propagation delay on the dendritic arbor will be very small. Thus, the Pettersen and Einevoll (2008) model cannot explain the $1/f$ spectral structure of LFPs and EEGs, which are most prominent for slow frequencies (see Figure 5.1).

We also investigated ways to explain the measurements of Logothetis et al. (2007), who reported that the extracellular medium was resistive and therefore did not display frequency dependence, in contradiction with the measurements of Gabriel et al. (1996b). We summarize and discuss our conclusions below.

In the experiments of Gabriel et al. (1996b), permittivity and conductivity were meausured in the medium in between two metal plates. This forms a capacitor, for which a (macroscopic) complex impedance is measured. This measure actually consists of two independent measurements, the real and imaginary parts of the impedance. These values are used to deduce the macroscopic permittivity and macroscopic conductivity of the medium. However, at the interface between the medium and the metal plates, the flow of electrons in the metal corresponds to a flow of charges in the tissue, and a variety of phenomena can occur, which can interfere with the measurement. The accumulation of charges that occurs at the interface between the electrode and the extracellular fluid implies a polarization impedance, which depends on the interaction between ions and the metal plate. Because this accumulation of charge implies a variation of concentration, the flow of ions may involve an important component of ionic diffusion.

In the experiments of Logothetis et al., a system of four electrodes was used, the two extreme electrodes inject current in the medium, while the two electrodes in the middle are used exclusively to measure the voltage. This system is supposed to be more accurate than that of Gabriel et al., because the electrodes that measure voltage are not subject to charge accumulation. However, the drawback of this method is non-linear effects. The magnitude of the injected current is such that the voltage at the extreme electrodes saturates. This voltage saturation also implies saturation of concentration (capacitive effect between electrodes), which limits ionic diffusion currents. Thus, the ratio between ionic diffusion currents and the currents due to the electric field is greatly diminished relative to the experiments of Gabriel et al.

We think that natural current sources are closer to the situation of Gabriel et al. for several reasons. First, the magnitude of the currents produced by biological sources is far too low for saturation effects. Second, the flow of charges across ion channels will produce perturbations of ionic concentration, which will be re-equilibrated by diffusion. The effects may not be as strong as the perturbations of concentrations induced by Gabriel et al.'s type of experiments, but ionic diffusion should play a role in both cases. This is precisely one of the aspects that should be evaluated in further experiments.

The experiments of Logothetis et al. (2007) were done using a four-electrode setup which neutralizes the influence of electrode impedance on voltage measurements (McAdams and Jossinet, 1992; Geddes, 1997). This system was used to perform high-precision impedance measurements, also avoiding ionic diffusion effects (Logothetis et al., 2007). Indeed, these experimental conditions, and the apparent resistive medium, could be reproduced by the present model if ionic diffusion was neglected. The present model therefore formulates the strong prediction that ionic diffusion is important, and that any measurement technique should allow ionic diffusion to reveal the correct frequency-dependent properties of impedance and electric parameters in biological tissue.

The main prediction of the present model is that ionic diffusion is an essential physical cause of the frequency dependence of LFPs. We have shown that the presence of ionic diffusion allows the model to account quantitatively for the macroscopic measurements of the frequency dependence of electric parameters in cortical tissue (Gabriel et al., 1996b). Ionic diffusion is responsible for a frequency dependence of the impedance as $1/\sqrt{\omega}$ for low frequencies (less than 1000 Hz), which accounts directly for the observed $1/f$ frequency scaling of LFP and EEG power spectra during wakefulness (Pritchard, 1992; Novikov et al., 1997; Bhattacharya and Petsche, 2001; Bédard et al., 2006a) (see Figure 5.14). Note that the EEG is more complex because it depends on the propagation of the signal across fluids, dura mater, skull, muscles and skin. However, this filtering is of low-pass type, and may not affect the low-frequency band, so there is a possibility that the $1/f$ scaling of EEG and LFPs have a common origin. The present model is consistent with the view that this apparent "$1/f$ noise" in brain signals is not generated by self-organized features of brain activity, but is rather a consequence of the genesis of the signal and its propagation through extracellular space (Bédard et al., 2006a). This is also in agreement with a recent study showing that the transfer function between simultaneously recorded intracellular and extracellular potentials is consistent with $1/f$ frequency filtering by extracellular media (Bédard et al., 2010).

It is important to note that the fact that ionic diffusion may be responsible for $1/f$ frequency scaling of LFPs is not inconsistent with other factors which may also influence frequency scaling. For example, the statistics of network activity – and more generally network state – can affect frequency scaling. This is apparent when comparing awake and slow-wave sleep LFP recordings in the same experiment, showing that the $1/f$ scaling is only seen in wakefulness but $1/f^2$ scaling is seen during sleep (Bédard et al., 2006a) (see Figure 5.14). In agreement with this, recent results indicate that the correlation structure of synaptic activity may influence frequency scaling at the level of the membrane potential, and that correlated network states scale with larger (more negative) exponents (Marre et al., 2007).

The critical question that remains to be solved is whether, under physiological conditions, ionic diffusion plays a role as important as that suggested here. We proposed a simple method to test this hypothesis (Bedard and Destexhe, 2009). The frequency dependence could be evaluated by using an extracellular electrode injecting current in conditions as close as possible to physiological conditions (a micropipette would be appropriate). By measuring the integration of the extracellular voltage following periodic current injection, one could estimate the "relaxation time" of the medium with respect to charge accumulation. If this relaxation time occurs at time scales relevant to neuronal currents (milliseconds) rather than the fast relaxation predicted by a purely resistive medium (picoseconds), then ionic diffusion will necessarily occur under physiological conditions, which would provide evidence in favor of the present mechanism.

Finally, our study predicts that the impedance tends to zero when the frequency tends to zero. Thus, for slow frequencies, the extracellular electric field can become very large, and perhaps creates significant ephaptic interactions. It is tempting to relate this to the observation that slow frequencies are more synchronized in general compared to fast frequencies (Bullock, 1997; Destexhe et al, 1999a). Hypersynchronized phenomena, such as epileptic or suppression burst EEG patterns, are usually of slow frequency. These observations also support the notion that frequency dependence is a fundamental property of the propagation of extracellular potentials.

Acknowledgments

This research was supported by the CNRS, the ANR (HR-Cortex) and the European Community (projects FACETS and BrainScales).

References

Bhattacharya, J. and Petsche, H. (2001). Universality in the brain while listening to music. *Proc. Biol. Sci.*, **268**, 2423–2433.

Bédard, C. and Destexhe, A. (2008). A modified cable formalism for modeling neuronal membranes at high frequencies. *Biophys. J.*, **94**, 1133–1143.

Bédard, C. and Destexhe, A. (2009). Macroscopic models of local field potentials and the apparent $1/f$ noise in brain activity. *Biophys. J.*, **96**, 2589–2603.

Bédard, C., Kröger, H. and Destexhe, A. (2004). Modeling extracellular field potentials and the frequency-filtering properties of extracellular space. *Biophys. J.*, **86**, 1829–1842.

Bédard, C., Kröger, H. and Destexhe, A. (2006a). Model of low-pass filtering of local field potentials in brain tissue. *Phys. Rev. E*, **73**, 051911.

Bédard, C., Kröger, H. and Destexhe, A. (2006b). Does the $1/f$ frequency scaling of brain signals reflect self-organized critical states? *Phys. Rev. Lett.*, **97**, 118102.

Bédard, C., Rodrigues, S., Roy, N., Contreras, D. and Destexhe, A. (2010). Evidence for frequency-dependent extracellular impedance from the transfer function between extracellular and intracellular potentials. *J. Comput. Neurosci.*, **29**, 389–403.

Beggs, J. and Plenz, D. (2003). Neuronal avalanches in neocortical circuits. *J. Neurosci.*, **23**, 11167–11177.

Braitenberg, V. and Shüz, A. (1998). *Cortex: Statistics and Geometry of Neuronal Connectivity* (2nd edition). Berlin: Springer-Verlag.

Bremer, F. (1938). L'activité électrique de l'écorce cérébrale. *Actual. Sci. Ind.* **658**, 3–46.

Bremer, F. (1949). Considérations sur l'origine et la nature des "ondes" cérébrales. *Electroencephalogr. Clin. Neurophysiol.*, **1**, 177–193.

Buckingham, M. J. J. (1985). *Noise in Electronic Devices and Systems.* New York: John Wiley & Sons.

Bullock, T. H. (1997). Signals and signs in the nervous system: the dynamic anatomy of electrical activity is probably information-rich. *Proc. Natl. Acad. Sci. USA*, **94**, 1–6.

Cajal, R. (1909). *Histologie du Système Nerveux de l'Homme et des Vertébrés.* Paris: Maloine.

Cole, K. S. and Cole, R. H. (1941). Dispersion and absorption in dielectrics. I. Alternating current characteristics. *J. Chem. Phys.*, **9**, 341–351.

Creutzfeldt, O., Watanabe, S. and Lux, H. D. (1966a). Relation between EEG phenomena and potentials of single cortical cells. I. Evoked responses after thalamic and epicortical stimulation. *Electroencephalogr. Clin. Neurophysiol.*, **20**, 1–18.

Creutzfeldt, O., Watanabe, S. and Lux, H. D. (1966b). Relation between EEG phenomena and potentials of single cortical cells. II. Spontaneous and convulsoid activity. *Electroencephalogr. Clin. Neurophysiol.*, **20**, 19–37.

Debye, P. and Hückel, E. (1923). The theory of electrolytes. I. Lowering of freezing point and related phenomena. *Phys. Z.*, **24**, 185206.

Destexhe, A. (1998). Spike-and-wave oscillations based on the properties of $GABA_B$ receptors. *J. Neurosci.*, **18**, 9099–9111.

Destexhe, A. and Paré, D. (1999). Impact of network activity on the integrative properties of neocortical pyramidal neurons in vivo. *J. Neurophysiol.*, **81**, 1531–1547.

Destexhe, A., Contreras, D. and Steriade, M. (1999a). Spatiotemporal analysis of local field potentials and unit discharges in cat cerebral cortex during natural wake and sleep states. *J. Neurosci.*, **19**, 4595–4608.

Destexhe, A., McCormick, D. A. and Sejnowski, T. J. (1999b). Thalamic and thalamocortical mechanisms underlying 3 Hz spike-and-wave discharges. *Prog. Brain Res.*, **121**, 289–307.

Diard, J. -P., Le Gorrec, B. and Montella, C. (1999). Linear diffusion impedance. General expression and applications. *J. Electroanal. Chem.*, **471**, 126–131.

Eccles, J. C. (1951). Interpretation of action potentials evoked in the cerebral cortex. *J. Neurophysiol.*, **3**, 449–464.

Foster, K. R. and Schwan, H. P. (1989). Dielectric properties of tissues and biological materials: a critical review. *Crit. Rev. Biomed. Eng.*, **17**, 25–104.

Gabriel, S., Lau, R. W. and Gabriel, C. (1996a). The dielectric properties of biological tissues: I. Literature survey. *Phys. Med. Biol.*, **41**, 2231–2249.

Gabriel, S., Lau, R. W. and Gabriel, C. (1996b). The dielectric properties of biological tissues: II. Measurements in the frequency range 10 Hz to 20 GHz. *Phys. Med. Biol.*, **41**, 2251–2269.

Gabriel, S., Lau, R. W. and Gabriel, C. (1996c). The dielectric properties of biological tissues: III. Parametric models for the dielectric spectrum tissues. *Phys. Med. Biol.*, **41**, 2271–2293.

Geddes, L. A. (1997). Historical evolution of circuit models for the electrode–electrolyte interface. *Ann. Biomed. Eng.*, **25**, 1–14.

Hille, B. (2001). *Ion Channels and the Excitable Membranes* (3rd edition). Sunderland, MA: Sinauer.

Hines, M. L. and Carnevale, N. T. (2000). The NEURON simulation environment. *Neural Comput.*, **9**, 1179–1209.

Hooge, F. N. (1962). $1/f$ noise is no surface effect. *Phys. Lett.*, **29A**, 139–140.

Hooge, F. N. and Bobbert, P. A. (1997). On the correlation function of $1/f$ noise. *Physica B*, **239**, 223–230.

Jensen, H. J. (1998). *Self-Organized Criticality: Emergent Complex Behavior in Physical and Biological Systems.* Cambridge: Cambridge University Press.

Johnston, D. and Wu, S. M.-S. (1999). *Foundation of Cellular Neurophysiology.* Cambridge, MA: MIT Press.

Katzner, S., Nauhaus, I., Benucci, A., Bonin, V., Ringach, D. L. and Carandini, M. (2009). Local origin of field potentials in visual cortex. *Neuron.*, **61**, 35–41.

Klee, M. and Rall, W. (1977). Computed potentials of cortically arranged populations of neurons. *J. Neurophysiol.*, **40**, 647–666.

Klee, M. R., Offenloch, K. and Tigges, J. (1965). Cross-correlation analysis of electroencephalographic potentials and slow membrane transients. *Science*, **147**, 519–521.

Koch, C. (1999). *Biophysics of Computation.* Oxford: Oxford University Press.

Koch, C. and Segev, I. (editors) (1998). *Methods in Neuronal Modeling* (2nd edition). Cambridge, MA: MIT Press.

Kronig, R. D. L. (1926). On the theory of dispersion of X-rays. *J. Opt. Soc. Am.*, **12**, 547.

Logothetis, N. K., Kayser, C. and Oeltermann, A. (2007). In vivo measurement of cortical impedance spectrum in monkeys: Implications for signal propagation. *Neuron*, **55**, 809–823.

Landau, L. D. and Lifshitz, E. M. (1984). *Electrodynamics of Continuous Media.* Oxford: Pergamon Press.

Marre, O., El Boustani, S., Baudot, P., Levy, M., Monier, C., Huguet, N., Pananceau, M., Fournier, J., Destexhe, A. and Frégnac, Y. (2007). Stimulus-dependency of spectral scaling laws in V1 synaptic activity as a read-out of the effective network topology. *Soc. Neurosci. Abstr.*, **33**, 790.6.

Maxwell, J. C. (1873). *A Treatise on Electricity and Magnetism*, Chapter 10, pp. 374–375. Oxford: Clarendon Press.

McAdams, E. T. and Jossinet, J. (1992). A physical interpretation of Schwan's limit current of linearity. *Ann. Biomed. Eng.*, **20**, 307–319.

Nicholson, C. (2005). Factors governing diffusing molecular signals in brain extracellular space. *J. Neural Transm.*, **112**, 29–44.

Nicholson, C. and Sykova, E. (1998). Extracellular space structure revealed by diffusion analysis. *Trends Neurosci.*, **21**, 207–215.

Niedermeyer, E. and Lopes da Silva, F. (editors) (1998). *Electroencephalography* (4th edition). Baltimore, MD: Williams and Wilkins.

Novikov, E., Novikov, A., Shannahoff-Khalsa, D., Schwartz, B. and Wright, J. (1997). Scale-similar activity in the brain. *Phys. Rev. E*, **56**, R2387–R2389.

Nunez, P. L. (1981). *Electric Fields of the Brain. The Neurophysics of EEG.* Oxford: Oxford University Press.

Nunez, P. L. and Srinivasan, R. (2005). *Electric Fields of the Brain* (2nd edition). Oxford: Oxford University Press.

Nyquist, H. (1928). Thermal agitation of electric charge in conductors. *Phys. Rev.*, **32**, 110–113.

Peters, A., Palay, S. L. and Webster, H. F. (1991). *The Fine Structure of the Nervous System.* Oxford: Oxford University Press.

Pethig, R. (1979). *Dielectric and Electronic Properties of Biological Materials.* New York: John Wiley & Sons.

Pettersen, K. H. and Einevoll, G. T. (2008). Amplitude variability and extracellular low-pass filtering of neuronal spikes. *Biophys. J.*, **94**, 784–802.

Pritchard, W. S. (1992). The brain in fractal time: $1/f$-like power spectrum scaling of the human electroencephalogram. *Int. J. Neurosci.*, **66**, 119–129.

Protopapas, A. D., Vanier, M. and Bower, J. (1998). Simulating large-scale networks of neurons. In: C. Koch and I. Segev (editors), *Methods in Neuronal Modeling* (2nd edition), Cambridge, MA: MIT Press, pp. 461–498.

Purcell, E. M. (1984). *Electricity and Magnetism.* New York: McGraw Hill.

Raju, G. G. (2003). *Dielectrics in Electric Fields.* New York: CRC Press.

Rall, W. and Shepherd, G. M. (1968). Theoretical reconstruction of field potentials and dendrodendritic synaptic interactions in olfactory bulb. *J. Neurophysiol.*, **31**, 884–915.

Ranck, J. B. (1963). Specific impedance of rabbit cerebral cortex. *Exp. Neurol.*, **7**, 144–152.

Rusakov, D. A. and Kullmann, D. M. (1998). Geometric and viscous components of the tortuosity of the extracellular space in the brain. *Proc. Natl. Acad. Sci. USA*, **95**, 8975–8980.

Steriade, M. (2003). *Neuronal Substrates of Sleep and Epilepsy.* Cambridge: Cambridge University Press.

Taylor, S. R. and Gileadi, E. (1995). The physical interpretation of the Warburg impedance. *Corrosion*, **51**, 664–671.

Vasilyev, A. M. (1983). *An Introduction to Statistical Physics.* Moscow: MIR Editions.

6

EEG and MEG: forward modeling

JAN C. DE MUNCK, CARSTEN H. WOLTERS AND
MAUREEN CLERC

6.1 Introduction

The electroencephalogram (EEG) represents potential differences recorded from the scalp as function of time (Niedermayer and Lopes da Silva, 1987). The generators of the EEG consist of time-varying ionic currents generated in the brain by biochemical sources. These current sources also generate a small but measurable magnetic induction field, which can be recorded with magnetoencephalographic (MEG) equipment (Hämäläinen et al., 1993). When EEG and MEG are studied in the time or frequency domain, several rhythms can be discriminated that contain valuable information about the collective behavior of the living human brain as a neural network. In this chapter EEG and MEG are discussed in the spatial domain. We consider that these signals are recorded from multiple sensors with known positions and study the spatial distribution of EEG and MEG (in the sequel abbreviated as MEEG) in relation to the spatial distribution of the underlying sources.

More precisely, we consider the mathematical problem of predicting the spatial distribution of MEEG, from several physiological assumptions on the current sources. This problem is commonly named the "forward problem." Solutions of the forward problem that are fast, accurate and practical are indispensable ingredients for the solution of the "inverse problem" or "backward problem," which is the problem of extracting as much information as possible about the cerebral current sources, on the basis of MEEG data. Both the forward and the inverse problems are formulated within the framework of a certain mathematical model, wherein the underlying physiological assumptions are precisely formulated. Regarding the forward problem, the modeling assumptions concern the geometry and conductivity of tissues such as brain, CSF, skull, and skin, whereas inverse models specify the assumed constraints (such as temporal behavior and noise characteristics) of the neuronal sources underlying the EEG and MEG. The inverse problem is the central topic of the next chapter of this book. Practical solutions of the MEEG

Handbook of Neural Activity Measurement, ed. Romain Brette and Alain Destexhe. Published by Cambridge University Press. © Cambridge University Press 2012.

inverse problem make it possible for example to localize the generators of epileptic events (e.g. Waberski et al., 1998; Ossenblok et al., 2007; Rullmann et al., 2009), to perform presurgical mapping of eloquent cortical areas (e.g. Roberts et al., 1998; Willemse et al., 2007) or to study which parts of the brain are involved in language processing (e.g. Friederici et al., 2000). Although each of these examples may require a different formulation of the inverse model, all are based on certain assumptions and limitations that are part of the forward problem.

In a living human brain there are many processes that are associated with ionic currents, implying that there could be a large variety of mechanisms contributing to the MEEG. However, considering that MEEG are generally recorded in the frequency range from 1 to 200 Hz, in reality there is one generator mechanism that dominates the others, giving rise to the current dipole model. In Section 6.2 the physiological basis of the current dipole model is presented together with a few other elementary biophysical assumptions. The requirements that forward models need to be fast, accurate and practical are mutually conflicting demands and this has given rise to many complementary approaches in the literature: analytic models, boundary element method (BEM), finite element method (FEM) and others. These are reviewed in Sections 6.3 to 6.6. Section 6.7 contains a general discussion and an outlook for further research.

6.2 The current dipole model and the quasi-static approximation

The living brain can be considered as a huge network of neurons in which information is passed through action potentials traveling over axons. These action potentials are caused by ionic currents that could in principle act as the generators of MEEG. However, because the electric current configuration associated with action potentials has a highly complex geometrical configuration with many positive and negative currents close to each other, the amplitude of potential and magnetic field will die out quickly with distance (see Figure 6.1A). As will be explained in Section 6.2.2 this behavior is a consequence of the fact that complex sources must be considered physically as higher order multipole sources. Moreover, another mechanism is present in the brain that generates currents of a *dipolar* nature, thereby overwhelming the contribution of action potentials to MEEG. This mechanism is provided by synaptic activation of pyramidal cells in layers III, IV and V whose dendrites have a regular parallel arrangement, perpendicular to the cortical sheet. When such a synapse is activated, dependent on the inhibitory or exhibitory nature of the synapse, a positive or negative electrical current will flow into the pyramidal cell and along the dendrite until it leaves the cell and returns through the intracellular space (Lopes da Silva and Van Rotterdam, 1982). When the synaptic currents of neighboring cells are

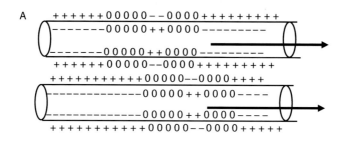

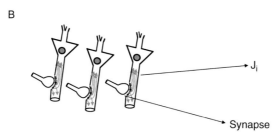

Figure 6.1 It is generally accepted that MEG and EEG are generated primarily by the synaptic currents of interacting pyramidal cells. Other time-varying charge distributions, such as action potentials, are associated with a much more complex charge distribution (sketched in A) than the synaptic currents at parallel pyramidal cells (B). In physical terms, synaptic currents give rise to a dipole source whereas action potentials are described with higher order multipole sources. Therefore, in the far field (like surface EEG), the signals are dominated by current dipole sources.

sufficiently synchronized, the current components coming into the cells from different directions will cancel, and from a distance one will only sense the current component parallel to the apical dendrites (see Figure 6.1B). In principle, the dendritic current could flow in both directions (upwards and downwards in Figure 6.1B), giving rise to a quadrupole source. However, since most of the current is flowing towards the cell body of the postsynaptic cell and because of a general asymmetry in the geometrical arrangement of the cortical layer, one component is dominant and therefore the macroscopic sources of MEEG are current dipoles.

Of course, this current dipole model is nothing more than a qualitative description. Few attempts have been undertaken to quantify the relative contributions to the possible sources of the EEG. One exception is given by Murakami and Okada (2006), who confirm that the main contribution is given by pyramidal cells in layers II/III and V. Therefore, it should be kept in mind that MEEG is only sensitive to an unknown portion of the total neural network. Furthermore, in the context of the current dipole model, current dipoles represent the macroscopic effect of dendritical currents, fed by simultaneously active synapses in layers II/III and V of the neocortex.

6.2.1 The mathematical physical foundation of the dipole model

Since with MEEG the characteristic distances are short and the time scales are long, the mathematical physical basis of the dipole model is given by the quasi-static approximation of Maxwell's equations (e.g. Jackson, 1962; Plonsey and Heppner, 1967). When linear material equations are assumed, and the time derivative of the magnetic induction **B** is neglected, Maxwell's equations are

$$\nabla \cdot (\varepsilon \mathbf{E}) = \rho \tag{6.1}$$

$$\nabla \times \frac{\mathbf{B}}{\mu_0} - \varepsilon \frac{\partial \mathbf{E}}{\partial t} = \mathbf{J} \tag{6.2}$$

$$\nabla \cdot \mathbf{B} = 0 \tag{6.3}$$

$$\nabla \times \mathbf{E} = 0. \tag{6.4}$$

Here, ε is the electrical permittivity, μ_0 the magnetic permeability, **E** the electric field, **B** the magnetic induction, ρ the charge density and **J** the total electric current. All these fields depend on space (**x**) and time (t). The symbol $\nabla \equiv \left(\frac{\partial}{\partial x}, \frac{\partial}{\partial y}, \frac{\partial}{\partial z} \right)^T$ is the nabla-operator and yields a column vector of partial derivatives with respect to **x**. The superscript ^{T} takes the transpose of a vector or matrix.

Equation (6.4) implies that in the quasi-static approximation the electric field can be expressed as (minus) the gradient of a scalar potential function $\psi(\mathbf{x})$,

$$\mathbf{E} = -\nabla \psi. \tag{6.5}$$

The physical meaning of ψ is the potential distribution measured with the EEG amplifier.

In order to apply the Maxwell equations to the prediction of the electric potential and the magnetic field caused by the dendritic currents, the total current is split into an intracellular or primary component $\mathbf{J}_i$, representing the parallel currents flowing inside the dendrites of the activated pyramidal cells, and an extracellular or volume current $\mathbf{J}_e$ representing the return current flowing through the extracellular space (see Figure 6.2). The physical mechanism underlying the intracellular currents $\mathbf{J}_i$ is the opening of ion channels, followed by the restoration of the concentration equilibrium. So, the energy sources for these currents are provided by the active ion pumps pushing ions into and out of the cell, against the concentration gradients. The extracellular current is a passive current, and satisfies Ohm's law, $\mathbf{J}_e = \underline{\sigma} \mathbf{E}$. So we have for the total current **J**

$$\mathbf{J} = \mathbf{J}_i + \mathbf{J}_e = \mathbf{J}_i + \underline{\sigma} \mathbf{E}, \tag{6.6}$$

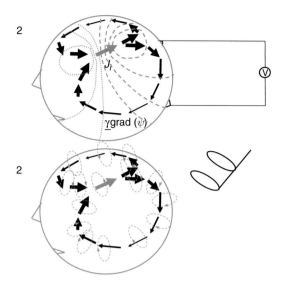

Figure 6.2 The dipole model is shown for EEG (A) and MEG (B). EEG and MEG
are generated by the same current sources, here represented by the current sources
$\mathbf{J}_i(\mathbf{x})$ (gray arrow). With EEG the observed potential differences at the scalp are
mediated through secondary Ohmic currents (black arrows) that depend on the
conductivity distribution $\underline{\gamma}(\mathbf{x})$. MEG is generated through Biot–Sarvart's law by
both primary and secondary currents. However, in the special case that the head,
including the skull and skin layers, is spherically symmetric, and also that the
MEG sensor has a perfect radial orientation, then the contribution of the volume
currents to the observed MEG is zero.

where

$$\underline{\sigma} = \begin{pmatrix} \sigma_{xx} & \sigma_{xy} & \sigma_{xz} \\ \sigma_{yx} & \sigma_{yy} & \sigma_{yz} \\ \sigma_{zx} & \sigma_{zy} & \sigma_{zz} \end{pmatrix}$$

is the conductivity tensor. Mathematically, the conductivity tensor can be inter-
preted as a symmetric matrix ($\sigma_{xy} = \sigma_{yx}$, $\sigma_{yz} = \sigma_{zy}$, $\sigma_{zx} = \sigma_{xz}$) that varies over
space ($\underline{\sigma} = \underline{\sigma}(\mathbf{x})$). Physically, it means that for different tissues the proportionality
between potential gradient and the resulting current is different. For instance, the
skull has a layered structure and therefore the same potential difference across the
skull may result in a weaker current than that potential difference applied along
the skull. Also, the white matter may show anisotropy due to the directionality of
the fibers of which it consists (see Figure 6.3). For isotropic parts of the brain, such
as the cerebrospinal fluid layer, $\underline{\sigma}$ reduces to a scalar function $\sigma(\mathbf{x})$.

In the context of the forward model, $\mathbf{J}_i$, ε, μ_0 and $\underline{\sigma}$ are given and the goal is to
determine the potential distribution at the scalp, as well as the magnetic induction
$\mathbf{B}$ outside this surface. Mathematically, this implies that a differential equation is to

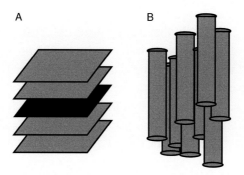

Figure 6.3 The conductivity distribution $\underline{\gamma}(\mathbf{x})$ appearing in MEEG forward models refers to the large scale "equivalent" or macroscopic conductivities of the different tissues constituting the head. Anisotropy of $\underline{\gamma}(\mathbf{x})$ can be the result of a microscopic directional sensitivity in space. For instance, in a layered structure (A) such as the skull, the same potential difference applied across the layers can result in a much lower current than the application of that potential difference parallel to the layers, in particular when the conductivity of one of the layers is small. In this case, $\underline{\gamma}(\mathbf{x})$ will have one small eigenvalue and two large ones. In a fibered structure (B), like the white matter, current will flow more easily parallel to the fibers than across them. Therefore, for such tissues $\underline{\gamma}(\mathbf{x})$ will have two small eigenvalues and a large one.

be derived for ψ and $\mathbf{B}$, wherein $\mathbf{J}_e$, $\mathbf{E}$ and ρ are eliminated. When time variations are decomposed in exponential functions of the type $e^{i\omega t}$, one finds for ψ and $\mathbf{B}$ respectively

$$\nabla \cdot \left(\underline{\gamma} \nabla \psi \right) = \nabla \cdot \mathbf{J}_i \qquad (6.7)$$

and

$$\nabla \times \left(\underline{\gamma}^{inv} \nabla \times \frac{\mathbf{B}}{\mu_0} \right) = \nabla \times \underline{\gamma}^{inv} \mathbf{J}_i, \qquad (6.8)$$

where

$$\underline{\gamma} \equiv \underline{\sigma} + i\omega\underline{\varepsilon}. \qquad (6.9)$$

So, $\underline{\gamma}$ describes the combined effects of Ohmic and displacement currents (De Munck and Van Dijk, 1991) and this is the most important quantity, apart from $\mathbf{J}_i$ that determines what is measured with MEEG. In this chapter, $\underline{\gamma}$ will be referred to as the conductivity, and it will be implicitly assumed that it may have a frequency dependence. In general, $\underline{\gamma}$ can be a tensor to describe local anisotropy. When anisotropy is included in the model, $\underline{\gamma}^{inv}$ is the matrix inverse of $\underline{\gamma}$. Since $\underline{\gamma}$ is a function of $\mathbf{x}$, it carries all the information about the geometry of the different tissues that constitute the head, as well as their possibly anisotropic conductivities. This becomes clear when a compartment model for the head is adopted, i.e. when it is modeled as a set of tissues (brain, cerebrospinal fluid layer, skull and skin) with different conductivities. In that case, $\underline{\gamma}$ takes the form of a piecewise constant function, being constant within each compartment and jumping from one value to another at the interfaces of two compartments. From Equations (6.7) and (6.8), it seems that the potential and the magnetic induction are completely separated. However, it should be kept in mind that the boundary conditions are specified in terms of ψ by the constraint that the head is electrically isolated and no current flows out of the head. In mathematical terms, this means that

$$\mathbf{n}(\mathbf{x}) \cdot \underline{\gamma}\nabla\psi(\mathbf{x}) = 0 \qquad \text{for} \quad \mathbf{x} \in S_{head}, \tag{6.10}$$

where $\mathbf{n}(\mathbf{x})$ is the normal vector of the surface representing the head. When the compartment model is adopted, $\underline{\gamma}(\mathbf{x})$ jumps from one value to another when $\mathbf{x}$ moves across a compartment boundary and the derivatives in (6.7) may not be defined. Therefore, at the interfaces of two compartments two continuity conditions need to be satisfied:

$$\left\{ \begin{array}{l} \psi(\mathbf{x}) \\ \mathbf{n}(\mathbf{x}) \cdot \underline{\gamma}\nabla\psi(\mathbf{x}) \end{array} \right. \quad \text{are continuous functions of } \mathbf{x}. \tag{6.11}$$

These two conditions complicate the use of (6.8) as an equation for the forward problem of MEG. Moreover, Equation (6.8) has the difficulty that at zero frequencies, $\underline{\gamma}$ vanishes outside the head, and therefore the inverse of $\underline{\gamma}$ is not defined there. For those reasons, in practice one always first computes ψ from (6.7) and integrates $\mathbf{B}$ from

$$\nabla \times \tfrac{\mathbf{B}}{\mu_0} = \mathbf{J}_i - \underline{\gamma}\nabla\psi, \tag{6.12}$$

instead of solving (6.8) directly. Note that (6.12) is derived from (6.2) by substituting (6.5) and (6.6). When the curl operator $\nabla\times$ is operated on (6.12), along with $\nabla \times \nabla \times \mathbf{B} = -\Delta\mathbf{B}$, where $\Delta \equiv \frac{\partial^2}{\partial x^2} + \frac{\partial^2}{\partial y^2} + \frac{\partial^2}{\partial z^2}$ is the Laplacian, one finds by integrating $\Delta\mathbf{B} = -\mu_0\nabla \times (\mathbf{J}_i - \underline{\gamma}\nabla\psi)$ that for $\mathbf{B}$ the following expression is derived

$$\mathbf{B}(\mathbf{x}) = \frac{\mu_0}{4\pi} \iiint\limits_{Head} \frac{\nabla' \times (\mathbf{J}_i(\mathbf{x}') - \underline{\gamma}\nabla'\psi(\mathbf{x}'))}{|\mathbf{x} - \mathbf{x}'|} d\mathbf{x}', \tag{6.13}$$

where ∇' represents the vector of partial derivatives with respect to the integration variable $\mathbf{x}'$. Note that the integral in (6.13) is done over the head instead of over the whole space, because outside the head both $\mathbf{J}_i(\mathbf{x})$ and $\underline{\gamma}(\mathbf{x})$ vanish.

6.2.2 The source term

In the previous section, the intracellular current $\mathbf{J}_i(\mathbf{x})$ was introduced, which represents the currents at the activated dendrites. In this section we take a closer look at the possible mathematical forms it can take in different source models. The simplest case is that the activated part of the brain is so small that it can be considered as a single point $\mathbf{x}_0$. For the mathematical description of $\mathbf{J}_i$ this implies that a Dirac delta function $\delta(\mathbf{x} - \mathbf{x}_0)$ should be used, which is infinite at $\mathbf{x}_0$, and zero everywhere else. At the point $\mathbf{x}_0$, a certain amount of current is flowing in the direction of the dendrites. When this current is represented with the symbol $\mathbf{m}_0$, one has for the current dipole model

$$\mathbf{J}_i = \mathbf{m}_0 \delta(\mathbf{x} - \mathbf{x}_0). \tag{6.14}$$

For reasons that become clearer in the sequel, $\mathbf{x}_0$ is called the dipole position and $\mathbf{m}_0$ the dipole moment. The source term of the equation for the potential (6.7) can now be expressed as

$$\nabla \cdot \mathbf{J}_i(\mathbf{x}) = \nabla \cdot (\mathbf{m}_0 \delta(\mathbf{x} - \mathbf{x}_0)) = \mathbf{m}_0 \cdot \nabla \delta(\mathbf{x} - \mathbf{x}_0) = -\mathbf{m}_0 \cdot \nabla_0 \delta(\mathbf{x} - \mathbf{x}_0). \tag{6.15}$$

Here ∇_0 represents the partial derivatives with respect to the source point.

To see the consequences of the dipole source model for the potential distribution and magnetic field, the very simplifying assumption is made that the conductivity is homogeneous and constant in space, i.e. $\boldsymbol{\gamma}(\mathbf{x}) = \mathbf{I}\gamma_\infty$ with $\mathbf{I}$ the identity matrix. Then (6.7) simplifies to

$$\Delta \psi_\infty^{dip} = \frac{-1}{\gamma_\infty} (\mathbf{m}_0 \cdot \nabla_0) \, \delta(\mathbf{x} - \mathbf{x}_0), \tag{6.16}$$

where ψ_∞^{dip} is the dipole "infinite medium" potential. The solution of (6.16) is somewhat complicated by the differential factor $\mathbf{m}_0 \cdot \nabla_0$. If this were a simple scalar M (current density), Equation (6.16) would read $\Delta \psi_\infty = -\frac{M}{\gamma_\infty} \delta(\mathbf{x} - \mathbf{x}_0)$, and its solution would be the monopole potential:

$$\psi_\infty^{mono}(\mathbf{x}) = \frac{M}{4\pi \gamma_\infty} \frac{1}{|\mathbf{x} - \mathbf{x}_0|}. \tag{6.17}$$

The physical interpretation of this solution would be the potential distribution, caused by a primary current sink at $\mathbf{x}_0$, and a secondary Ohmic current of ions that replace the ions that disappear in the sink at $\mathbf{x}_0$. However, since there is no current source, charge would accumulate in the cell, which would be unphysiological. Suppose that the accumulated charge leaves the cell at a neighboring point $\mathbf{x}_0 + \mathbf{h}$, then the resulting potential distribution would be the sum of two monopolar sources, sink at $\mathbf{x}_0$ and a source at $\mathbf{x}_0 + \mathbf{h}$:

$$\psi_\infty^{bip}(\mathbf{x}) = \frac{M}{4\pi\gamma_\infty}\frac{1}{|\mathbf{x}-\mathbf{x}_0|} - \frac{M}{4\pi\gamma_\infty}\frac{1}{|\mathbf{x}-\mathbf{x}_0-\mathbf{h}|}. \tag{6.18}$$

When current source and sink are very close to each other and relatively far away from the measurement point, this implies that $|\mathbf{h}|$ is small with respect to $|\mathbf{x}-\mathbf{x}_0|$ and hence we can approximate the second term in (6.18) by a Taylor approximation:

$$\begin{aligned}
\psi_\infty^{bip}(\mathbf{x}) &= \frac{M}{4\pi\gamma_\infty}\frac{1}{|\mathbf{x}-\mathbf{x}_0|} - \frac{M}{4\pi\gamma_\infty}\frac{1}{|\mathbf{x}-\mathbf{x}_0|} - \frac{M\mathbf{h}}{4\pi\gamma_\infty}\cdot\nabla_0\frac{1}{|\mathbf{x}-\mathbf{x}_0|} + \cdots \\
&\approx \frac{-M\mathbf{h}}{4\pi\gamma_\infty}\cdot\nabla_0\frac{1}{|\mathbf{x}-\mathbf{x}_0|}.
\end{aligned} \tag{6.19}$$

When the limit of $\mathbf{h}\to 0$ is taken while keeping the product $M\mathbf{h}$ finite, the bipole potential (where the two sources are separated) approaches the dipole potential (where the two sources coincide)

$$\psi_\infty^{dip}(\mathbf{x}) = \frac{-\mathbf{m}_0}{4\pi\gamma_\infty}\cdot\nabla_0\cdot\frac{1}{|\mathbf{x}-\mathbf{x}_0|} = \frac{1}{4\pi\gamma_\infty}\frac{\mathbf{m}_0\cdot(\mathbf{x}-\mathbf{x}_0)}{|\mathbf{x}-\mathbf{x}_0|^3}, \tag{6.20}$$

which is the solution of (6.16) with

$$\mathbf{m}_0 = M\mathbf{h}. \tag{6.21}$$

Therefore, when the electrode is in the far field, i.e. much further away from the source than the source dimensions (as is the case with surface EEG), the approximation made in (6.19) is valid and the potential distribution can be considered as the result of a current source very close to a current sink. The direction of the dipole determines the relative positions of the current source and sink, whereas the magnitude of the dipole is determined by their relative distance $|\mathbf{h}|$ and current densities M.

When the potential distributions of the monopole source (6.17) and the dipole case (6.20) are compared, one observes that the dipole potential falls off with distance much faster than the monopole potential ($\sim|\mathbf{x}-\mathbf{x}_0|^{-2}$ compared to $\sim|\mathbf{x}-\mathbf{x}_0|^{-1}$).

This finding can be generalized by considering more complex combinations of current sources and sinks (see Nolte and Curio, 1997) than a simple bi-pole. When two dipoles of the same magnitude are combined, one will find in an analysis similar to (6.19) that the leading term will contain a second-order derivative of the monopole potential, implying that the resulting potential will disappear with distance in a manner proportional to $|\mathbf{x}-\mathbf{x}_0|^{-3}$. For instance, to model the potential distribution due to action potentials, such complex source configurations (see Figure 6.1A) would be appropriate, and the above multipole argument would lead to the conclusion that the effect of action potential hardly contributes to the surface

EEG. In order to record action potentials, one needs to use a needle electrode that is put in the near field, i.e. close to the axonal source.

We would like to remark that not every combination of sources and sinks is physically feasible, considering the fact that charge cannot accumulate. No matter what conductivity profile $\underline{\gamma}(\mathbf{x})$, there is the constraint that any combination of sources should be compensated by a combination of sinks. Mathematically, this constraint can be expressed with help of the boundary condition (6.10). When the equation for the potential (6.7) is integrated over the head volume, one finds

$$\iiint\limits_{Head} \nabla \mathbf{J}_i d\mathbf{x} = \iiint\limits_{Head} \nabla \cdot \left(\underline{\gamma}\nabla\psi\right) d\mathbf{x} = \oiint\limits_{S_{Head}} \left(\underline{\gamma} \cdot \nabla\psi\right) \mathbf{n} dS = 0, \qquad (6.22)$$

where the latter equality follows from the boundary condition (6.10). This equation represents a necessary condition for (6.7), known as the so-called *compatibility condition* (Braess, 2007), to have a solution. Therefore, the single monopole source is not only unphysiological, because of the charge accumulation, it is also mathematically incompatible with the isolation condition at the boundary. Nevertheless, to find a solution of the forward problem, it is often convenient to consider an unphysiological monopole source, and to find the corresponding dipole solution by taking the derivative with respect to the source point. To do this in a mathematically consistent way, one could replace the isolation condition (6.10) by

$$\mathbf{n}(\mathbf{x}) \cdot \underline{\gamma}\nabla\psi(\mathbf{x}) = C \qquad \text{for} \quad \mathbf{x} \in S_{head}, \qquad (6.23)$$

where C is some constant, which is adapted such that

$$\iiint\limits_{Head} \nabla \mathbf{J}_i d\mathbf{x} = \oiint\limits_{S_{Head}} \left(\underline{\gamma}\nabla\psi\right) \cdot \mathbf{n} dS = C \oiint\limits_{S_{Head}} dS = C A_{Head}. \qquad (6.24)$$

When two monopolar sources with opposite amplitude are added, or when the derivative with respect to the source position is taken, the right hand side of (6.24) disappears and the resulting potential will satisfy the isolation condition (6.10).

With this idea in mind, one can verify that the dipole potential can always be expressed as the gradient of the monopole potential, also when the conductivity varies over space. Similar to (6.20) one finds,

$$\psi^{dip}(\mathbf{m}_0, \mathbf{x}_0; \mathbf{x}) = -\mathbf{m}_0 \cdot \nabla\psi^{mono}(\mathbf{x}_0; \mathbf{x}), \qquad (6.25)$$

where ψ^{mono} is the potential due to a monopole with unit current density, which is injected at point $\mathbf{x}_0$ into a conductor with conductivity profile $\underline{\gamma}(\mathbf{x})$, and boundary conditions (6.23).

One might object against the current dipole model that it is a non-physiological model in situations where large parts of the cortex are activated. In such cases it seems more realistic to assume that the source term is given by a distribution of current dipoles over a curved thin layer, representing the active part of the cortex. Because of the superposition principle, the resulting layer potential can be expressed as the dipole potentials, integrated over all surface elements in the active layer,

$$\psi^{layer}(\mathbf{x}) = -\iint\limits_{S_{layer}} \mathbf{m}(\mathbf{x}') \cdot \nabla'\psi^{mono}(\mathbf{x}';\mathbf{x})dS', \qquad (6.26)$$

where here $\mathbf{m}(\mathbf{x}')$ is the dipole density at position $\mathbf{x}'$ of the activated cortex.

In the special case that the dipoles are directed perpendicular to the cortical layer (such as is the case for pyramidal cells) and that the density is homogeneous, it appears that the potential due to the activated layer only depends on the shape of the layer contour and not on the shape of the layer itself. Under these conditions, we have that $\mathbf{m}(\mathbf{x}') = M^{dip}\mathbf{n}(\mathbf{x}')$ and one finds

$$\psi^{layer}(\mathbf{x}) = M^{dip}\iint\limits_{S_{layer}} \nabla'\psi^{mono}(\mathbf{x}',\mathbf{x}) \cdot \mathbf{n}(\mathbf{x}')dS'. \qquad (6.27)$$

When in addition S_{layer} is a closed surface, Gauss' theorem can be applied, and under the assumption that the complete layer is placed in a region with constant conductivity, the kernel $\Delta'\psi^{mono}(\mathbf{x}',\mathbf{x})$ vanishes and therefore also the integral in (6.27):

$$\psi^{ClosedLayer}(\mathbf{x}) = M^{dip}\oiint\limits_{S_{layer}} \nabla'\psi^{mono}(\mathbf{x}',\mathbf{x}) \cdot \mathbf{n}(\mathbf{x}')dS'$$

$$= M^{dip}\iiint \Delta'\psi^{mono}(\mathbf{x}',\mathbf{x}) \cdot \mathbf{n}(\mathbf{x}')dS' = 0. \qquad (6.28)$$

Therefore, if S_{layer} represents part of the cortex, we may conclude that the resulting potential only depends on the contour of the active layer. As a consequence, both dipolar source layers, depicted for example in Figure 6.4, result in exactly the same potential distribution.

Several studies have investigated the effect of source extension on the dipole potential (De Munck et al., 1988a, 1988b) and magnetic induction (Cuffin, 1985). From simulations, where an extended source (dipole layer) was taken as forward model, and a mathematical point dipole was taken as inverse model, it has appeared that source extension hardly has an effect on the potential distribution. In other words, if a point dipole is placed near the gravity point of an extended source, both source configurations produce very similar potential distributions. Therefore, one can conclude that one does not need to worry too much about the effect of source

Figure 6.4 A more realistic source model than a current dipole would consist of a dipole layer, representing an activated part of the cortex. When the dipoles are oriented perpendicularly to the layer (such as the pyramidal cells in the cortex) and the dipole density is homogeneous, it appears that the potential (and magnetic field) caused by the dipole layer only depends on the outer contour of the layer. Therefore, in terms of potential, the flat layer on the left and the curved layer at the middle generate exactly the same potential distribution. Furthermore, it appears from simulation studies that a single dipole placed at the "gravity point" of the flat layer almost produces the same potential distribution. In this sense the dipole model should be interpreted as an "equivalent dipole" for a more realistic source configuration.

extension, provided that the dipole position is interpreted as the gravity point of the dipole layer. The other side of the coin is, however, that the high similarity of the potential distributions indicates that it is hard to extract the size of the activated brain area from potential measurements, if both dipole density and source position are unknown a priori.

For completeness we also present the infinite medium solutions for MEG. In the case that the conductivity is constant one finds that the second term in (6.13) disappears because the curl of a gradient field is null. For $\mathbf{B}_\infty$ one finds

$$\mathbf{B}_\infty(\mathbf{x}) = \frac{\mu_0}{4\pi} \iiint_{Head} \frac{\nabla' \times \mathbf{J}_i(\mathbf{x}')}{|\mathbf{x} - \mathbf{x}'|} d\mathbf{x}'. \qquad (6.29)$$

Note that the magnetic field due to a current source in an infinite medium is independent of the conductivity. This implies that Ohmic extracellular current does not contribute to the infinite medium solution of MEG. For the dipolar case (substitute (6.14)) one finds

$$\mathbf{B}_\infty^{dip}(\mathbf{x}) = \frac{\mu_0}{4\pi} \mathbf{m}_0 \times \frac{\mathbf{x} - \mathbf{x}_0}{|\mathbf{x} - \mathbf{x}_0|^3}. \qquad (6.30)$$

If one compares the dipole expressions for the potential (6.20) and the magnetic induction (6.30), one observes that both fall off with $|\mathbf{x} - \mathbf{x}_0|^{-2}$, and that both expressions are very similar. The dot product in (6.20) is replaced by a cross product in (6.30). This difference implies that the field patterns of MEG and EEG are rotated with respect to each other by 90°.

6.2.3 EEG and MEG sensors

The purpose of the forward model is to compute a prediction of the measured EEG and MEG for a given conductor and source model. This prediction needs to be as accurate as possible because any compromise on accuracy may lead to systematic errors in the application of the forward solutions in inverse modeling. One important aspect that determines the predicted fields is the fact that, with both EEG and MEG, reference sensors are used. With EEG such reference is a matter of principle; with MEG the reference is optional and is meant to improve the signal to noise ratio. Furthermore, MEG is measured with a coil of finite diameter and in principle the extension of the loop has to be accounted for. With EEG, the potential distribution is so smooth compared to the electrode size that an infinitely small sensor can be assumed.

EEG is a measurement of potential *differences* between two positions on the skin. The fact that no absolute potentials are measured (and cannot be measured) is reflected by the fact that, mathematically, the equation for the potential (6.7) and its boundary condition (6.10) has no unique solution. It can be verified simply that if $\psi(\mathbf{x})$ is a solution, so is $\psi(\mathbf{x}) + c$, where c is an arbitrary constant. When a potential difference is considered, $\psi(\mathbf{x}) - \psi(\mathbf{x}_{ref})$ or $\psi(\mathbf{x}) - \psi_{Average}$, this constant cancels. As a consequence, in a practical application of the forward model one always has to account for the position of the reference electrode, or the positions of the other electrodes when EEG data are referred to average reference.

With MEG one should consider which sensor type is used: a magnetometer or a gradiometer (Hämäläinen et al., 1993). Magnetometers consist of a single super-conducting coil through which the magnetic flux is measured. An MEG signal recorded with a magnetometer is "absolute" or reference free. However, the signal to noise ratio of magnetometers is often too low to provide useful MEG signals and in such cases gradiometers can be used. Gradiometers consist of a pair of coils, the pickup and compensation coil, with identical area and exactly opposite orientation. The effect of gradiometers is that homogeneous far field sources (traffic, motion of the building, earth's magnetic field) are suppressed because they produce an exactly opposite signal in both coils, whereas near field sources (brain) are relatively enhanced because they produce different signals at both coils. When gradiometers are used, this has to be accounted for in forward models by computing the **B**-field differences at both coils. This aspect of the forward problem of MEG is mathematically equivalent to the reference problem of EEG, with the difference that for EEG it is a fundamental need, whereas for MEG the reference problem appears because of practical signal to noise ratio considerations.

If the MEG signal were recorded with an infinitely small loop with orientation $\mathbf{n}_{coil}$, the predicted MEG signal (the flux) at that sensor would be computed as

$A_{coil}\mathbf{n}_{coil} \cdot \mathbf{B}(\mathbf{x})$, where A_{coil} is the coil area. However, the pickup and compensation coils have a finite size and in principle it is better to account for the fact that the **B**-field varies over the coil area (Cuffin and Cohen, 1983) and compute the magnetic flux ψ_M through the coil. When the coil is represented by the curve Γ, the magnetic flux can be computed as

$$\psi_M(\mathbf{x}) = \iint_{Coil} \mathbf{B}(\mathbf{x}) \cdot d\mathbf{S}. \tag{6.31}$$

Considering that $\nabla \cdot \mathbf{B} = 0$, Equation (6.3), the magnetic induction can be expressed as the curl of a vector potential $\mathbf{A}(\mathbf{x})$,

$$\mathbf{B}(\mathbf{x}) = \nabla \times \mathbf{A}(\mathbf{x}) \tag{6.32}$$

and (6.31) can be simplified as

$$\psi_M(\mathbf{x}) = \iint_{Coil} (\nabla \times \mathbf{A}(\mathbf{x})) \cdot d\mathbf{S} = \oint_{\Gamma} \mathbf{A}(\mathbf{x}) \cdot d\gamma. \tag{6.33}$$

An expression for $\mathbf{B}(\mathbf{x})$ in terms of the derivative of $\mathbf{A}(\mathbf{x})$, such as in Equation (6.32) can always be found. For instance, one can verify that for the magnetic field in an infinite medium (6.30), one finds for the vector potential

$$\mathbf{A}_\infty^{dip}(\mathbf{x}) = \frac{\mu_0}{4\pi} \frac{\mathbf{m}_0}{|\mathbf{x} - \mathbf{x}_0|}, \tag{6.34}$$

as can be verified by substituting (6.34) into (6.32). When using the finite element Method the computation of the magnetic flux can be nicely integrated in the computations of the electric potentials (see Section 6.5).

6.3 Analytical solutions

The mathematical challenge of the forward problem is to find a solution of (6.7) which is accurate and for which a fast computational algorithm exists. These are generally conflicting demands and that explains to some extent why so many approaches to the forward problem exist in the literature. The accuracy problem has a physical and a mathematical aspect. The physical aspect is to choose a model for the conductivity distribution $\gamma(\mathbf{x})$ (infinite medium, isolated homogeneous sphere, concentric spheres, piecewise constant compartment models, etc.) Each of these models describes the head with varying levels of simplification, thereby introducing different kinds of systematic errors in the solution of the forward problem. All these conductor models will be reviewed and discussed in the remainder of this chapter. The mathematical aspect of the forward problem refers to the numerical accuracy with which the differential equation (6.7) can be solved for a given conductor model. Often different numerical approaches exist to compute the solution of (6.7), with different trade-offs between speed and accuracy.

For the simplest conductor model, the infinite medium, expressions for the potential and magnetic field were derived in the previous section. Although this conductor model is far from realistic, these solutions are of great theoretical interest and also play a role in more advanced models. The first step towards more realistic source models is to make assumptions in such a way that the isolation of the conductor and the differences in conductivity are accounted for, but such that (6.7) can still be solved analytically. One of the advantages of analytically solvable models is that the problem of numerical accuracy is eliminated. Another advantage is that the algorithms used to compute the solution are generally quite fast. But the main advantage is that these solutions provide insight into the main characteristics of the forward problem solution, such as the main differences between EEG and MEG, the dependence on conductivity parameters and the spatial resolution.

6.3.1 Models that allow closed form expressions

6.3.1.1 The electric potential in the homogeneous sphere

It appears that a closed form solution of the EEG forward problem exists when $\gamma(\mathbf{x})$ is modeled as a homogeneous isotropic sphere ($\gamma(\mathbf{x}) = \gamma_{hom}$ for $|\mathbf{x}| \leq R$ and $\gamma(\mathbf{x}) = 0$ for $|\mathbf{x}| > R$, where R is the radius of the conductor). The solution was found in the early 1950s (Wilson and Bayley, 1950; Frank, 1952) with the application of electrocardiography in mind. However, the solution has often been used for EEG modeling. The solution method of Frank (1952) is based on the differentiation of the monopole potential. In the vector notation adopted in this chapter, the monopole solution can be presented as

$$\psi_{hom}^{mono}(\mathbf{x}) = \frac{M}{4\pi\gamma_{hom}} \left(\frac{1}{|\mathbf{x} - \mathbf{x}_0|} + \frac{R/r_0}{\left|\mathbf{x} - \frac{R^2}{r_0^2}\mathbf{x}_0\right|} \right.$$
$$\left. - \frac{1}{R}\log\left(\frac{\left|r_0\mathbf{x} - \frac{R^2}{r_0}\mathbf{x}_0\right| + R^2 + \mathbf{x}\cdot\mathbf{x}_0}{2R^2} \right) \right), \quad (6.35)$$

where $r_0 \equiv |\mathbf{x}_0|$ is the distance of the source to the center of the sphere. Here the measurement point $\mathbf{x}$ is at an arbitrary point inside the conductor or at its boundary. The first term of (6.35) can be interpreted as the monopole potential found in an infinite medium. One can easily verify that the other two terms vanish as $R \to \infty$. The second term can be interpreted as a mirror source, positioned at $\mathbf{x}_0' = \frac{R^2}{r_0^2}\mathbf{x}_0$ and with relative strength proportional to R/r_0 (see Figure 6.5). Such a term could

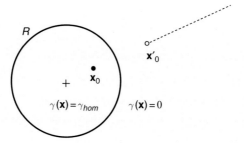

Figure 6.5 The simplest non-trivial dipole model for EEG consists of a current dipole embedded in a homogeneously conducting sphere of radius R. The dipole is positioned at $\mathbf{x}_0$ and the measurement electrode is at an arbitrary point $\mathbf{x}$ inside the sphere. It appears that the solution of the boundary value problem is identical to the sum of three sources that are embedded in an infinite medium: the original current source at $\mathbf{x}_0$, a mirror current source at $R^2/r_0^2\,\mathbf{x}_0$, and a line source, extending from the mirror source to infinity. These three sources should be properly weighted to obtain the correct solution. Similar to these virtual sources of the homogeneous sphere model, in the boundary element method they can be interpreted as a representation of the dipole potential in terms of virtual sources that take the form of triangular dipole or monopole layers. Proper weighting of these sources gives the solution of the boundary value problem in a complex geometry.

be expected, considering that compared to the reflection of light in a plane mirror, we are just studying solutions of Maxwell's equation at an infinitely long wavelength (the quasi-static approximation). The third term, representing the potential due to a line source segment which extends from the origin to $\mathbf{x}_0$ is somewhat unexpected, because it does not seem to have a simple equivalent in terms of light reflection. Note that in this spherical case, the solution of the forward problem is interpreted retrospectively in terms of virtual mirror sources that are embedded in an infinite medium of constant conductivity. With other solution methods, such as the boundary element method, a representation of the solution in terms of virtual sources also plays a fundamental role, and different choices for representing the solution yield algorithms to solve the forward problem with different properties.

The dipole potential in vector notation can now be obtained by applying the differential operator $-\mathbf{m}_0 \cdot \nabla_0$ as shown in Section 6.2.2. One finds

$$\psi_{hom}^{dip}(\mathbf{x}) = \frac{\mathbf{m}_0 \cdot}{4\pi\gamma_{hom}} \left(\frac{\mathbf{x} - \mathbf{x}_0}{|\mathbf{x} - \mathbf{x}_0|^3} + \frac{R^3\mathbf{x} - Rr^2\mathbf{x}_0}{\left|r_{0\mathbf{x}} - \frac{R^2}{r_0}\mathbf{x}_0\right|^3} - \frac{1}{R} \frac{(r^2\mathbf{x}_0 - R^2\mathbf{x})\Big/\left|r_{0\mathbf{x}} - \frac{R^2}{r_0}\mathbf{x}_0\right| + \mathbf{x}}{\left|r_{0\mathbf{x}} - \frac{R^2}{r_0}\mathbf{x}_0\right| + R^2 + \mathbf{x} \cdot \mathbf{x}_0} \right).$$

$$(6.36)$$

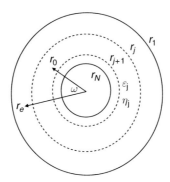

Figure 6.6 When the head is modeled as a set of concentric spheres with different (possibly anisotropic) conductivties, the potential distribution can be obtained
using the technique of separation of variables and the application of special functions of mathematical physics. The geometry of that model is depicted here.

6.3.1.2 The magnetic induction outside a concentric sphere model

For MEG, closed form solutions exist for a more general class of models. When
the head is spherically symmetric (see Figure 6.6) it appears that the **B**-field is
independent of the conductivity profile. A detailed and comprehensive derivation of
this result is given in Sarvas (1987). First, an alternative expression for **B** is derived,
starting from (6.13). When using $|\mathbf{x} - \mathbf{x}'|^{-1} \nabla' \times \mathbf{a}(\mathbf{x}') = \nabla' \times (\mathbf{a} |\mathbf{x} - \mathbf{x}'|^{-1}) + \mathbf{a} \times$
$\nabla' |\mathbf{x} - \mathbf{x}'|^{-1}$ in addition to isolation condition (6.10) one finds

$$\mathbf{B}(\mathbf{x}) = \mathbf{B}_\infty(\mathbf{x}) - \frac{\mu_0}{4\pi} \iiint\limits_{Head} \underline{\gamma} \nabla' \psi(\mathbf{x}') \times \nabla' \frac{1}{|\mathbf{x} - \mathbf{x}'|} d\mathbf{x}'. \qquad (6.37)$$

Note that only the second term in (6.37) depends on the conductivity profile. If the
additional assumption is made that the conductivity is isotropic and piecewise constant, the volume integral in (6.37) can be converted into a sum of surface integrals
on which Stokes' theorem can be applied. One finds

$$\mathbf{B}(\mathbf{x}) = \mathbf{B}_\infty(\mathbf{x}) - \frac{\mu_0}{4\pi} \sum_i \gamma_i \iiint\limits_{V_i} \nabla' \psi(\mathbf{x}') \times \nabla' \frac{1}{|\mathbf{x} - \mathbf{x}'|} d\mathbf{x}'$$

$$= \mathbf{B}_\infty(\mathbf{x}) - \frac{\mu_0}{4\pi} \sum_j (\gamma_j^+ - \gamma_j^-) \oiint\limits_{S_j} \psi(\mathbf{x}') \frac{\mathbf{x} - \mathbf{x}'}{|\mathbf{x} - \mathbf{x}'|^3} \times \mathbf{n}(\mathbf{x}') dS', \qquad (6.38)$$

where γ_j^+ and γ_j^- are the conductivities just inside and outside the surface S_j that
separates two compartments with different conductivities. The vectors $\mathbf{n}(\mathbf{x}')$ are
the local normals of the surfaces S_j. The most important advantage of (6.38) compared to (6.37) is that with (6.38) the magnetic induction can be computed from

knowledge of the potential ψ instead of $\nabla\psi$. Moreover, only knowledge of ψ at the interfaces between two compartments and not inside the compartments is required. Equation (6.38) was originally derived by Geselowitz (1970).

The computation of the **B**-field in the spherical case is based on knowledge of the radial component B_r of **B** (i.e. $B_r = \mathbf{B} \cdot \frac{\mathbf{x}}{r}$, with $r \equiv |\mathbf{x}|$), combined with the constraints on **B** that are provided by the Maxwell equations (6.2) and (6.3), applied outside the head. When the radial component of **B** is computed in the case of spherical symmetry, it appears that the second term of (6.38) vanishes. The reason is that $\mathbf{x} - \mathbf{x}'$, $\mathbf{n}(\mathbf{x}') = \frac{\mathbf{x}'}{r'}$ and $\frac{\mathbf{x}}{r}$ are in the same plane and hence linearly dependent, so that $((\mathbf{x} - \mathbf{x}') \times \mathbf{n}) \cdot \mathbf{x}/r = \det(\mathbf{x} - \mathbf{x}', \mathbf{n}, \mathbf{x}) = 0$. As a result,

$$B_r = \mathbf{B}^{dip}_{sphere}(\mathbf{x}) \cdot \frac{\mathbf{x}}{r} = \frac{\mu_0}{4\pi}\mathbf{m}_0 \times \frac{\mathbf{x} - \mathbf{x}_0}{|\mathbf{x} - \mathbf{x}_0|^3} \cdot \frac{\mathbf{x}}{r}, \tag{6.39}$$

implying that the radial component of the magnetic induction can be computed as if the dipole were in an infinite medium. Volume currents do not contribute to this component. However, this does not imply, as is sometimes wrongly stated, that secondary currents do not contribute at all to MEG. When the orientation of the MEG sensors is not perpendicular to the scalp or when the sensor area cannot be neglected, there is a contribution of these currents. This contribution is however, independent of the conductivities, because it appears that the non-radial components of $\mathbf{B}^{dip}_{sphere}$ can be computed from B_r.

Outside the head, no currents are present, so that for the low-frequency limit Maxwell's equations result in $\nabla \cdot \mathbf{B} = 0$ and $\nabla \times \mathbf{B} = 0$. These equations imply that **B** can be expressed as the gradient of a magnetic potential function $U(\mathbf{x})$, for which $\Delta U(\mathbf{x}) = 0$. Sarvas (1987) showed how this potential can be found from $\mathbf{B} \cdot \frac{\mathbf{x}}{r}$, giving

$$\mathbf{B}^{dip}_{sphere}(\mathbf{x}) = \frac{\mu_0}{4\pi F^2}\left(F\mathbf{m}_0 \times \mathbf{x}_0 - ((\mathbf{m}_0 \times \mathbf{x}_0) \cdot \mathbf{x})\nabla F\right), \tag{6.40}$$

with

$$\begin{cases} F \equiv |\mathbf{x} - \mathbf{x}_0|\left(|\mathbf{x} - \mathbf{x}_0||\mathbf{x}| + |\mathbf{x}|^2 - \mathbf{x} \cdot \mathbf{x}_0\right) \\ \nabla F = \left(\frac{|\mathbf{x} - \mathbf{x}_0|^2}{r} + \frac{(\mathbf{x} - \mathbf{x}_0)\cdot\mathbf{x}}{|\mathbf{x} - \mathbf{x}_0|} + 2|\mathbf{x} - \mathbf{x}_0| + 2r\right)\mathbf{x} - \left(|\mathbf{x} - \mathbf{x}_0| + 2r + \frac{(\mathbf{x} - \mathbf{x}_0)\cdot\mathbf{x}}{|\mathbf{x} - \mathbf{x}_0|}\right)\mathbf{x}_0. \end{cases} \tag{6.41}$$

Similar expressions were derived earlier by Ilmoniemi et al. (1985). From Equation (6.40) it appears that when $\mathbf{m}_0$ is parallel to $\mathbf{x}_0$, i.e. when the dipole has a radial orientation, the magnetic field outside the conductor vanishes (see also Grynszpan and Geselowitz, 1973). In particular, this implies that a dipole at the center of the sphere does not generate a magnetic induction outside the sphere.

The derivation above is strictly speaking only valid for isotropic conductivities, but with a slightly more general argumentation De Munck and Van Dijk (1991) have shown that the conclusion is true for any (anisotropically) spherically symmetrical conductor. A further generalization can be found in Ilmoniemi (1995). The analytical solution for the magnetic field in a spherical head model provides the insights that MEG has the advantage of being insensitive to electrical conductivity differences but it also has the disadvantage that many sources (deep and radial sources) are undetectable with MEG. The impact thereof in experimental studies, when dealing with non-spherical head models, remains to be explored, although one can expect that the deeper and more radial a current source, the lower its signal to noise ratio will be (De Jongh et al., 2005).

6.3.1.3 The electric potential in an anisotropic infinite medium

It seems that for EEG the most advanced model that allows a closed form solution is the homogeneously conducting sphere as presented in Equation (6.36). Models accounting for multiple compartments or having a single compartment of a more complex shape either require infinite series of special functions or numerical approximations. When the medium is anisotropic, a closed form expression for the potential distribution can be derived for the case that the medium is of infinite extent and homogenous. This case only seems of theoretical interest, but it appears to play a practical role when the finite element method (FEM) is concerned.

The assumption of a homogeneous conductive medium of infinite extent implies that the conductivity tensor $\underline{\gamma}$ is independent of $\mathbf{x}$: $\underline{\gamma}(\mathbf{x}) = \underline{\gamma}_\infty$, where $\underline{\gamma}_\infty$ is a symmetric matrix. The solution of (6.7) can be found by adopting new coordinates $\mathbf{x}'$ that depend on the principal axes of $\underline{\gamma}_\infty$ (see e.g. Peters and Elias, 1988; Wolters et al., 2007a). Because $\underline{\gamma}_\infty$ is symmetric and positive definite (conductivities are positive quantities in all directions), it can be represented in terms of another matrix $\underline{\mathbf{w}}$ as follows:

$$\underline{\gamma}_\infty = \underline{\mathbf{w}}\underline{\mathbf{w}}^T, \tag{6.42}$$

where $\underline{\mathbf{w}}^T$ is the transposed matrix of $\underline{\mathbf{w}}$. The precise choice of $\underline{\mathbf{w}}$ is unimportant here, what matters is that a matrix $\underline{\mathbf{w}}$ exists such that (6.42) is true. The situation here is almost equivalent to the theory of generalized least squares estimation, where $\underline{\mathbf{w}}$ is a prewhitening matrix. In the present case, $\underline{\mathbf{w}}$ can be thought of as a combination of the diagonal matrix of the (square roots of the) three principal conductivities and the rotation matrix required to rotate the $\mathbf{x}$-coordinate system to the principal directions of $\underline{\gamma}_\infty$. Here $\underline{\mathbf{w}}$ is used to define new coordinates as follows:

$$\mathbf{x} = \underline{\mathbf{w}}^T \mathbf{x}'. \tag{6.43}$$

Using (6.43) derivatives to $\mathbf{x}$ as they appear in the $\nabla = d/\sqrt{d\mathbf{x}}$-operator can be converted to derivatives to $\mathbf{x}'$, with the side effect that the conductivity tensor in (6.7) disappears. One finds

$$\nabla' = \frac{d}{d\mathbf{x}'} = \frac{d\mathbf{x}}{d\mathbf{x}'}\frac{d}{d\mathbf{x}} = \underline{\mathbf{w}}^T\frac{d}{d\mathbf{x}} = \underline{\mathbf{w}}^T\nabla, \qquad (6.44)$$

and

$$\Delta' = \nabla' \cdot \nabla' = \nabla \cdot \underline{\mathbf{w}}\underline{\mathbf{w}}^T\nabla = \nabla \cdot \underline{\gamma}_\infty\nabla. \qquad (6.45)$$

Note that in (6.44) and (6.45) it was essential to assume that $\underline{\gamma}_\infty$ and $\underline{\mathbf{w}}$ are independent of $\mathbf{x}$. If (6.45) is inserted in (6.7) and a monopole source is assumed one finds

$$\Delta'\psi_{anis}^{'mono} = -M\delta(\underline{\mathbf{w}}^T\mathbf{x}' - \underline{\mathbf{w}}^T\mathbf{x}_0') = -\frac{M}{\det(\underline{\mathbf{w}}^T)}\delta(\mathbf{x}' - \mathbf{x}_0') = -\frac{M}{\sqrt{\det(\underline{\gamma}_\infty)}}\delta(\mathbf{x}' - \mathbf{x}_0'). \qquad (6.46)$$

It appears that in the primed coordinate system, the equation is, apart from a scaling factor caused by the transition from one coordinate system to the other, identical to that of a homogeneous isotropic coordinate system, of which the solution is the simple $1/R$-potential. Therefore, one finds for the monopole potential that

$$\psi_{anis}^{mono}(\mathbf{x}) = \frac{M}{4\pi\sqrt{\det(\underline{\gamma}_\infty)}}\frac{1}{|\underline{\mathbf{w}}^{T\,inv}(\mathbf{x} - \mathbf{x}_0)|}$$

$$= \frac{M}{4\pi\sqrt{\det(\underline{\gamma}_\infty)}}\frac{1}{\sqrt{(\mathbf{x} - \mathbf{x}_0) \cdot \underline{\gamma}_\infty^{inv}(\mathbf{x} - \mathbf{x}_0)}}. \qquad (6.47)$$

Interestingly, in the final expression $\underline{\mathbf{w}}$ has completely disappeared and the end result is expressed directly in terms of $\underline{\gamma}_\infty$. Finally, for the dipole potential it is found by straightforward differentiation that

$$\psi_{anis}^{dip}(\mathbf{x}) = \frac{1}{4\pi\sqrt{\det(\underline{\gamma}_\infty)}}\frac{\mathbf{m}_0 \cdot \underline{\gamma}_\infty^{inv}(\mathbf{x} - \mathbf{x}_0)}{\sqrt{(\mathbf{x} - \mathbf{x}_0) \cdot \underline{\gamma}_\infty^{inv}(\mathbf{x} - \mathbf{x}_0)}^3}. \qquad (6.48)$$

6.3.2 *Models that can be solved with series expansions*

A general mathematical technique for solving equations like (6.7) is to adopt a coordinate transformation such that the surfaces of constant $\gamma(\mathbf{x})$ coincide with a coordinate surface (see e.g. Morse and Feshbach, 1953). For instance, when the model consists of concentric spheres, one would convert the Cartesian coordinates

$\mathbf{x} = (x, y, z)$ using

$$
\begin{cases}
x = r \sin(\theta) \cos(\varphi) \\
y = r \sin(\theta) \sin(\varphi) \\
z = r \cos(\theta).
\end{cases}
\tag{6.49}
$$

Note that, contrary to the coordinate transform used in Section 6.3.1.3, the transform in (6.49) is highly non-linear. Therefore, the transformation of the derivatives is far more complicated than those used in (6.45).

When the model is spherically symmetric, this implies that $\underline{\gamma}(\mathbf{x})$ is a function of r alone and that the conductivities in the θ- and φ-directions are equal. Therefore, such a model is specified in spherical coordinates by a radial conductivity $\varepsilon(r)$ and tangential conductivity $\eta(r)$. This idea was applied to conductor modeling by Hosek et al. (1978) for the isotropic four-sphere model.

When (6.7) is expressed in spherical coordinates, one obtains

$$
\frac{1}{\eta} \frac{\partial}{\partial r} \left(r^2 \varepsilon \frac{\partial \psi}{\partial r} \right) + \frac{1}{\sin \theta} \frac{\partial}{\partial \theta} \left(\sin \theta \frac{\partial \psi}{\partial \theta} \right) + \frac{1}{\sin^2 \theta} \frac{\partial^2 \psi}{\partial \varphi^2}
$$
$$
= M \frac{\delta(r - r_0)\delta(\theta - \theta_0)\delta(\varphi - \varphi_0)}{\eta \sin \theta},
\tag{6.50}
$$

where a monopole source has been adopted with $(r_0, \theta_0, \varphi_0)$ as spherical coordinates. One observes that the left hand side of (6.50) consists of three terms that each depend on a single coordinate (although this is not completely true for the third term). This result makes it possible to find solutions of the homogeneous equation (right hand side of (6.50) set to zero) that are of the form $R(r)T(\theta)F(\varphi)$. The radial functions $R(r)$ will contain the "hard" part of the problem and depend on the conductivity profiles $\varepsilon(r)$ and $\eta(r)$ as well as the boundary condition (6.10). The angular functions $T(\theta)$ and $F(\varphi)$ appear to be the special functions of mathematical physics, called associated Legendre functions and sine/cosine functions respectively. These functions are very general in mathematical physics and also appear for example in the solution of the Schrödinger equation for the hydrogen model in quantum mechanics (e.g. Merzbacher, 1961). They are periodic in θ and φ and are numbered with indices n and m. The solution of (6.50) will then consist of an appropriate linear combination of all solutions $R_n(r)T_{nm}(\theta)F_m(\varphi)$.

The use of a monopole source makes the problem axially symmetric with the line through the source as symmetry line. Therefore the solution will only depend on the angle ω between $\mathbf{x}$ and $\mathbf{x}_0$. It appears that the monopole potential can be expressed as

$$
\psi_{sphere}^{mono}(\mathbf{x}; \mathbf{x}_0) = \frac{M}{4\pi} \sum_{n=0}^{\infty} (2n + 1) R_n(r_0, r) P_n(\cos \omega),
\tag{6.51}
$$

with

$$\cos \omega = \frac{\mathbf{x} \cdot \mathbf{x}_0}{r r_0}. \tag{6.52}$$

Here the functions $P_n()$ are the familiar Legendre polynomials, which can be computed from certain recursion relations. The dipole potential can be computed by taking the derivative of (6.51) with respect to $\mathbf{x}_0$. Because

$$\begin{cases} \nabla_0 \cos \omega = \frac{1}{r_0} \left(\frac{\mathbf{x}}{r} - \cos \omega \frac{\mathbf{x}_0}{r_0} \right) \\ \nabla_0 r_0 = \frac{\mathbf{x}_0}{r_0}, \end{cases} \tag{6.53}$$

one finds

$$\psi_{sphere}^{dip}(\mathbf{x}; \mathbf{x}_0) = \frac{\mathbf{m}_0}{4 \pi r_0} \cdot \left(\left(\sum_{n=0}^{\infty} (2n+1) R_n' P_n \right) \mathbf{x}_0 \right.$$
$$\left. + \left(\sum_{n=0}^{\infty} (2n+1) R_n P_n' \right) \left(\frac{\mathbf{x}}{r} - \cos \omega \frac{\mathbf{x}_0}{r_0} \right) \right), \tag{6.54}$$

where R_n' and P_n' are the derivatives of $R_n()$ with respect to r_0 and $P_n()$ with respect to $\cos \omega$ respectively.

To find $R_n()$ two linearly independent solutions $H_n^{(1)}(r)$ and $H_n^{(2)}(r)$ of the equation

$$\frac{d}{dr} \left(r^2 \varepsilon(r) \frac{d H_n^{(i)}}{dr} \right) + n(n+1) \eta(r) H_n^{(i)} = 0, \qquad i = 1, 2, \tag{6.55}$$

need to be combined as

$$R_n(r_0, r) = \frac{1}{r^2 \varepsilon(r) W(H_n^{(1)}, H_n^{(2)})} \begin{cases} H_n^{(2)}(r_0) H_n^{(1)}(r) & r_0 > r \\ H_n^{(1)}(r_0) H_n^{(2)}(r) & r_0 < r, \end{cases} \tag{6.56}$$

where

$$W(H_n^{(1)}, H_n^{(2)}) \equiv H_n^{(1)}(r) \frac{d}{dr} H_n^{(2)}(r) - H_n^{(2)}(r) \frac{d}{dr} H_n^{(1)}(r) \tag{6.57}$$

is the Wronski determinant. For the case that the conductivity profiles $\varepsilon(r)$ and $\eta(r)$ are piecewise constant functions (see Figure 6.6),

$$\begin{cases} \varepsilon(r) = \varepsilon_j \\ \eta(r) = \eta_j \end{cases} \quad \text{for } r_{j+1} < r < r_j, \tag{6.58}$$

with $0 = r_{N+1} < r_N < \ldots < r_1$. The solutions of (6.55) in each shell j ($r_{j+1} < r < r_j$) consist of linear combinations of the functions r^{v_j} and r^{v_j-1}, with

$$v_j = \frac{1}{2} \left(-1 + \sqrt{1 + 4n(n+1) \eta_j / \varepsilon_j} \right). \tag{6.59}$$

Note that for isotropic shells, $\varepsilon_j = \eta_j$, we have that $v_j = n$, so that we obtain the familiar power functions r^n and r^{-n-1} as solutions.

At the interfaces between two shells the conductivity profile is discontinuous, and interface conditions (6.11) must be satisfied, implying that $H_n^{(i)}(r)$ and $\varepsilon(r)\frac{d}{dr}H_n^{(i)}(r)$ are continuous. When these two quantities are known for $r = r_{j+1}$, the 2×2 matrices $M_j(r_j, r_{j+1})$ can be used to obtain their values at $r = r_j$

$$\begin{pmatrix} H_n^{(i)}(r_j) \\ \varepsilon_j \frac{d}{dr}H_n^{(i)}(r_j) \end{pmatrix} = M_j(r_j, r_{j+1}) \begin{pmatrix} H_n^{(i)}(r_{j+1}) \\ \varepsilon_{j+1}\frac{d}{dr}H_n^{(i)}(r_{j+1}) \end{pmatrix}. \tag{6.60}$$

The expression for $M_j(r_j, r_{j+1})$ is given by

$$M_j(a, b) = \left(\tfrac{a}{b}\right)^{v_j}\left(S_j^+(a, b) + \left(\tfrac{a}{b}\right)^{-2v_j-2} S_j^-(a, b)\right), \tag{6.61}$$

with

$$S_j^-(a, b) = \frac{1}{2v_j + 1}\begin{pmatrix} v_j\frac{a}{b} & -\frac{a}{\varepsilon_j} \\ -\varepsilon_j\frac{v_j(v_j+1)}{b} & v_j + 1 \end{pmatrix}$$

and

$$S_j^+(a, b) = \frac{1}{2v_j + 1}\begin{pmatrix} v_j + 1 & \frac{b}{\varepsilon_j} \\ \varepsilon_j\frac{v_j(v_j+1)}{a} & v_j\frac{b}{a} \end{pmatrix}. \tag{6.62}$$

When (6.60) is applied repeatedly, the solution for the piecewise constant case can be found for the complete range $0 < r < r_1$ from knowledge at the boundaries. From (6.56) it appears that $H_n^{(2)}(r)$ should satisfy the isolation condition at $r = r_1$, whereas $H_n^{(1)}(r)$ is in some way constrained at $r = 0$. Therefore, the coefficients of the Legendre series ((6.51) and (6.54)) basically consist of a matrix product of 2×2 matrices, of which a few matrix elements are selected. In De Munck and Peters (1993) the remaining details of the computation of $R_n()$ and its derivatives are given.

An important aspect of infinite series is the number of terms that need to be included to obtain an accurate approximation. With the formulation of De Munck and Peters (1993) some insight into this matter can be achieved because it appears that the asymptotic behavior of the coefficients $R_n()$ of the Legendre series can be obtained for $n \to \infty$. The way this can be achieved can be understood from (6.61). When $a > b$ then $M_j(a, b) \to \left(\tfrac{a}{b}\right)^{v_j} S_j^+(a, b)$ for $n \to \infty$ and because S_j^+ is rank 1, the matrix product of S_j^+ can be largely simplified. Moreover, many of the factors $\left(\frac{a}{b}\right)^{\nu_j}$ in numerator and denominator of (6.56) cancel. As a result, a sum is found of the form

$$\psi(\Lambda, \omega) = \sum_{n=0}^{\infty} (f_n - nf^{(-1)} - f^{(0)} - n^{-1}f^{(1)} - \cdots)\Lambda^n P_n(\cos\omega)$$

$$+ f^{(-1)} \frac{\Lambda\cos\omega - \Lambda^2}{D^3} + f^{(0)} \frac{1 - D}{D} + f^{(1)} \log\frac{2}{D + 1 - \Lambda\cos\omega}$$

$$(6.63)$$

where

$$D \equiv \sqrt{1 - 2\Lambda\cos\omega + \Lambda^2}. \qquad (6.64)$$

When layers between the source point and the field point are isotropic, one finds $\Lambda = r_0/r$, otherwise different expressions must be used that depend on the anisotropy of the layers between the source point and the measurement point. The last three terms in (6.63) can be recognized as the dipole, monopole and line source potential for an infinite medium and normalized parameters. The computation of (6.63) is independent of the choice of the parameters $f^{(-1)}$, $f^{(0)}$ and $f^{(1)}$ because the sums of terms that are subtracted exactly equal the closed form expressions that are added. However, from the analysis of the asymptotic behavior of the coefficients needed in (6.54), the f-parameters are chosen such that

$$f_n \to nf^{(-1)} + f^{(0)} + n^{-1}f^{(1)} + \cdots \quad \text{for} \quad n \to \infty. \qquad (6.65)$$

Therefore, the sum of closed form expressions already makes a very good approximation of the end result, and in the remaining series of differences much fewer terms are needed to achieve the same accuracy. In De Munck et al. (1991) this idea is elaborated for the special case of the isotropic three-sphere model, where the electrode is located at the outer surface. Finally, Berg and Scherg (1994) and Zhang (1995) have presented heuristic but efficient methods to compute the potential in the multi-sphere model in a fast way, based on a linear combination of mirror source potentials, for which fast analytical solutions are used.

6.3.3 Elementary differences between EEG and MEG

Although the models described so far are far from realistic, they are advanced enough to demonstrate a few characteristics of MEG and EEG. Figure 6.7 shows a modeling setup where a tangential dipole is embedded in a concentric spherical conductor, consisting of a brain, skull and skin compartment. The conductivities

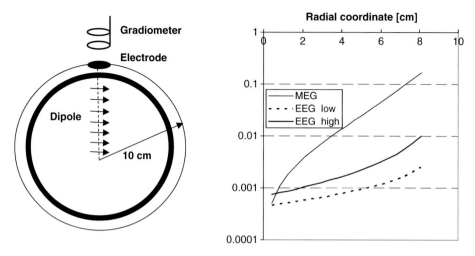

Figure 6.7 At the left a tangential dipole is depicted with varying radial coordinate. The dipole is embedded in a spherically symmetric volume conductor, consisting of a brain, a poorly conducting skull and a skin compartment. At the right the resulting magnetic induction is plotted on a logarithmic scale as a function of depth. At the origin, the **B**-field vanishes, due to symmetry. Contrary to the electric potential (which is referenced to the average of 64 other equally spaced electrodes), the magnetic induction is independent of the skull conductivity. When the skull conductivity is relatively low, the return current is more restricted to the brain compartment, and the potential differences at the skin are smaller. Also the electric potential becomes smaller for deeper dipoles, but it does not vanish at the center.

of brain and skin are taken equal, and the conductivity of the skull is either low (1% of brain conductivity) or high (10% of brain conductivity). On the right the potential at the z-axis (referenced to the average of 64 regularly spaced electrodes) is plotted as function of the dipole depth on a logarithmic scale. One observes that the potential reduces with depth, but does not vanish for a dipole at the sphere's center. When the skull conductivity is low, the potential differences at the skin are relatively small. This is the shielding effect of the skull as it occurs in EEG. The magnetic induction is plotted for a gradiometer located at a small distance from the sphere directly above the dipole. The **B**-field is independent of the skull conductivity and therefore only one MEG curve is shown.

Further differences between EEG, MEG magnetometers and MEG gradiometers are explored in another simulation experiment, presented in Figure 6.8. On the left the modeling setup is depicted. A tangential dipole on the z-axis is used with a radial coordinate of 8.5 cm, embedded in a spherical head model with 10 cm radius. The head model is either a three-sphere model, a homogeneous sphere model, or an infinite medium. EEG electrodes are at the upper part of the sphere and their

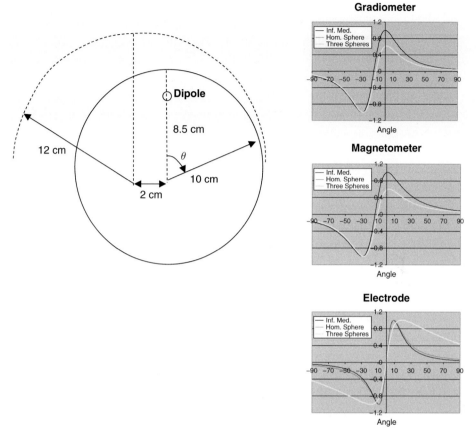

Figure 6.8 On the left a dipole in a spherical head model is depicted. The radius of the head is 10 cm and the dipole is 8.5 cm off-center, pointing into the plane of the drawing. EEG electrodes are located at the upper hemisphere of the head. The magnetic induction is measured with magnetometers or gradiometers that are located on a sphere with radius 12 cm, and which is shifted 2 cm to the left. The right part of the figure shows the **B**-fields (gradiometer and magnetometer) and at the bottom the electric potential is shown as a function of the angle θ. One observes that, in this configuration, the magnetometers and gradiometers generate almost identical signals. The small asymmetry of 2 cm leftward shift causes a substantial difference in the magnetic fields present in an infinite medium compared to the spherically symmetric conductors. For the electric potential one observes that the largest disturbance of the field is caused by the conductivity jump at the skull, and not at the head–air interface. Finally, the magnetic field tends to zero much faster than the electric potential. For MEG the homogeneous model and the three sphere models give identical distributions and therefore in the magnetomer and the gradiometer panels only two curves are visible.

positions are indicated with a θ-coordinate, extending from $-90°$ to $+90°$. MEG sensors (magnetometers or gradiometers) are fixed perpendicularly to a hemisphere (radius 12 cm), which is shifted 2 cm to the left with respect to the head. On the right, the potentials and fields are plotted as a function of θ, for different head models.

These graphs show in the first place that gradiometers and magnetometers are not very different in spatial sensitivity. Nevertheless, from other simulations it appears that ignoring the compensation coil in inverse modeling can produce (significant) systematic errors in estimated dipole positions. Therefore, we strongly recommend that compensation coils are always included in the forward model, when gradiometers are used, despite the similarity of the curves.

If no shift between sensors and volume conductor were present then the curves for the infinite medium and the concentric sphere models would be identical. The graphs show furthermore that the small shift of 2 cm has a relatively large effect on the recorded MEG signal. In practice the situation is even worse, because the MEG sensors are aligned with the head shaped MEG helmet, and not with a hypothetical sphere. We therefore also recommend that volume current effects be accounted for in MEG inverse modeling.

One also observes from these curves that EEG is highly sensitive for the smearing effect of the skull, whereas for MEG this effect is absent. The effect of the skull has a major effect on the spatial resolution of EEG. It appears that without a skull, resolutions of MEG and EEG would be comparable. One could even conclude that without a skull, the resolution of EEG would be better than MEG, and this has probably to do with the fact the MEG sensors are necessarily further away from the sources than the EEG electrodes. Finally, it appears from the EEG graphs that the skull has a much larger effect on the EEG signal than the fact that the head is an isolated sphere. This may seem somewhat surprising because the jump in conductivity at the boundary of the head is much larger than the conductivity jump at the skull-brain interface. A simple explanation for this "paradox" may be that the skull is much closer to the source than the boundary of the head.

6.3.4 More advanced models that are analytically solvable

One might expect that by a simple combination of coordinate scaling (as used in Section 6.3.1.3) and the spherical coordinate system used in (6.49) it would be possible to arrive at an analytical solution for the potential distribution in an ellipsoid coordinate system. However, the approach does not appear to work because it leads to a non-orthogonal coordinate system for which the Laplacian is not separable. A general strategy that does work is to adopt an orthogonal coordinate system consisting of confocal ellipses and hyperbolas. Spheroidal coordinates (ξ, ζ, φ)

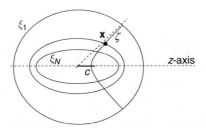

Figure 6.9 Prolate or oblate spheroidal coordinates can be obtained by starting from a set of two-dimensional ellipses, with the same foci. In this figure, these foci are on the positive and negative sides of the z-axis, at a distance c from the origin. In two-dimensional the coordinates of a point are defined by the ellipse (indicated by ξ) that passes through $\mathbf{x}$ and the asymptotic hyperbole (indicated by ζ). When the ellipses are rotated about the long axis (in this case the z-axis) or the short axis, a third coordinate is provided by the required rotation angle φ. These three coordinates are mutually orthogonal, at every point in space. When a rotation over the long axis is used, one obtains prolate spheroidal coordinates, in the other case oblate coordinates are obtained. The former is appropriate to describe elongated objects with axial symmetry, the latter is for flat objects.

are obtained when the ellipses are rotated about their long axis (prolate spheroidal coordinates) or its short axis (oblate spheroidal coordinates, see Figure 6.9)

$$
\begin{cases}
x = c\sqrt{\xi^2 \pm 1}\,\sin(\zeta)\cos(\varphi) \\
y = c\sqrt{\xi^2 \pm 1}\,\sin(\zeta)\sin(\varphi) \\
z = c\xi\cos(\zeta).
\end{cases}
\tag{6.66}
$$

Here c is half the focal distance and z is the symmetry axis. For prolate spheroids, the $-$ sign should be used and for oblate spheroids the $+$ sign. The "radial" conductivity, (i.e. the conductivity perpendicular to the spheroidal surface passing through $\mathbf{x}$) is now indicated by $\varepsilon(\xi)$ and the "tangential" (parallel to the spheroids) conductivity by $\eta(\xi)$. The analysis is very similar to the spherical case, one difference being that the separation of variables is less complete and therefore the radial functions $X_{nm}^{(i)}(\xi)$ depend on both n and m. With the piecewise constant approximation it can be found that these functions satisfy (see Equation (6.55))

$$
\frac{d}{d\xi}\left((\xi^2 \pm 1)\frac{dX_{nm}^{(i)}}{d\xi}\right) - \left(\nu_j(\nu_j + 1) \mp \frac{\mu_j^2}{\xi^2 \pm 1}\right)\eta(\xi)X_{nm}^{(i)} = 0, \qquad i = 1, 2,
\tag{6.67}
$$

where we have used that $\nu_j(\nu_j + 1) = n(n + 1)\eta_j/\varepsilon_j$ and $\mu_j \equiv m\sqrt{\eta_j/\varepsilon_j}$. The solutions of (6.67) are generalized associated Legendre functions with non-integer coefficients ν_j and μ_j. Very similar to the spherical case, these functions can be combined to obtain an infinite series of spherical harmonics as end result. Contrary to concentric spheres, with the confocal spheroids, the double sum over n and m cannot be simplified to a single sum over n. As the range of m-terms is $2n+1$, the analytical solution for the confocal spheroids is computationally much more expensive than the solution for spheres. However, it might be feasible to derive a convergence acceleration formula for spheroids, in the same way as was done in Equation (6.63).

For the MEG case no analytically closed form solution is available for conductors with a confocal spheroidal conductivity profile. Strategies to obtain $\mathbf{B}(\mathbf{x})$ in the form of infinite series of special functions are based on the integration of the electric potential according to Equation (6.38). In Cuffin and Cohen (1977) this principle is applied for the cases of a homogeneous prolate or oblate spheroid. Their formulae are illustrated graphically for dipoles on one of the coordinate axes, in which case the sum over m disappears from the expression.

More general ellipsoidal coordinates can be obtained by considering surfaces defined by Morse and Feshbach (1953)

$$\frac{x^2}{\xi^2 - a^2} + \frac{y^2}{\xi^2 - b^2} + \frac{z^2}{\xi^2} = 1, \quad \text{with } 0 \leq b \leq a. \tag{6.68}$$

By varying ξ_1 in the range of $a < \xi_1 < \infty$, ξ_2 in the range of $b < \xi_2 < a$, $\xi_2 < \infty$ and ξ_3 in the range of $0 < \xi_3 < b$ three sets or orthogonal quadratic surfaces are obtained that can be used to define elliptic coordinates

$$\begin{cases} x = \sqrt{\frac{(\xi_1^2 - a^2)(\xi_2^2 - a^2)(\xi_3^2 - a^2)}{a^2(a^2 - b^2)}} \\ y = \sqrt{\frac{(\xi_1^2 - b^2)(\xi_2^2 - b^2)(\xi_3^2 - b^2)}{b^2(b^2 - a^2)}} \quad \text{with } 0 < \xi_3 < b < \xi_2 < a < \xi_1. \\ z = \frac{\xi_1 \xi_2 \xi_3}{ab} \end{cases} \tag{6.69}$$

It appears that these coordinates also lead to a separable system of second-order differential equations, opening the possibility of finding an analytical solution. This was pursued in Kariotou (2004). The ordinary differential equations are the so-called Lamé equations (Morse and Feshbach, 1953) and multiplication of their solutions for ξ_1, ξ_2 and ξ_3 gives the ellipsoidal harmonics. However, these functions can only be computed explicitly for orders up to 3, implying that the series expansion has to be truncated very early, unless numerical approximations are used. Dassios et al. (2007) and Gutiérrez and Nehorai (2008) have used this solution method to find the corresponding $\mathbf{B}$-field outside the head. Because of the lower order approximation that is mathematically possible, it seems to us that such solutions are mainly of theoretical interest. Other coordinate systems in which Equation (6.7) is separable can be found in Moon and Spencer (1988).

It is obvious that the applicability of analytical solutions using orthogonal coordinate systems is limited in practical situations because these models can never be tailored to the geometries of the brain, CSF, skull and skin of individual subjects. However, it appears to be feasible by analytical means to combine spherical multiple compartments with shifts, as was done by Rudy and Plonsey (1979) and Cuffin (1991) in the eccentric sphere model. Furthermore, analytical models can be used to approximate realistic geometries applying different (sphere) models with different centers and radii for different sensors. The challenge is then to adjust the sphere centers and radii in such a way that for each sensor the sphere model gives acceptable accuracy. This approach was followed by Huang et al. (1999) and Ermer et al. (2001).

6.4 The boundary element method

When the volume conductor does not have a "simple" shape, such as a sphere or an ellipsoid, one has to leave the mathematical ideal to solve the volume conduction equation exactly, and rely on numerical methods to approximate the solution with finite precision. One of the earliest methods for this was based on what is nowadays called the boundary element method. In Barnard et al. (1967a) a set of integral equations is derived describing the potential distribution on the torso due to a current dipole representing the electrical activity of the heart. The mathematical physical model is exactly the same as we need to describe the potential distribution due to a dipole in the brain. In Barnard et al. (1967b) and Lynn and Timlake (1968a), methods are described with great mathematical rigor to solve these integral equations numerically.

The underlying assumptions to make this approach possible imply that the conductivity profile $\underline{\gamma}(\mathbf{x})$ is isotropic ($\underline{\gamma}(\mathbf{x})$) $= \underline{\mathbf{I}}\gamma(\mathbf{x})$) and *piecewise constant*. The latter assumption means that the conductors consist of compartments (tissues) of different conductivities, and that within each compartment the conductivity does not depend on $\mathbf{x}$. A schematic example of such a conductor is plotted in Figure 6.10. The example of Figure 6.10 has the additional property that all compartments are nested, i.e. interfaces between two compartments do not intersect and form

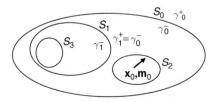

Figure 6.10 An example of a nested geometry.

closed surfaces. When the compartments are represented as volumes V_i and the conductivities of volume i as γ_i, Equation (6.7) simplifies to

$$\gamma_i \Delta \psi(\mathbf{x}) = s(\mathbf{x}) \qquad \mathbf{x} \in V_i, \qquad (6.70)$$

where the right hand side represents the source term, which is usually a superposition of dipoles. It is important to note that Equation (6.70) does not hold for $\mathbf{x}$ on the compartment interfaces, which can be phrased mathematically as the assumption that V_i are open subsets of $\mathbb{R}^3$. At the compartment interfaces, denoted by ∂V_i, the continuity conditions (6.11) must be satisfied. In terms of the compartment model:

$$\begin{cases} \psi^+(\mathbf{x}) = \psi^-(\mathbf{x}) \\ \gamma^+ \mathbf{n}(\mathbf{x}) \cdot \nabla \psi^+(\mathbf{x}) = \gamma^- \mathbf{n}(\mathbf{x}) \cdot \nabla \psi^-(\mathbf{x}) \end{cases} \qquad \mathbf{x} \in \partial V_i. \qquad (6.71)$$

Here the $^+$ and $^-$ superscripts refer to variables evaluated just outside or inside the boundary ∂V_i. The strategy for obtaining a numerical approximation of (6.70) and (6.71) is to derive first an equivalent integral equation for ψ, such that ψ only appears on the interfaces, and then to discretize the resulting boundary integral equation. This approach is sketched roughly below for the so-called double layer BEM approach.

6.4.1 The double layer BEM

The starting point of the derivation is Green's second identity, which implies that for any functions $f(\mathbf{x})$ and $g(\mathbf{x})$ that satisfy certain mild constraints and an arbitrary volume V with smooth boundary ∂V, one has

$$\iiint_V \left(f \Delta' g - g \Delta' f \right) d\mathbf{x}' = \oiint_{\partial V} \left(f \nabla' g - g \nabla' f \right) \cdot \mathbf{n}(\mathbf{x}') dS'. \qquad (6.72)$$

Here the primes denote that the derivatives are taken with respect to $\mathbf{x}'$, which is the integration variable. Since (6.72) is very generally valid, the following choices can be made to obtain an integral equation: (1) $f(\mathbf{x}') = \psi(\mathbf{x}')$, (2) $g(\mathbf{x}) = G(\mathbf{x}, \mathbf{x}') \equiv \frac{1}{|\mathbf{x}-\mathbf{x}'|}$ and (3) $V = V_i$. For the first of these functions one has, using (6.70), that $\Delta' f(\mathbf{x}') = s(\mathbf{x}')/\gamma_i$.

For the second function, one has $\Delta' g(\mathbf{x}') = -4\pi\delta(\mathbf{x}-\mathbf{x}')$ and the volume integral of $f \Delta' g$ yields $-4\pi f(\mathbf{x})$ when $\mathbf{x}$ is inside V_i, and 0 when it is outside. In the former case, it is found that

$$\gamma_i \psi(\mathbf{x}) = \frac{1}{4\pi} \iiint_{V_i} \frac{s(\mathbf{x}')}{|\mathbf{x} - \mathbf{x}'|} d\mathbf{x}' + \frac{\gamma_i}{4\pi} \oiint_{\partial V_i} \left(\psi \nabla' \frac{1}{|\mathbf{x} - \mathbf{x}'|} - \frac{1}{|\mathbf{x} - \mathbf{x}'|} \nabla' \psi \right) \cdot \mathbf{n}(\mathbf{x}') dS'$$

$$\mathbf{x} \in V_i. \qquad (6.73)$$

This equation is applied for all volumes V_i and the results are summed. Since $\mathbf{x}$ is inside one of the volumes and outside all others, the sum of the left hand side equals $\gamma(\mathbf{x})\psi(\mathbf{x})$. The first term on the right hand side of (6.73) can be recognized as the potential in an infinite medium of unit conductivity, due to a current source density distribution $s(\mathbf{x})$, restricted to the volume V_i. When all such terms are summed, the effects of all current sources from all volumes are added, and one simply obtains the unrestricted infinite medium potential. On the right hand side, it is noted that each boundary ∂V_i occurs twice in the summation, once as the inside of one volume, and once as the outside of its neighboring volume. Therefore, considering the second boundary condition of (6.71), the last terms in the sum of (6.73) cancel precisely . The resulting sum is expressed using a numbering over surfaces S_j, which are the compartment interfaces, and one obtains

$$\gamma(\mathbf{x})\psi(\mathbf{x}) = \psi_{\infty}^s(\mathbf{x}) + \frac{1}{4\pi}\sum_j \left(\gamma_j^+ - \gamma_j^-\right) \oiint_{S_j} \psi\nabla' \frac{1}{|\mathbf{x}-\mathbf{x}'|}\cdot\mathbf{n}(\mathbf{x}')dS' \qquad \mathbf{x}\in V_i.$$

$$(6.74)$$

Note that the index j of the conductivity in this equation refers to the surface index. Therefore, in this notation, the same physical tissue conductivity can be referred to by different symbols. For instance, in the example of Figure 6.10 the inside of S_0 is the same compartment as the outsides of S_1 and S_2 and therefore one has $\gamma_0^- = \gamma_1^+ = \gamma_2^+$. To avoid some confusion in the sequel, the index j is always used to refer to surfaces whereas i always refers to volume.

Equation (6.74) is almost the equation that is needed. It expresses the potential at the point $\mathbf{x}$ as the infinite medium potential plus a correction term that can be interpreted as a dipole layer potential, where the dipole density has an amplitude $\frac{\gamma_j^+ - \gamma_j^-}{4\pi}\psi(\mathbf{x}')$ and the dipoles are directed perpendicular to the surfaces S_j (cf. (6.26)). Equation (6.74) looks very similar to (6.38), which expresses the magnetic induction outside the head in terms of the potential distribution on each of the compartment interfaces. However, (6.74) cannot be used as a boundary integral equation because it is only valid for $\mathbf{x}$ outside the boundaries ∂V_i. One of the difficulties is that for $\mathbf{x}$ on a boundary, $\gamma(\mathbf{x})$ is not defined. To obtain an expression for $\mathbf{x}\in S_k$ one can restart from Green's identity (6.72), assume that $\mathbf{x}\in\partial V$ and use that the volume integral of $f\Delta'g$ equals $f\Omega(\partial V_i, \mathbf{x})$ when $\mathbf{x}$ is at a boundary point of V_i, where $\Omega(\partial V_i, \mathbf{x})$ is the solid angle of the surface ∂V_i, subtended at $\mathbf{x}$. When V_i does not contain any sharp corners, one can replace this quantity by 2π, and find similarly to (6.74)

$$\frac{\gamma_k^+ + \gamma_k^-}{2}\psi(\mathbf{x}) = \psi_{\infty}^s(\mathbf{x}) + \frac{1}{4\pi}\sum_j \left(\gamma_j^+ - \gamma_j^-\right)\oiint_{S_j}\psi\nabla'\frac{1}{|\mathbf{x}-\mathbf{x}'|}\cdot\mathbf{n}(\mathbf{x}')dS' \qquad \mathbf{x}\in S_k.$$

$$(6.75)$$

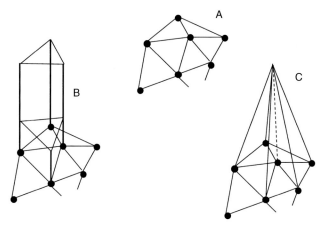

Figure 6.11 The same triangular mesh (A) can lead two different interpolation functions with a different smoothness and number of data points. In B the
unknowns are taken from each triangle and the function is interpolated by a piecewise constant function. In C the unknowns are attributed to each node of the mesh
and tent shaped functions lead to a piecewise linear interpolation scheme.

So for $\gamma(\mathbf{x})$ at the boundary the average conductivity at both sides of the boundary
is taken.

The appealing aspect of (6.75) is that the dimensionality of the problem has
been reduced from a three-dimensional problem to a two-dimensional problem,
because in (6.75) the unknown potential only appears at the boundary. The numerical approximation of (6.75) is based on the description of the surfaces S_j in terms of
triangular elements and a definition of local base functions $h_n(\mathbf{x})$ (see Figure 6.11):

$$\psi(\mathbf{x}) \approx \sum_{n=1}^{N} \psi_n h_n(\mathbf{x}). \tag{6.76}$$

Several choices for $h_n(\mathbf{x})$ can be made. One can assume that each $h_n(\mathbf{x})$ takes on
a constant value on each triangle or, alternatively, one can use the same triangular
grid and assume that $h_n(\mathbf{x})$ varies linearly from one triangle to the next. In the
former case $\psi(\mathbf{x})$ is approximated by a piecewise constant function. Higher order
spline functions have also been explored (e.g. Frijns et al., 2000).

The second step of the BEM consists of substituting (6.76) in (6.74). This results
in an infinite number of equations (for all $\mathbf{x}$) and a finite number N of unknowns
ψ_n. To deal with this situation, one can choose for the collocation or the Galerkin
approach. With the collocation method N equations are obtained by letting $\mathbf{x}$ run
over the triangle centers (piecewise constant case) or the nodes (piecewise linear
case) and ignoring all other $\mathbf{x}$-positions. The resulting system of N equations with
N unknowns

$$\sum_{n=1}^{N} A_{mn}\psi_n = b_m \qquad (6.77)$$

has matrix elements

$$A_{mn} = \frac{\gamma_{k_m}^{+} + \gamma_{k_m}^{-}}{2}\delta_{nm} - \frac{1}{4\pi}\left(\gamma_{k_n}^{+} - \gamma_{k_n}^{-}\right)\iint_{\Delta_n} h_n(\mathbf{x}_m)\nabla'\frac{1}{|\mathbf{x}_m - \mathbf{x}'|}\cdot\mathbf{n}(\mathbf{x}')\mathrm{d}S',$$

$$n = 1,\ldots,N,$$

$$b_m = \psi_{\infty}^{s}(\mathbf{x}_m). \qquad (6.78)$$

When a piecewise constant approximation is used, the integrals in (6.78) represent the solid angles of each of the triangles Δ_m, observed by the node points $\mathbf{x}_n$ (center of triangle Δ_n). Therefore, these integrals can be computed using analytically closed expressions for the solid angle of a triangle (Van Oosterom and Strackee, 1983). A derivation of the solid angle formula can be found for example in Eriksson (1990). Furthermore, in the piecewise constant approximation the number of unknowns is equal to the number of triangles. For linear interpolation analytical formulae for the integration can also be derived (e.g. De Munck, 1992). In this case the number of unknowns in (6.77) equals the number of nodes, which is only about half the number of triangles.

The solution of the system (6.77) of equations can be performed by direct LU decomposition, provided one adds a constraint like $\sum_{n=1}^{N} w_n\psi_n = 0$, to remove the ambiguity caused by the reference potential.

With the Galerkin approach, which was already proposed by Lynn and Timlake (1968a) and Barnard et al. (1967b), the left and right hand sides of Equation (6.74) are multiplied with the base functions $h_m(\mathbf{x})$, and integrated over the surfaces. For each m this yields an equation with N unknowns that can also be solved by standard means. The disadvantage of the Galerkin approach is that the computation of the matrix elements is more time consuming than in the collocation approach, but the great advantage is that it yields more accurate results (Mosher et al., 1999).

Many more variants of the BEM have been described in the literature. In the early days of applying computers to solve potential problems, computer memory was relatively small and systems like (6.77) could only be solved iteratively (see e.g. Golub and Van Loan, 1989). In a highly sophisticated analysis of the problem, Lynn and Timlake (1968b) managed to speed up these iterations by deriving an analytic expression for the small eigenvalues of the matrix A, associated with large conductivity jumps. Another attempt to deal with the large conductivity jump of the skull, is the so-called isolated problem approach (Hämäläinen and Sarvas, 1989; Meijs et al., 1989; Fuchs et al., 1998). In this approach, first the potential is computed on a model which is isolated at the brain, and then this potential is propagated through the skull to the skin. However, Mosher et al. (1999) reported that the benefits of this method are limited: although it improves the EEG forward solution for superficial dipoles, it degrades the MEG forward solution.

6.4.2 The single layer BEM

In the previous section a boundary integral equation was derived that is equivalent to Equation (6.7) when the conductivity is homogeneous and piecewise constant. Retrospectively, the potential was interpreted as the sum of the infinite medium potential and a double layer correction term. In Kybic et al. (2005b) this interpretation in terms of virtual sources was used as a formal starting point of the derivation of the integral equation, leading to alternative integral equations with computationally attractive properties. An important notion in this theory is the representation of function $u(\mathbf{x})$ that is harmonic in every volume V_i (meaning that $\Delta u(\mathbf{x}) = 0$), except at the boundaries, where $u(\mathbf{x})$ and/or its normal derivative $\partial_n u(\mathbf{x}) \equiv \frac{du(\mathbf{x})}{d\mathbf{x}} \cdot \mathbf{n}(\mathbf{x})$ can be discontinuous. When a square bracket notation $[u]$ is introduced to denote such jumps

$$
\begin{cases}
[u](\mathbf{x}) \equiv u^+(\mathbf{x}) - u^-(\mathbf{x}) \\
[\partial_n u](\mathbf{x}) \equiv \lim_{\alpha \downarrow 0} \left(\frac{du(\mathbf{x}+\alpha\mathbf{n})}{d\alpha} - \frac{du(\mathbf{x}-\alpha\mathbf{n})}{d\alpha} \right)
\end{cases}
\tag{6.79}
$$

the following representation theorem can be derived (Kybic et al., 2005b)

$$
\frac{u^+(\mathbf{x}) + u^-(\mathbf{x})}{2} = \frac{1}{4\pi} \sum_j \left(-\oiint_{S_j} [u] \, \partial_{n'} G(\mathbf{x} - \mathbf{x}') dS' + \oiint_{S_j} [\partial_{n'} u] \, G(\mathbf{x} - \mathbf{x}') dS' \right),
\tag{6.80}
$$

where $\mathbf{x}$ is located on one of the boundaries and

$$
G(\mathbf{x} - \mathbf{x}') \equiv \frac{1}{|\mathbf{x} - \mathbf{x}'|}.
\tag{6.81}
$$

Note that these definitions of $[u]$ and $G()$ are slightly different from those of Kybic et al. (2005b). Equation (6.80) shows that, very generally, the contribution to the average $u(\mathbf{x})$ consists of a double layer part with dipole density $[u]$ and a single layer part with charge density $[\partial_n u]$.

It appears that for a certain choice of $u(\mathbf{x})$, the integral equation (6.74) can be derived with a single step from (6.80). For that purpose, one defines $\psi_{\infty,i}(\mathbf{x})$ as the potential in an infinite medium of unit conductivity, caused by the current source $s_i(\mathbf{x})$, which is $s(\mathbf{x})$ inside V_i and zero elsewhere. In other words, $s_i(\mathbf{x})$ is the restriction of $s(\mathbf{x})$ to V_i and $\psi_{\infty,i}(\mathbf{x})$ is the resulting potential. Then the sum of these functions equals the infinite medium potential $\psi_\infty(\mathbf{x})$ and satisfies

$\Delta\psi_\infty(\mathbf{x}) = s(\mathbf{x})$. Furthermore both ψ_∞ and its normal derivatives are continuous for all $\mathbf{x}$. Therefore, and because $\Delta\psi = s_i$ (6.70), the function

$$u(\mathbf{x}) = u_d(\mathbf{x}) = \gamma(\mathbf{x})\psi(\mathbf{x}) - \sum_i \psi_{\infty,i}(\mathbf{x}) \tag{6.82}$$

will be harmonic everywhere, except at the compartment interfaces. This choice of $u()$ satisfies the conditions needed for the application of (6.80) and considering (6.71) for the continuity conditions one finds that $[\partial_n u_d] = 0$ and $[u_d] = (\gamma_j^+ - \gamma_j^-)\psi$ (for $\mathbf{x} \in S_j$) so that substitution of this $u()$ indeed yields (6.74).

So, the double layer BEM is obtained by choosing a function $u()$ such that the single layer term in (6.80) disappears. Similarly, the single layer BEM can be derived by choosing another $u()$ such that the double layer term vanishes. When

$$u(\mathbf{x}) = u_s(\mathbf{x}) = \psi(\mathbf{x}) - \sum_i \frac{\psi_{\infty,i}(\mathbf{x})}{\gamma_i}, \tag{6.83}$$

it can be verified that this $u_s(\mathbf{x})$ also is harmonic for $\mathbf{x}$ outside the compartment interfaces ($\mathbf{x} \notin S_j$) and that $[u_s](\mathbf{x}) = 0$, because of the continuity of ψ and the infinite medium potentials. Therefore, one can apply (6.80) to this $u()$ and now the first term on the right hand side of (6.83) vanishes. One finds

$$\psi(\mathbf{x}) = \sum_i \frac{\psi_{\infty,i}(\mathbf{x})}{\gamma_i} + \frac{1}{4\pi}\sum_j \oiint_{S_j} [\partial_{n'}u_s]\,G(\mathbf{x} - \mathbf{x}')dS'$$

$$= \sum_i \frac{\psi_{\infty,i}(\mathbf{x})}{\gamma_i} + \frac{1}{4\pi}\sum_j \oiint_{S_j} \xi(\mathbf{x}')G(\mathbf{x} - \mathbf{x}')dS', \tag{6.84}$$

where the abbreviation

$$\xi(\mathbf{x}) \equiv [\partial_n u_s](\mathbf{x}) \tag{6.85}$$

has been used.

This function can be found by solving an integral equation, obtained by taking the derivative of (6.80) in the direction $\mathbf{n}$ (the normal at the boundary point $\mathbf{x} \in S_k$)

$$\partial_n u^+(\mathbf{x}) + \partial_n u^-(\mathbf{x}) = \frac{-1}{2\pi}\sum_j \oiint_{S_j} [u]\,\partial_n\partial_{n'}G(\mathbf{x} - \mathbf{x}')dS' + \oiint_{S_j} [\partial_{n'}u]\,\partial_n G(\mathbf{x} - \mathbf{x}')dS'. \tag{6.86}$$

With the choice made in (6.83), the first term in (6.86) with the double derivative vanishes. Then, considering (6.71), for $[\gamma\partial_n u_s]$ one finds

$$[\gamma\,\partial_n u_s](\mathbf{x}) = -\sum_i \left[\frac{\gamma\,\partial_n\psi_{\infty,i}}{\gamma_i}\right](\mathbf{x}) = -\left(\gamma_k^+ - \gamma_k^-\right)\partial_n\sum_i \frac{\psi_{\infty,i}(\mathbf{x})}{\gamma_i} \qquad \mathbf{x} \in S_k, \tag{6.87}$$

This part can be computed directly and will act as the right hand side of a system of equations. On the other hand, one finds

$$
\begin{aligned}
\left[\gamma \partial_n u_s\right](\mathbf{x}) &= \gamma_k^+ \partial_n u_s^+ - \gamma_k^- \partial_n u_s^- \\
&= \frac{\left(\gamma_k^+ + \gamma_k^-\right)\left(\partial_n u_s^+ - \partial_n u_s^-\right)}{2} + \frac{\left(\gamma_k^+ - \gamma_k^-\right)\left(\partial_n u_s^+ + \partial_n u_s^-\right)}{2}.
\end{aligned}
$$

$$(6.88)$$

Finally, combining (6.87) and (6.88), (6.85) and (6.86) gives

$$
\frac{1}{2}\frac{\gamma_k^+ + \gamma_k^-}{\gamma_k^+ - \gamma_k^-}\xi(\mathbf{x}) = -\partial_n \sum_i \frac{\psi_{\infty,i}(\mathbf{x})}{\gamma_i} - \frac{1}{4\pi}\sum_j \oiint_{S_j} \xi(\mathbf{x}')\partial_n G(\mathbf{x}-\mathbf{x}')\mathrm{d}\mathbf{x}, \qquad \mathbf{x} \in S_k,
$$

$$(6.89)$$

which is an integral equation in $\xi(\mathbf{x})$.

In sum, with the single layer BEM formulation, one first solves Equation (6.89) and substitutes the result in (6.84). The discretization of (6.89) can be done in much the same way as described for the double layer formulation. Since the kernels are quite different, both approaches could have different numerical advantages and disadvantages. From a simulation study presented in Kybic et al. (2005b) it appears that the single layer formulation generally performs better than the double layer approach.

6.4.3 The symmetric BEM

The representation theorems (6.80) and (6.86) allow still another formulation, which appears to have even more favorable properties than the single layer BEM. The idea is to make a choice of $u(\mathbf{x})$ such that the integral equation becomes self adjoint, leading to a symmetric system matrix A in the numerical approximation (6.77). This goal can be achieved by choosing a set of functions $u_i(\mathbf{x})$ correspond-ing to each compartment i, such that $u_i(\mathbf{x}) = \psi(\mathbf{x}) - \psi_{\infty,i}(\mathbf{x})/\gamma_i$ for $\mathbf{x}$ in V_i and $u_i(\mathbf{x}) = -\psi_{\infty,i}(\mathbf{x})/\gamma_i$ for $\mathbf{x}$ outside V_i. Since these functions are also harmonic, except on the boundaries, one can evoke the representation formulae (6.80) and (6.86). This idea was proposed in Kybic et al. (2005b) and we refer to that paper for all details.

With the symmetric formulation of the BEM one deals with a combined system of equations involving both the discretized potentials and the normal currents. In this sense, the strategy is the same as for the multi-sphere model (Equation (6.60)), where a matrix vector formulation was used to relate potential and normal current from one shell to the next one. The 2×2 matrices $M_j()$ occurring in (6.60) are however non-symmetric.

A property of the symmetric BEM is that the system matrix only contains non-zero blocks corresponding to adjacent surfaces. For example, in the case of the conductor depicted in Figure 6.10, the block of numbers corresponding to combinations of S_4 and S_1 would vanish.

6.4.4 Numerical comparison of BEM variants

To give an impression of the differences in the numerical accuracies of the different BEM formulations a simulation study was performed. The potential distribution is computed for an isotropic concentric three-shell model with normalized conductivities (1.0, 0.0125, 1.0) using the methods described in Section 6.3.2. For the numerical computations the spheres are discretized using 1280 triangles per sphere. Using either the double layer, or the single layer or the symmetric formulation of the BEM, the potential was computed on a set of 642 EEG electrodes on the outer sphere as a function of the dipole radial coordinate. EEG data simulated by both methods were compared after converting data to average reference. Dipole orientation was kept fixed at a half radial, half tangential orientation.

The differences between numerical and analytical models were expressed by a relative difference measure (RDM) and a magnification factor (MAG), where

$$\text{RDM} = \left| \frac{\psi^{Ana}}{\left|\psi^{Ana}\right|} - \frac{\psi^{Num}}{\left|\psi^{Num}\right|} \right| \tag{6.90}$$

and

$$\text{MAG} = \frac{\left|\psi^{Num}\right|}{\left|\psi^{Ana}\right|} \tag{6.91}$$

and where

$$|\psi| = \sqrt{\sum_n \psi_n^2}. \tag{6.92}$$

Note that for example Meijs et al. (1989), who introduced these error measures to source analysis, used different definitions for RDM than we do. There is thus certain confusion about the nomenclature in the literature that sometimes makes it difficult to compare results from different authors.

Figure 6.12 presents the results of the simulation. It appears that for the double layer formulation the numerical error (in terms of both RDM and MAG) increases dramatically when the dipole gets closer to the compartment boundary. To some extent, this effect can be reduced by using the Galerkin approach (Mosher et al., 1999). However, with the single layer and, more generally, the symmetric BEM formulation the reduction of numerical errors is much more effective.

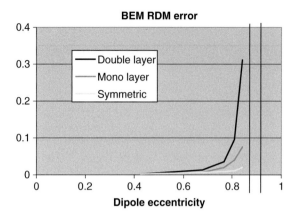

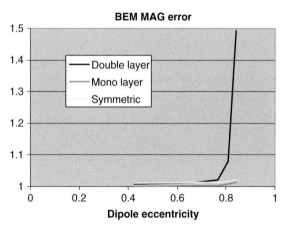

Figure 6.12 (See plate section for color version.) A comparison of the numerical accuracy of different BEM variants is shown. As a true model a three-shell concentric sphere model was used. The RDM and MAG errors are shown as functions of dipole eccentricity (normalized with respect to the outer radius of volume conductor r_1) for the double layer method, the single layer method and the symmetric method. The vertical black lines in the upper plot represent the locations r_3 and r_2 of the compartment boundaries.

6.4.5 Non-nested geometries

Boundary element methods (BEM) require a geometrical model describing the interfaces between tissues with different conductivities. The head models classically used for solving the forward and inverse problems of MEEG consist of nested volumes. However, in many cases, such geometrical models are too restrictive (Bénar and Gotman, 2002; Oostenveld and Oostendorp, 2002). For instance, when the subject has a hole in the skull (e.g. after surgery) this has a major effect on

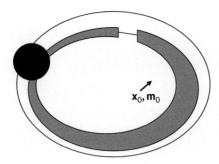

Figure 6.13 An example of a non-nested geometry. Such geometries are required when modeling holes in the skull or modeling the eye balls.

the potential distribution. Such geometry cannot be modeled with nested compartments. Also the shape of the volume conductor at the eye balls cannot be modeled with a nested geometry (see Figure 6.13).

Apart from the use of the multiple deflation technique (Lynn and Timlake, 1968b), there is no theoretical need to assume nested geometries in the foundations of the BEM, and the BEM presented in the previous sections do not rely on this assumption (Barnard et al., 1967a). Kybic et al. (2006) present an implementation of the symmetric BEM for a very general volume oriented geometry.

6.4.6 The fast multipole method for large problems

Any solution of the EEG forward problem based on the BEM and other numerical methods described in further sections requires the inversion of a linear system of equations (6.77). In the BEM the matrix elements can be interpreted as the interaction of boundary elements through singular integral operators, and even in the symmetric formulation of the BEM most of these interactions are non-zero. When sophisticated head models such as those described in Section 6.4.4 are employed and the number of degrees of freedom of the system grows large (over 10^4), solving the linear system by direct LU decomposition becomes practically unfeasible.

Instead, iterative methods (e.g. Golub and Van Loan, 1989), which only access the matrix through matrix-vector products, are used. That is, the solution of the system of equations is based exclusively on computations of the form $\sum A_{mn}u_n^{(k)}$, for appropriate choices of $u_n^{(k)}$. This way of solving the system of equations was also followed in the early days of the BEM (Barnard et al., 1967b; Lynn and Timlake 1968a, 1968b). Recently, iterative methods have been revisited as a useful strategy for solving the discretized BEM equations (Hackbusch and Nowak, 1989). The underlying idea can be explained by the notion that for a singular matrix, of which a decomposition $A = CB^T$ is known, the matrix product $A\mathbf{u}$ can be more efficiently computed as $C(B^T\mathbf{u})$, when the number of columns of B and C are limited. In the most extreme case, when B and C are column vectors, the computational cost reduces from $O(N^2)$ to $O(N)$. The same principle can be applied when A has a blocked structure, and for each block a singular decomposition is known

$$
\begin{pmatrix}
C_{11}B_{11}{}^T & C_{12}B_{12}{}^T & \vdots \\
C_{21}B_{21}{}^T & C_{22}B_{22}{}^T & \vdots \\
\cdots & \cdots &
\end{pmatrix}
\begin{pmatrix}
\mathbf{u}_1 \\
\mathbf{u}_2
\end{pmatrix}.
\tag{6.93}
$$

This idea is exploited in the fast multipole method (FMM) or the panel clustering method of Hackbusch and Nowak (1989). The matrix A itself is not singular but, using a multipole approximation, A can be subdivided into singular partitions (not necessarily contiguous blocks as in (6.93)) and some small non-singular parts. Also the vector $\mathbf{u}$ is split into different parts, corresponding to the splitting of A. The matrix vector product $A\mathbf{u}^{(k)}$ is then computed using the decomposition of the singular partitions. Since this decomposition can be done very efficiently, without computing A itself, this procedure results in a very substantial reduction in overall complexity of calculating $A\mathbf{u}^{(k)}$. Typically, a decrease from $O(N^2)$ to $O(N)$ can be achieved. The two main advantages of the FMM are the reduced memory consumption, and the accelerated computations. There is extensive literature dealing with the FMM (Epton and Dembart, 1995; Beatson and Greengard, 1997; Dembart and Yip, 1998; Rahola, 1998; Cheng et al., 1999) for gravitational or electromagnetic scattering calculations. The FMM has been applied to the electrostatic Maxwell problem and the symmetric BEM, but with only one interface (Of et al., 2002), and to the multi-compartment model of EEG and MEG in (Kybic et al., 2005a).

6.5 The finite element method

The merits of the BEM are that for two decades forward models have been constructed that describe a realistic geometry of homogeneous isotropic compartments and that can be solved with an acceptable amount of memory and computer time. However, computer technology and especially adapted mathematical algorithms have developed rapidly in recent years so that three-dimensional approaches such as the finite element method (FEM), the subject of this section, have become practically feasible.

The finite element method (FEM) is a very well founded mathematical method for the numerical solution of (especially so-called elliptic) partial differential equations as they appear in science and engineering. It is therefore particularly well suited to solving Equation (6.7) for very general models of $\underline{\gamma}(\mathbf{x})$. One of the reasons why the BEM has been used much more than the FEM in MEEG research is that it was often assumed that FEM needs much more computer power. However, as shown recently (Weinstein et al., 2000; Wolters et al., 2002; Gencer and Acar, 2004; Wolters et al., 2004; Hallez et al., 2005; Schimpf, 2007; Vallaghé et al., 2009), the FEM forward problem for EEG and MEG can be reduced to an amount of computational work that increases completely linearly with the number of FE nodes. Therefore, the use of even millions of FE nodes no longer constitutes a real practical problem (Rullmann et al., 2009). Special adaptations of the FEM were developed to make it suitable for point dipoles and point monopoles. Here only a brief description of these strategies is presented (for a thorough description, see Drechsler et al., 2009; Lew et al., 2009b), thereby emphasizing the differences and similarities with the BEM.

The starting point of the FEM is to define a solution of (6.7) with boundary condition (6.10) in terms of a so-called weak formulation. When $h(\mathbf{x})$ is a (piecewise) differentiable function satisfying certain constraints, one can multiply (6.7) with $h(\mathbf{x})$ and integrate over the conductor volume V. Using boundary condition (6.10) one obtains

$$
\begin{aligned}
\iiint_V (\nabla \cdot \mathbf{J}_i)\, h(\mathbf{x}) dx &= \iiint_V \left(\nabla \cdot \left(\underline{\gamma} \cdot \nabla \psi \right) \right) h(\mathbf{x}) dx \\
&= \oiint_{\partial V} h(\mathbf{x}) \left(\underline{\gamma} \cdot \nabla \psi \right) \cdot d\mathbf{S} - \iiint_V \left(\underline{\gamma} \cdot \nabla \psi \right) \cdot \nabla h(\mathbf{x})\, dx \\
&= - \iiint_V \left(\underline{\gamma} \cdot \nabla \psi \right) \cdot \nabla h(\mathbf{x})\, dx.
\end{aligned}
\tag{6.94}
$$

The weak formulation implies that a function $\psi(\mathbf{x})$ is sought that satisfies (6.94), for all functions $h()$ of a certain class. By using the weak formulation, it becomes possible to establish a firm mathematical basis under the numerical approximation of the differential equation as discussed in Section 6.5.1 for the subtraction potential approach. The Galerkin approach described in Section 6.4.1 for the BEM, where a system of equations was obtained by projecting an integral equation on a set of base functions, can also be interpreted as a weak formulation.

The discretization of (6.94) is analogous to what has been used in (6.76) for the BEM. But because in (6.94) the derivative is needed, the base functions need to be at least piecewise linear. Furthermore, base functions are derived from three-dimensional (e.g. tetrahedra or hexahedra) instead of two-dimensional elements (triangles or quadrangles on a surface). When the approximation

$$\psi(\mathbf{x}) \approx \sum_{n=1}^{N} \psi_n h_n(\mathbf{x}) \tag{6.95}$$

is substituted in (6.94) one obtains, taking for $h()$ one of the base functions $h_m()$,

$$-\sum_n \psi_n \iiint_V \left(\underline{\gamma}\nabla h_n(\mathbf{x})\right) \cdot \nabla h_m(\mathbf{x})\, dx = \iiint_V \nabla \cdot \mathbf{J}_i\, h_m(\mathbf{x}) dx, \quad m = 1, \dots, N.$$

$$\tag{6.96}$$

Equation (6.96) can be interpreted as a matrix-vector equation $A\psi = \mathbf{b}$ with matrix elements

$$A_{mn} = -\iiint_V \left(\underline{\gamma}\nabla h_n(\mathbf{x})\right) \cdot \nabla h_m(\mathbf{x}) dx$$

$$b_m = \iiint_V \nabla \cdot \mathbf{J}_i\, h_{\mathrm{m}}(\mathbf{x}) dx. \tag{6.97}$$

Similar to the BEM, the goal is to solve the matrix vector equation and to use the coefficients to compute the potential at any point in space using (6.95). There are a few substantial differences however. First of all, with FEM the complete three-dimensional space is filled with elements, whereas for the BEM there are only elements at the boundaries. Therefore, with FEM the dimension N is much larger (typically 10^5 to 10^6) than with BEM (typically 10^3 to 10^4) and most often iterative solution methods are used that are based on matrix vector products $A\psi^{(k)}$.

The matrix elements A_{mn} for FEM are much simpler to compute than for BEM (Equation (6.78)) and, more important, nearly all of them are zero. Only when the supports of $h_n()$ and $h_m()$ (region where these functions deviate from zero) overlap is the corresponding matrix element A_{nm} non-zero. When computing $A\psi^{(k)}$ in an iterative procedure (see e.g. Wolters et al., 2002; Lew et al., 2009b) all zero elements can be skipped and therefore in the implementation of the FEM one needs a dedicated data structure that keeps track of the non-zero elements. Very efficient multigrid method based iterative techniques can reduce the amount of computational work to $O(N)$, i.e. linear in the number of unknowns (Hackbusch, 1994; Wolters et al., 2002; Lew et al., 2009b). Finally, the matrix A in (6.97) can be shown to be symmetric positive definite (positive definiteness follows from the ellipticity of the underlying bilinear form, see Wolters et al., 2007), which makes even more efficient multigrid solver techniques possible (Hackbusch, 1994; Wolters et al., 2002; Lew et al., 2009b). The BEM symmetry depends on the use of mixed single and double layer formulation. With regard to the tissue conductivities, the FEM allows a much larger flexibility than the BEM since each element can contain its own conductivity tensor.

The magnetic field can be computed using the FEM on the basis of the integral presented in (6.13). As explained in Section 6.2.3, in practical applications one needs the magnetic flux $\psi_M(\mathbf{x})$ through the coil. When forward modeling is based on the FEM, it can be shown (e.g. Wolters et al., 2004) that the flux at MEG sensor i, $\psi_M(\mathbf{x}_i)$, can be computed as a sum of the flux of the primary dipolar source and the flux of the secondary sources

$$\psi_M^{dip}(\mathbf{x}_i) = \Psi_{prim}(\mathbf{x}_i) + \Psi_{sec}(\mathbf{x}_i) = \oint_{\Gamma_i} \mathbf{A}_{\infty}^{dip}(\mathbf{x}) \cdot d\boldsymbol{\gamma} - \iiint_{Head} \left(\underline{\boldsymbol{\gamma}} \nabla' \psi(\mathbf{x}') \right) \mathbf{c}_i(\mathbf{x}') d\mathbf{x}'$$

(6.98)

where

$$\mathbf{c}_i(\mathbf{x}') \equiv \frac{\mu_0}{4\pi} \oint_{\Gamma_i} \frac{1}{|\mathbf{x} - \mathbf{x}'|} d\boldsymbol{\gamma}.$$

(6.99)

Here Γ_i denotes the loop of the ith MEG sensor. The first term of Equation (6.98) can be computed in practice using Equations (6.33) and (6.34). The second term requires the integration of the function $\mathbf{c}_i(\mathbf{x}')$ over the whole head. This integration can be computed efficiently in discretized form, by means of the matrix-vector product

$$\boldsymbol{\psi}_{M,sec} = S\boldsymbol{\psi}$$

(6.100)

where S is a matrix with matrix elements

$$S_{in} = - \iiint_{Head} \left(\underline{\boldsymbol{\gamma}} \nabla' h_n(\mathbf{x}') \right) \mathbf{c}_i(\mathbf{x}') d\mathbf{x}', \quad i = 1, \ldots, I_{MEG}; n = 1, \ldots, N.$$

(6.101)

The secondary flux matrix S is only dependent on the volume conductor (geometry and conductivity) and the MEG sensors. When it is assumed that the MEG sensors are fixed with respect to the head the matrix S needs to be computed only once. Then it can be used for fast FE-based MEG forward solutions within the inverse problem as shown in Section 6.5.3. Modern MEG devices are supplied with accurate head motion correction systems (Uutula et al., 2001, Taulu et al., 2005, Taulu and Simola, 2006; Wehner et al., 2008) and therefore the assumption of fixed MEG sensors does not pose severe constraints.

6.5.1 The dipole singularity

One of the numerical difficulties of applying the FEM to MEG and EEG forward modeling is related to the use of mathematical point dipoles, which give a singularity in the right hand side of (6.96). When using partial integration and when the dipole model (6.15) is afterwards substituted in the expression for the right hand side, b_m (Equation (6.97)), one obtains

$$b_m = -\mathbf{m}_0 \cdot \nabla_0 \iiint\limits_V \delta\,(\mathbf{x} - \mathbf{x}_0)\,h_m(\mathbf{x})d\mathbf{x}$$

$$= \begin{cases} 0 & \text{if } \mathbf{x}_0 \notin \text{ support } h_m(\mathbf{x}) \\ k_{m0} & \text{if } \mathbf{x}_0 \in \text{ support } h_m(\mathbf{x}) \end{cases} \tag{6.102}$$

where k_{m0} is a constant depending on the dipole orientation and strength and on the gradient of the basis functions of the element that contains the source. This *partial integration direct potential approach* was first presented in Yan et al. (1991) and was used and/or compared to other dipole modeling approaches (Awada et al., 1997; Schimpf et al., 2002; Lew et al., 2009b; Vallaghé and Papadopoulo, 2010). One can observe from (6.102) that, in the case of linear interpolation functions $h_m(\mathbf{x})$, one obtains a right hand side vector b_m that depends on the dipole position in a non-smooth way. As long as the dipole position stays within the support of a certain element, the right hand side is independent of $\mathbf{x}_0$, and when it crosses an element border, b_m will jump from one vector to another. On a first view, this characteristic of the forward solution seems to make Equation (6.102) unattractive in inverse modeling applications. However, it appears that the implementation of (6.102) in combination with linear basis functions leads to quite accurate solutions as long as the source is close to the barycenter of an element and/or the mesh resolution in the source area is sufficiently high (Lew et al., 2009b; Vallaghé and Papadopoulo, 2010). The approach only leads to larger errors when the dipole position $\mathbf{x}_0$ is close to the boundary of an element and the mesh resolution in the source area is not sufficiently high (Awada et al., 1997 (two-dimensional case); Schimpf et al., 2002; Lew et al., 2009b). A possible approach to increase the accuracy and practicability of the partial integration potential approach with regard to the inverse problem might thus be to precompute a leadfield for dipole sources in the barycenters of the elements and use leadfield interpolation strategies for all other dipole sources (De Munck et al., 2001; Yvert et al., 2001) or to use higher order basis functions. With regard to computational complexity, as will be presented in Section 6.5.2, the partial integration direct potential approach is particularly wellsuited because nearly all entries on the right hand side of (6.102) are zero. Therefore, an advantage of this approach is that highest FE resolutions can easily be realized in practical inverse source analysis scenarios.

Another approach to solving the singularity problem is to use a smoother model for the primary sources which, from a physiological perspective, might even be more realistic than the possibly too focal point-like mathematical dipole (Murakami and Okada, 2006). The FE method is also flexible with regard to such considerations and first approaches such as the representation of the primary sources through Whitney elements (Tanzer et al., 2005), Raviart–Thomas elements (Pursiainen, 2008; Pursiainen et al., 2012) or through a blurred dipole within the *Venant direct potential approach* (Buchner et al., 1997) have been presented. As the comparison of the Venant FE approach with the partial integration FE approach and the subtraction FE approach (see below) with analytical solutions for mathematical dipoles in multilayer anisotropic sphere models showed, high accuracies can be achieved despite the slight differences in source modeling (Lanfer, 2007; Lanfer et al., 2007; Lew et al., 2009b). The Venant direct potential approach is well tested overall and has already been successfully applied in several inverse source analysis scenarios (Buchner et al., 1997; Waberski et al., 1998; Wolters, 2003; Fuchs et al., 2007; Rullmann et al., 2009). One of the difficulties with dipole smoothing is that it introduces several parameters, such as the level of smoothing, and the optimal tuning of those parameters in combination with chosen FE mesh resolution towards a realistic physiological situation is a difficult task. As a last remark, similar to the partial integration direct potential approach, the right hand side vector for the Venant approach has only some few non-zero entries so that, when using the newest methodological developments (Gencer et al., 2004; Wolters et al., 2004) it is especially attractive with regard to computational complexity and therefore allows highest FE resolutions in practical FE source analysis scenarios (Rullmann et al., 2009).

The last solution method discussed here is to eliminate the infinities associated with delta functions right from the start and formulate the FEM problem in terms of a correction to the infinite medium potential. The gross strategy of this so-called *subtraction potential approach* is comparable to what has been used in Equation (6.63) for the analytical solution of the multi-sphere model, and also to the isolated problem approach mentioned with the BEM.

With the subtraction approach of the FEM the unknown potential $\psi(\mathbf{x})$ is split into an easily computable part $\psi_\infty(\mathbf{x})$ and a correction term $\psi_{corr}(\mathbf{x})$,

$$\psi(\mathbf{x}) = \psi_\infty(\mathbf{x}) + \psi_{corr}(\mathbf{x}). \tag{6.103}$$

Substituting (6.103) in (6.7) yields the following boundary value problem for $\psi_{corr}(\mathbf{x})$

$$\begin{cases} \nabla \cdot \left(\underline{\gamma} \nabla \psi_{\text{corr}} \right) & = \nabla \cdot \mathbf{J}_i - \nabla \cdot \left(\underline{\gamma} \nabla \psi_\infty \right) & \mathbf{x} \in V \\ \mathbf{n}(\mathbf{x}) \cdot \underline{\gamma} \nabla \psi_{\text{corr}}(\mathbf{x}) & = -\mathbf{n}(\mathbf{x}) \cdot \underline{\gamma} \nabla \psi_\infty(\mathbf{x}) & \mathbf{x} \in \partial V. \end{cases} \tag{6.104}$$

When the current source is a dipole located at $\mathbf{x}_0$ and the conductivity tensor used to compute $\psi_\infty(\mathbf{x})$ equals $\underline{\gamma}_\infty = \underline{\gamma}(\mathbf{x}_0)$ in a non-empty domain around $\mathbf{x}_0$ (the so-called *homogeneity condition* (see Wolters et al., 2007a; Drechsler et al., 2009; Lew et al., 2009b) the dipole singularity in (6.104) disappears and one obtains

$$\begin{cases} \nabla \cdot \left(\underline{\gamma} \nabla \psi_{corr} \right) = \nabla \cdot \left((\underline{\gamma}_\infty - \underline{\gamma}) \nabla \psi_\infty \right) & \mathbf{x} \in V \\ \mathbf{n}(\mathbf{x}) \cdot \underline{\gamma} \nabla \psi_{corr}(\mathbf{x}) = -\mathbf{n}(\mathbf{x}) \cdot \underline{\gamma} \nabla \psi_\infty(\mathbf{x}) & \mathbf{x} \in \partial V. \end{cases} \tag{6.105}$$

One can interpret ψ_{corr} in Equation (6.105) as the potential distribution caused by a current source density $\nabla \cdot \left(\left(\underline{\gamma}_\infty - \underline{\gamma} \right) \nabla \psi_\infty \right)$, in a conductor described by $\underline{\gamma}(\mathbf{x})$ when on the boundary a current $-\mathbf{n}(\mathbf{x}) \cdot \underline{\gamma} \nabla \psi_\infty(\mathbf{x})$ is injected.

It has been proved (Wolters et al., 2007a) that (6.105) fulfils the compatibility condition given in Equation (6.22). Furthermore, because of the homogeneity condition, $\underline{\gamma}_\infty - \underline{\gamma}$ is precisely zero where $\nabla \psi_\infty$ is infinite. Therefore, the singularity on the right hand side of Equation (6.7) is eliminated and equation (6.105) has appropriately smooth right hand sides, so that existence and uniqueness of the solution can be proved for the correction potential (Wolters et al., 2007a). Then, the numerical analysis also has a firm mathematical basis and, under certain conditions on the material coefficients, one can even prove that the FEM discretizations of (6.105) converge to the true solution with certain quantitative orders and one can give a reason why numerical errors increase with increasing source eccentricity (Wolters et al., 2007a; Drechsler et al., 2009).

The underlying idea beneath the elimination of the singularity for the FEM is comparable to that discussed in Section 6.3.3, Equation (6.63), where the goal was to speed up the summation of spherical harmonics. In both cases an infinite medium potential is added and subtracted. The difference is that in the case of the FEM the conductivity of $\psi_\infty(\mathbf{x})$ corresponds precisely to the conductivity at the position of the dipole ($\underline{\gamma}_\infty$). In the analytical solution of the multi-sphere model, the optimal conductivity is a combination of all conductivities of the shells between source and electrode (De Munck and Peters, 1993).

The fact that the conduction profile is the same as in the original problem, and that only the source term and boundary condition are different, results in a discretization of (6.105) with the same sparse system matrix A as defined in (6.97), but a modified right hand side,

$$b_m = - \iiint_V \left(\left(\underline{\gamma}_\infty - \underline{\gamma}(\mathbf{x}) \right) \nabla \psi_\infty(\mathbf{x}) \right) \cdot \nabla h_m(\mathbf{x}) d\mathbf{x}$$

$$+ \oiint_{\partial V} \left((\underline{\gamma}_\infty \nabla \psi_\infty(\mathbf{x})) \cdot \mathbf{n}(\mathbf{x}) \right) h_m(\mathbf{x}) dS. \tag{6.106}$$

Because of the homogeneity condition the singularity in b_m (first term) is eliminated. The price one has to pay here is that the vector b_m is now everywhere filled with non-zero entries that are expensive to compute and which additionally makes the transfer matrix approach presented in Section 6.5.3 more expensive. Solution strategies to these problems that are currently under investigation are indicated in Section 6.5.3.

Two methods are described in the literature (Wolters et al., 2007a; Drechsler et al., 2009) to implement Equation (6.106). In one method, called *projected subtraction* (Wolters et al., 2007a), $\psi_\infty(\mathbf{x})$ is first approximated in terms of the base functions $h_n(\mathbf{x})$ using (6.95) and (6.106) is expressed as an integral over the gradients of the base functions. The advantage of this method is that it is fast and flexible, but less accurate than the *full subtraction* method (Drechsler et al., 2009), where the integrals are computed numerically.

6.5.2 Numerical comparison of FEM variants

Similar to Section 6.4.4 the differences in the numerical accuracies of different FEM variants are shown for an illustrative simulation study. For the analytical model a concentric four-sphere model was used (relative radii: 1.0, 0.93, 0.87 and 0.85), where the second shell (skull) had an anisotropy ratio of 10 following Marin et al. (1998) and, following the measurements for the cerebrospinal fluid (CSF) of Baumann et al. (1997), a factor 5.4 times higher conductivity than for the brain was used for the third (CSF) shell. For the numerical computations the spheres are discretized using either 161 086 (left side) or 360 056 tetrahedra (right side) in total, see Figure 6.14. The four FEM variants discussed above (partial integration, Venant dipole, projected subtraction and full subtraction) are compared in Figure 6.15, using the same error measures as those that were used for the BEM. Test electrodes were distributed regularly over 748 points at the outer sphere and data were converted to average reference before computing error measures. The orientation of the dipole was tangential (errors are very similar for radial sources and are therefore not shown here).

Figure 6.15 shows that for both meshes A and B the errors for the direct approaches are oscillating, in particular for mesh B. This is because in the direct methods the right hand side vector $\mathbf{b}$ (see Equation (6.102)) jumps when the dipole source moves from one element (for partial integration) or one node (for Venant) to the other and mesh B contains very large elements in the middle of the sphere. Mesh B is optimized for the subtraction approach because there the FEM is only used to compute corrections onto the infinite medium potential, and because these corrections dominate at higher eccentricities relatively few elements are needed at lower eccentricities. As a result, in terms of accuracy, the subtraction approaches are able to take more advantage of the increased number of elements in mesh B than the direct approaches. Furthermore, it appears that the full subtraction approach is an order of magnitude more accurate than the projected subtraction approach, for

A   B

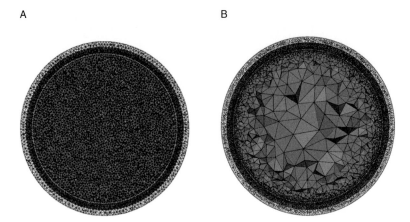

Figure 6.14 (See plate section for color version.) The finite element method is based on a three-dimensional discretization of space. In the examples presented here constrained Delaunay tetrahedralizations are used to describe the concentric sphere model. In A the discretization is homogeneous and consists of 161 086 nodes. Example B consists of 360 056 nodes, most of which are put at larger eccentricities in order to minimize the error of the FEM-computed correction potential within the subtraction approaches. While example A is optimized for the needs of the direct FEM approaches, example B is optimized for the FEM subtraction approaches, because in those methods the largest part of the potential variations in the inner compartments is explained by the infinite medium potential and the largest corrections are located at higher eccentricities. (Reprinted from Drechsler et al. (2009), with permission from Elsevier.)

dipole sources close to the conductivity discontinuity as also shown in Drechsler et al. (2009).

From more extensive validation studies it appears that, when compared to the partial integration and Venant direct methods, the subtraction approach achieved superior results over all tested source eccentricities, if the homogeneity condition was sufficiently fulfilled, i.e. as long as there was a sufficient number of elements with conductivity $\underline{\gamma}_\infty$ between the source and the next conductivity discontinuity (Wolters et al., 2007a; Drechsler et al., 2009; Lew et al., 2009b). Both direct approaches were found to be less sensitive with regard to the homogeneity condition, a sufficiently high mesh resolution in the source area is however also essential for those approaches. As also shown by Lew et al. (2009b), with optimized FE meshes for the needs of the direct potential approaches and for dipole sources up to 1 mm below the CSF compartment, topography and magnitude errors were also only at about 1% as long as all elements between the load FE nodes were brain elements (this is just one element for the partial integration and up to about 30 elements for the Venant approach). Therefore, in summary, in a volume conductor where CSF and brain anisotropy are modeled, the accurate and very fast direct methods are currently still the methods of choice (see, e.g. Rullmann et al., 2009),

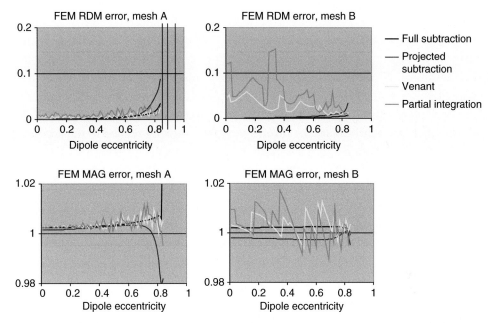

Figure 6.15 A comparison of the numerical accuracy of the four different FEM variants: partial integration, Venant dipole, projected subtraction and full subtraction. In the left two graphs the mesh depicted in Figure 6.14 A is used, the right two graphs are based on mesh B of Figure 6.14. The vertical black lines in the upper plot represent the locations of the compartment boundaries.

whereas in the multi-compartment (skin, skull spongiosa and compacta, CSF, brain) head model with homogeneous and isotropic brain tissue modeling, the full subtraction approach, if sufficiently speeded up (see Section 6.5.3), might be an even better choice. Low errors (localization errors below 1.7 mm, orientation errors far below 1° and magnitude errors far below 1% for sources up to 2 mm below the CSF compartment in the FE model A from Figure 6.14 with about 161k nodes) were also found in Venant FE method based dipole fit scenarios (using Nelder–Mead simplex optimization in continuous parameter space) in a four-compartment anisotropic sphere model where the reference data were computed using the quasi-analytical methods (Wolters, 2008). It therefore currently seems that the slight RDM and MAG error curve oscillations of the Venant approach are of less practical importance even if not specifically treated by means of higher FE basis functions or leadfield interpolation techniques (Vorwerk, 2011).

6.5.3 The use of FEM in inverse models

The inverse problem of EEG and MEG is to localize the dipole sources underlying observed potential differences and recorded magnetic inductions. Many different models and algorithms have been presented in the literature (see chapter 7), all with different assumptions and applicability. What they have in common is that they are all based on solutions of the forward problem as presented in this chapter. Furthermore, all inverse methods are based on a comparison, by means of a cost or gain function, of the data with predicted fields caused by a certain configuration of dipole sources. Since for almost any inverse problem multiple model predictions are required, the performance of inverse models is directly related to the accuracy and speed with which the forward problem can be solved.

A simple strategy to deal efficiently with multiple inverse algorithms applied to a single subject is to precompute a table containing the potentials/magnetic fields at the sensors, for a dense grid of possible dipole positions in the head (De Munck et al., 2001; Ermer et al., 2001; Yvert et al., 2001). Because dipole moments act linearly on the required fields, one has to store three fields for three orthogonal unit dipoles at each point of the grid. Therefore, when the number of sensors is I and the number of grid points is K, the size of the table is $3IK$ floats or doubles.

For the BEM or analytical methods the precomputation of the table works well, but for the FEM, without a further reduction of the computational work, it would be too expensive in practice. The problem is that for an accurate computation of a single FE forward solution using optimal multigrid techniques of $O(N)$ complexity (Hackbusch, 1994; Wolters et al., 2002; Lew et al., 2009b), a computation time in the seconds range, say 20 s, is often needed, dependent surely on mesh-size, algorithm and hardware. Therefore, if $K = 30\,000$ (which corresponds to a mesh size of about 3 mm) the computation of the table would take more than 160 h. However, it appears that there is an easy solution for the problem of computation which not only works efficiently for the precomputation of such a table, but works even more generally for all inverse source analysis scenarios (Weinstein et al., 2000; Gencer and Acar, 2004; Wolters et al., 2004; Pursiainen, 2008; Pursiainen et al., 2012). The main idea is to precompute transfer matrices

$$\begin{cases} T_{EEG} = RA^{-1} & \text{for EEG} \\ T_{MEG} = SA^{-1} & \text{for MEG} \end{cases} \tag{6.107}$$

with R being a $(I_{EEG}-1) \times N$ restriction matrix that maps the full three-dimensional FE potential vector onto the $I_{EEG} - 1$ non-reference electrodes, such that $\psi_{isense} = \sum R_{isens,\,inode} \psi_{inode}$. Similarly, S is the $I_{MEG} \times N$ secondary flux matrix from (6.101). The transfer matrices T_{EEG} and T_{MEG} can be efficiently computed using the symmetry and the sparseness of the system matrix A by means of solving $I_{EEG} - 1$ or I_{MEG} large sparse linear equation systems for EEG or MEG, respectively. In the above example, this would take I_{MEG} times 20 s for the precomputation of T_{MEG}. In this way we have obtained direct FE forward operators, i.e.

$$\begin{cases} T_{EEG}\mathbf{b} = RA^{-1}\mathbf{b} = R\mathbf{\psi} = \mathbf{\psi}_{isens} & \text{for EEG} \\ T_{MEG}\mathbf{b} = SA^{-1}\mathbf{b} = S\mathbf{\psi} = \mathbf{\psi}_{M,sec} & \text{for MEG.} \end{cases} \quad (6.108)$$

Note that the magnetic flux of the primary source, $\mathbf{\psi}_{M,prim}$, still has to be added for a complete MEG forward solution and that for the subtraction approach the singularity potential should be accounted for (Wolters et al., 2004; Drechsler et al., 2009). For each source, we thus only have to compute the FE right hand side vector $\mathbf{b}$ and to multiply it to the precomputed transfer matrices. The fewer non-zero entries $\mathbf{b}$ has, the faster the FE forward solution will be. For the direct potential approaches, only a few multiplications have to be performed per sensor, independent of the FE resolution, so that FE forward solutions in the millisecond range are possible for FE models with even millions of unknowns offering highest accuracies for very moderate computation times (Rullmann et al., 2009). Figure 6.16 shows the dramatic decrease in computation time for the Venant FE approach. For the subtraction approach, $\mathbf{b}$ is expensive to compute and non-sparse, so that its multiplication to the transfer matrices consists of number of sensors many nearly full scalar products of dimension N. However, the process is easily parallelizable and, for the precomputation of leadfield tables, new numerical techniques are currently under investigation to store many right hand side vectors in data sparse so-called *H-matrix* format which can then be easily multiplied with $O(N \log N)$

Numerical/modeling accuracy versus computation time

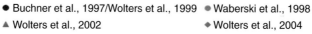

● Buchner et al., 1997/Wolters et al., 1999   ● Waberski et al., 1998
▲ Wolters et al., 2002   ◆ Wolters et al., 2004
■ Rullmann et al., 2009

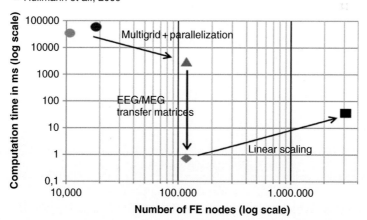

Figure 6.16 FE developments from 1997 to 2009. The figure illustrates the dramatical decrease in computation time for a single dipole FE forward computation (Venant approach) which enabled a huge increase in the practically feasible "number of FE nodes" from some thousand to millions of nodes.

complexity to the precomputed transfer matrices (Wolters et al., 2004; Drechsler et al., 2009).

6.6 Other forward methods

The problem of solving realistic volume conductor models has, apart from with the BEM and the FEM, also been addressed with other numerical techniques, like the finite difference method (FDM) (e.g. Saleheen and Kwong, 1997; Mohr and Vanrumste, 2003; Mohr, 2004; Hallez et al., 2005) and the finite volume method (FVM) (e.g. Pruis et al., 1993; Rosenfeld et al., 1996; Mohr, 2004; Cook and Koles, 2006). Similar to the FEM, the FDM and the FVM are methods where a sparse set of equations is derived, where the unknowns refer to the potentials at the nodes of a three-dimensional mesh. With the FDM, these equations are obtained by approximating derivatives as $f''(x) \approx \frac{1}{h^2}(f(x-h) - 2f(x) - f(x+h))$. When this is done in three-dimensional the potential at each node is only related to its neighbors on the mesh and therefore the system of equations is sparse (Saleheen and Kwong, 1997). Because the FDM is based on the approximation of partial derivatives along the Cartesian coordinates, the structure of the three-dimensional mesh is necessarily much more regular than for example the meshes one can choose with the FEM or FVM.

Besides the above mentioned numerical methods, perturbation methods have also been explored (Nolte and Curio, 1999; Nolte et al., 2001). The starting point of Nolte's analysis is the double layer integral of the BEM, Equation (6.74). Exploiting the fact that the shape of the head only deviates little from a sphere, one can accurately express the shape of the head as a linear combination of spherical harmonics (Van 't Ent et al., 2001) using only a very few coefficients. When the same expansion is used with unknown coefficients for the potential distribution a very small linear system of equations is obtained that can be solved numerically. The speed and accuracy obtained with this method compare well with the double layer BEM, but the price one has to pay is the quite involved expressions for the matrix elements.

6.7 Discussion and conclusion

After discussing so many details of possible approaches for the forward problem, do we now know which one is the method of choice? Because of the wide availability of MR scans and software tools to derive realistic models from them, the role of the analytically solvable models is mainly as a golden standard when testing numerical methods. BEM approaches implicitly assume that the conductivity distribution is piecewise constant. In principle this limitation could be overcome by taking smaller and smaller compartments, but this approach is hampered by an enormous (non-linear) increase in computational cost. Although the fast multipole technique can be invoked to speed up BEM computations with complex geometries (Kybic et al., 2005a), when there is too much spatial variation in conductivity or when the anisotropy of the conductivity becomes important, the only option is to use three-dimensional approaches like the FEM. The question therefore is, what is known about in vivo conductivity and what do simulation studies teach us of their importance in forward modeling?

The skull itself is inhomogeneous and consists of low-conducting compacta and much better conducting spongiosa (Akhtari et al., 2002). Furthermore, cerebrospinal fluid (CSF) has a much higher conductivity than brain tissues and its value does not seem to have an intersubject variability (Baumann et al., 1997). Finally, brain tissues are strongly anisotropic (Tuch et al., 2001).

Recent publications show that, depending on parameters such as, for example, source location and orientation and sensor arrangement, the sensitivity of source analysis to skull inhomogeneity caused by the spongiosa and compacta (Ollikainen et al., 1999; Sadleir and Argibay, 2007; Dannhauer et al., 2011), to the highly-conducting CSF (Ramon et al., 2004; Wolters et al., 2006; Wendel et al., 2008) and to brain anisotropy (Wolters et al., 2001, 2006; Haueisen et al., 2002; Hallez et al., 2005; Güllmar et al., 2006, 2010; Bangera et al., 2010) might be significant so that the use of three-compartment models might result in systematic errors in the inverse analysis. For the human skull, a global sensitivity analysis showed that for three- and four-layer conductor models, the EEG is most sensitive to the radial skull-to-scalp conductivity ratio (Vallaghé and Clerc, 2009). This is in agreement with the results of Dannhauer et al. (2011). However, as shown by Ollikainen et al. (1999), Sadleir and Argibay (2007) and Dannhauer et al. (2011), the concept of skull inhomogeneity (especially the different radial conductivities at different skull positions caused by different local amounts of spongiosa and compacta) might be much more important than the concept of skull anisotropy. Based on these results, Dannhauer et al. (2011) recommend an explicit modeling of the skull's three-layeredness.

Adequate strategies to measure, register and segment multimodal T1-, T2- and diffusion-tensor- (DT-) weighted magnetic resonance images (MRI) are available (see, e.g. Wolters et al., 2006; Rullmann et al., 2009; Olesch et al., 2010; Dannhauer et al., 2011) so that the setup of multi-compartment inhomogeneous and anisotropic volume conductor models is becoming feasible. However, an inhomogeneous, anisotropic volume conductor model will be in need of the individually varying conductivity parameters for skin, skull compacta, skull spongiosa as well as the linear scaling parameter for the brain anisotropy following the models of Tuch et al. (2001) and Wang et al. (2008), i.e. altogether four parameters instead of only three for the three-compartment isotropic approaches.

Considering that, besides the mainly invariant CSF conductivity at body temperature (Baumann et al., 1997), substantial inter-individual differences in most of the remaining conductivity parameters can be expected, it is an important question how much these parameters can be measured or estimated from the individual subject under investigation. One possibility is to derive tailor made models by injecting small artificial currents into the head during EEG setup and analyzing the resulting voltages (e.g. Gonçalves et al., 2003a, 2003b). Tailored conductivity parameters can then be achieved by fitting conductivity parameters to these voltage responses using the same conductivity model as used for MEEG forward modeling. Another possibility consists in measuring (additionally to the MEEG datasets of interest) a block of combined somatosensory evoked potential (SEP) and field (SEF) data and then using the SEP/SEF data for fitting not only the source parameters but additionally some few conductivity parameters for a setup of an individually (not only with regard to the geometry, but also with regard to the conductivity) calibrated head model, which can then be used in the later analysis of the data of interest (Huang et al., 2007; Vallaghé et al., 2007; Lew et al., 2009a; Wolters et al., 2010).

Thus, in summary, how does the two-dimensional BEM approach compare to the three-dimensional FDM, FVM and FEM approaches? The answer to this question depends partly on the computational aspects. For a piecewise constant and isotropic model the question is open. Besides some first comparison studies of BEM and FEM (Clerc et al., 2002; Lanfer et al., 2007; Wolters, 2008; Vorwerk, 2011), no detailed memory/speed/accuracy comparisons using state-of-the-art approaches have yet been made. Moreover, the best choice also depends on the application: is the forward model used in an inverse modeling setting, how many forward calculations are required, etc.? The answers to these questions determine how much computational work should be done in advance to obtain the optimal BEM or FEM variant. Furthermore, if model setup computations like the LU decomposition of the system matrix in the case of the BEM, are part of the comparison, then one should also compare non-numerical preparation time such as the effort it takes to obtain a triangular mesh for the BEM (e.g. Wagner et al., 1995), a structured voxel grid for the FVM (Hallez, 2008) or the FEM (Rullmann et al., 2009; Vallaghé and Papadopoulo, 2010), or a geometry-adapted hexahedral (Wolters et al., 2007b), an ordinary (e.g. Wagner et al., 1995) or constrained Delaunay tetrahedral (Drechsler et al., 2009; Lew et al., 2009b) model for the FEM. Finally, because of the complexity of the problem, most researchers are expert in either the BEM or the FEM, which is another obstacle in a complete comparison of the optimal BEM with the optimal FEM under given testing conditions.

In order to combine the modeling advantages of both methods, it is possible to couple the BEM with the FEM, using domain decomposition techniques: some of the domains are assigned a BEM solver, others a FEM solver, and communication between domains occurs through the boundary conditions (Olivi et al., 2010). The BEM-FEM coupling is more versatile, than each of the solvers separately, but it comes with the price of a lower computation speed and it is difficult to decide which head tissue compartment should then later be modeled with the BEM.

The comparison of different forward problem solution methods also depends on the ease with which one can derive an accurate tailored mesh from given anatomical MRI scans. For the BEM a reasonably accurate mesh can be obtained by fitting a parameterized model to the gross delineation of the individual's head geometry (Van 't Ent et al., 2001). An example of such a mesh is shown in the left hand panel of Figure 6.17. An FE tetrahedralization of the three-dimensional domain has originally been considered a much harder problem than the two-dimensional surface triangularization for the BEM. However, nowadays, accurate and efficient constrained delaunay tetrahedralization (CDT) meshing approaches can be used for FE based source analysis (Drechsler et al., 2009; Lew et al., 2009b). Sometimes, when very accurate BEM geometries are required, two-dimensional BEM surface triangularizations are derived from prior three-dimensional volume tetrahedralizations (Wagner et al., 2000). More importantly, the FE approach can also be based on regular hexahedra elements (Rullmann et al., 2009), geometry-adapted hexahedra elements (Wolters et al., 2007b) or implicit meshes using level sets (Vallaghé and Papadopoulo, 2010). Given an appropriate voxel segmentation of the head tissues, such FE approaches are then easy to realize as long as they can use the resolution of the underlying voxel segmentation directly. An example of a FEM mesh is given in the right hand panel of Figure 6.17.

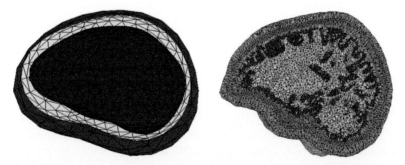

Figure 6.17 (See plate section for color version.) On the left a typical BEM mesh is shown. It consists of three surfaces representing the outer boundaries of the brain, skull, and skin compartment. In this example, the brain surface consists of about 700 nodes, the skull of about 400 nodes, and the skin of about 200 nodes. The right shows a sagittal cut through a five-compartment tetrahedral finite element volume conductor model of the human head (Reprinted from Wolters et al., 2006, with permission from Elsevier.)

Finally, the comparison of forward problem solvers depends on the availability of reliable conductivity parameters and the ease with which these parameters can be obtained from in vivo measurements. The practical and computational aspects of such conductivity estimation protocols are just being investigated and no comparisons have been presented yet (Wolters and De Munck, 2007).

References

Akhtari, M., Bryant, H. C., Mamelak, A. N., Flynn, E. R., Heller, L., Shih, J. J. et al. (2002). Conductivities of three-layer live human skull. *Brain Topogr.*, **14**(3), 151–167.

Awada, K. A., Jackson, S. E., Williams, J. T., Wilton, D. R., Baumann, S. B. and Papanicolaou, A. C. (1997). Computational aspects of finite element modelling in EEG source localization. *IEEE Trans. Biomed. Eng.*, **44** (8), 736–752.

Bangera, N. B., Schomer, D. L., Dehghani, N., Ulbert, I., Cash, S., Papavasiliou, S., Eisenberg, S. R., Dale, A. M. and Halgren, E. (2010). Experimental validation of the influence of white matter anisotropy on the intracranial EEG forward solution. *J. Comput. Neurosci.*, **29**, 371–387.

Barnard, A. C. L., Duck, J. M. and Lynn M. S. (1967a). The application of electromagnetic theory to electrocardiology. I. Derivation of the integral equations. *Biophys. J.*, **7**, 443–462.

Barnard, A. C. L., Duck, J. M. and Lynn M. S. and Timlake, W. P. (1967b). The application of electromagnetic theory to electrocardiology. II. Numerical solution of the integral equations. *Biophys. J.*, **7**, 463–491.

Baumann, S. B., Wozny, D. R., Kelly, S. K. and Meno, F. M. (1997). The electrical conductivity of human cerebrospinal fluid at body temperature. *IEEE Trans. Biomed. Eng.*, **44** (3), 220–223.

Beatson, R. K. and Greengard, L. (1997). A short course on fast multipole methods. In: M. Ainsworth, J. Levesley, W. Light and M. Marletta (editors), *Wavelets, Multilevel Methods and Elliptic PDEs*, Oxford: Oxford University Press, pp. 1–37.

Bénar, C. G. and Gotman, J. (2002). Modeling of post-surgical brain and skull defects in the EEG inverse problem with the boundary element method. *Clin. Neurophysiol.*, **113**, 48–56.

Berg, P. and Scherg, M. (1994). A fast method for forward computation of multiple-shell spherical head models. *Electroenceph. Clin. Neurophysiol.*, **90** (1), 58–64.

Braess, D. (2007). *Finite Elements: Theory, Fast Solvers and Applications in Solid Mechanics*. Cambridge: Cambridge University Press.

Buchner, H., Knoll, G., Fuchs, M., Rienäcker, A., Beckmann, R., Wagner, M., Silny, J and Pesch, J. (1997). Inverse localization of electric dipole current sources in finite element models of the human head. *Electroenceph. Clin. Neurophysiol.*, **102**, 267–278.

Cheng, H., Greengard, L. and Rokhlin, V. (1999). A fast adaptive multipole algorithm in three dimensions. *J. Comput. Phys.*, **155**, 468–498.

Clerc, M., Dervieux, A., Keriven, R., Faugeras, O., Kybic, J. and Papadopoulo, T. (2002) Comparison of BEM and FEM Methods for the E/MEG problem. *Proceedings of the 13th Int.Conf. on Biomagnetism, Jena, Germany, August 10–14*, pp. 688–690.

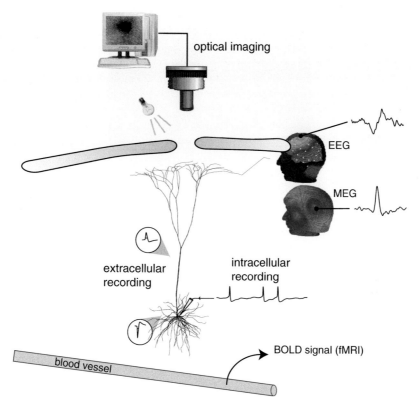

Plate 1.2 Correlates of neural activity and their measurement. A pyramidal cortical cell is displayed in the middle. Its membrane potential can be recorded with an intracellular electrode (*intracellular recording*, Chapter 3). Current flowing through the neuronal membrane creates extracellular potentials, which can be measured with an extracellular electrode (*extracellular recording*, Chapters 4 and 5). These potentials can also be measured with electrodes on the scalp (*EEG*, Chapters 6 and 7). Similarly, neural activity produces magnetic fields, measured with *MEG* (Chapters 6 and 7). Membrane potential can also be seen with a camera after opening the scalp and applying voltage-sensitive dyes onto the surface of the cortex (voltage-sensitive dye imaging, Chapter 9, an *optical imaging* technique). Calcium enters the cell when it spikes, which can also be recorded optically with a different technique (calcium imaging, Chapter 10). More indirectly, neural activity impacts metabolism, in particular the blood vessels, which produces signals that can be recorded with intrinsic signal optical imaging (Chapter 8, also an optical imaging method) and functional magnetic resonance imaging (*fMRI*, Chapter 11).

Plate 2.12 Epicortical grid arrays. Electrode sites (silver dots) are embedded into silicone rubber. The grid is placed on the surface of the brain. Dark lines are sulci of the brain.

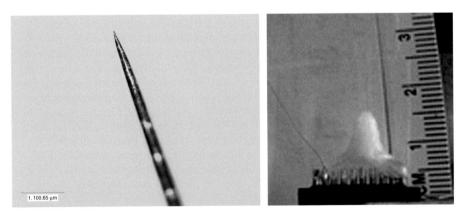

Plate 2.13 Photograph of a tungsten microelectrode (left) and a wire electrode array from NB Labs (www.nblabslarry.com) (right).

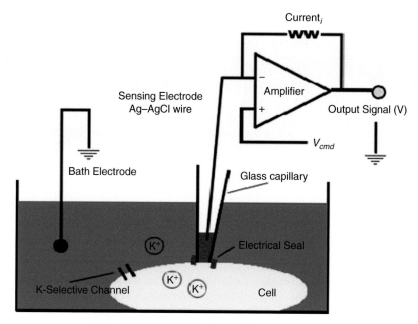

Plate 2.17 Principle of a patch clamp setup.

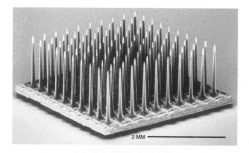 

Plate 2.18 Typical designs of micromachined electrode arrays: Utah electrode array-UEA (left) (Nordhausen et al., 1996, © Elsevier) and Michigan array (right) (Wise et al., 2004, © 2004 IEEE).

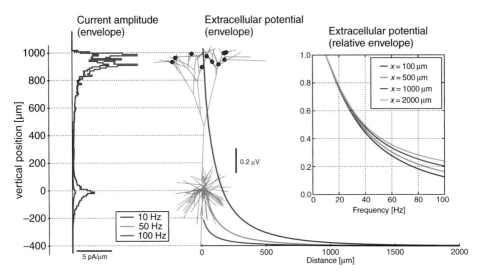

Plate 4.3 Illustration of frequency filtering of LFP for the passive layer-5 pyramidal model neuron in Fig. 4.2A receiving simultaneous sinusoidal input currents
$I_s(t) = I_0 \cos(2\pi f t)$ at 10 apical synapses (red dots in middle panel). The middle panel shows the envelope (amplitude) of the sinusoidally varying extracellular
potential plotted at different lateral positions at the level of the soma (x-direction).
The left panel shows the envelope of the linear current source density of the *return
current* along the depth direction (z-direction) for $f = 10$ Hz and $f = 100$ Hz.
The right panel shows the relative magnitude of envelopes of the extracellular potential as a function of frequency for different lateral distances from the
soma. Here curves are normalized to unity for the lowest frequency considered,
$f = 10$ Hz.

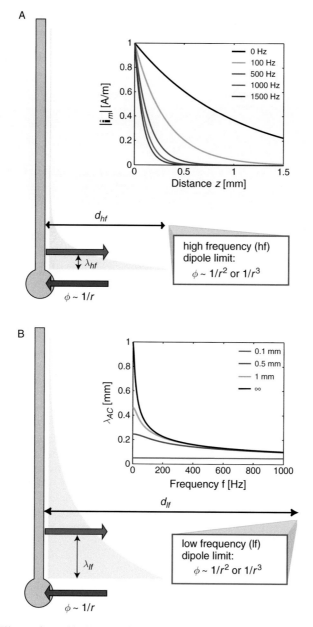

Plate 4.6 Illustration of ball-and-stick neuron and its frequency-dependent dipole sizes and corresponding far-field limits. A. For high frequencies (hf), the center of gravity (blue arrow) of the dendritic return current is close to soma. Therefore, the AC length constant λ_{hf} is small and transition to the far-field limit occurs around a distance d_{hf}, relatively close to the neuron. Inset: transmembrane return-current profile along an infinite dendritic stick for different frequencies (Pettersen and Einevoll, 2008). Parameters: stick diameter 2 mm, membrane and axial resistivities $R_m = 30000\ \Omega\,cm^2$, $R_i = 150\ \Omega\,cm^2$, membrane capacitance $C_m = 1\ \mu F/cm^2$. λ_{AC}^{∞} is 317 mm, 145 mm, 103 mm, and 84 mm for 100 Hz, 500 Hz, 1000 Hz, and 1500 Hz, respectively. B. For low frequencies (lf) the AC length constant λ_{lf} is relatively large and the far-field limit is reached for a larger distance d_{lf} than for the higher frequency in A. Inset: AC length constant $\lambda_{AC}(\omega)$ as a function of frequency for ball-and-stick models of different length; parameter values for diameter, resistivity, and capacitance are the same as in A.

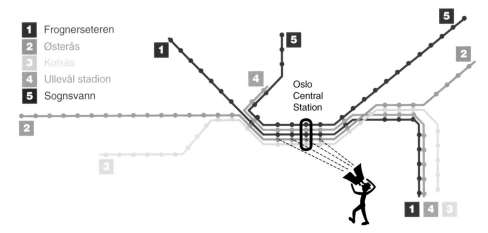

Map of the Oslo subway system. With its branchy structure of different lines ("dendrites") stretching out from the hub at Oslo Central Station ("soma"), the subway system resembles a neuron. If we pursue this analogy, the subway stations (marked with dots) may correspond to "neuronal compartments" and the net number of passengers entering or leaving the subway system at each station to the net "transmembrane current" at this compartment. If more passengers enter than leave the subway system at a point in time, it means that the number of people in the subway system, i.e. the "intracellular membrane potential," increases. The intracellular soma membrane potential, crucial for predicting the generation of neuronal action potentials (which luckily have no clear analogy in normal subway traffic), would then correspond to the number of passengers within the subway station at Oslo Central Station. The extracellular potential on the other hand would be more similar to what could be measured by an eccentric observer counting passengers flowing in and out of a few neighboring subway stations (with binoculars on the top of a large building maybe). While the analogy is not perfect, it should illustrate that intracellular and extracellular potentials are correlated, but are really two different things. (Adapted from Pettersen and Einevoll, 2009.)

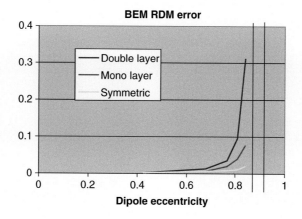

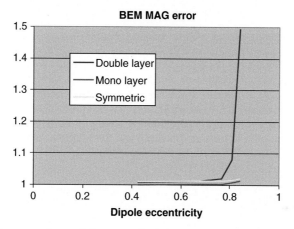

Plate 6.12 A comparison of the numerical accuracy of different BEM variants is shown. As a true model a three shell concentric sphere model was used. The RDM and MAG errors are shown as functions of dipole eccentricity (normalized with respect to the outer radius of volume conductor r_1) for the double layer method, the single layer method and the symmetric method. The vertical black lines in the upper plot represent the locations r_3 and r_2 of the compartment boundaries.

A   B

Plate 6.14 The finite element method is based on a three-dimensional discretization of space. In the examples presented here constrained Delaunay tetrahedralizations are used to describe the concentric sphere model. In A the discretization
is homogeneous and consists of 161 086 nodes. Example B consists of 360 056
nodes, most of which are put at larger eccentricities in order to minimize the error
of the FEM-computed correction potential within the subtraction approaches.
While example A is optimized for the needs of the direct FEM approaches,
example B is optimized for the FEM subtraction approaches, because in those
methods the largest part of the potential variations in the inner compartments is
explained by the infinite medium potential and the largest corrections are located
at higher eccentricities. (Reprinted from Drechsler et al. (2009), with permission
from Elsevier.)

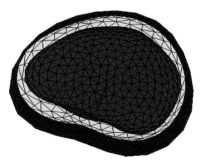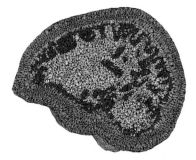

Plate 6.17 On the left a typical BEM mesh is shown. It consists of three surfaces
representing the outer boundaries of the brain, skull, and skin compartment. In
this example, the brain surface consists of about 700 nodes, the skull of about 400
nodes, and the skin of about 200 nodes. The right shows a sagittal cut through
a five-compartment tetrahedral finite element volume conductor model of the
human head (Reprinted from Wolters et al., 2006, with permission from Elsevier).

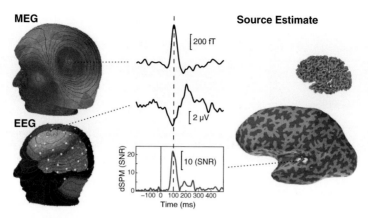

Plate 7.1 Example of MEG and EEG signals and source estimates. Left: maps of event-related magnetic fields (top) and electric scalp potentials (bottom), depicted with MRI-based reconstruction of the subject's head. The isocontour lines indicate the spatial pattern of the recorded signals at 100 ms after the onset of an auditory tone stimulus. The locations of the MEG sensors within a helmet-shaped array are indicated by the squares; the white dots indicate the locations of the EEG electrodes on the scalp. The top two waveforms show the time course of the signal recorded by one MEG and one EEG sensor, respectively. Right: map of the distributed source estimate. The noise-normalized minimum-norm estimate (dynamic statistical parametric map, dSPM) for the auditory evoked response at 100 ms latency is shown on the cerebral cortex. The realistic shape of the cortex was reconstructed from anatomical MRI; the cortical surface was inflated (bottom) to visualize the source estimates within sulci. The lowermost waveform shows the estimated time course of one source element in the auditory cortex in the lateral sulcus.

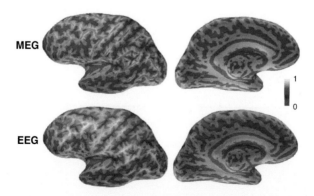

Plate 7.4 Sensitivity maps of MEG and EEG sensor arrays. The color code indicates the normalized root-mean-square signal value for 102 MEG magnetometers (left) and 70 EEG electrodes (right), as generated by cortically constrained dipole elements oriented perpendicular to the cortical surface. Lateral (top) and medial (bottom) views of the inflated left hemisphere are shown. Note that MEG is most sensitive to superficial tangentially oriented source located within the walls of sulci, whereas the sensitivity map for EEG is more uniform.

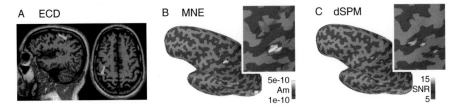

Plate 7.5 Example of different source estimates for the same MEG data. A. The equivalent current dipole (ECD) for the averaged contralateral response at 20 ms after an electrical stimulation of the right median nerve at the wrist. The green dot indicates the location of the ECD, superimposed on axial (left) and sagittal (right) slices of anatomical MRI. The green line extending from the dot indicates the orientation of the current dipole moment. The location of the ECD is consistent with the primary somatosensory cortex in the posterior wall of the central sulcus. B. The minimum-norm estimate (MNE) computed on the cortical surface. The location of the ECD is indicated with a green dot. In this distributed source estimate the reconstructed activity spreads from the posterior wall of the central sulcus also to the neighboring regions. The striped pattern is typical in inflated views of the cortex, especially when an orientation constraint for the dipole elements is imposed. The maximum in the MNE is at the crown of the gyrus, slightly more superficial than the ECD. C. The noise-normalized MNE, known as dynamic statistical parametric map (dSPM), shows a somewhat different spatial distribution than the MNE. In this case, the maximum in dSPM is deeper compared with the ECD and the maximum in MNE. In general, however, the overall localization of activity is consistent across these three different source estimates.

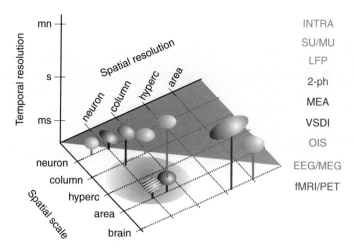

Plate 9.1 Imaging and non-imaging techniques are classified according to their resolutions, both spatial and temporal, and their spatial scale (INTRA intracellular recording, SU/MU single-unit/multi-unit recording, LFP local field potential, 2-ph two-photon imaging, MEA multi-electrode array, VSDI voltage-sensitive dye imaging, OIS optical imaging of intrinsic signals, EEG/MEG electroencephalography/magnetoencephalography, fMRI/PET functional magnetic resonance imaging/positron emission tomography). The mesoscopic scale is represented by the oval shaded area. (Modified from Chemla and Chavane, 2010b).

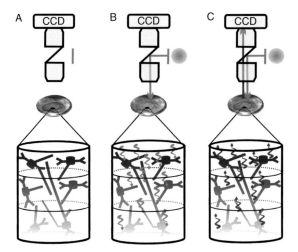

Plate 9.2 The principle of VSDI in three steps. The imaging chamber allows direct access to the primary visual cortex V1 represented as a patch of cortex with its six layers. A. The dye, applied on the surface of the cortex, penetrates through the cortical layers of V1. B. All neuronal and non-neuronal cells are now stained with the dye and when the cortex is illuminated, the dye molecules act as molecular tranducers that transform changes in membrane potential into optical signals. C. The fluorescent signal (red arrow) is recorded by a CCD camera. (Modified from Chemla and Chavane, 2010b).

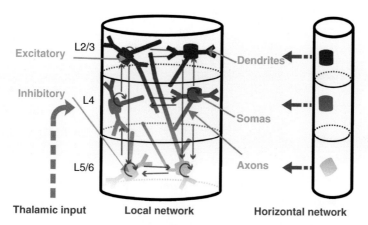

Plate 9.3 Contributions of the optical signal. Once neurons are stained by the VSD, every neuronal membrane contributes to the resulting fluorescent signal, but from where and in what proportion? Answering these four questions could clarify the origins of optical signals: 1) which cells? 2) which parts of the cell? 3) which layers? 4) which presynaptic origins? (Modified from Chemla and Chavane, 2010b).

A Cortical column model

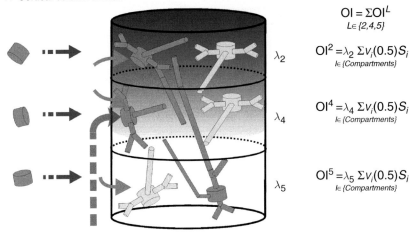

$$OI = \Sigma OI^L$$
$$L \in \{2,4,5\}$$

$$OI^2 = \lambda_2 \, \Sigma v_i(0.5) S_i$$
$$i \in \{Compartments\}$$

$$OI^4 = \lambda_4 \, \Sigma v_i(0.5) S_i$$
$$i \in \{Compartments\}$$

$$OI^5 = \lambda_5 \, \Sigma v_i(0.5) S_i$$
$$i \in \{Compartments\}$$

B Experimental vs. Modeled VSD signal

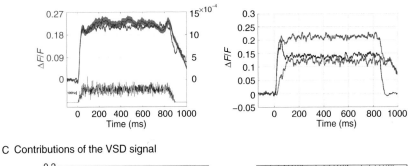

C Contributions of the VSD signal

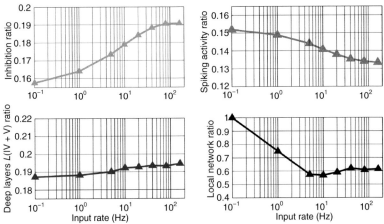

Plate 9.4 Biophysical model implementation. A. Model representation with thalamic input (red), background activity (green) and lateral connections (blue). The model offers the possibility of computing the VSD signal with a linear formula (right panel) taking into account the dye concentration (λ parameter), and the membrane

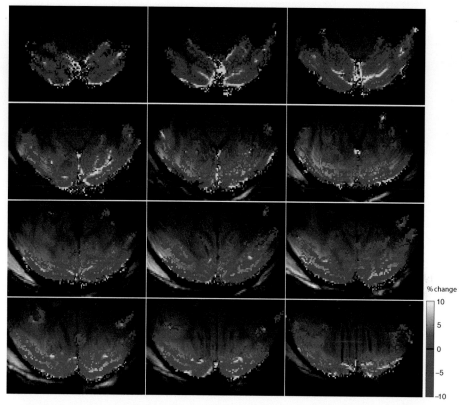

% change

10

5

0

−5

−10

Plate 11.3 Functional activation acquired in the visual cortex of an anesthetized monkey at 7 T at a spatial resolution of $0.5 \times 0.5 \times 2\,mm^3$. The map shows the percentage signal change in response to a black-and-white rotating checkerboard presented to both eyes. Functional activation is seen in V1–V5. The scanner and experimental procedures are described in Logothetis et al. (1999) and Pfeuffer et al. (2004).

Caption for Plate 9.4 (cont.)
potential (V_i) at each compartment. B. *Left:* Temporal evolution of the VSD signal in response to an input of 800 ms and for a given input strength (80% contrast for the experimental and 300 Hz for the model). Comparison between modeled (red) and experimental (gray, mean ± sem) responses. *Right:* Model prediction of the relative contributions of local (black line) and lateral (blue line) inputs to the total VSD signal (red line). C. The different contributions of the VSD signal: inhibition vs. excitation, axonic vs. dendritic activity, deep vs. superficial layers and local vs. global activity. (Modified from Chemla and Chavane, 2010a).

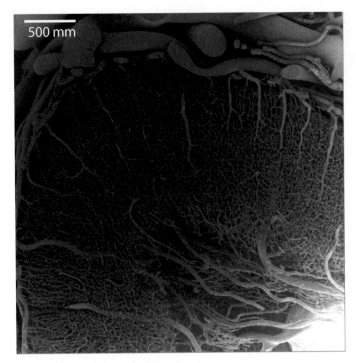

Plate 11.5 Corrosion cast of monkey V1 showing the microvasculature of the cortex (from Weber et al., 2008). Arteries and large draining are located in the pial layer. Penetrating arterioles and venules can be seen in addition to the capillary network. Intracortical arteries are shown in red and intracortical veins in blue. The vascular density in white matter is lower than in gray matter. (Adapted from Weber et al., 2008, with permission.)

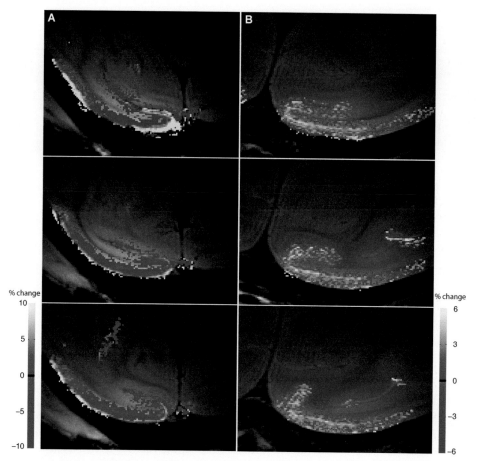

Plate 11.7 Specificity of GE- and SE-BOLD fMRI acquired at 4.7 T. The GE-BOLD map at an in-plane resolution of $333 \times 333\,\mu m^2$ (A) shows highest activity at the cortical surface and near vessels in the sulcus (Goense et al., 2007; Logothetis, 2008), whereas the SE-BOLD map in-plane resolution of $250 \times 180\,\mu m^2$ shows a BOLD signal that is better confined to the gray matter (B). Laminar structure can be seen in the SE-fMRI signal. Partial volume effects in the slice direction were not detrimental due to the anatomy of monkey V1 and by positioning the slices perpendicular to the cortical surface.

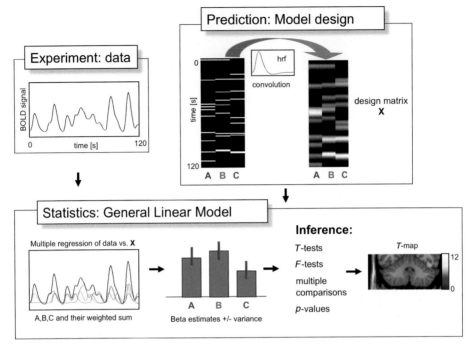

Plate 11.8 Illustration of the statistical parametric mapping approach. The aim is to determine whether and to what extent the BOLD signal time-series of a given voxel ("data") can be explained by experimental conditions, here denoted as A, B and C. Here, each condition was presented ten times for 1 s at random intervals within a 120 s period, as shown in the left panel of the model design. The stimulus time series are convolved by the hrf to obtain the expected effect of each condition on the BOLD signal of the voxels. A subsequent GLM analysis estimates the weights by which each predictor (regressor) contributes to the data. In a last step (inference), the contributions of the different predictors (betas) are compared against zero or against each other. In a whole-brain analysis, the resulting statistical T-values are projected in a color code onto each voxel, providing a statistical spatial map.

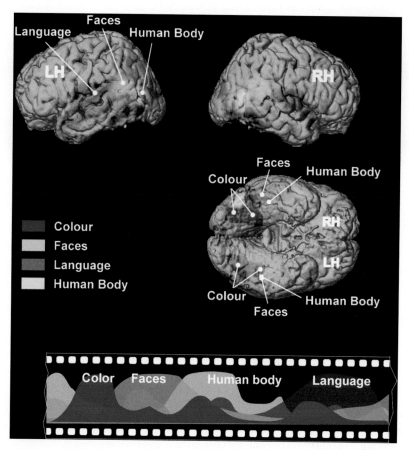

Plate 11.9 Activation maps obtained using a GLM analysis of volunteers' brains watching a James Bond movie. Voxels that achieved significant beta-estimates for one of the four feature regressors were color coded to identify the feature and superimposed on a surface rendering of the anatomical image of a single brain. Statistics was calculated on eight volunteers, and activity is thresholded at $p<0.001$ using FWE correction. (Modified with permission from Bartels and Zeki, 2003.)

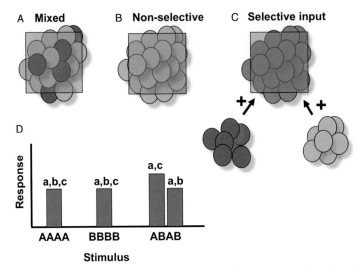

Plate 11.10 Ambiguities in interpreting fMRI adaptation results. Panels A–C illustrate neural functional organizations within a voxel (in blue). Direct recordings of neurons (not relying on adaptation) would differentiate stimulus-selective (A) from non-selective neurons (B, C) neurons. The adaptation effects that fMRI-A relies on can however lead to ambiguous net effects (D). A mixed neural population (A) would adapt its spike rates more to same-stimulus repeats (AAAA or BBBB) than to stimulus alternations (ABAB). Neurovascular coupling may render the BOLD effect of this differential spike rate adaptation more apparent in the downstream site (e.g. V5/MT) whose input is affected rather than in the site where it originates (e.g. V1, Kohn and Movshon, 2003), especially when the mechanism of adaptation involves primarily intraneural spike rate adaptation of the output neurons (with potentially little BOLD signal effect) rather than network adaptation (substantial BOLD effect). Hence the ambiguous outcome of (A) illustrated in (D). Non-selective sites should not reveal differential adaptation (B) – unless they get input from selective sites (C). In this case, adaptation can render neurons or sites that are not selective to a feature to appear selective following adaptation (Tolias et al., 2005a, 2005b). (Modified from Bartels et al., 2008b.)

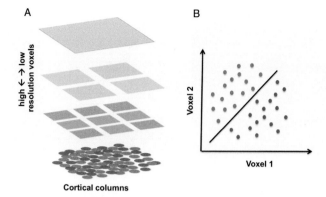

Cortical columns

Plate 11.11 Effects of cortical functional organization on pattern classification and high-resolution imaging. A. Schematic diagram illustrating how a region (left) packet with feature-specific functional columns (circles, e.g. green responding to leftwards and red to rightwards motion) may lead to less feature bias in voxel responses (squares) as the voxel size increases (from bottom to top). Top level: the areal net response is not biased to red/green (e.g. left/right motion). Second level: medium high resolution may lead to voxel-wise yet subthreshold biases, allowing multivariate statistics to inform about the presence of feature-specific information in this region. Third level: high-resolution imaging may reveal the actual functional organization of the region using voxel-wise, standard univariate statistics. B. Multivariate statistics can combine weak signal biases from many voxels. The example shows the signal strength of one voxel plotted against that of one other voxel, each dot representing the response to one trial, of two stimulus classes. While the mean response to the red stimulus class may not be statistically different from that to the green stimulus class in any of the two voxels, their combined responses allow classification of the two stimulus classes.

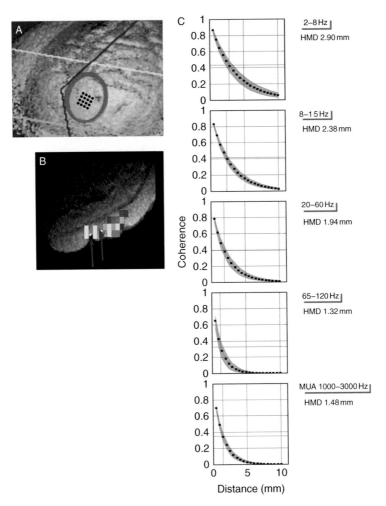

Plate 11.13 Coherence between signals recorded at different electrodes as a function of electrode distance for the different neural signals. A. Diagram of the 4 × 4 electrode array used for measuring the neural signals. B. Position of two electrodes (arrows) in combined MRI and physiology experiments. C. Average coherence-over-distance for the different neural signals, which shows the decrease in coherence with interelectrode distance. The gray shading shows 1–99% confidence intervals. The distance at which the coherence is halved is indicated by the red lines and shown next to the panels. The loss of coherence with distance was 1–3 mm and was comparable for all neural frequency bands. (Adapted from Goense and Logothetis, 2008.)

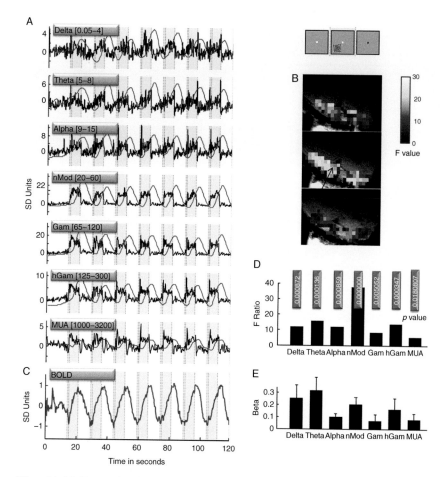

Plate 11.14 Dependencies between BOLD and neural signals in V1 of an awake monkey. A. Band-separated neural signals (black) in response to a 6° rotating checkerboard (inset). Green shading indicates the times the stimulus was presented. The different band-limited power signals were convolved with a theoretical HRF and used as regressors (red). The response to the stimulus was especially pronounced in the neuromodulatory and gamma bands. B. Functional activation maps superimposed on GE anatomical images. The location of the electrode is indicated by the arrow. C. The BOLD time course acquired at a temporal resolution of 250 ms shows clear modulation to the stimulus. The output of the GLM analysis and F-test yielded significant p-values for all neural bands (D) indicating that all bands contributed to the BOLD response. E. Beta values lacked dramatic differences across bands.

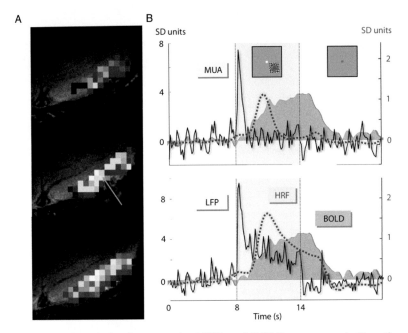

Plate 11.15 Dissociation between the MUA and BOLD response. A. Functional activation maps in response to a 6° peripheral rotating checkerboard stimulus (inset in B). The arrow indicates the location of the electrode. B. The average MUA, LFP in the 20–60 Hz range and BOLD time courses show that while the neuromodulatory component of the LFP stayed elevated for the duration of the stimulus, the MUA rapidly returned to baseline after a transient onset response. The prolonged time course of the BOLD response suggests a more sustained driving mechanism of the BOLD response as opposed to the transient MUA signal. The dotted line shows the regressor, i.e. the neural signal convolved with the theoretical hrf, which indicates that the MUA-derived regressor cannot capture the sustained part of the BOLD response. (Adapted from Goense and Logothetis, 2008.)

Cook, M. J. D. and Koles, Z. J. (2006). A high-resolution anisotropic finite-volume head model for EEG source analysis. *Proc. 28th Ann. Conf. IEEE Eng. Med. Biol. Soc.*, pp. 4536–4539.

Cuffin, B. N. (1985), A comparison of moving dipole inverse solutions using EEG's and MEG's. *IEEE Trans. Biomed Eng.*, **32** (11), 905–910.

Cuffin, B. N. (1991). Eccentric spheres models of the head. *IEEE Trans. Biomed. Eng.*, **38** (9), 871–878.

Cuffin, B. N. and Cohen, D. (1977). Magnetic fields of a dipole in special volume conductor shapes. *IEEE Trans. Biomed. Eng.*, **24**(4), 372–381.

Cuffin, B. N. and Cohen, D. (1983). Effects of detector coil size and configuration on measurements of the magnetoencephalogram. *J. Appl. Phys.*, **54**, 3589–3594.

Dannhauer, M., Lanfer, B., Wolters, C. H. and Knösche, T. R. (2011). Modeling of the human skull in EEG source analysis. Human Brain Mapping, **32** (9), 1383–1399.

Dassios, G., Giapalaki, S. N., Kandili, A. N. and Kariotou, F. (2007). The exterior magnetic field for the multilayer ellipsoidal model of the brain. *Q. J. Mech. Appl. Math.*, **60** (1), 1–25.

Dembart, B. and Yip, E. (1998). The accuracy of fast multipole methods for Maxwell's equations. *IEEE Comput. Sci. Eng.*, **5** (3), 48–56.

De Jongh, A., De Munck, J. C., Gonçalves, S. I. and Ossenblok, P. P. W. (2005). Differences in MEG/EEG epileptic spike yields explained by regional differences in signal to noise ratios. *J. Clin. Neurophysiol.*, **22** (2), 153–158.

De Munck, J. C. (1988). The potential distribution in a layered anisotropic spheroidal volume conductor. *J. Appl. Phys.*, **64** (2), 464–470.

De Munck, J. C. (1992). A linear discretization of the volume conductor boundary integral equation using analytically integrated elements. *IEEE Trans. Biomed. Eng.*, **39**, 986–989.

De Munck, J. C. and Peters, M. J. (1993). A fast method to compute the potential in the multi-sphere model. *IEEE Trans. Biomed. Eng.*, **40** (11), 1166–1174.

De Munck, J. C. and Van Dijk, B. W. (1991). Symmetry considerations in the quasi-static approximation of volume conductor theory. *Phys. Med. Biol.*, **36** (4), 521–529.

De Munck, J. C., Van Dijk, B. W. and Spekreijse, H. (1988a). An analytic method to determine the effect of source modelling errors on the apparent location and direction of biological sources. *J. Appl. Phys.*, **63** (3), 944–956.

De Munck, J. C., Van Dijk, B. W. and Spekreijse, H. (1988b). Mathematical dipoles are adequate to describe realistic generators of human brain activity. *IEEE Trans. Biomed. Eng.*, **35** (11), 960–966.

De Munck, J. C., Hämäläinen, M. H. and Peters, M. J. (1991). The use of the asymptotic expansion to speed up the computation of a series of spherical harmonics. *Clin. Phys. Physiol. Meas.*, **12A**, 83–87.

De Munck, J. C., De Jongh, A. and Van Dijk, B. W. (2001). The localization of spontaneous brain activity: an efficient way to analyze large data sets. *IEEE Trans. Biomed. Eng.*, **48** (11), 1221–1228.

Drechsler, F., Wolters, C. H., Dierkes, T., Si, H. and Grasedyck, L. (2009). A full subtraction approach for finite element method based source analysis using constrained Delaunay tetrahedralisation. *NeuroImage*, **46** (4), 1055–1065.

Epton, M. A. and Dembart, B. (1995). Multipole translation theory for the three-dimensional Laplace and Helmholtz equations. *SIAM J. Sci. Comput.*, **16** (4), 865–897.

Eriksson, F. (1990). On the measure of solid angles. *Math. Mag.*, **63** (3), 184–187.

Ermer, J. J., Mosher, J. C., Baillet, S. and Leahy, R. M. (2001). Rapidly Re-computable EEG forward models for realistic head shapes. *Phys. Med. Biol.*, **46** (4), 1265–1281.

Frank, E. (1952). Electric potential produced by two point current sources in a homogeneous conducting sphere. *J. Appl. Phys.*, **23** (11), 1225–1228.

Friederici, A., Wang, Y., Herrmann, C., Maess, B. and Oertel, U. (2000). Localization of early syntactic processes in frontal and temporal cortical areas: an MEG study. *Human Brain Mapping*, **11**, 1–11.

Frijns, J. H. M., De Loo, S. L. and Schoonhoven, R. (2000). Improving accuracy of the boundary element method by the use of second-order interpolation function. *IEEE Trans. Biomed. Eng.*, **47** (10), 1336–1346.

Fuchs, M., Drenckhahn, R., Wischmann, H. A. and Wagner, M. (1998). An improved boundary element method for realistic volume conductor modeling. *IEEE Trans. Biomed. Eng.*, **45** (8), 980–997.

Fuchs, M., Wagner, M. and Kastner, J. (2007). Development of volume conductor and source models to localize epileptic foci. *J. Clin. Neurophysiol.*, **24**, 101–119.

Gencer, N. G. and Acar, C. E. (2004). Sensitivity of EEG and MEG measurements to tissue conductivity. *Phys. Med. Biol.*, **49**, 701–717.

Geselowitz, D. B. (1970). On the magnetic field generated outside an inhomogeneous volume conductor by integral current sources. *IEEE Trans. Magn.*, **6**, 346–347.

Golub, G. H. and Van Loan, C. F. (1989). *Matrix Computations* (2nd edition). Baltimore, MD: Johns Hopkins University Press.

Gonçalves, S. I., De Munck, J. C., Verbunt, J. P. A., Heethaar R. M. and Lopes da Silva, F. H. (2003a). In vivo measurement of brain and skull resistivities using an EIT based method and the combined analysis of SEP/SEF data. *IEEE Trans. Biomed. Eng.*, **50** (9), 1124–1127.

Gonçalves, S. I., De Munck, J. C., Verbunt, J. P. A., Bijma, F., Heethaar, R. M. and Lopes da Silva, F. H. (2003b). In vivo measurement of the brain and skull resistivities using an EIT based method and realistic models for the head. *IEEE Trans. Biomed. Eng.*, **50** (6), 754–767.

Grynszpan, F. and Geselowitz, D. B. (1973). Model studies of the magnetocardiogram. *Biophys. J.*, **13**, 911–925.

Güllmar, D., Haueisen, J., Eiselt, M., Giessler, F., Flemming, L., Anwander, A., Knösche, T. R., Wolters, C. H., Dümpelmann, M., Tuch, D. S., and Reichenbach, J. R. (2006). Influence of anisotropic conductivity on EEG source reconstruction: investigations in a rabbit model. *IEEE Trans. Biomed. Eng.*, **53** (9), 1841–1850.

Güllmar, D., Haueisen, J. and Reichenbach, J. R. (2010). Influence of anisotropic electrical conductivity in white matter tissue on the EEG/MEG forward and inverse solution. A high resolution whole head simulation study. *NeuroImage*, **51**, 145–163.

Gutiérrez, D. and Nehorai, A. (2008). Array response kernels for EEG and MEG in multilayer ellipsoidal geometry. *IEEE Trans. Biomed. Eng.*, **55** (3), 1103–1111.

Hackbusch, W. (1994). *Iterative Solution of Large Sparse Systems of Linear Equations*, Applied Mathematical Sciences, Vol. 95. New York: Springer Verlag.

Hackbusch, W. and Nowak, Z. K. (1989). On the fast matrix multiplication in the boundary element method by panel clustering. *Numer. Math.*, **54**, 463–491.

Hallez, H. (2008). Incorporation of anisotropic conductivities in EEG source analysis. *Dissertation*, Universiteit Gent, Biomedical Engineering, ISBN 978-90-8578-229-2.

Hallez, H., Vanrumste, B., Van Hese, P., D'Asseler, Y., Lemahieu, I. and Van de Walle, R. (2005). A finite difference method with reciprocity used to incorporate anisotropy in electroencephalogram dipole source localization, *Phys. Med. Biol.*, **50**, 3787–3806.

Hämäläinen, M. S. and Sarvas, J. (1989). Realistic conductivity geometry model of the human head for interpretation of neuromagnetic data. *IEEE Trans. Biomed. Eng.*, **36**, 165–171.

Hämäläinen, M. S., Hari, R., Ilmoniemi, R. J., Knuutila, J. and Lounasmaa, O. V. (1993). Magnetoencephalography theory, instrumentation, and applications to noninvasive studies of the working human brain. *Rev. Mod. Phys.*, **65**, 413–497.

Haueisen, J., Tuch, D. S., Ramon, C., Schimpf, P. H., Wedeen, V. J., George, J. S. and Belliveau, J. W. (2002). The influence of brain tissue anisotropy on human EEG and MEG. *NeuroImage*, **15**, 159–166.

Hosek, R. S., Sances, A., Jodat, R. W. and Larson, S. J. (1978). The contributions of intracerebral currents to the EEG and evoked potentials. *IEEE Trans. Biomed. Eng.*, **25** (5), 405–413.

Huang, M. X., Mosher, J. C. and Leahy, R. M. (1999). "A sensor-weighted overlapping-sphere head model and exhaustive head model comparison for MEG. *Phys. Med. Biol.*, 423–440.

Huang, M. X., Song, T., Hagler, D. J., Podgorny, I., Jousmaki, V., Cui, L., Gaa, K., Harrington, D. L., Dale, A. M., Lee, R. R., Elman, J. and Halgren, E. (2007). A novel integrated MEG and EEG analysis method for dipolar sources. *NeuroImage*, **37**, 731–748.

Ilmoniemi, R. J. (1995). Radial anisotropy added to a spherically symmetric conductor does not affect the external magnetic field due to internal sources. *Europhys. Lett.*, **30** (5), 313–326.

Ilmoniemi, R. J., Hämäläinen, M. S. and Knuutila, J. (1985). The forward and inverse problems in the spherical model. In: H. Weinberg, G. Stroink and T. Katila (editors), *Biomagnetism: Applications and Theory*. Oxford: Pergamon Press, pp. 278–282.

Jackson, J. D. (1962). *Classical Electrodynamics*. New York: Wiley.

Kariotou, F. (2004). Electroencephalography in ellipsoidal geometry. *J. Math. Anal. Appl.*, **290**, 324–342.

Kybic, J., Clerc, M., Faugeras, O., Keriven, R. and Papadopoulo, T. (2005a). Fast multipole acceleration of the MEG/EEG boundary element method. *Phys. Med. Biol.*, **50**, 4695–4710.

Kybic, J., Clerc, M., Abboud, T., Faugeras, O., Keriven, R. and Papadopoulo, T. (2005b). A common formalism for the integral formulations of the forward EEG problem. *IEEE Trans. Med. Im.*, **24**, 12–28.

Kybic, J., Clerc, M., Faugeras, O., Keriven, R. and Papadopoulo, T. (2006). Generalized head models for MEG/EEG: boundary element method beyond nested volumes. *Phys. Med. Biol.*, **51**, 1333–1346.

Lanfer, B. (2007) Validation and comparison of realistic head modelling techniques and application to tactile somatosensory evoked EEG and MEG data. *Diploma Thesis in Physics*, Fachbereich Physik, Westfaelische Wilhelms-Universitaet Muenster.

Lanfer, B., Wolters, C. H., Demokritov, S. O. and Pantev, C. (2007). Validating finite element method based EEG and MEG forward computations. *Proc. 41. Jahrestagung der DGBMT, Deutsche Gesellschaft fuer Biomedizinische Technik im VDE, Aachen, Germany, September 26–29*, pp. 140–141. ISSN: 0939-4990, www.bmt2007.de.

Lew, S., Wolters, C. H., Anwander, A., Makeig, S. and MacLeod, R. S. (2009a). Improved EEG source analysis using low resolution conductivity estimation in a four-compartment finite element head model. *Human Brain Mapping*, **30** (9), 2862–2878.

Lew, S., Wolters, C. H., Röer, C., Dierkes, T. and MacLeod, R. S. (2009b). Accuracy and run-time comparison for different potential approaches and iterative solvers in finite element method based EEG source analysis. *Appl. Numer. Math.*, **59** (8), 1970–1988.

Lopes da Silva, F. H. and Van Rotterdam, A. (1982). Biophysical aspects of EEG and MEG generation. In: E. Niedermeyer and F. H. Lopes da Silva (editors), *Electroencephalography: Basic Principles, Clinical Applications and Related Fields*, (4th edition). Munich: Urban & Schwarzenberg, pp. 93–109.

Lynn, M. S. and Timlake, W. P. (1968a). The numerical solution of singular integral equations of potential theory. *Numer. Math.*, **11**, 77–98.

Lynn, M. S. and Timlake, W. P. (1968b). The use of multiple deflations in the numerical solution of singular systems of equations, with applications to potential theory. *SIAM J. Numer. Anal.*, **5** (2), 303–322.

Marin, G., Guerin, C., Baillet, S., Garnero, L. and Meunier, G. (1998). Influence of skull anisotropy for the forward and inverse problem in EEG: simulation studies using the FEM on realistic head models. *Human Brain Mapping*, **6**, 250–269.

Meijs, J. W. H., Weier, O. W., Peters, M. J. and Van Oosterom, A. (1989). On the numerical accuracy of the boundary element method. *IEEE Trans. Biomed. Eng.*, **36**, 1038–1049.

Merzbacher, E. (1961). *Quantum Mechanics*. New York: Wiley.

Mohr, M. (2004). Simulation of bioelectric fields: the forward and inverse problem of electroencephalographic source analysis. *Dissertation*, Friedrich-Alexander-Universitaet Erlangen-Nuernberg, Arbeitsberichte des Instituts fuer Informatik, Band 37, Nummer 6, ISSN 1611–4205.

Mohr, M. and Vanrumste, B. (2003). Comparing iterative solvers for linear systems associated with the finite difference discretisation of the forward problem in electroencephalographic source analysis. *Med. Biol. Eng. Comput.*, **41**, 75–84.

Moon, P. and Spencer, D. E. (1988). *The Field Theory Handbook*. Berlin: Springer-Verlag.

Morse, P. M. and Feshbach, H. (1953). *Methods of Theoretical Physics*, Vol I and II. New York: McGraw-Hill.

Mosher, J. C., Leahy, R. M. and Lewis, P. S. (1999). EEG and MEG: forward solutions for inverse methods. *IEEE Trans. Biomed. Eng.*, **46** (3), 245–259.

Murakami, S. and Okada, Y. (2006). Contributions of principal neocortical neurons to magnetoencephalography and electroencephalography signals. *J. Physiol.*, **575** (3), 925–936.

Niedermayer, E. and Lopes da Silva, F. (editors) (1987). *Electroencephalography. Basic Principles, Clinical Applications and Related Fields*, (2nd edition). Baltimore, MD: Urban and Schwarzenberg.

Nolte, G. and Curio, G. (1997). On the calculation of magnetic fields based on multipole modeling of focal biological current sources. *Biophys. J.*, **73**, 1253–1262.

Nolte, G. and Curio, G. (1999). Perturbative solutions of the electric forward problem for realistic volume conductors. *J. Appl. Phys.*, **86** (5), 2800–2812.

Nolte, G., Fieseler, T. and Curio, G. (2001). Perturbative analytical solutions of the magnetic forward problem for realistic volume conductors. *J. Appl. Phys.*, **89** (4), 2360–2370.

Of, G., Steinbach, O. and Wendland, W. L. (2002). A fast multipole boundary element method for the symmetric boundary integral formulation. In: *Proceedings of IABEM, Austin, TX, 2002*.

Olivi, E., Clerc, M., Papadopoulo, T. and Vallaghé, S. (2010). Domain decomposition for coupling finite and boundary element methods in EEG. In: *Proceedings of International Conference on Biomagnetism*.

Olesch, J., Ruthotto, L., Kugel, H., Skare, S., Fischer, B. and Wolters, C. H. (2010). A variational approach for the correction of field-inhomogeneities in EPI sequences. SPIE Medical Imaging, Image Processing, 7623(1), 8 pages, San Diego, CA. doi: 10.1117/12.844375.

Ollikainen, J., Vaukhonen, M., Karjalainen, P. A. and Kaipio, J. P. (1999). Effects of local skull inhomogeneities on EEG source estimation. *Med. Eng. Phys.*, **21**, 143–154.

Oostenveld, R. and Oostendorp, T. F. (2002). Validating the boundary element method for forward and inverse EEG computations in the presence of a hole in the skull. *Human Brain Mapping*, **17** (3), 179–192.

Ossenblok, P., De Munck, J. C., Drolsbach, W., Colon, A. and Boon P. (2007). The advantages of interictal MEG compared to EEG in case of frontal lobe epilepsy. *Epilepsia*, **48** (11), 2139–2149.

Plonsey, R. and Heppner, D. (1967). Considerations of quasistationarity in electrophysiological systems. *Bull. Math. Biophys.*, **29**, 657–664.

Peters, M. J. and Elias, P. J. H. (1988). On the magnetic field and the electric potential generated by biological sources in an anisotropic volume conductor. *Med. Biol. Eng. Comput.*, **26**, 617–623.

Pruis, G. W., Guilding, B. H. and Peters, M. J. (1993). A comparison of different numerical methods for solving the forward problem of EEG and MEG. *Physiol. Meas.*, **14**, A1–A9.

Pursiainen, S. (2008). Computational methods in electromagnetic biomedical inverse problems. *Thesis*, Helsinki University of Technology, Faculty of Information and Natural Sciences, Institute of Mathematics.

Pursiainen, S., Lucka, F. and Wolters, C. H. (2012). Complete electrode model in EEG: relationship and differences to the point electrode model. *Phys. Med. Biol.*, **57**, 999–1017.

Rahola, J. (1998). Experiments on iterative methods and the fast multipole method in electromagnetic scattering calculations. Technical Report TR/PA/98/49, CERFACS, 98.

Ramon, C., Schimpf, P., Haueisen, J., Holmes, M. and Ishimaru, A. (2004). Role of soft bone, CSF and gray matter in EEG simulations. *Brain Topography*, **16** (4), 245–248.

Roberts, T., Poeppel, D. and Rowley, H. (1998). Magnetoencephalography and magnetic source imaging. *Neuropsych. Behav. Neurol.*, **11**, 49–64.

Rosenfeld, M., Tanami, R. and Abboud, S. (1996). Numerical solution of the potential due to dipole sources in volume conductors with arbitrary geometry and conductivity. *IEEE Trans. Biomed. Eng.*, **43** (7), 679–0689.

Rudy, R. and Plonsey, R. (1979). The eccentric spheres model as the basis for a study of the role of geometry and inhomogeneities in electrocardiography. *IEEE Trans. Biomed. Eng.*, **26** (7), 392–399.

Rullmann, M., Anwander, A., Dannhauer, M., Warfield, S. K., Duffy, F. H. and Wolters, C. H. (2009). EEG source analysis of epileptiform activity using a 1mm anisotropic hexahedra finite element head model. *NeuroImage*, **44**, 399–410.

Sadleir, R. J. and Argibay, A. (2007). Modeling skull electrical properties. *Ann. Biomed. Eng.*, **35** (10), 1699–1712.

Saleheen, H. I. and Kwong, T. N. (1997). New finite difference formulations for general inhomogeneous anisotropic bioelectric problems. *IEEE Trans. Biomed. Eng.*, **44** (9), 800–809.

Sarvas, J. (1987). Basic mathematical and electromagnetic concepts of the basic biomagnetic inverse problem. *Phys. Med. Biol.*, **32** (1), 11–22.

Schimpf, P. H. (2007). Application of quasi-static magnetic reciprocity to finite element models of the MEG lead-field. *IEEE Trans. Biomed. Eng.*, **54** (11), 2082–2088.

Schimpf, P. H., Ramon, C. and Haueisen, J. (2002). Dipole models for the EEG and MEG. *IEEE Trans. Biomed. Eng.*, **49** (5), 409–418.

Tanzer, O., Järvenpää, S., Nenonen, J. and Somersalo, E. (2005). Representation of bio-electric current sources using Whitney elements in finite element method. *Phys. Med. Biol.*, **50**, 3023–3039.

Taulu, S. and Simola, J. (2006). Spatiotemporal signal space separation method for rejecting nearby interference in MEG measurements. *Phys. Med. Biol.*, **51**, 1759–68.

Taulu, S., Simola, J. and Kajola, M. (2005). Applications of the signal space separation method. *IEEE Trans. Signal Process.*, **53**, 3359–72.

Tuch, D. S., Wedeen, V. J., Dale, A. M., George, J. S. and Belliveau, J. W. (2001). Conductivity tensor mapping of the human brain using diffusion tensor MRI. *Proc. Natl. Acad. Sci. USA*, **98** (20), 11697–11701.

Uutela, K., Taulu, S. and Hämäläinen, M. (2010). Detecting and correcting for head movements in neuromagnetic measurements. *NeuroImage*, **14**, 1424–1431.

Vallaghé, S. and Clerc, M. (2009). A global sensitivity analysis of three- and four-layer EEG conductivity models. *IEEE Trans. Biomed. Eng.*, **56** (4), 988–995.

Vallaghé, S. and Papadopoulo, T. (2010). A trilinear immersed finite element method for solving the EEG forward problem. *SIAM J. Sci. Comput.*, **32**, 2379–2394.

Vallaghé, S., Clerc, M. and Badier, J.-M. (2007). In vivo conductivity estimation using somatosensory evoked potentials and cortical constraints on the sources. *Proceedings of 4th IEEE International Symposium on Biomedical Imaging: from Nano to Macro*, pp. 1036–1039.

Vallaghé, S., Papadopoulo, T. and Clerc, M. (2009). The adjoint method for general EEG and MEG sensor-based lead field equations. *Phys. Med. Biol.*, **54**, 135–147.

Van Oosterom, A. and Strackee, J. (1983). The solid angle of a plane triangle. *IEEE Trans. Biomed. Eng.*, **30** (2), 125–126.

Van 't Ent, D., De Munck, J. C. and Kaas, A. L. (2001). An automated procedure for deriving realistic volume conductor models for MEG/EEG source localization. *IEEE Trans. Biomed. Eng.*, **48** (12), 1434–1443.

Vorwerk, J. (2011). Comparison of numerical approaches to the EEG forward problem, *Master's thesis*, Mathematics, Münster.

Waberski, T., Buchner, H., Lehnertz, K., Hufnagel, A., Fuchs, M., Beckmann, R. and Rienäcker, A. (1998). The properties of source localization of epileptiform activity using advanced headmodelling and source reconstruction. *Brain Topography*, **10** (4), 283–290.

Wagner, M., Fuchs, M., Wischmann, H. A., Ottenberg, K. and Dössel, O. (1995). Cortex segmentation from 3D MR images for MEG reconstructions. In: C. Baumgartner et al. (editors), *Biomagnetism: Fundamental Research and Clinical Applications*. Amsterdam: Elsevier/IOS Press, pp. 433–438.

Wagner, M., Fuchs, M., Drenckhahn, R., Wischmann, H.-A., Koehler, T. and Theissen, A. (2000). Automatic generation of BEM and FEM meshes from 3D MR data. *NeuroImage*, **3**, S168.

Wang, K., Zhu, S., Mueller, B., Lim, K. O., Liu, Z. and He, B. (2008). A new method to derive white matter conductivity from diffusion tensor MRI. *IEEE Trans. Biomed. Eng.*, **55** (10), 2481–2486.

Weinstein, D., Zhukov, L. and Johnson, C. (2000). Lead-field bases for electroencephalography source imaging. *Ann. of Biomed. Eng.*, **28** (9), 1059–1066.

Wehner, D. T., Hämäläinen, M. S., Mody, M. and Ahlfors, S. P. (2008). Head movements of children in MEG: quantification, effects on source estimation, and compensation *NeuroImage*, **40** (2), 541–550.

Wendel, K., Narra, N. G., Hannula, M., Kauppinen, P. and Malmivuo, J. (2008). The influence of CSF on EEG sensitivity distributions of multilayered head models. *IEEE Trans. Biomed. Eng.*, **55** (4), 1454–1456.

Willemse, R. B., De Munck, J. C., Van 't Ent, D., Ris, P., Baayen, J. C., Stam, C. J. and Vandertop, W. P. (2007). Magnetoencephalographic study of posterior tibial nerve stimulation in patients with intracranial lesions around the central sulcus. *Neurosurgery*, **61** (6), 1209–1218.

Wilson, F. N. and Bayley, R. H. (1950). The electric field of an eccentric dipole in a homogeneous spherical conducting medium. *Circulation*, **1**, 84–92.

Wolters, C. H. (2003). Influence of tissue conductivity inhomogeneity and anisotropy on EEG/MEG based source localization in the human brain. *MPI of Cognitive Neuroscience Leipzig*, MPI Series in Cognitive Neuroscience, No. 39, ISBN 3-936816-11-5 (http://lips.informatik.uni-leipzig.de/pub/2003-33).

Wolters, C. H. (2008). Finite element method based electro- and magnetoencephalography source analysis in the human brain. *Habilitation*, Fachbereich für Mathematik und Informatik, Westfälische Wilhelms-Universität Münster.

Wolters, C. H. and De Munck, J. C. (2007). Volume conduction, *Scholarpedia*, **2** (3), 1738, www.scholarpedia.org/article/Volume_conduction.

Wolters, C. H., Beckmann, R. F., Rienäcker, A. and Buchner, H. (1999). Comparing regularized and non-regularized nonlinear dipole fit methods: A study in a simulated sulcus structure. *Brain Topography*, **12** (1), 3–18.

Wolters, C. H., Anwander, A., Koch, M., Reitzinger, S., Kuhn, M. and Svensen, M. (2001). Influence of head tissue conductivity anisotropy on human EEG and MEG using fast high resolution finite element modeling, based on a parallel algebraic multigrid solver. In: T. Plesser and V. Macho (editors), *Contributions to the Heinz-Billing Award*, pp. 111–157. Gesellschaft für wissenschaftliche Datenverarbeitung mbH. Forschung und wissenschaftliches Rechnen, ISSN: 0176-2516, www.billingpreis.mpg.de.

Wolters, C. H., Kuhn, M., Anwander, A. and Reitzinger, S. (2002). A parallel algebraic multigrid solver for fininite element methods based source localization methods in the human brain. *Comput. Vis. Sci.*, **5** (3), 165–177.

Wolters, C. H., Graesdyck, L. and Hackbush, W. (2004). Efficient computation of lead field bases and influence matrix for the FEM-based EEG and MEG inverse problem. *Inverse Problems*, **20**, 1099–1116.

Wolters, C. H., Anwander, A., Weinstein, D., Koch, M., Tricoche, X. and MacLeod, R. S. (2006). Influence of tissue conductivity anisotropy on EEG/MEG field and return current computation in a realistiuc head model: a simulation and visualization study using high-resolution finite element modeling. *NeuroImage*, **30** (3), 813–826.

Wolters C. H., Köstler, H., Möller, C., Härtlein, J., Grasedyck, L. and Hackbusch, W. (2007a). Numerical mathematics of the subtraction method for the modeling of a current dipole in EEG source reconstruction using finite element head models. *SIAM J. Sci. Comput.*, **30** (1), 24–45.

Wolters, C. H., Anwander, A., Berti, G. and Hartmann, U. (2007b). Geometry-adapted hexahedral meshes improve accuracy of finite element method based EEG source analysis. *IEEE Trans. Biomed. Eng.*, **54** (8), 1446–1453.

Wolters, C. H., Lew, S., MacLeod, R. S. and Hämäläinen, M. S. (2010). Combined EEG/MEG source analysis using calibrated finite element head models. *Proc. 44th Annual Meeting of the DGBMT, Rostock-Warnemünde, Germany*, http://conference.vde.com/bmt-2010, to appear.

Yan, Y., Nunez, P. L., Hart, R. T. (1991). Finite element model of the human head: scalp potentials due to dipole sources. *Med. Biol. Eng. Comput.*, **29**, 475–481.

Yvert, B., Crouzeix-Cheylus, A. and Pernier, J. (2001). Fast realistic modelling in bioelectromagnetism using lead field interpolation. *Human Brain Mapping*, **14**, 48–63.

Zhang, Z. (1995). A fast method to compute the surface potentials generated by dipoles within multilayer anisotropic spheres. *Phys. Med. Biol.*, **40**, 335–349.

MEG and EEG: source estimation

SEPPO P. AHLFORS AND MATTI S. HÄMÄLÄINEN

7.1 Introduction

In magnetoencephalography (MEG) and electroencephalography (EEG), scalp potentials and extracranial magnetic fields generated by electrical activity in the brain are detected non-invasively (Berger, 1929; Cohen, 1972) (for an overview of the methodology see, e.g., Hamalainen et al., 1993; Niedermeyer and Lopes da Silva, 1999; Michel et al., 2009; Hansen et al., 2010). MEG and EEG signals are superpositions of contributions from sources at different locations in the brain. *Source estimation* (also known as *inverse modeling*) refers to the problem of determining the spatiotemporal patterns of neural activity on the basis of the recorded signals (Figure 7.1). The specific goal in source estimation can be stated in two closely related ways: (a) to identify the locations of the sources of the measured signals as a function of time, or (b) to disentangle the contributions from different brain regions in the measured time-varying signals. The often used term *source localization* refers to the former, whereas *spatiotemporal imaging* emphasizes the latter, reflecting the use of MEG and EEG source estimation in the analysis of the dynamical activity in networks of brain areas.

A given source in the brain generates a characteristic spatial pattern of signals in arrays of MEG and EEG sensors. These patterns can be calculated by using a *forward model* (see Chapter 6). In source estimation, the measured spatial patterns of signals are analyzed in order to make inferences about the distribution of the sources in the brain. Source analysis requires knowledge of the sensitivity patterns of the MEG and EEG sensors, known as the *lead fields*, which are determined by the forward model. As the lead fields of individual sensors are different, information about the spatial distribution of the brain sources can be deduced from the signals detected by an array of sensors.

The forward model provides a solution to the Maxwell equations for current sources located in the brain using approximations of the conductor geometry

Handbook of Neural Activity Measurement, ed. Romain Brette and Alain Destexhe. Published by Cambridge University Press. © Cambridge University Press 2012.

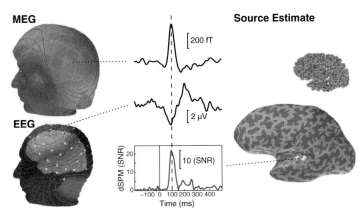

Figure 7.1 (See plate section for color version.) Example of MEG and EEG signals and source estimates. Left: maps of event-related magnetic fields (top) and electric scalp potentials (bottom), depicted with MRI-based reconstruction of the subject's head. The isocontour lines indicate the spatial pattern of the recorded signals at 100 ms after the onset of an auditory tone stimulus. The locations of the MEG sensors within a helmet-shaped array are indicated by the squares; the white dots indicate the locations of the EEG electrodes on the scalp. The top two waveforms show the time course of the signal recorded by one MEG and one EEG sensor, respectively. Right: map of the distributed source estimate. The noise-normalized minimum-norm estimate (dynamic statistical parametric map, dSPM) for the auditory evoked response at 100 ms latency is shown on the cerebral cortex. The realistic shape of the cortex was reconstructed from anatomical MRI; the cortical surface was inflated (bottom) to visualize the source estimates within sulci. The lowermost waveform shows the estimated time course of one source element in the auditory cortex in the lateral sulcus.

(shape, tissue conductivities) and the sensor configuration (engineering design, calibration). The forward model mainly concerns physics and geometry, and its accuracy depends on how well the physical and physiological details are known. In contrast, inverse modeling makes use of estimation theory and statistics. For example, in a Bayesian approach, the inverse problem is formulated as determining the posterior probability density of brain activation patterns, given the combination of the measured signals and a priori knowledge about the system (Tarantola, 1987; Kaipio and Somersalo, 2004). Various quantities of interest, such as the maximum a posteriori (MAP) estimate and Bayesian confidence regions for the source model parameters, are derived from this probability density function.

MEG and EEG source estimation is a powerful tool for non-invasive studies of human brain activity. Spatiotemporal patterns of spontaneous and event-related activity can be detected in the time scale of milliseconds, corresponding to electrophysiological neural activity, and the spatial scale of a few millimeters to centimeters, corresponding to functional areas in the cerebral cortex (Hari and

Lounasmaa, 1989). MEG and EEG source estimation has been applied to a wide range of sensory and cognitive paradigms, involving populations with normal and abnormal brain function (see, e.g., Lounasmaa et al., 1996; Salmelin and Kujala, 2006). The most important clinical application is presurgical evaluation of patients with epilepsy or brain tumors, where MEG and EEG can help to identify the location of epileptogenic foci and eloquent functional areas, such as the primary sensory and motor cortices and language areas (Knowlton et al., 1997; Papanicolaou et al., 1999; Nakasato and Yoshimoto, 2000; Ebersole and Hawes-Ebersole, 2007; Stufflebeam et al., 2009). For cognitive studies, MEG and EEG source estimation provides a means to study sequences of activation in networks of interacting functional cortical areas (see, e.g., Salmelin et al., 1994; Gross et al., 2001; Bar et al., 2006; Supp et al., 2007; Gow et al., 2008).

In the following we will describe the general relationship between neural activity and the MEG and EEG source estimates (Section 7.2), provide an overview of various MEG and EEG source estimation approaches (Section 7.3), discuss their general limitations and interpretation (Section 7.4), and briefly relate MEG and EEG to other neuroimaging techniques (Section 7.5).

7.2 Relationship between neural activity and the MEG and EEG source estimates

The relationship between the neural activity and the measurable MEG and EEG signals is discussed in detail by De Munck et al. in Chapter 6. Here we concentrate on source estimates derived from the measured signals. The emphasis is on how the estimated source currents, rather than the measured magnetic fields and scalp potentials, are related to the neural activity (Figure 7.2).

7.2.1 Source estimates: primary current distribution

The basic concept for linking macroscopic MEG and EEG source estimates and microscopic neural activity is the *primary current* density $\mathbf{J}^P(\mathbf{x}, t)$ (Plonsey, 1969; Tripp, 1983). The goal of the MEG and EEG source estimation is to determine the primary current distribution over space ($\mathbf{x}$) and time (t). Primary currents represent, at a macroscopic level, the vector sum of local active currents within a small region of the brain. In addition to the primary currents, there are passive, Ohmic *volume currents*, which depend on the primary currents and the conductive properties of the head. MEG and EEG signals are linear functions of the primary currents:

$$v_i(t) = \int \mathbf{L}_i(\mathbf{x}) \cdot \mathbf{J}^P(\mathbf{x}, t)d\mathbf{x}, \tag{7.1}$$

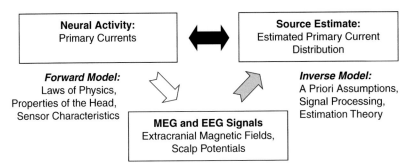

Figure 7.2 Schematic relationship between neural activity and the MEG and EEG source estimates. The sources of MEG and EEG signals are described in terms of primary currents. The forward model (white arrow) describes the dependence of the MEG and EEG signals on the primary currents, incorporating physical and geometrical properties of the head and the sensors. The inverse problem (gray arrow) refers to the estimation of the primary current distribution on the basis of the measured MEG and EEG signals. Source estimation is based on statistical models and various types of a priori assumptions. In the present chapter we describe several approaches to the MEG and EEG inverse problem and discuss how the source estimates are related to the actual neural activity (black arrow).

where $\mathbf{L}_i(\mathbf{x})$ is the lead field describing the spatial sensitivity pattern of the ith MEG or EEG sensor given the details of the conductivity distribution in the head which contains the primary currents.

Since MEG and EEG detect magnetic fields and electric potentials typically at a distance of several centimeters from the sources, the primary currents within a volume of the order of a few cubic millimeters can be approximated as a *current dipole*. The current dipole can be taken as the basic element for all the source estimation methods discussed here. Sometimes, higher terms from the current multipole expansion can be useful to describe spatially extended sources (Nolte and Curio, 2000; Jerbi et al., 2004). However, the quadrupole and higher moments can always be approximated with combinations of dipoles close to each other. We emphasize that the use of current dipoles as a basic element for source reconstruction is conceptually different from the equivalent current dipole (ECD) model discussed in Section 2.3: ECD is the dipole that best explains a given measured signal pattern, whereas a dipole element is a discretized macroscopic representation of the net primary current within a small region of the brain (Figure 7.3). Thus, the current dipole element plays a role analogous to a volume element (voxel) in magnetic resonance images (MRI). Note, however, that in MEG and EEG source estimation the source amplitude for a single dipole element usually cannot be determined independently of all the other dipole elements.

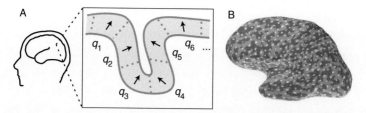

Figure 7.3 Representation of the sources of MEG and EEG signals using current dipole elements. A. In a discretized model of the primary current distribution, the source space is divided into macroscopic dipole elements. The dipole elements, depicted here by the small arrows corresponding to elements q_j, are often constrained to lie in the cerebral cortex, sometimes also constrained to be oriented perpendicular to the cortical surface. B. Locations of dipole elements (white dots) shown on a lateral view of the inflated cortical surface.

Using discrete current dipole elements, Equation (7.1) can be approximated as

$$v_i(t) = \sum_{j=1}^{n} a_{ij} \, q_j(t), \quad i = 1, \ldots, m, \qquad (7.2a)$$

where m is the number of sensors and n is the number of source elements. In matrix notation

$$\mathbf{v}(t) = \mathbf{A} \, \mathbf{q}(t), \qquad (7.2b)$$

where $\mathbf{v}(t) = (v_1(t), \ldots, v_n(t))^T$ is the signal vector, $\mathbf{q}(t) = (q_1(t), \ldots, q_n(t))^T$ is a vector whose elements $q_j(t)$ correspond to the activation (source strength) of the jth dipole element representing the primary current distribution $\mathbf{J}^p(\mathbf{x})$ in the brain, and $\mathbf{A} = [\mathbf{a}_1, \ldots, \mathbf{a}_n]$ is the forward matrix. The columns $\mathbf{a}_j = (a_{1j}, \ldots, a_{mj})^T$ of $\mathbf{A}$ are the signal patterns of source elements of unit amplitude. The rows of $\mathbf{A}$ are the discretized versions $\mathbf{L}_i(\mathbf{x}_j)$ of the continuous lead fields. For notational simplicity, we assume that there is only one source element at each location, indexed by j, as is the case, for example, when the source orientation is constrained to be perpendicular to the cortical surface (see Section 7.4.2(i) below); however, often it would be necessary to modify the equations to express the vector components of a current dipole at a given location explicitly.

A source estimate can be written formally as

$$\hat{\mathbf{q}}(t) = W(\mathbf{v}(t)), \qquad (7.3)$$

where W is an operator that, when applied to the data, gives an estimate of $\mathbf{q}(t)$. In general, the solution to the inverse problem is not unique (Helmholtz, 1853): many different source patterns $\mathbf{q}(t)$ can produce the same measured signal values $\mathbf{v}(t)$. Different choices for W will result in different estimates $\hat{\mathbf{q}}(t)$.

7.2.2 Silent sources

Non-uniqueness in MEG and EEG source estimation is equivalent to the existence of "silent" source patterns $\mathbf{q}_0$, which do not result in a measurable extracranial signal (i.e. $\mathbf{v}_0 = \mathbf{A}\mathbf{q}_0 \approx \mathbf{0}$): if the difference between two solutions $\mathbf{q}_0 = \mathbf{q}_2 - \mathbf{q}_1$ is silent, then both explain the data equally well ($\mathbf{v}_2 = \mathbf{A}\mathbf{q}_2 = \mathbf{A}\mathbf{q}_1 + \mathbf{A}\mathbf{q}_0 \approx \mathbf{A}\mathbf{q}_1 = \mathbf{v}_1$). From a practical point of view, any source for which the recordable signal is indistinguishable from the measurement noise is a silent source. Source configurations that differ from each other by a silent source belong to the same equivalence class of solutions. Importantly, the non-uniqueness cannot be eliminated merely by increasing the number of sensors in the MEG and EEG sensor arrays: even if the scalp potential and the extracranial magnetic field were known with arbitrary high accuracy, there is always a possibility of source configurations that do not generate scalp potentials or extracranial magnetic fields at all.

Silent sources can originate at both microscopic and macroscopic levels. "Closed" source patterns (Lorente de No, 1947) produce no measurable signal in either MEG or EEG. For these sources, the integral of the microscopic source currents over a region corresponding to a single macroscopic source element vanishes. In this case, little can be done in source estimation: the microscopically canceling patterns are undetectable by MEG and EEG, since the primary current itself is zero. All source estimation methods discussed here assume that the source space consists of macroscopic current dipole elements; no inferences are made about patterns of microscopically canceling source currents. It is important, however, to consider microscopically closed sources when MEG and EEG signals are compared with other brain imaging methods, as the activity related to those sources may contribute to, for example, functional MRI signals (cf. Section 7.5).

Macroscopically, non-zero primary current distributions can also be silent. These silent source patterns may consist of one or more non-zero source elements that are each silent in MEG and/or EEG, or they may contain complex patterns of source elements for which the signals cancel out. A particular example of a silent source element is a radially oriented current dipole in spherically symmetric conductivity geometry, which is silent in MEG (Baule and McFee, 1965; Grynszpan and Geselowitz, 1973). Mathematically, the silent source configurations correspond to the null space of the forward operator $\mathbf{A}$. Because of the different sensitivity patterns of MEG and EEG sensors, sources that are silent in MEG may be detectable in EEG and vice versa (Cohen and Cuffin, 1983). The detectability of individual dipole elements is determined by the lead fields of the sensors (Figure 7.4). In general, for both MEG and EEG the sensitivity to individual dipoles diminishes rapidly as a function of distance between the source and the sensor (Cohen and Cuffin, 1983; Hillebrand and Barnes, 2002; Goldenholz et al., 2009). The

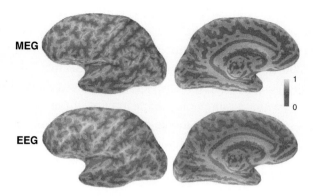

Figure 7.4 (See plate section for color version.) Sensitivity maps of MEG and EEG sensor arrays. The color code indicates the normalized root-mean-square signal value for 102 MEG magnetometers (left) and 70 EEG electrodes (right), as generated by cortically constrained dipole elements oriented perpendicular to the cortical surface. Lateral (top) and medial (bottom) views of the inflated left hemisphere are shown. Note that MEG is most sensitive to superficial tangentially oriented source located within the walls of sulci, whereas the sensitivity map for EEG is more uniform.

lead field components of MEG sensors are insensitive to dipole elements oriented perpendicularly with respect to the inner surface of the skull (radial dipoles) (Melcher and Cohen, 1988; Haueisen et al., 1995; Ahlfors et al., 2010a). In contrast, EEG lead fields are sensitive to both tangential and radial sources.

Cancelation of signals occurs also for more complex configurations of sources (Ahlfors et al., 2010b). For example, a macroscopically closed source pattern, which produces little measurable signal in either MEG or EEG may occur when the two walls of a sulcus are equally active (cf. Figure 7.3A). The cancelation of sources in the opposing sides of sulci, but not at the bottom of sulci or crowns of gyri, may result in selective cancelation of tangential versus radial source elements (Eulitz et al., 1997; Ahlfors et al., 2010b), suggesting that comparison of MEG and EEG could provide useful insights about widespread coherent cortical activity (Freeman et al., 2009). Selective cancelation of signals due to background brain activity is also an important factor in determining the relative signal to noise ratio for the source of interest in MEG versus EEG (de Jongh et al., 2005; Goldenholz et al. 2009; Ahlfors et al., 2010b).

7.3 Source estimation methods

In source estimation, a priori assumptions are necessary to select one (the "optimal") solution from the equivalence class of all the possible solutions that can explain the measured data (Backus, 1971). Note that the a priori assumptions made in specific inverse estimation methods are different from the general a priori-like assumptions that are part of the forward model and, therefore, common to all MEG and EEG source estimation methods. Assumptions made at the forward modeling stage include the choice of the source space (see Section 7.4.2(i)) and the use of the current dipole as the basic element for source estimation (Section 7.2.1). In contrast, the specific assumptions characterizing different source estimation methods concern the expected spatiotemporal distribution of the amplitudes of the source elements q_j. These assumptions can be about the global properties of the primary current distribution (e.g. energy, L1-norm, or mean local smoothness), about the extent of regional activity, or about preferred locations suggested, for example, by fMRI data.

Below we describe various inverse estimation methods in terms of two broad categories (Baillet et al., 2001): parametric source localization methods and distributed source reconstruction methods. In the parametric source localization methods, which are closely related to the concept of an equivalent current dipole (ECD), foci of activation are determined using a test function evaluated for each possible source element q_j. The distributed source reconstruction methods are closer to what would commonly be considered as imaging of neural activity, providing estimates $\hat{\mathbf{q}}$ for the spatial pattern of the primary currents.

7.3.1 Parametric source localization

In parametric source localization approaches, the activation is typically assumed to consist of a small number of spatially separate focal sources represented by current dipoles. The definition of "focal" is related to the spatial resolution of MEG and EEG (see Section 7.4.4 below); in practice, a source whose extent is $\sim 1\ cm^3$ or less can be considered focal. A central goal of source localization methods is to identify the locations of the foci. Here we discuss three approaches: the ECD model, the multiple signal classification (MUSIC) technique, and the beamformer method. In each of these, a test function is evaluated for possible source elements within the source space. Typically the source space is the cranial volume or the cortical surface. A minimum (or maximum) point of the test function indicates a likely location of an activation focus. Maps of the test functions can be considered "pseudo-images" of activity; however, these maps should not be interpreted as direct estimates of the primary current $\mathbf{q}$.

(i) Equivalent current dipoles (ECD) In the ECD approach, the MEG and EEG signals are assumed to be generated by one or more focal sources that can be approximated by current dipoles. The location $\mathbf{x}$ and the dipole moment vector $\mathbf{Q}$ of a single ECD are determined by minimizing a non-linear test function, usually the weighted sum of squared differences between the measured v_i^{meas} and modeled $v_i^{ECD}(\mathbf{x}, \mathbf{Q})$ signal values:

$$\arg \min_{\mathbf{x}, \mathbf{Q}} \sum_{i=1}^{m} [v_i^{meas} - v_i^{ECD}(\mathbf{x}, \mathbf{Q})]^2. \qquad (7.4)$$

Generalization of Equation (7.4) to multiple ECDs is straightforward; however, solving for multiple ECDs is difficult in practice. In the single-ECD case, it may be computationally feasible to perform an exhaustive search of the six-dimensional parameter space at some specified resolution. With multi-ECD models, an iterative search is usually the only computationally feasible approach. The total number of unknown parameters becomes prohibitively large as the number of ECDs increases: for each ECD there are three non-linear parameters for the location and three linear parameters for the dipole moment. A common problem is that the iterative search algorithms easily converge to a local minimum, which may provide a poor representation of the actual primary current distribution. Sophisticated algorithms based on global optimization (Uutela et al., 1998) or multi-start simplex (Huang, et al. 1998) methods can mitigate this problem.

The ECD approach has been the most popular MEG and EEG source estimation method, and works especially well for many types of sensory evoked response where the activity can be modeled with a small number of focal sources (Brenner et al., 1978; Fender, 1987; Scherg and Ebersole, 1994; Hari and Forss, 1999; Martinez et al., 1999). The ECD model has been successfully applied also to higher-level cognitive event-related activity (see, e.g., Salmelin et al., 1994; Simos et al., 2002).

(ii) Multiple signal classification (MUSIC) method The MUSIC method enables localization of multiple foci without having to perform a search in a high-dimensional parameter space (Mosher et al., 1992). In MUSIC, using a time window of data, the m-dimensional space of signals recorded by m sensors is divided into signal and noise subspaces. In MUSIC, likely foci of activation are determined by evaluating a test (scanning) function that is a measure of the projection of the measured data to the noise subspace. For a dipole with the forward vector $\mathbf{a}_j$, the test function is

$$\|\mathbf{P}_s^{\perp} \mathbf{a}_j\|^2 / \|\mathbf{a}_j\|^2, \qquad (7.5)$$

where $\mathbf{P}_s^{\perp}$ is the projection matrix onto the noise subspace. The advantage of MUSIC over multi-ECD fit is that only a single scalar value of the cost function needs to be evaluated for each location in the source space, whereas in the multi-dipole fit, all possible combinations of dipoles need to be evaluated.

In principle, MUSIC will correctly detect multiple foci of activation, located at the local minima of the scanning function, as long as the source time courses are linearly independent. However, in the presence of noise, strongly correlated sources cause the algorithm to fail (Supek and Aine, 1993; Sekihara et al., 1997; Mosher et al., 1999). In principle, to detect correlated sources, higher-order versions of the methods could be constructed; however, this is computationally challenging. Another difficulty in the original MUSIC approach is to find the multiple local minima of the scanning function. This problem has been alleviated by the introduction of recursively applied and projected (RAP) MUSIC, which modifies the scanning function after each source has been found by projecting out the contribution of the sources already identified (Mosher and Leahy, 1999). As a result, at each iteration it is sufficient to find the global minimum of the scanning function to locate a new dipole.

(iii) Beamformers In the beamformer approach, the test function is a spatial filter that is constructed to extract optimally the contribution of a focus of activation at a given location in the source space (Van Veen et al., 1997; Gross et al., 2001; Vrba and Robinson, 2001; Hillebrand et al., 2005; Sekihara and Nagarajan, 2008). By applying the linear beamformer operator constructed for each individual source element to the recorded data, one obtains a pseudo-image of the source activity. In adaptive beamforming, the spatial filter is constructed using the data covariance matrix, whereas the beamformer operator can also be non-adaptive, based on a priori assumptions and the properties of the forward matrix. The minimum-variance adaptive beamformer output is computed as

$$\hat{q}_j(t) = \mathbf{w}_j^T \mathbf{v}(t), \tag{7.6a}$$

where the weight vector is

$$\mathbf{w}_j^T = \mathbf{C}_d^{-1}\mathbf{a}_j / \mathbf{a}_j^T \mathbf{C}_d^{-1}\mathbf{a}_j; \tag{7.6b}$$

$\mathbf{C}_d$ is the data covariance matrix. This spatial filter has the desired property of unit gain: $\mathbf{w}_j^T \mathbf{a}_j = 1$ for the source of interest at location j. Similarly to MUSIC, beamformer methods based on the covariance matrix may fail to detect activity that is correlated across multiple foci (Sekihara et al., 2002). Beamformer methods have been particularly useful for localizing non-phase locked source activity, for example changes in the spectral content of MEG signals during the performance of sensorimotor and cognitive tasks (see, e.g., Taniguchi et al., 2000; Gross et al., 2001; Cheyne et al., 2003; Fawcett et al., 2004; Herdman et al., 2004; Itier et al., 2006).

(iv) Determining the time course of the activations Once the locations of the foci of activation have been identified, there are various ways to determine the source amplitudes as a function of time. In the *spatiotemporal multi-dipole model* (Scherg and Von Cramon, 1986), the source locations are held fixed and the amplitude time courses are solved using a linear least-squares method. This is analogous to the distributed source reconstruction approach described below in Section 7.3.2, in which the amplitudes for a fixed set of source elements are determined. Note, however, that usually the spatiotemporal multidipole model is an over-determined problem (typically with less than ten sources), whereas the distributed source reconstruction is an under-determined problem (often with thousands of source elements). For the spatiotemporal multi-dipole model, several methods have been proposed for determining the model order, i.e. the appropriate number of ECDs in a multi-dipole model (De Munck, 1990). This is often difficult, and one reason why the resulting multi-dipole model may depend on the experience of the person analyzing the data. In the beamformer approach, the time course for each source is solved using a spatial filter optimized for that source location. However, a combination of solutions for multiple source locations will not generally add up to explain the measured field pattern; in this way the beamformer approach differs from both the spatiotemporal multi-dipole model and the distributed source reconstructions.

7.3.2 *Distributed source reconstruction*

For distributed source models, the source locations within the source space are pre-specified, and the amplitudes of the sources are determined on the basis of the MEG and EEG data (Hamalainen and Ilmoniemi, 1984, 1994; Dale and Sereno, 1993; Ioannides et al., 1993; Jeffs et al., 1987). The reconstruction problem is usually under-determined: data from tens or hundreds of sensors are used to estimate thousands of source amplitudes.

In distributed source reconstruction methods, the source estimate $\hat{\mathbf{q}}$ typically minimizes a cost function with two terms:

$$\arg \min_{\mathbf{q}} [f_1(\mathbf{v} - \mathbf{Aq}) + f_2(\mathbf{q})]. \tag{7.7}$$

The first term $f_1(\mathbf{v} - \mathbf{Aq})$ ensures that solution $\hat{\mathbf{q}}$ is consistent with the measured data $\mathbf{v}$ within the limits of experimental uncertainties. Commonly,

$$f_1(\mathbf{v} - \mathbf{Aq}) = ||(\mathbf{v} - \mathbf{Aq})||_{\mathbf{C}}^2 = (\mathbf{v} - \mathbf{Aq})^T \mathbf{C}^{-1}(\mathbf{v} - \mathbf{Aq}), \tag{7.8}$$

where $\mathbf{C}$ is the noise covariance matrix. The second term $f_2(\mathbf{q})$ imposes an additional criterion required to select one solution from the set of the possible solutions that can explain the measured data.

It is often possible to give Equation (7.7) a probabilistic (Bayesian) interpretation, in which $f_1(\mathbf{v} - \mathbf{Aq})$ represents the logarithm of the likelihood function (Equation (7.8) corresponds to Gaussian noise), and $f_2(\mathbf{q})$ incorporates a priori information about the sources (Clarke, 1989; Baillet and Garnero, 1997; Dale et al., 2000; Phillips et al., 2002). In this interpretation, the general solution is the a posteriori probability distribution rather than one particular solution (Schmidt et al., 1999; Nummenmaa et al., 2007); from the a posteriori distribution one can then extract specific solutions, for example, the maximum a posteriori (MAP) estimate. The Bayesian statistical approach provides a unified framework for incorporating prior information into the MEG/EEG inverse solutions (Tarantola, 1987; Kaipio and Somersalo, 2004). The Bayesian interpretation can be problematic, however, as in practice $f_2(\mathbf{q})$ may describe some heuristic constraint, often computationally convenient, rather than actual a priori knowledge.

Several choices for the prior term $f_2(\mathbf{q})$ have been proposed. In the weighted minimum L2-norm case

$$f_2(\mathbf{q}) = ||\mathbf{q}||_{\mathbf{R}}^2 = \mathbf{q}^T \mathbf{R}^{-1} \mathbf{q}, \tag{7.9}$$

where $\mathbf{R}$ is the a priori source covariance matrix. The minimum energy (minimum-L2-norm) constraint is obtained with $\mathbf{R} = \mathbf{I}$ (identity matrix) (Hamalainen and Ilmoniemi, 1994). The LORETA method uses a smoothness constraint by incorporating the Laplacian operator (i.e. second spatial derivative) in $\mathbf{R}$ (Pascual-Marqui et al., 1994). An inverse gain weighting (lead field normalization) in $\mathbf{R}$ deemphasizes superficial sources (see Section 7.4.3 below) (Fuchs et al., 1999; Lin et al., 2006c). Also, a priori information about likely source locations, as obtained, for example, from fMRI data, can be incorporated in $\mathbf{R}$ (Liu et al., 1998; Ahlfors and Simpson, 2004). When both f_1 and f_2 have a quadratic form (Equations (7.8) and (7.9)), the weighted minimum-L2-norm solution can be expressed in closed form:

$$\hat{\mathbf{q}} = \mathbf{W} \, \mathbf{v} = \mathbf{R}\mathbf{A}^T (\mathbf{A}\mathbf{R}\mathbf{A}^T + \mathbf{C})^{-1}\mathbf{v}. \tag{7.10}$$

Below, this solution is called the minimum-norm estimate (MNE) (Hamalainen and Ilmoniemi, 1994). From Equation (7.10), we can see that the estimate can be thought of as a weighted sum of transformed lead field vectors $\mathbf{l}_i^R = \mathbf{R} \, \mathbf{l}_i$, where $\mathbf{l}_i$ are the rows of the forward matrix $\mathbf{A} = [\mathbf{l}_1^T, \ldots, \mathbf{l}_m^T]^T$

$$\hat{\mathbf{q}} = \Sigma_i \beta_i \mathbf{l}_i^R, \tag{7.11}$$

where the scalar weight factor β_i is the ith element of $\boldsymbol{\beta} = (\mathbf{A}\mathbf{R}\mathbf{A}^T + \mathbf{C})^{-1}\mathbf{v}$. In the case of $\mathbf{R} = \mathbf{I}$, the estimate $\hat{\mathbf{q}}$ is always in the subspace spanned by the sensor lead fields $\mathbf{l}_i$ (Hamalainen and Ilmoniemi, 1994).

For the minimum L1-norm constraint (Matsuura and Okabe, 1997; Uutela et al., 1999)

$$f_2(\mathbf{q}) = ||\mathbf{q}||_1 = \Sigma_j |q_j|. \tag{7.12}$$

The solution to Equation (7.7) with the condition Equation (7.12) can be obtained by linear programming methods (Uutela et al., 1999). The minimum-L1-norm estimate, also known as the minimum-current estimate (MCE), typically shows spatially more focal peaks in the estimated source distribution, compared with the minimum-L2-norm solution (see also Section 7.4.3(i)). An alternative way to obtain distributed sources with sharp foci is to use iterative methods, such as the FOCUSS method which converges to a distribution with multiple foci (Gorodnitsky et al., 1995; see also Ioannides et al., 1990). In a variation of the cortically constrained MCE called VESTAL (Huang et al., 2006), the usually discontinuous MCE source waveforms are projected onto the subspace derived from the spatiotemporal MEG/EEG data matrix; this results in smooth waveforms resembling those produced by the minimum-L2-norm approach while preserving the focal quality of the L1-norm solutions. The recently proposed L1L2-norm estimate provides spatially focal but temporally smooth estimates by using a temporal basis function expansion for the source waveforms and a combined L1L2-norm regularizer for the basis function coefficients (Ding and He, 2008; Ou et al., 2009a).

The distributed source estimates can be converted to noise-normalized maps called dynamical statistical parametric maps (dSPM) by dividing the estimated amplitude for each element $\hat{q}_j$ of the source estimate by the estimated magnitude of noise sources at the corresponding location (Dale et al., 2000):

$$s_j = \hat{q}_j / \sqrt{(\mathbf{w}_j \mathbf{C} \, \mathbf{w}_j^T)}, \qquad (7.13)$$

where $\mathbf{w}_j$ is the jth row of $\mathbf{W}$. The dSPMs are similar to the statistical parametric maps commonly used to display regions of significant activation in fMRI. A slightly different normalization is used in a method called sLORETA in which the matrix $\mathbf{C}$ in Equation (7.13) is replaced by the data covariance matrix $\mathbf{C}_d = \mathbf{C} + \mathbf{A}\mathbf{R}\mathbf{A}^T$ (Pascual-Marqui, 2002; Wagner et al., 2004). Since each source element q_j is divided by a constant specific to that element, the noise normalization does not change the waveform of the estimated time course for individual elements. However, the relative magnitude between elements will be changed by the normalization; therefore, the spatial patterns are different in the original source estimates (MNE) and noise-normalized (dSPM) maps (Figure 7.5) (see also discussion of the depth profile in Section 7.4.3(ii)). The noise-normalized maps can be particularly useful for identifying regions of interest in which further statistical analysis of the source estimates will be performed, for example between experimental conditions or subject groups (Ciesielski et al., 2005; Wehner et al., 2007).

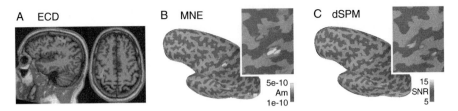

Figure 7.5 (See plate section for color version.) Example of different source estimates for the same MEG data. A. The equivalent current dipole (ECD) for the
averaged contralateral response at 20 ms after an electrical stimulation of the right
median nerve at the wrist. The green dot indicates the location of the ECD, superimposed on axial (left) and sagittal (right) slices of anatomical MRI. The green
line extending from the dot indicates the orientation of the current dipole moment.
The location of the ECD is consistent with the primary somatosensory cortex in
the posterior wall of the central sulcus. B. The minimum-norm estimate (MNE)
computed on the cortical surface. The location of the ECD is indicated with a
green dot. In this distributed source estimate the reconstructed activity spreads
from the posterior wall of the central sulcus also to the neighboring regions. The
striped pattern is typical in inflated views of the cortex, especially when an orientation constraint for the dipole elements is imposed. The maximum in the MNE
is at the crown of the gyrus, slightly more superficial than the ECD. C. The noise-
normalized MNE, known as dynamic statistical parametric map (dSPM), shows a
somewhat different spatial distribution than the MNE. In this case, the maximum
in dSPM is deeper compared with the ECD and the maximum in MNE. In general, however, the overall localization of activity is consistent across these three
different source estimates.

7.4 Interpretation of the source estimates

As described in the previous section, a priori assumptions are used in MEG and
EEG source estimation methods to select one solution $\hat{\mathbf{q}}$ from the equivalence class
of all those source distributions that are consistent with the measured extracranial
signals. Experimental uncertainties extend this equivalence class, thereby increasing the ambiguity in source estimation. In this section we examine some basic
limitations in MEG and EEG source estimation, including effects of measurement
noise and other uncertainties. We also consider the general effects of different a
priori assumptions on the source estimates, and discuss the question of the spatial
resolution of MEG and EEG.

7.4.1 Effects of measurement noise

Noise in the recorded MEG and EEG signals consists of instrumentation noise,
environmental noise, and subject noise. The subject noise includes magnetic fields
and electric potentials generated by the eyes, the heart, muscles, and magnetic
material in the subject. In addition, if activity that is time locked to a stimulus

(event-related/evoked fields and potentials) is studied, the ongoing background brain activity is often considered to be noise with respect to the signals of interest. Environmental and non-brain subject noise in MEG can be substantially reduced by shielded rooms (Cohen and Halgren, 2003) and various types of spatial filtering techniques, such as signal space projection (SSP) or source signal separation (SSS) (Uusitalo and Ilmoniemi, 1997; Taulu and Simola, 2006). The effect of noise on the source estimates can be evaluated statistically, for example by computing confidence limits for the ECD parameters (Hamalainen et al., 1993; Fuchs et al., 2004; Darvas et al., 2005). In the case of a single ECD, the largest uncertainty is in the depth direction (Hari et al., 1988; Mosher et al., 1993). Uncertainties in the ECD parameters are often interdependent, for example, if the ECD is erroneously localized deeper than the actual source, the estimated amplitude of the dipole moment will be larger than the true amplitude. For distributed source reconstruction methods, to prevent over-fitting of the data and amplification of noise, higher levels of noise require more regularization, leading to loss of spatial details in the source estimates (Ahlfors et al., 1992; Hamalainen and Ilmoniemi, 1994).

7.4.2 Uncertainties in forward modeling

Next we discuss how three types of uncertainties in forward modeling affect the source estimates: (i) misspecification of the source space, (ii) uncertainty in the description of the physical properties of the head (shape, conductivity), and (iii) uncertainty in the positioning of the sensors with respect to the head.

(i) Source space The source space, specifying the allowed locations for MEG and EEG source estimates, is usually restricted either to the intracranial volume or to the cerebral gray matter (cortex). The source space should include all brain regions contributing to the measured signals; otherwise, the source estimates may be distorted by signals originating from sources outside the source space. For example, if the cerebral cortex is chosen as the source space, the presence of cerebellar activity may result in erroneously estimated activity in the inferior occipitotemporal areas, or signals from the eyes may affect the source estimates in the orbitofrontal cortex. In general, one cannot assume that activity in regions that are excluded from the source space will not affect the source estimates.

Restricting the source space to the cerebral cortex, as identified from anatomical MRI (Dale and Sereno, 1993), is physiologically justifiable: the primary currents are mainly due to postsynaptic currents in cortical pyramidal cells, whereas action potentials in the white matter produce little if any measurable signal in MEG and EEG (Niedermeyer and Lopes da Silva, 1999). Since the characteristic spatial resolution (see Section 7.4.4) is of the same order of magnitude as the spacing of gyral folding patterns (~1 cm), merely restricting the source locations to the gray matter instead of the whole cerebral volume may have little effect on the source estimates. In contrast, additional constraints based on properties of the cortical sheet, such as neighborhood relationships or the orientation of the local surface normal, can substantially affect the source estimates. It is important, however, to keep in mind that the spatial resolution of MEG and EEG source estimates is a "volume" property even when the source space is constrained to the cortical surface. Therefore, the source estimate may misleadingly suggest activation in regions that are functionally unrelated to the actual source, for example on the opposite wall of a sulcus.

The orientations of the sources are expected to be perpendicular to the cortical surface, since for symmetry reasons the signals due to the components of the primary currents parallel to the cortical surface are likely to cancel out macroscopically (Niedermeyer and Lopes da Silva, 1999). Distributed source reconstructions with a cortical orientation constraint (Dale and Sereno, 1993; Lin et al., 2006a) commonly show striped patterns of estimated activity on the cortical surface (cf. Figure 7.5), corresponding to source elements whose orientation is similar to that of the actual source. The orientation constraint for the source elements in the in-between regions is likely to be inconsistent with the measured MEG or EEG data; therefore, these elements show little activation in the source estimate. For practical computations, adequate spatial sampling of the source space is achieved in most cases when the distance been adjacent source locations is ~3–10 mm intervals, in accordance with the characteristic spatial resolution of MEG and EEG (cf. Section 7.4.4). Some mislocalization of the sources due to the sparseness of the spatial sampling may occur when a strict orientation constraint is used. To alleviate this potential problem, a "loose" orientation constraint was introduced, in which the perpendicular orientation is given more a priori weighting, but without completely excluding the other orientations (Lin et al., 2006a).

(ii) Head model The accuracy of the head model used for the forward calculation has, in general, a larger effect on EEG than on MEG. The spherically symmetric conductor model has been popular for both MEG and EEG, but the boundary element model (BEM) is commonly used when anatomical MRI data for individual subjects are available (see Chapter 6). For EEG, a four-compartment model is commonly used (brain, CSF, skull, and scalp), and the source localization accuracy is moderately dependent on the accuracy of the conductivity values and the tissue boundaries (Vaughan, 1974; Scherg and Von Cramon, 1986). If the assumed skull/soft tissue conductivity ratio is too high or too low, the EEG forward solution for a given source will be too focal or too widespread, respectively. This can result in a systematic error of the ECD location being too superficial or too deep. Note that from the point of view of source estimation, the uncertainties in the conductivity and thickness of the skull are closely interdependent, i.e. a thicker skull with higher conductivity is practically equivalent to a thinner skull with lower conductivity (Goncalves et al., 2000).

One problem with the use of the sphere model for EEG is the difficulty in co-registering the results with the individual subject's anatomy. For MEG, this is straightforward, as the anatomical MRI can be co-registered with the actual position of the MEG sensor array and the source estimates can be superimposed directly on the MR images. In contrast, for EEG the true locations of the electrodes on the scalp have to be transformed to the surface of the spherical head model. Therefore, for accurate registration of EEG source estimates with individual anatomy, a conductivity model that takes into account the shape of the head is necessary.

For MEG, the spherical model is both computationally simpler and more accurate than for EEG. The magnetic field of a current dipole in spherically symmetric connectivity geometry can be computed using a closed-form expression, which is independent of the actual conductivity values (Sarvas, 1987). In general, the MEG source estimates are insensitive to small changes in the location of the center of the spherical model (Tarkiainen et al., 2003). Comparison of simulated source estimates obtained with the spherical model with anatomically more realistic BEM models has indicated that the largest errors in ECD location are expected in the frontal regions (Hamalainen and Sarvas, 1989; Tarkiainen et al., 2003).

(iii) Sensor positioning Besides the specification of the source space and the head model, errors in the forward model are also caused by uncertainty in the positioning of the sensors with respect to the head. The positions of the EEG electrodes on the scalp may vary from session to session, but are fixed during a single recording session, whereas the relative position of the head and the MEG sensor array may also change during the recording. Due to the smoothness of the spatial patterns of both the scalp potentials and the extracranial magnetic fields, the source estimates are only moderately sensitive to small errors in sensor locations. This is different from fMRI, in which head-movement-related shifts of high-contrast boundaries in sequentially obtained images may be misinterpreted as changes in the brain activation. In an MEG study with 7–13 year old children performing a cognitive task, the average change in the relative position of the MEG sensors and head was found to be ~12 mm during a session of four 5-minute runs (Wehner et al., 2008). The source localization error due to this amount of head movement was of the same order of magnitude as is typically caused by other sources of uncertainty, such as measurement noise and the head model. The peak amplitude in the MNE was found to be slightly reduced when the head movements were not compensated for (Wehner et al., 2008).

The amount of error in the MEG source estimates due to head movement can depend on the location of the brain activity. Whereas translational changes in the head position transfer almost directly to general localization errors, rotation of the head commonly results in largest uncertainties in the frontal lobe. As the back of the head usually leans against the helmet-shaped MEG sensor array, movement-related errors are smallest in occipital areas. Even without head movements, the positioning of the MEG sensor array can have a prominent effect on the source estimates: if the sensors are far from the activity of interest, the noise-normalized maps of distributed source reconstructions can show substantially reduced activation (Marinkovic et al., 2004).

7.4.3 Explicit and implicit consequences of specific a priori assumptions

Whenever possible, one should try to evaluate whether the a priori assumptions defining the different source estimation approaches are consistent with the underlying brain activity using independent means, for example fMRI data, intracranial recordings, or computational modeling. In some situations it may be possible to evaluate a priori assumptions on the basis of the MEG and EEG data themselves: for example, a low value of the goodness-of-fit may indicate that a given set of data cannot be fully explained by an ECD. For the under-determined distributed source reconstructions, however, the solution can always be made consistent with the recorded data at the specified level of regularization, and there is little intrinsic information about whether the a priori cost function $f_2(\mathbf{q})$ in Equation (7.7) is justified in terms of providing a meaningful source estimate.

Obviously, the properties of the source estimates will reflect the a priori assumptions made: a single-ECD model gives one dipole location, constraint on the spatial derivative will result in a smooth distributed solution, etc. It is important, however, to make a distinction between explicit a priori assumptions and their secondary, indirect consequences. Here we discuss how two properties of distributed source reconstructions, smoothness (or focality) and depth profile, are related to various a priori assumptions.

(i) Smoothness Of the distributed source estimates, neither MNE nor MCE imposes any explicit constraint on the spatial extent of the estimate. The apparent difference in the focality between MNE and MCE solutions is a secondary consequence of the fact that nearby source elements typically have similar signal patterns, i.e. similar forward solutions. As an example, consider two dipole elements close to each other such that the signal patterns are indistinguishable in the presence of noise, thereby contributing equally to the first term $f_1(\mathbf{v} - \mathbf{Aq})$ in Equation (7.7). The allocation of amplitudes between the two elements in the distributed source estimate is determined by the constraint $f_2(\mathbf{q})$. If the Euclidean norms of the signal patterns for the two source elements are similar, the L2-norm in MNE is minimized when the elements are allocated similar amplitudes, providing a spatially smooth solution. In contrast, the L1-norm of the source pattern in MCE is minimized by a focal solution in which all activity is assigned to the source element that explains the measured signals with a smaller dipole amplitude.

Thus, the apparent smoothness of the MNE solutions is a consequence of the combination of two things: nearby dipole elements with like orientation typically generate similar characteristic signal patterns (i.e. the lead fields are smooth), and the minimum-L2-norm constraint prefers solutions in which the total amount of source amplitude is divided among sources with similar signal patterns. Properly regularized MNE solutions are typically smooth over both space and time. Because of the smooth spatial distribution, the estimated time course for any given source element is also typically smooth in MNE. In contrast, in the minimum-L1-norm estimates, noise may switch the allocation of the amplitudes at subsequent time instants between dipole elements that have similar signal patterns, resulting in discontinuous source reconstructions over time at a given source location.

(ii) Depth Various types of distributed source reconstructions differ also in their depth profile, even when there is no explicit constraint for the depth dependence. The unconstrained L2-MNE (for which $\mathbf{R} = \mathbf{I}$ in Equation (7.9)) assigns largest amplitudes to superficial source elements. The bias toward superficial sources can be compensated for using an "inverse gain" (also known as "lead field normalization") weighting, which is inversely proportional to the sensitivity of the sensor array to a source at a given location (cf. Figure 7.4) (Fuchs et al., 1999; Lin et al., 2006c). This is not an explicit constraint on the depth of the source, but utilizes the fact that the gain factor is generally larger for source elements that are close to the sensors. The practical value of the inverse-gain a priori constraint appears to rely on the preference of many experimenters to interpret the MEG and EEG source estimates in terms of focal activations. It is usually desirable for a distributed source estimate to show the maximum amplitude close to the location of a focal source (Greenblatt et al., 2005).

In the case of noise-normalized distributed solutions, division by the noise estimate, which typically has a similar depth profile to that for the sources of interest, compensates for the bias towards superficial sources (Dale et al., 2000; Pascual-Marqui, 2002; Wagner et al., 2004). Interestingly, with noise-normalized estimates, an explicit inverse-gain constraint has little effect, as the influence of the constraint tends to cancel out when applied also to the noise (Lin et al., 2006c). A similar cancelation effect can also occur for other types of constraints, such as fMRI weighting.

7.4.4 What is the spatial resolution of MEG and EEG?

The deceptively simple question regarding the spatial resolution of MEG and EEG source estimates turns out to be quite difficult to answer. Not only does the quality of the source estimates depend on the signal to noise ratio of the recorded data as well as various sources of uncertainties in forward modeling (see Sections 7.4.1–7.4.2), but it also depends on the overall pattern of activity across the brain.

Distributed MEG and EEG source reconstructions are often visualized on anatomical MRI, such that the maps of the source estimates graphically look like fMRI activation maps. It is important, however, to realize the very different nature of these maps and the different interpretation they require. Unlike fMRI, in which the signals obtained for each voxel can be determined largely independently of the activation in other voxels, the MEG and EEG source estimates critically depend on the global pattern of the primary currents. This can be characterized by the cross-talk and point-spread functions for dipole source elements (obtained from the resolution function, i.e. the product of the inverse and forward operators), which are typically widespread and contain multiple local maxima (Grave de Peralta-Menendez and Gonzalez-Andino, 1998; Liu et al., 1998). Notably, the MEG and EEG source estimates are not simply spatially smoothed, low-pass filtered versions of the true primary current distribution.

In interpreting distributed source estimates, it is important to keep in mind that a maximum point in the estimate does not necessarily coincide with the focus of activation, and that "ghost" sources may appear due to specific signal patterns generated by combinations of sources. This can lead to alternative interpretations of the measured data: for example, whether the large maximum in frontal EEG electrodes for auditory evoked responses is generated by radially oriented sources in the frontal cortex or by bilateral tangential sources in the auditory cortex at the Sylvian fissure (Vaughan, 1974). Similarly, the bilateral auditory sources generate a "dipolar" magnetic field pattern of the parietal lobe in MEG (Hamalainen et al., 1995). Often, the combination of MEG and EEG data can resolve these types of ambiguities.

One consequence of the limited spatial resolution resulting from the smoothness of the sensor lead fields is that there is, in general, only little information in the MEG and EEG data about the actual extent of a patch of activity within a functional region. Therefore, the apparent extent of a local source is usually determined more by the a priori assumptions in the source estimation method rather than by the actual extent of the activation. We prefer to use the term "extended source" for local patches of activity, and reserve "distributed source" for patterns consisting of multiple patches across the cortex (Ahlfors et al., 2010b). When interpreting distributed source reconstructions, the choice of thresholding used in the visualization will affect the apparent local extent of the active regions, scaling proportionally to the amplitude of the source. Conversely, when locations of ECDs are visualized on anatomical MRI, one has to keep in mind that the true activation is not necessarily focal. Confidence volume for ECDs should be indicated together with the dipole location, but again, the confidence volume (which depends on the signal to noise ratio) may have little to do with the actual extent of the source. However, if one can assume that the source strength per cortical area is constant (Lu and Williamson, 1991), then the ECD amplitude can be used to estimate the extent of the activation.

If the activation can be assumed to be confined to a small region of the brain at a given time instant, such that the source is well represented by a single ECD, a spatial resolution of a few millimeters can be achieved. For example, the relative location of the representation of individual fingers within the primary somatosensory cortex can be determined (Suk et al., 1991; Elbert et al., 1995; Nakamura et al., 1998). With artificial current dipoles, a mean localization accuracy of 3 mm in MEG and 7–8 mm in EEG in a skull phantom (Leahy et al., 1998), and 8 mm in MEG and 10 mm in EEG in the human brain (Cohen et al., 1990) has been reported.

7.5 Comparison with other techniques and future developments

A unique property of MEG and EEG source estimation, compared with other brain imaging techniques, is that it allows non-invasive mapping of electrical activity in the human brain with millisecond time resolution. Intracranial electrophysiological recordings can provide a high spatial resolution, but they require surgery and can typically be obtained only from a restricted part of the brain. Functional MRI (fMRI) is non-invasive and can provide a millimeter-scale spatial resolution over the whole brain without the type of ambiguity related to non-uniqueness in MEG and EEG source estimates. However, fMRI and other hemodynamic methods, such as PET and optical near-infrared imaging, only indirectly measure neural activity and do not have the same time resolution as MEG and EEG. Transcranial magnetic stimulation (TMS), by means of interrupting and/or stimulating neural activity, can provide information about the relationship between brain processes and behavior complementary to that obtained from imaging methods.

Multimodal imaging, in which data from different types of techniques are merged, holds the promise of achieving a more comprehensive view of brain activation than is possible with any single method alone. In MEG studies, scalp

EEG is commonly recorded simultaneously. Technological innovations have also made it possible to record EEG simultaneously with TMS or fMRI (Ilmoniemi et al., 1997; Bonmassar et al., 1999; Paus, 1999; Gotman et al., 2006; Ritter and Villringer, 2006). It is also possible to record intracranial EEG or optical signals simultaneously with MEG (Sander et al., 2007; Santiuste et al., 2008; Dalal et al., 2009; Ou et al., 2009b). Simultaneous detection of MEG and MR signals using a common SQUID detector has also been introduced (Volegov et al., 2004; Clarke et al., 2007). The whole-head MEG sensor arrays have in part inspired novel designs of multi-coil MRI arrays and their use for fast MR data acquisition (Lin et al., 2006b). Direct detection of neural currents using MR is an exciting possibility; however, it is still unclear whether the required sensitivity can be achieved for human studies (Petridou et al., 2006; Witzel et al., 2008).

Even when not acquired simultaneously, the different types of recordings can provide useful complementary information about the brain function. In regard to MEG and EEG source estimation, the good spatial resolution of fMRI (and PET) can be used to guide the MEG and EEG source estimates (Heinze et al., 1994; Simpson et al., 1995; Liu et al., 1998; Ahlfors et al., 1999; Dale et al., 2000). Specifically, fMRI can suggest likely locations of MEG and EEG sources among all the possible sources that, due to the non-uniqueness of the inverse problem, can explain the measured MEG and EEG signals equally well (Ahlfors and Simpson, 2004). A better understanding of the relationship between electrophysiological and hemodynamic signals, emerging from simultaneous recordings in animal models (see Chapter 11) as well as from biophysical modeling (Tagamets and Horwitz, 1998; Murakami et al., 2003; David et al., 2006; Jones et al., 2007) is likely to facilitate the integration of the different types of non-invasive human brain imaging data.

In the future, advances in the interpretation of MEG and EEG signals are expected to result from the integration of source estimation with sophisticated signal processing methods. The inherent limitations of MEG and EEG due to non-uniqueness and cancelation effects should be taken into account throughout the analysis and interpretation of the data in relation to models of brain dynamics. For example, distributed source estimation can be integrated with the analysis of oscillatory activity (Gross et al., 2001; Lin et al., 2004; Astolfi et al., 2005; Pantazis et al., 2009). Thus, the source estimation of non-invasive MEG and EEG recordings provides a valuable tool for studies of large-scale brain networks.

Acknowledgments

We would like to acknowledge support from the National Institutes of Health (NS057500, EB009048, and P41RR14075).

References

Ahlfors, S. P. and Simpson, G. V. (2004). Geometrical interpretation of fMRI-guided MEG/EEG inverse estimates. *NeuroImage*, **22**, 323–332.

Ahlfors, S. P., Ilmoniemi, R. J. and Hamalainen, M. S. (1992). Estimates of visually evoked cortical currents. *Electroencephalogr. Clin. Neurophysiol.*, **82**, 225–236.

Ahlfors, S. P., Simpson, G. V., Dale, A. M., Belliveau, J. W., Liu, A. K., Korvenoja, A., Virtanen, J., Huotilainen, M., Tootell, R. B., Aronen, H. J. and Ilmoniemi, R. J. (1999). Spatiotemporal activity of a cortical network for processing visual motion revealed by MEG and fMRI. *J. Neurophysiol.*, **82**, 2545–2555.

Ahlfors, S. P., Han, J., Belliveau, J. W. and Hamalainen, M. S. (2010a). Sensitivity of MEG and EEG to source orientation. *Brain Topogr.*, **23**, 227–232.

Ahlfors, S. P., Han, J., Lin, F. H., Witzel, T., Belliveau, J. W., Hamalainen, M. S. and Halgren, E. (2010b). Cancellation of EEG and MEG signals generated by extended and distributed sources. *Human Brain Mapping*, **31**, 140–149.

Astolfi, L., Cincotti, F., Mattia, D., Babiloni, C., Carducci, F., Basilisco, A., Rossini, P. M., Salinari, S., Ding, L., Ni, Y., He, B. and Babiloni, F. (2005). Assessing cortical functional connectivity by linear inverse estimation and directed transfer function: simulations and application to real data. *Clin. Neurophysiol.*, **116**, 920–932.

Backus, G. E. (1971). Inference from inadequate and inaccurate data. In: W. H. Reid (editor), *Mathematical Problems in the Geophysical Sciences 2. Inverse Problems, Dynamo Theory, and Tides*. Providence, RI: American Mathematical Society, pp. 1–105.

Baillet, S. and Garnero, L. (1997). A Bayesian approach to introducing anatomo-functional priors in the EEG/MEG inverse problem. *IEEE Trans. Biomed. Eng.*, **44**, 374–385.

Baillet, S., Mosher, J. C. and Leahy, R. M. (2001). Electromagnetic brain mapping. *IEEE Signal Process. Mag.*, **18**, 14–30.

Bar, M., Kassam, K. S., Ghuman, A. S., Boshyan, J., Schmidt, A. M., Dale, A. M., Hamalainen, M. S., Marinkovic, K., Schacter, D. L., Rosen, B. R. and Halgren, E. (2006). Top-down facilitation of visual recognition. *Proc. Natl. Acad. Sci. USA*, **103**, 449–454.

Baule, G. and McFee, R. (1965). Theory of magnetic detection of the heart's electrical activity. *J. Appl. Phys.*, **36**, 2066–2073.

Berger, H. (1929). Uber das Elektrenkephalogramm des Menschen. *Arch. Psychiat. Nervenkr.*, **87**, 527–570.

Bonmassar, G., Anami, K., Ives, J. and Belliveau, J. W. (1999). Visual evoked potential (VEP) measured by simultaneous 64-channel EEG and 3T fMRI. *NeuroReport*, **10**, 1893–1897.

Brenner, D., Lipton, J., Kaufman, L. and Williamson, S. J. (1978). Somatically evoked magnetic fields of the human brain. *Science*, **199**, 81–83.

Cheyne, D., Gaetz, W., Garnero, L., Lachaux, J. P., Ducorps, A., Schwartz, D. and Varela, F. J. (2003). Neuromagnetic imaging of cortical oscillations accompanying tactile stimulation. *Brain Res. Cogn. Brain. Res.*, **17**, 599–611.

Ciesielski, K. T., Hamalainen, M. S., Lesnik, P. G., Geller, D. A. and Ahlfors, S. P. (2005). Increased MEG activation in OCD reflects a compensatory mechanism specific to the phase of a visual working memory task. *NeuroImage*, **24**, 1180–1191.

Clarke, C. J. S. (1989). Probabilistic methods in a biomagnetic inverse problem. *Inverse Problems*, **5**, 999–1012.

Clarke, J., Hatridge, M. and Mossle, M. (2007). SQUID-detected magnetic resonance imaging in microtesla fields. *Annu. Rev. Biomed. Eng.*, **9**, 389–413.

Cohen, D. (1972). Magnetoencephalography: detection of the brain's electrical activity with a superconducting magnetometer. *Science*, **175**, 664–666.

Cohen, D. and Cuffin, B. N. (1983). Demonstration of useful differences between magnetoencephalogram and electroencephalogram. *Electroencephalogr. Clin. Neurophysiol.*, **56**, 38–51.

Cohen, D. and Halgren, E. (2003). Magnetoencephalogaphy. In: G. Adelman (editor), *Encyclopedia of Neuroscience* (2nd edition). Boston, MA: MIT Press.

Cohen, D., Cuffin, B. N., Yunokuchi, K., Maniewski, R., Purcell, C., Cosgrove, G. R., Ives, J., Kennedy, J. G. and Schomer, D. L. (1990). MEG versus EEG localization test using implanted sources in the human brain. *Ann. Neurol.*, **28**, 811–817.

Dalal, S. S., Baillet, S., Adam, C., Ducorps, A., Schwartz, D., Jerbi, K., Bertrand, O., Garnero, L., Martinerie, J. and Lachaux, J. P. (2009). Simultaneous MEG and intracranial EEG recordings during attentive reading. *NeuroImage*, **45**, 1289–1304.

Dale, A. M. and Sereno, M. I. (1993). Improved localization of cortical activity by combining EEG and MEG with MRI cortical surface reconstruction: a linear approach. *J. Cogn. Neurosci.*, **5**, 162–176.

Dale, A. M., Liu, A. K., Fischl, B. R., Buckner, R. L., Belliveau, J. W., Lewine, J. D. and Halgren, E. (2000). Dynamic statistical parametric mapping: combining fMRI and MEG for high-resolution imaging of cortical activity. *Neuron*, **26**, 55–67.

Darvas, F., Rautiainen, M., Pantazis, D., Baillet, S., Benali, H., Mosher, J. C., Garnero, L. and Leahy, R. M. (2005). Investigations of dipole localization accuracy in MEG using the bootstrap. *NeuroImage*, **25**, 355–368.

David, O., Kiebel, S. J., Harrison, L. M., Mattout, J., Kilner, J. M. and Friston, K. J. (2006). Dynamic causal modeling of evoked responses in EEG and MEG. *NeuroImage*, **30**, 1255–1272.

de Jongh, A., De Munck, J. C., Goncalves, S. I. and Ossenblok, P. (2005). Differences in MEG/EEG epileptic spike yields explained by regional differences in signal-to-noise ratios. *J. Clin. Neurophysiol.*, **22**, 153–158.

De Munck, J. C. (1990). The estimation of time varying dipoles on the basis of evoked potentials. *Electroencephalogr. Clin. Neurophysiol.*, **77**, 156–160.

Ding, L. and He, B. (2008). Sparse source imaging in electroencephalography with accurate field modeling. *Human Brain Mapping*, **29**, 1053–1067.

Ebersole, J. S. and Hawes-Ebersole, S. (2007). Clinical application of dipole models in the localization of epileptiform activity. *J. Clin. Neurophysiol.*, **24**, 120–129.

Elbert, T., Pantev, C., Wienbruch, C., Rockstroh, B. and Taub, E. (1995). Increased cortical representation of the fingers of the left hand in string players. *Science*, **270**, 305–307.

Eulitz, C., Eulitz, H. and Elbert, T. (1997). Differential outcomes from magneto- and electroencephalography for the analysis of human cognition. *Neurosci. Lett.*, **227**, 185–188.

Fawcett, I. P., Barnes, G. R., Hillebrand, A. and Singh, K. D. (2004). The temporal frequency tuning of human visual cortex investigated using synthetic aperture magnetometry. *NeuroImage*, **21**, 1542–1553.

Fender, D. (1987). Source localization of brain electrical activity. In: A. Gevins and A. Remond (editors), *Handbook of Electroencephalography and Clinical Neurophysiology*, Amsterdam: Elsevier, pp. 355–403.

Freeman, W. J., Ahlfors, S. P. and Menon, V. (2009). Combining fMRI with EEG and MEG in order to relate patterns of brain activity to cognition. *Int. J. Psychophysiol.*, **73**, 43–52.

Fuchs, M., Wagner, M., Kohler, T. and Wischmann, H. A. (1999). Linear and nonlinear current density reconstructions. *J. Clin. Neurophysiol.*, **16**, 267–295.

Fuchs, M., Wagner, M. and Kastner, J. (2004). Confidence limits of dipole source reconstruction results. *Clin. Neurophysiol.*, **115**, 1442–1451.

Goldenholz, D. M., Ahlfors, S. P., Hämäläinen, M. S., Sharon, D., Ishitobi, M., Vaina, L. M. and Stufflebeam, S. M. (2009). Mapping the signal-to-noise-ratios of cortical sources in magnetoencephalography and electroencephalography. *Human Brain Mapping*, **30**, 1077–1086.

Goncalves, S., de Munck, J. C., Heethaar, R. M., Lopes da Silva, F. H. and van Dijk, B. W. (2000). The application of electrical impedance tomography to reduce systematic errors in the EEG inverse problem – a simulation study. *Physiol. Meas.*, **21**, 379–393.

Gorodnitsky, I. F., George, J. S. and Rao, B. D. (1995). Neuromagnetic source imaging with FOCUSS: a recursive weighted minimum norm algorithm. *Electroencephalogr. Clin. Neurophysiol.*, **95**, 231–251.

Gotman, J., Kobayashi, E., Bagshaw, A. P., Benar, C. G. and Dubeau, F. (2006). Combining EEG and fMRI: a multimodal tool for epilepsy research. *J. Magn. Reson. Imaging*, **23**, 906–920.

Gow, Jr., D. W., Segawa, J. A., Ahlfors, S. P. and Lin, F. H. (2008). Lexical influences on speech perception: a Granger causality analysis of MEG and EEG source estimates. *NeuroImage*, **43**, 614–623.

Grave de Peralta-Menendez, R. and Gonzalez-Andino, S. L. (1998). A critical analysis of linear inverse solutions to the neuroelectromagnetic inverse problem. *IEEE Trans. Biomed. Eng.*, **45**, 440–448.

Greenblatt, R. E., Ossadtchi, A. and Pflieger, M. E. (2005). Local linear estimators for the bioelectromagnetic inverse problem. *IEEE Trans. Signal Process.*, **53**, 3403–3412.

Gross, J., Kujala, J., Hamalainen, M., Timmermann, L., Schnitzler, A. and Salmelin, R. (2001). Dynamic imaging of coherent sources: studying neural interactions in the human brain. *Proc. Natl. Acad. Sci. USA*, **98**, 694–699.

Grynszpan, F. and Geselowitz, D. B. (1973). Model studies of the magnetocardiogram. *Biophys. J.*, **13**, 911–925.

Hamalainen, M. S. and Ilmoniemi, R. J. (1984). Interpreting measured magnetic fields of the brain: estimates of current distributions. Helsinki University of Technology.

Hamalainen, M. S. and Ilmoniemi, R. J. (1994). Interpreting magnetic fields of the brain: minimum norm estimates. *Med. Biol. Eng. Comput.*, **32**, 35–42.

Hamalainen, M. S. and Sarvas, J. (1989). Realistic conductivity geometry model of the human head for interpretation of neuromagnetic data. *IEEE Trans. Biomed. Eng.*, **36**, 165–171.

Hamalainen, M., Hari, R., Ilmoniemi, R. J., Knuutila, J. and Lounasmaa, O. V. (1993). Magnetoencephalography – theory, instrumentation, and applications to noninvasive studies of the working human brain. *Rev. Mod. Phys.*, **65**, 413–497.

Hamalainen, M., Hari, R., Lounasmaa, O. V. and Williamson, S. J. (1995). Do auditory stimuli activate human parietal brain regions? *NeurolReport*, **6**, 1712–1714.

Hansen, P., Kringelbach, M. and Salmelin, R. (2010). *MEG: An Introduction to Methods*. New York: Oxford University Press.

Hari, R. and Forss, N. (1999). Magnetoencephalography in the study of human somatosensory cortical processing. *Philos. Trans. R. Soc. London Ser. B*, **354**, 1145–1154.

Hari, R. and Lounasmaa, O. V. (1989). Recording and interpretation of cerebral magnetic fields. *Science*, **244**, 432–436.

Hari, R., Joutsiniemi, S. and Sarvas, J. (1988). Spatial resolution of neuromagnetic records: theoretical calculations in a spherical model. *Electroencephalogr. Clin. Neurophysiol.*, **71**, 64–72.

Haueisen, J., Ramon, C., Czapski, P. and Eiselt, M. (1995). On the influence of volume currents and extended sources on neuromagnetic fields: a simulation study. *Ann. Biomed. Eng.*, **23**, 728–739.

Heinze, H. J., Mangun, G. R., Burchert, W., Hinrichs, H., Scholz, M., Munte, T. F., Gos, A., Scherg, M., Johannes, S., Hundeshagen, H., et al. (1994). Combined spatial and temporal imaging of brain activity during visual selective attention in humans. *Nature*, **372**, 543–546.

Helmholtz, H. (1853). Ueber einige Gesetze der Vertheilung elektrischer Strome in korperlichen Leitern, mit Anwendung auf die thierisch-elektrischen Versuche. *Ann. Phys. Chem.*, **89**, 211–233, 353–377.

Herdman, A. T., Wollbrink, A., Chau, W., Ishii, R. and Pantev, C. (2004). Localization of transient and steady-state auditory evoked responses using synthetic aperture magnetometry. *Brain Cogn.*, **54**, 149–151.

Hillebrand, A. and Barnes, G. R. (2002). A quantitative assessment of the sensitivity of whole-head MEG to activity in the adult human cortex. *NeuroImage*, **16**, 638–650.

Hillebrand, A., Singh, K. D., Holliday, I. E., Furlong, P. L. and Barnes, G. R. (2005). A new approach to neuroimaging with magnetoencephalography. *Human Brain Mapping*, **25**, 199–211.

Huang, M., Aine, C. J., Supek, S., Best, E., Ranken, D. and Flynn, E. R. (1998). Multi-start downhill simplex method for spatio-temporal source localization in magnetoencephalography. *Electroencephalogr. Clin. Neurophysiol.*, **108**, 32–44.

Huang, M. X., Dale, A. M., Song, T., Halgren, E., Harrington, D. L., Podgorny, I., Canive, J. M., Lewis, S. and Lee, R. R. (2006). Vector-based spatial-temporal minimum L1-norm solution for MEG. *NeuroImage*, **31**, 1025–1037.

Ilmoniemi, R. J., Virtanen, J., Ruohonen, J., Karhu, J., Aronen, H. J., Naatanen, R. and Katila, T. (1997). Neuronal responses to magnetic stimulation reveal cortical reactivity and connectivity. *NeuroReport*, **8**, 3537–3540.

Ioannides, A. A., Bolton, J. P. R., Hasson, R. and Clarke, C. J. S. (1990). Localised and distributed source solutions for the biomagnetic inverse problem. II. In: S. J. Williamson et al. (editors), *Advances in Biomagnetism*, New York: Plenum, pp. 591–594.

Ioannides, A. A., Hellstrand, E. and Abraham-Fuchs, K. (1993). Point and distributed current density analysis of interictal epileptic activity recorded by magnetoencephalography. *Physiol. Meas.*, **14**, 121–130.

Itier, R. J., Herdman, A. T., George, N., Cheyne, D. and Taylor, M. J. (2006). Inversion and contrast-reversal effects on face processing assessed by MEG. *Brain Res*, **1115**, 108–120.

Jeffs, B., Leahy, R. and Singh, M. (1987). An evaluation of methods for neuromagnetic image reconstruction. *IEEE Trans. Biomed. Eng.*, **34**, 713–723.

Jerbi, K., Baillet, S., Mosher, J. C., Nolte, G., Garnero, L. and Leahy, R. M. (2004). Localization of realistic cortical activity in MEG using current multipoles. *NeuroImage*, **22**, 779–793.

Jones, S. R., Pritchett, D. L., Stufflebeam, S. M., Hamalainen, M. and Moore, C. I. (2007). Neural correlates of tactile detection: a combined magnetoencephalography and biophysically based computational modeling study. *J. Neurosci.*, **27**, 10751–10764.

Kaipio, J. and Somersalo, E. (2004). *Statistical and Computational Inverse Problems*. New York: Springer.

Knowlton, R. C., Laxer, K. D., Aminoff, M. J., Roberts, T. P., Wong, S. T. and Rowley, H. A. (1997). Magnetoencephalography in partial epilepsy: clinical yield and localization accuracy. *Ann. Neurol.*, **42**, 622–631.

Leahy, R. M., Mosher, J. C., Spencer, M. E., Huang, M. X. and Lewine, J. D. (1998). A study of dipole localization accuracy for MEG and EEG using a human skull phantom. *Electroencephalogr. Clin. Neurophysiol.*, **107**, 159–173.

Lin, F. H., Witzel, T., Hamalainen, M. S., Dale, A. M., Belliveau, J. W. and Stufflebeam, S. M. (2004). Spectral spatiotemporal imaging of cortical oscillations and interactions in the human brain. *NeuroImage*, **23**, 582–595.

Lin, F. H., Belliveau, J. W., Dale, A. M. and Hamalainen, M. S. (2006a). Distributed current estimates using cortical orientation constraints. *Human Brain Mapping*, **27**, 1–13.

Lin, F. H., Wald, L. L., Ahlfors, S. P., Hamalainen, M. S., Kwong, K. K. and Belliveau, J. W. (2006b). Dynamic magnetic resonance inverse imaging of human brain function. *Magn. Reson. Med.*, **56**, 787–802.

Lin, F. H., Witzel, T., Ahlfors, S. P., Stufflebeam, S. M., Belliveau, J. W. and Hamalainen, M. S. (2006c). Assessing and improving the spatial accuracy in MEG source localization by depth-weighted minimum-norm estimates. *NeuroImage*, **31**, 160–171.

Liu, A. K., Belliveau, J. W. and Dale, A. M. (1998). Spatiotemporal imaging of human brain activity using functional MRI constrained magnetoencephalography data: Monte Carlo simulations. *Proc. Natl. Acad. Sci. USA*, **95**, 8945–8950.

Lorente de No, R. (1947). *A Study of Nerve Physiology*. Studies of the Rockefeller Institute 132, Chapter 16.

Lounasmaa, O. V., Hamalainen, M., Hari, R. and Salmelin, R. (1996). Information processing in the human brain: magnetoencephalographic approach. *Proc. Natl. Acad. Sci. USA*, **93**, 8809–8815.

Lu, Z. L. and Williamson, S. J. (1991). Spatial extent of coherent sensory-evoked cortical activity. *Exp. Brain Res.*, **84**, 411–416.

Marinkovic, K., Cox, B., Reid, K. and Halgren, E. (2004). Head position in the MEG helmet affects the sensitivity to anterior sources. *Neurol. Clin. Neurophysiol.*, **30**, 1–6.

Martinez, A., Anllo-Vento, L., Sereno, M. I., Frank, L. R., Buxton, R. B., Dubowitz, D. J., Wong, E. C., Hinrichs, H., Heinze, H. J. and Hillyard, S. A. (1999). Involvement of striate and extrastriate visual cortical areas in spatial attention. *Nature Neurosci.*, **2**, 364–369.

Matsuura, K. and Okabe, Y. (1997). A robust reconstruction of sparse biomagnetic sources. *IEEE Trans. Biomed. Eng.*, **44**, 720–726.

Melcher, J. R. and Cohen, D. (1988). Dependence of the MEG on dipole orientation in the rabbit head. *Electroencephalogr. Clin. Neurophysiol.*, **70**, 460–472.

Michel, C. M., Koenig, T., Brandeis, D., Gianotti, L.R.R. and Wackermann, J. (2009). *Electrical Neuroimaging*. Cambridge: Cambridge University Press.

Mosher, J. C. and Leahy, R. M. (1999). Source localization using recursively applied and projected (RAP) MUSIC. *IEEE Trans. Signal Process.*, **47**, 332–340.

Mosher, J. C., Lewis, P. S. and Leahy, R. M. (1992). Multiple dipole modeling and localization from spatio-temporal MEG data. *IEEE Trans. Biomed. Eng.*, **39**, 541–557.

Mosher, J. C., Spencer, M. E., Leahy R. M. and Lewis, P. S. (1993). Error bounds for EEG and MEG dipole source localization. *Electroencephalogr. Clin. Neurophysiol.*, **86**, 303–321.

Mosher, J. C., Baillet, S. and Leahy, R. M. (1999). EEG source localization and imaging using multiple signal classification approaches. *J. Clin. Neurophysiol.*, **16**, 225–238.

Murakami, S., Hirose, A. and Okada, Y. C. (2003). Contribution of ionic currents to magnetoencephalography (MEG) and electroencephalography (EEG) signals generated by guinea-pig CA3 slices. *J. Physiol.*, **553**, 975–985.

Nakamura, A., Yamada, T., Goto, A., Kato, T., Ito, K., Abe, Y., Kachi, T. and Kakigi., R. (1998). Somatosensory homunculus as drawn by MEG. *NeuroImage*, **7**, 377–386.

Nakasato, N. and Yoshimoto, T. (2000). Somatosensory, auditory, and visual evoked magnetic fields in patients with brain diseases. *J. Clin. Neurophysiol.*, **17**, 201–211.

Niedermeyer, E. and Lopes da Silva, F. (1999). *Electroencephalography: Basic Principles, Clinical Applications, and Related Fields.* Philadelphia, PA: Lippincott, Williams and Wilkins.

Nolte, G. and Curio, G. (2000). Current multipole expansion to estimate lateral extent of neuronal activity: a theoretical analysis. *IEEE Trans. Biomed. Eng.*, **47**, 1347–1355.

Nummenmaa, A., Auranen, T., Hamalainen, M. S., Jaaskelainen, I. P., Lampinen, J., Sams, M. and Vehtari, A. (2007). Hierarchical Bayesian estimates of distributed MEG sources: theoretical aspects and comparison of variational and MCMC methods. *NeuroImage*, **35**, 669–685.

Ou, W., Hamalainen, M. S. and Golland, P. (2009a). A distributed spatio-temporal EEG/MEG inverse solver. *NeuroImage*, **44**, 932–946.

Ou, W., Nissila, I., Radhakrishnan, H., Boas, D. A., Hamalainen, M. S. and Franceschini, M. A. (2009b). Study of neurovascular coupling in humans via simultaneous magnetoencephalography and diffuse optical imaging acquisition. *NeuroImage*, **46**, 624–632.

Pantazis, D., Simpson, G. V., Weber, D. L., Dale, C. L., Nichols, T. E. and Leahy, R. M. (2009). A novel ANCOVA design for analysis of MEG data with application to a visual attention study. *NeuroImage*, **44**, 164–174.

Papanicolaou, A. C., Simos, P. G., Breier, J. I., Zouridakis, G., Willmore, L. J., Wheless, J. W., Constantinou, J. E., Maggio, W. W. and Gormley, W. B. (1999). Magnetoencephalographic mapping of the language-specific cortex. *J. Neurosurg*, **90**, 85–93.

Pascual-Marqui, R. D. (2002) . Standardized low-resolution brain electromagnetic tomography (sLORETA): technical details. *Methods Find. Exp. Clin. Pharmacol.*, **24D**, 5–12.

Pascual-Marqui, R. D., Michel, C. M. and Lehmann, D. (1994). Low resolution electromagnetic tomography: a new method for localizing electrical activity in the brain. *Int. J. Psychophysiol.*, **18**, 49–65.

Paus, T. (1999). Imaging the brain before, during, and after transcranial magnetic stimulation. *Neuropsychologia*, **37**, 219–224.

Petridou, N., Plenz, D., Silva, A. C., Loew, M., Bodurka, J. and Bandettini, P. A. (2006). Direct magnetic resonance detection of neuronal electrical activity. *Proc. Natl. Acad. Sci. USA.*, **103**, 16015–16020.

Phillips, C., Rugg, M. D. and Friston, K. J. (2002). Systematic regularization of linear inverse solutions of the EEG source localization problem. *NeuroImage*, **17**, 287–301.

Plonsey, R. (1969). *Bioelectric Phenomena.* New York: McGraw-Hill.

Ritter, P. and Villringer, A. (2006). Simultaneous EEG-fMRI. *Neurosci. Biobehav. Rev.*, **30**, :823–838.

Salmelin, R. and Kujala, J. (2006). Neural representation of language: activation versus long-range connectivity. *Trends Cogn. Sci.*, **10**, 519–525.

Salmelin, R., Hari, R., Lounasmaa, O. V. and Sams, M. (1994). Dynamics of brain activation during picture naming. *Nature*, **368**, 463–465.

Sander, T. H., Liebert, A., Mackert, B. M., Wabnitz, H., Leistner, S., Curio, G., Burghoff, M., Macdonald, R. and Trahms, L. (2007). DC-magnetoencephalography and time-resolved near-infrared spectroscopy combined to study neuronal and vascular brain responses. *Physiol. Meas.*, **28**, 651–664.

Santiuste, M., Nowak, R., Russi, A., Tarancon, T., Oliver, B., Ayats, E., Scheler, G. and Graetz, G. (2008). Simultaneous magnetoencephalography and Intracranial EEG registration: technical and clinical aspects. *J. Clin. Neurophysiol.*, **25**, 331–339.

Sarvas, J. (1987). Basic mathematical and electromagnetic concepts of the biomagnetic inverse problem. *Phys. Med. Biol.*, **32**, 11–22.

Scherg, M. and Ebersole, J. S. (1994). Brain source imaging of focal and multifocal epileptiform EEG activity. *Neurophysiol. Clin.*, **24**, 51–60.

Scherg, M. and Von Cramon, D. (1986). Evoked dipole source potentials of the human auditory cortex. *Electroencephalogr. Clin. Neurophysiol.*, **65**, 344–360.

Schmidt, D. M., George, J. S. and Wood, C. C. (1999). Bayesian inference applied to the electromagnetic inverse problem. *Human Brain Mapping*, **7**, 195–212.

Sekihara, K. and Nagarajan, S. S. (2008). *Adaptive Spatial Filters for Electromagnetic Brain Imaging*. Berlin: Springer.

Sekihara, K., Poeppel, D., Marantz, A., Koizumi, H. and Miyashita, Y. (1997). Noise covariance incorporated MEG-MUSIC algorithm: a method for multiple-dipole estimation tolerant of the influence of background brain activity. *IEEE Trans. Biomed. Eng.*, **44**, 839–847.

Sekihara, K., Nagarajan, S. S., Poeppel, D. and Marantz, A. (2002). Performance of an MEG adaptive-beamformer technique in the presence of correlated neural activities: effects on signal intensity and time-course estimates. *IEEE Trans. Biomed. Eng.*, **49**, 1534–1546.

Simos, P. G., Breier, J. I., Fletcher, J. M., Foorman, B. R., Castillo, E. M. and Papanicolaou, A. C. (2002). Brain mechanisms for reading words and pseudowords: an integrated approach. *Cereb. Cortex*, **12**, 297–305.

Simpson, G. V., Pflieger, M. E., Foxe, J. J., Ahlfors, S. P., Vaughan, Jr., H. G., Hrabe, J., Ilmoniemi, R. J. and Lantos, G. (1995). Dynamic neuroimaging of brain function. *J. Clin. Neurophysiol.*, **12**, 432–449.

Stufflebeam, S. M., Tanaka, N. and Ahlfors, S. P. (2009). Clinical applications of magnetoencephalography. *Human Brain Mapping*, **30**, 1813–1823.

Suk, J., Ribary, U., Cappell, J., Yamamoto, T. and Llinas, R. (1991). Anatomical localization revealed by MEG recordings of the human somatosensory system. *Electroencephalogr. Clin. Neurophysiol.*, **78**, 185–196.

Supek, S. and Aine, C. J. (1993). Simulation studies of multiple dipole neuromagnetic source localization: model order and limits of source resolution. *IEEE Trans. Biomed. Eng.*, **40**, 529–540.

Supp, G. G., Schlogl, A., Trujillo-Barreto, N., Muller, M. M. and Gruber, T. (2007). Directed cortical information flow during human object recognition: analyzing induced EEG gamma-band responses in brain's source space. PLoS ONE 2:e684.

Tagamets, M. A. and Horwitz, B. (1998). Integrating electrophysiological and anatomical experimental data to create a large-scale model that simulates a delayed match-to-sample human brain imaging study. *Cereb. Cortex*, **8**, 310–320.

Taniguchi, M., Kato, A., Fujita, N., Hirata, M., Tanaka, H., Kihara, T., Ninomiya, H., Hirabuki, N., Nakamura, H., Robinson, S. E., Cheyne, D. and Yoshimine, T. (2000). Movement-related desynchronization of the cerebral cortex studied with spatially filtered magnetoencephalography. *NeuroImage*, **12**, 298–306.

Tarantola, A. (1987). *Inverse Problem Theory. Methods for Data Fitting and Model Parameter Estimation*. Amsterdam: Elsevier.

Tarkiainen, A., Liljestrom, M., Seppa, M. and Salmelin, R. (2003). The 3D topography of MEG source localization accuracy: effects of conductor model and noise. *Clin. Neurophysiol.*, **114**, 1977–1992.

Taulu, S. and Simola, J. (2006). Spatiotemporal signal space separation method for rejecting nearby interference in MEG measurements. *Phys. Med. Biol.*, **51**, 1759–1768.

Tripp, J. H. (1983). Physical concepts and mathematical models. In: S. J. Williamson, G-L. Romani, L. Kaufman and I. Modena (editors), *Biomagnetism: An Interdisciplinary Approach*, New York: Plenum Press, pp. 101–139.

Uusitalo, M. A. and Ilmoniemi, R. J. (1997). Signal-space projection method for separating MEG or EEG into components. *Med. Biol. Eng. Comput.*, **35**, 135–140.

Uutela, K., Hamalainen, M. and Salmelin, R. (1998). Global optimization in the localization of neuromagnetic sources. *IEEE Trans. Biomed. Eng.*, **45**, 716–723.

Uutela, K., Hamalainen, M. and Somersalo, E. (1999). Visualization of magnetoencephalographic data using minimum current estimates. *NeuroImage*, **10**, 173–180.

Van Veen, B. D., van Drongelen, W., Yuchtman, M. and Suzuki, A. (1997). Localization of brain electrical activity via linearly constrained minimum variance spatial filtering. *IEEE Trans. Biomed. Eng.*, **44**, 867–880.

Vaughan, Jr., H. G. (1974). The analysis of scalp-recorded brain potentials. In: R. F. Thompson and M. M. Patterson (editors). *Bioelectric Recording Techniques*, New York: Academic Press, pp. 157–207.

Volegov, P., Matlachov, A. N., Espy, M. A., George, J. S. and Kraus, Jr., R. H. (2004). Simultaneous magnetoencephalography and SQUID detected nuclear MR in microtesla magnetic fields. *Magn. Reson. Med.*, **52**, 467–470.

Vrba, J. and Robinson, S. E. (2001). Signal processing in magnetoencephalography. *Methods*, **25**, 249–271.

Wagner, M., Fuchs, M. and Kastner, J. (2004). Evaluation of sLORETA in the presence of noise and multiple sources. *Brain Topogr.*, **16**, 277–280.

Wehner, D. T., Ahlfors, S. P. and Mody, M. (2007). Effects of phonological contrast on auditory word discrimination in children with and without reading disability: a magnetoencephalography (MEG) study. *Neuropsychologia*, **45**, 3251–3262.

Wehner, D. T., Hamalainen, M. S., Mody, M. and Ahlfors, S. P. (2008). Head movements of children in MEG: quantification, effects on source estimation, and compensation. *NeuroImage*, **40**, 541–550.

Witzel, T., Lin, F. H., Rosen, B. R. and Wald, L. L. (2008). Stimulus-induced rotary saturation (SIRS): a potential method for the detection of neuronal currents with MRI. *NeuroImage*, **42**, 1357–1365.

Intrinsic signal optical imaging

RON D. FROSTIG AND CYNTHIA H. CHEN-BEE

8.1 Introduction

The most popular technique for investigating the functional organization and plasticity of the cortex involves the use of a single microelectrode. It offers the advantage of recording action potentials and subthreshold activity directly from cortical neurons with high spatial (point) and temporal (millisecond) resolution sufficient to follow real-time changes in neuronal activity at any location along a volume of cortex, with the disadvantage that recordings are invasive to the cortex. In order to assess the functional representation of a sensory organ (e.g. a finger, a whisker), neurons are recorded from different cortical locations and the functional representation of the organ is then defined as the cortical region containing neurons responsive to stimulation of that organ (i.e. neurons that have receptive fields localized at the sensory organ). A change in the spatial distribution of neurons responsive to a given sensory organ and/or in their amplitude of response is typically taken as evidence for plasticity in the functional representation of that sensory organ (Merzenich et al., 1984). As a cortical functional representation could comprise thousands to millions of neurons distributed over a volume of cortex, the use of a single microelectrode to map a functional representation and its plasticity requires many recordings across a large cortical region, recordings that can only be obtained in a serial fashion and require many hours to complete, thus the animal is typically anesthetized. Because the number of recordings will be limited due to cortical tissue damage incurred during the experiment and time constraints, the characterization of a functional representation is dependent on the extrapolation between recording locations. Also, because of its invasiveness, this technique is not ideal for the assessment of the same functional representation before and after a manipulation within the same animal. The recent development of simultaneous recordings from a chronically implanted array of microelectrodes (see review by Nicolelis and Ribeiro, 2002) provides some solution to some of the challenges

Handbook of Neural Activity Measurement, ed. Romain Brette and Alain Destexhe. Published by Cambridge University Press. © Cambridge University Press 2012.

described above, and has great promise for its potential for long-term recordings from the same animal, although recordings still damage the cortical tissue, the sample size is still small, and extrapolation is still needed between recording locations. Therefore, for better understanding of large-scale cortical functional organization and its plasticity, microelectrode arrays are still limited in fulfilling the need for visualizing mass-action of millions of neurons at different points in time before, during and after an experimental manipulation.

The ability to visualize the functional organization of the living brain, especially the cortex, has been a long-held dream of neuroscientists as eloquently expressed by Sherrington's "enchanted loom" metaphor. The realization of this hope is now manifested in several functional imaging methods, each with its own advantages and limitations. Ideally, a functional imaging technique should have the ability (1) to sample neuronal spiking activity, (2) non-invasively, (3) simultaneously from large cortical regions, (4) with high, three-dimensional spatial resolution, and (5) high temporal resolution, (6) in the awake animal, and (7) without the use of extrinsic contrast agents, or dyes. This ideal technique would enable the direct, non-invasive (hence, long-term and non-damaging) assessment of a cortical functional representation (comprising many neurons collectively occupying a volume of cortex) with sufficient spatial and temporal resolution to track real-time changes in the functional representation in the awake, behaving animal. In recent years, there has been a growing interest also in studying evoked subthreshold activity and therefore the ability to record subthreshold activation should also be added to the above list of abilities. Unfortunately, such an ideal imaging technique does not yet exist but each of the most widely used current techniques for studying cortical functional organization and plasticity offers a particular subset of the above-described advantages.

This chapter is devoted to intrinsic signal optical imaging (ISOI), a functional imaging technique that has revolutionized our understanding of cortical functional organization and its plasticity since it was launched more than 20 years ago (Grinvald et al., 1986; Frostig et al., 1990; Ts'o et al., 1990). ISOI lacks the ability to measure neuronal activity directly (spiking or subthreshold) and track data in millisecond temporal resolution, but is capable of visualizing functional representations via the simultaneous sampling of a large cortical region with high spatial resolution. In this chapter we will concentrate on ISOI as a cortical activity mapping technique and the many examples of its applications that originate from functional imaging of rat barrel cortex, the largest subdivision of primary somatosensory cortex dedicated to processing input from the large facial whiskers. We find the combined use of ISOI with rat barrel cortex particularly appealing because (a) this animal model offers many advantages for the study of cortical functional organization and its plasticity, (b) it can be non-invasive to the cortical tissue – by imaging through the skull – and thus offers the opportunity to image chronically from the same animal, and (c) it is capable of visualizing functional representations in real time (albeit providing only an aerial, i.e. two-dimensional, view of functional representations).

8.2 Background and theory

8.2.1 *Principles of cortical functional organization: a brief introduction*

Because the target of most ISOI investigations (including ours) is the cerebral cortex, there is a need to understand properly what is imaged and how imaging is optimized for cortical functional organization. For a better understanding of the current chapter, we should mention at least briefly the following fundamental principles of cerebral cortex functional organization. (1) *A gross parcelation of the cortex into modality-specific areas.* Based on histological techniques and later on various electrophysiological techniques (e.g. evoked potentials, microelectrode recordings), it became clear that different areas of the primary cortex are dedicated to different modalities (e.g. visual, auditory, somatosensory, motor, etc.). (2) *Inputs to the cortex are topographically organized.* Inputs to the different cortical areas are organized in a manner that closely follows the spatial organization of the peripheral receptors of their respective sensory epithelia, namely the inputs of the retina to the primary visual area, the cochlea to the primary auditory cortex, and the skin to the primary somatosensory cortex. Accordingly, the representations of the sensory inputs in the primary sensory areas together form topological maps; while they may not preserve the exact scale and orientation of their respective sensory epithelium, these maps do preserve the relative spatial relationships between peripheral points. These mappings are named according to their respective sensory modality followed by the suffix "topy": for example, retinotopy for vision, tonotopy for hearing, and somatotopy for touch. It follows from this organization principle that using electrophysiological or imaging tools one can "map" the functional representation of different peripheral organs in the cortex (e.g. the representation of a finger, a hand, etc.). (3) *Columnar organization.* The cerebral cortex is a layered structure, whose orientation axes are described by the terms horizontal or tangential to indicate the direction within a given layer, and by the term vertical to indicate the perpendicular direction. Cells that are stacked together vertically, typically across all cortical layers, are interconnected along the vertical axis of the cortex. Such connectivity underlies the columnar organization of primary cortex, where these interconnected cells share similar response preference to sensory stimulations. Exceptions to these fundamental principles exist for each of the sensory modalities. As a general rule, however, the functional organization of the cortex typically exhibits parcelation into modality-specific areas, topographical mapping of input to each area, and presence of functional columns within each topographical map, properties which can be visualized effectively with functional imaging.

8.2.2 Advantages of applying ISOI to the rat barrel cortex

A popular animal model for studying cortical functional organization is the rat barrel cortex. One of the main advantages offered by this animal model is the relative ease in imaging cortical activity. Unlike imaging experiments in cats and monkeys, the surgical preparation for imaging in rats is minimally invasive to the animal (Masino et al., 1993) because there is no longer a need to (a) control for respiration and heart rate artifacts via paralysis and artificial ventilation in addition to the synchronization of data collection to heart and respiration, (b) remove the skull and dura in order to image cortical activity, therefore leaving the brain intact and in optimal condition throughout the experiment, and (c) build and affix an elaborate "recording chamber" to the skull with screws and cement as typically done for imaging in cats and monkeys. Instead, only removal of the scalp and muscle tissue along with thinning of the skull with a drill above the imaged area is needed for better light penetration. In addition, because the skull and the dura are left intact, the opportunity is available to image the same rat repeatedly over an extended time period (chronic imaging), which can be advantageous for the study of cortical plasticity within the same animal. It is worthwhile noting that, in contrast to rats, the imaging of mice somatosensory cortex can be achieved through the intact skull (no thinning needed) because it is sufficiently transparent (Prakash et al., 2000).

The other main advantage is related to the use of rat barrel cortex as an animal model of cortical function and plasticity. The body of literature accumulated thus far on the rat barrel cortex suggests that this system shares many features with the sensory cortex in higher mammals. These include the topographical organization of whisker inputs to the cortex – each whisker has its own anatomical representation of aggregate of cells commonly called a "barrel" in the contralateral barrel cortex, and therefore the matrix-like organization of the large whiskers on the snout is mapped in a one-to-one fashion to a matrix of layer IV barrels. Further, within this topographical map, each whisker has its own functional column of cells that respond with the shortest latency and strongest amplitude to the stimulation of a whisker. Therefore, the matrix-like organization of the large whiskers on the snout is functionally mapped in a one-to-one fashion to a matrix-like columnar organization that is in register with the layer IV barrels. Findings on these exquisite structural and functional maps of whiskers provide a rich source of information from which hypotheses can be formulated about cortical functional organization and plasticity (Feldman and Brecht, 2005; Petersen, 2007). The popularity of the barrel cortex for the study of cortical functional organization and plasticity stems in part from the relative ease in manipulating the whiskers: precise stimulation can be delivered to a specific whisker or a specific combination of whiskers; sensory deprivation can be easily achieved by trimming and/or plucking of whiskers; and sensory deprivation can be easily reversed because of whisker regrowth following the termination of whisker trimming and/or plucking. Thus, because of the above-mentioned advantages, the combination of ISOI with rat barrel cortex can be a powerful means for investigating adult cortical functional organization and its plasticity.

8.2.3 *Intrinsic signals*

Utilization of ISOI is based on the finding that when the brain is illuminated neuronal activity causes changes in the intensity of the light that is reflected from the brain. Accordingly, patterns of evoked activity can be detected and quantified by measuring reflectance patterns from the living brain (Grinvald et al., 1986). These stimulus-evoked optical changes are referred to as *intrinsic signals* to differentiate them from other optical signals obtained using extrinsic probes such as voltage-sensitive dyes or calcium indicators. The intrinsic signals originate from activity-dependent changes in several intrinsic brain sources including: oxygen consumption affecting hemoglobin saturation level (i.e. oxygenated versus deoxygenated hemoglobin, or HbO_2 versus Hbr, respectively), changes in blood volume affecting tissue light absorption, changes in blood flow affecting hemoglobin saturation level, and changes in tissue light scattering related to physiological events accompanying neuronal activity such as ion and water movement in and out of neurons, changes in the volume of neuronal cell bodies, capillary expansion, and neurotransmitter release – several of these sources can be utilized for imaging brain activity (for a recent review see Vanzetta and Grinvald, 2008).

The above mentioned sources associated with the vascular system are known as *hemodynamic* sources, and their dependence on neuronal activity is referred to as *neurovascular coupling*. The intrinsic signals captured with ISOI, a hemodynamic-based imaging technique, can be exploited in two main ways, each aimed at pursuing different research questions. First, the captured intrinsic signals can be used to *map* brain activity, and only one wavelength is typically required for illumination of the brain. The choice of the illumination wavelength depends on which underlying hemodynamic source will dominate the intrinsic signals during imaging. For example, if the chosen illumination wavelength is in the orange, red or near-infrared part of the spectrum, the map of brain activity obtained with ISOI is based on intrinsic signals dominated by stimulus-induced oxymetry (transition between oxygenated and deoxygenated hemoglobin), whereas if the chosen wavelength is in the green part of the spectrum, the map of brain activity is based on intrinsic signals dominated by stimulus-induced changes in blood volume. When used for mapping purposes, no mathematical models or parameter assumptions are needed to analyze the data. Second, the captured intrinsic signals can be used to *separate and characterize individual underlying hemodynamic sources of ISOI* such as Hbr and HbO2. The contribution of each underlying hemodynamic source using spectral imaging techniques is a topic of intense research and such research is technically and mathematically more intricate, utilizing spectral imaging techniques that rely on sampling brain activation using several wavelengths of illumination and mathematical modeling along with parameter assumptions for data analysis. The quantitative description of the spatiotemporal interaction between all these sources has become increasingly complex and a general consensus has yet to be reached (for a review see Vanzetta and Grinvald, 2008). Therefore, this chapter provides only some general comments about the underlying hemodynamic sources because it focuses on using intrinsic signals for mapping purposes (one illumination wavelength, no mathematical modeling), mapping that can be successfully achieved independently of whether the exact spatiotemporal dynamics of the underlying hemodynamic sources are known.

During imaging, the camera is positioned above the cortex and captures an aerial, or two-dimensional, collection of light reflection from the illuminated cortex; thus, the collected imaging data contain the cumulative activity integrated orthogonally to the cortical surface. Penetration of light into the cortex depends on the wavelength of the illumination: longer wavelengths penetrate deeper into the cortical tissue. However, the amplitude of intrinsic signals is stronger with short wavelength illumination and becomes progressively weaker with longer wavelength illumination. Therefore red illumination is a good compromise between depth and signal amplitude and signal to noise considerations. At present it is still difficult, however, to estimate to what extent the red light penetrates in vivo and consequently to what extent different cortical layers, especially the deeper layers, contribute to the collected data. Due to the exponential loss of photons with cortical depth and due to a further exponential loss of photons in the return path from cortical depth to the surface, at the very minimum ISOI is inherently biased for functional measurements of activity in the upper layers of the cortex, although the upper layers of the cortex include large dendritic trees of neurons located in deep cortical layers and therefore activity in the upper layers can be influenced or modulated by neurons in deep cortical layers.

It should be noted that in recent years the term intrinsic signals has also been expanded to describe new classes of intrinsic signals such as stimulus-evoked fast scattering responses and intrinsic auto-fluorescence responses – all of which can also be used for mapping activity in the brain. The description of these signals and their advantages and limitations for mapping neuronal activity in the brain are beyond the scope of the present chapter and will not be mentioned further (for detailed recent reviews see Husson and Issa, 2009; Rector et al., 2009; Shibuki et al., 2009).

8.2.4 *"Spread" versus "preference"; "point" versus "large-scale" stimulation; "global" versus "mapping" signal; "specific" versus "non-specific" signals – a guide to the perplexed*

Two popular and complementary types of evoked activity are typically imaged when characterizing the functional organization of the cortex: **spread** *versus* **preference**. Imaging the first type of activity, evoked spread, allows us to address a fundamental question about the functional organization of the cortex: what is the total cortical region activated by a specific stimulus delivery? In addressing such a question, the imaging of the total cortical spread of evoked activity is used and the result is also known as the *functional representation* of a specific stimulus. Because preference of responses to a specific stimulus is typically organized in columns, imaging the second type of activity (evoked preference) allows us to address another fundamental question: what cortical regions respond preferentially to a given stimulus delivery? In addressing this question, the imaging of cortical regions with the preferred evoked activity is used to map specific columnar systems. Besides differing in the research question being addressed, the two types of evoked activity that can be imaged also differ in the type of stimulus delivery used during data collection, the imaging signal being exploited, and the data analysis used for visualizing the functional map.

When mapping the **spread** of cortical activity, the cortex is imaged in response to delivering a single type of stimulus (e.g. set of whiskers as a tactile stimulus, set of pure tones as an auditory stimulus, set of visual orientations as a visual stimulus). Because of the topographical nature of cortical organization, the location(s) of peak activity evoked by the stimulus delivery will depend on the specific topographical properties for that cortical area (i.e. somatotopy for tactile, tonotopy for auditory, and retinotopy for visual). The spatial extent of activation beyond the peak location(s), along with the peak(s), constitutes the evoked cortical activity spread. An important and popular case of mapping the spread of cortical activity is when the stimulus delivery is spatially focused, or in a **point**-like fashion (e.g. single whisker, single pure tone, discrete $1° \times 1°$ visual area). The resultant cortical activity spread following point-stimulation is known as the **cortical point-spread** analogous to the term used in the field of optics. Characterizing the point-spread addresses the following question about the functional organization of the cortex:

how much cortex is activated by delivering a point-like stimulation to the peripheral sensory epithelium? The cortical point-spread concept is important because it is a measure of the basic divergence unit of cortical activation, as opposed to the fundamental concept of the receptive field of a cortical neuron, which is a basic unit of convergence (i.e. measure of how much periphery can activate a single neuron in the cortex).

The cortical response in the rat barrel cortex evoked by single whisker stimulation is an appealing model for mapping cortical point-spreads because stimulation of a single whisker is a simple and straightforward means of achieving point-like stimulation, in addition to all the advantages offered by studying the barrel cortex as discussed in Section 8.2.2. Figure 8.1A shows the first successful mapping of a cortical activity area in the barrel cortex following single whisker stimulation (Grinvald et al., 1986). The figure contains individual traces of intrinsic signals captured by a photodiode array mounted on a microscope, with each trace corresponding to a photodiode within the array. Each trace has a clear baseline prior to stimulation and, as expected, a single signal peak exists whose location can be readily seen. Beyond the peak there exists an approximately symmetrical gradient of progressively diminishing amplitude and increasing latency to the peak of signal traces. This signal spread, along with the peak signal, is taken as a measure of the cortical point-spread of the stimulated whisker or the whisker's functional representation; we use these terms interchangeably. Also as expected, a similar pattern of signal spread is obtained when stimulating different single whiskers, except for the shifting of peak signal location to co-localize above the "appropriate" layer IV barrel. The data presented in Figure 8.1A constituted the first proof that intrinsic signals could be used to map the functional representations of different whiskers. The point-spread of a single orientation or a single pure tone can also be mapped as an analog to the point-spread of a single whisker. When using a photodiode array, note that only intrinsic signals can be plotted, and high spatial resolution images of activity maps had to wait until video and CCD cameras (without a microscope) were introduced to the field of neuroscience.

After successfully mapping a cortical spread, specifically a point-spread, in rat barrel cortex, research attention was then directed towards using the same intrinsic signal recording setup to map the **preference** of cortical activity such as the orientation columns (example of a cortical columnar system) known to exist over the entire visual cortex (Grinvald et al., 1986). Orientation columns refer to columns of cells that respond preferentially to the specific orientation (e.g. vertical, horizontal) of a line, border, or grating in the visual world. Because columnar systems typically cover the entire cortical area being imaged, mapping the entire extent of a columnar system requires the use of a **large-scale** stimulus that spans over a large territory of the peripheral sensory epithelium; for example, simultaneous delivery

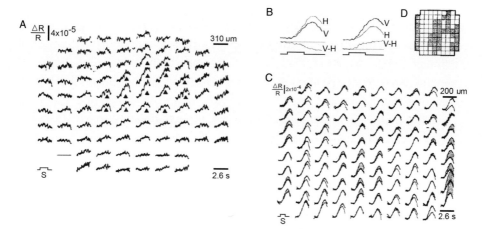

Figure 8.1 Mapping of intrinsic signals using a photodiode array. A. Mapping of a single whisker functional organization (point-spread) in the barrel cortex using intrinsic signals. The spatial distribution of intrinsic signals was obtained in response to mechanical stimulation of a single whisker. Simultaneous intensity changes in reflected light were measured with 96 photodiodes, averaged over 200 trials. Wavelengths of 665 to 750 nm were used. The duration of the stimulus (S), the size of the cortical area viewed by each detector, the time scale, and the fractional change in light intensity reaching the diode are provided at the corners of the panel. Signals marked by triangles are the largest 15 signals selected by computer integration. B–D. Visualization of orientation columns in cat visual cortex. B. Demonstration of the signal sensitivity to the orientation of the stimulus. Right traces: the response to vertical drifting gratings (V) and to horizontal drifting gratings (H). The difference between these two traces is shown in the third trace (V-H). An electrode penetration at this site recorded from cells which responded only to vertical gratings. Left traces: as for right, but the optical signals are from diodes that monitored cortical area in which units responded only to horizontal stimuli. C. The raw data. Two superimposed patterns of reflected light signals in response to horizontal (fine lines) and vertical (bold lines) drifting gratings. These experiments were interlaced. Positions where the response to horizontal stimuli was larger than the response to vertical stimuli are shaded in gray. The pattern of optical signals may seem puzzling as optical signals are seen everywhere, also in areas where single-unit responses were not recorded for a given stimulus orientation. D. A two-state map of the orientation columns. For each pixel, the signals for the vertical and horizontal stimuli were integrated. Shaded diodes are those giving a larger response to horizontal than to vertical stimuli (corresponding to those positions shaded in gray in panel C). A and C are examples of the global signal and the results of their subtraction in B and D are the mapping signals. (First published in Grinvald et al., 1986.)

of stimulation to all the large whiskers, or many sound frequencies, or delivery of a full screen visual stimulation. It is important to keep in mind the type of stimulus delivery used during imaging ("point" versus "large-scale") as the literature can be confusing on this subject. In recent years there has also been a growing popularity for the use of naturalistic stimuli (movies of natural scenes, natural calls, etc.) but this type of stimulation has yet to be applied with ISOI. Electrophysiological studies demonstrated that, with some exceptions, there is a smooth transition of orientation preference as one records neuronal response to different orientations along tangential (horizontal) distances in the cortex. Notably, many neurons that respond to a given orientation would respond in a weaker manner to similar orientations and not at all to dissimilar orientations. Therefore, by extrapolating results from electrophysiological studies, if it were feasible to implant an expansive and very dense array of microelectrodes, one would expect that delivery of an orientation stimulus (e.g. horizontal gratings) over the entire full screen (i.e. "large-scale" stimulus) should evoke a response over the entire visual cortex that would contain many local activity peaks (above the columns with preferred response to horizontal orientation) surrounded by less active (above columns preferring similar orientations such as near-horizontal) or non-responsive cortical regions (above columns with preferred response to dissimilar orientations such as vertical and near-vertical). Likewise, one would have similar expectations for imaging of intrinsic signals: when presenting a full screen grating stimulation of a specific orientation and recording intrinsic signals from the visual cortex surface, the spatial locations of the responding orientation columns should be easily identified by local peaks of intrinsic signal responses on a background of weakly or non-responding areas. However, as Figure 8.1D demonstrates, it quickly became clear that merely using a single stimulus (e.g. horizontal orientation) to image local cortical areas that are maximally responsive to it was not sufficient to obtain the horizontal orientation column map. Contrary to expectations, large-scale (e.g. full screen) stimulation of a single orientation led to a strong and uniform presence of intrinsic signal over a larger area of cortex without any obvious multiple signal peaks. Furthermore, similar findings of a large and uniform signal area were obtained for other orientations. Such large and non-specific general signal without clear peaks as captured with imaging can be referred to as the **global** signal (to differentiate it from the actual underlying cortical activity) and appears to occur irrespective of which orientation stimulus is presented. It should be noted that the signal imaged and used in mapping a single whisker cortical point-spread (Figure 8.1A) is also an example of a global signal.

How then can intrinsic signals captured with imaging be used to map cortical preference activity such as orientation columns? Further study demonstrated that the global signal amplitude was slightly stronger at cortical locations corresponding to the activated orientation columns, but in order to localize these slightly stronger signals one had to subtract (or divide) from them the global signal evoked by its opposite orientation (e.g. vertical – horizontal, $45° - 135°$, etc.). In other words, while already containing mapping information for the location of orientation columns, global signals first required the subtracting (or dividing) out of reference global signals (e.g. global signal evoked by an opposite orientation stimulus) in order for them effectively to provide such mapping information. The resultant differential signal, whether achieved by subtraction or division, between the global signals of two opposite orientation stimuli (e.g. vertical and horizontal; Figure 8.1B, D) is routinely used for mapping purposes and can be referred to as the **mapping** signal (an alternative to differential imaging is the use of "single condition" maps where the signal for one stimulus is not divided by the signal for its opposite stimulus but by a "cocktail" that includes cortical activation by all other stimuli, in this case all other orientations including oblique orientations). The mapping signal is therefore a component of the global signal, whose amplitude and spatial pattern corresponds better to those for evoked suprathreshold responses (action potentials) for the same stimulus. Indeed, this correspondence to suprathreshold neuronal activity helped to establish intrinsic signal optical imaging further as a mapping imaging method that could also be used to map columnar systems. Mapping signals were later also used to image other columnar systems such as the ocular dominance system, a system characterized by columns of neurons with preferred response to one eye interspersed with columns of neurons with preferred response to the other eye. Therefore, imaging of the ocular dominance system is obtained by subtracting or dividing global signals obtained from stimulating one eye by those of the other eye. In summary, unlike mapping a cortical activity spread, the mapping of cortical preference activity requires the use of multiple types of stimulus deliveries (e.g. horizontal versus vertical, horizontal versus "cocktail") all of them "large-scale," and involves the use of the mapping signal because the global signal for the stimulus of interest is analyzed relative to the global signal for another (orthogonal or "cocktail") stimulus (stimuli). Similar to mapping of an entire columnar system, mapping the point-spread of preference activity requires the use of multiple types of stimulus deliveries (e.g. horizontal versus vertical) as well as the use of the mapping signal except that all stimuli are delivered in a "point" instead of "large-scale" fashion. We should point out explicitly that the global signal for a given stimulus is subtracted or divided by the signal for a resting condition (e.g. prestimulus or control data) when the goal is to map the cortical activity spread (e.g. point-spread of a single whisker) rather than preference; such maps can be considered "true" single condition maps because the analysis does not involve the signal for any stimuli other than the stimulus of interest. Further, unlike the mapping of a columnar system using orthogonal orientations or the "cocktail" approach, which are ad hoc solutions based on prior knowledge about the visual cortex, the mapping of the global signal is more satisfying because it is equivalent to the way single-unit or LFP evoked responses are analyzed compared to their prestimulus controls.

The global signal captured with imaging is generally viewed as being **non-specific** because of its inability to provide visualization of a specific columnar system, as opposed to the mapping signal being viewed as **specific** because of its ability to do so. However, we should stress that the "non-specific" global signal is not necessarily less informative or less useful an imaging signal for mapping the functional organization of the cortex. As discussed above, the global signal actually contains the specific mapping information for the columnar system; we just cannot extract this information directly without the use of the differential analysis methods discussed above. Also, whether an imaging signal is considered more informative or useful depends on the research question at hand; if one in interested in mapping a columnar system, then the mapping signal is more useful for mapping cortical function because of its ability to target these columns; however, if characterizing a point-spread of a stimulus is the question of interest, then the global signal is more useful. Therefore, there are instances when the global signal is more informative or useful for functional imaging compared to the mapping signal. This is an important point as some ISOI research projects as well as most functional magnetic resonance imaging (fMRI) research projects routinely use only the global signal for functional mapping (although fMRI research typically relies on a different phase of the signal, discussed below).

It should be emphasized that the type of intrinsic imaging signal being exploited (global versus mapping) is typically associated with a given type of analysis used to generate images of functional maps. When the global signal is used, imaging data evoked by a single stimulus (e.g. single whisker) are analyzed relative to pre-stimulus or control data. Such an approach is suited for mapping cortical spreads including point-spreads such as the functional representation of a single whisker (interestingly, however, in our hands the use of control or no stimulation data, rather than prestimulus data, was ineffective for mapping the functional representation of single whiskers). When the mapping signal is used, imaging data evoked by a given stimulus (e.g. horizontal) are analyzed relative to data evoked by an orthogonal stimulus (e.g. vertical) or a "cocktail" of stimuli (e.g. combined activation of all other orientations). This approach is suited for the mapping of functional columns that exhibit a preferred (stronger) response to that given stimulus and typically requires that all delivered stimuli are "large-scale," if characterization of the columnar system over the entire imaged area is of interest. As mentioned above, the mapping signal (where data of a given stimulus are processed relative to data of another stimulus) can also be used to map a point-spread (specifically the point-spread of a columnar system) as long as all the stimuli are delivered in a "point" rather than "large-scale" fashion, in which case the functional map obtained is a point-spread of a columnar system. Hence, the former approach for data analysis is employed by those ISOI research projects and most fMRI research projects that

Table 8.1 *Two commonly imaged types of evoked cortical activity: spread versus preference.*

	Type of evoked cortical activity	
	Spread	Preference
Research question	What is the total cortical region activated by a given stimulus delivery? Relevant for mapping *functional representations* (e.g. single whisker functional representation)	What cortical regions respond strongest (i.e. preferentially) to a given stimulus delivery? Relevant for mapping *columnar systems* (e.g. orientation columns)
Stimulus delivery	Single stimulus type (e.g. one whisker)	Multiple stimulus types (e.g. one visual orientation, such as horizontal, and then an orthogonal orientation)
Imaging signal	Global	Mapping
Data analysis	Stimulus data are analyzed relative to pre-stimulus or control data	Differential analysis where data for one stimulus are analyzed relative to data for the other stimulus

exploit the global signal, whereas the latter approach is employed by ISOI research projects that exploit the mapping signal.

Taken together, mapping of activity spread versus preference, both of them useful for better understanding of cortical organization, is achieved depending on the type of stimulus being delivered (single stimulus versus multiple stimuli; "point" versus "large-scale"), and type of imaging signal being exploited ("global" versus "mapping") which in turn dictates the type of data analysis used to visualize the functional maps (relative to pre stimulus or control data versus relative to data of another stimulus). See Table 8.1 for a summary.

The original ISOI experiments were conducted with the use of photodiode arrays, which provided excellent temporal resolution but limited spatial resolution. The arrays were less than optimal for imaging intrinsic signals because the signals were relatively slow and thus did not require such high temporal resolution, and the imaging of these signals could benefit from increased spatial resolution. Furthermore, images of cortical activity could not be readily generated from data collected by the photodiode arrays. The arrays were eventually replaced by CCD (or video) cameras (see Figure 8.2) that offered improved spatial resolution and the opportunity to create clear images of cortical activation (Frostig et al., 1990; Ts'o et al., 1990). For examples of images obtained with CCD cameras, we refer to various figures throughout the remainder of the chapter. The continuous technical

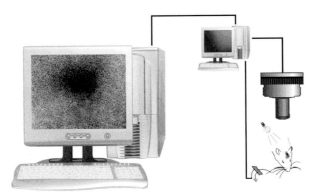

Figure 8.2 Setup for in vivo intrinsic signal optical imaging of cortical functional representations. Images are taken through the animal's thinned skull. The cortex is illuminated with a red (630 nm) light source. The CCD camera captures a sequence of images before, during and after a computer controlled mechanical stimulation to a whisker. The sequence of digital images is sent to a computer that also controls the experiment. After averaging of 32–128 image sequences, analyzed data can be displayed as an image using an 8-bit linear grayscale on a computer monitor. The black patch represents activity evoked by 5 Hz rostrocaudal stimulation to a single whisker. Typically, one computer acquires the data and controls the experiment, while the data are sent to a second computer for analysis.

evolution of CCD and recently CMOS cameras that offer even higher spatial and temporal resolutions promises continuous parallel improvements and refinement of research into cortical organization and its plasticity.

After such detailed discussion of imaging evoked activity, one has to wonder what is the relationship of the intrinsic imaging signal (global and mapping) to the underlying neuronal activation? A detailed discussion of this topic is provided in the next section.

8.3 Relationship between intrinsic signals and underlying neuronal activation

As alluded to in the previous section, a good correlation has been found between ISOI mapping signals and suprathreshold neuronal activation when both are evoked by an identical sensory stimulus. Indeed, such correlation has been the default working hypothesis of many research groups utilizing ISOI.

However, as the ISOI *global* signal and *subthreshold* neuronal activation attracted increasing research interest over the years, it gradually became more pertinent to consider other possible relationships between the type of intrinsic imaging signal being exploited (global versus mapping) and neuronal activation. In particular, questions arose in response to the observed nature of the global signal. As

Figure 8.3 Three-dimensional visualization of stimulus-evoked activity in rat bar-rel cortex as assessed with intrinsic signal optical imaging. The activity level of a 4.7 mm by 3.8 mm cortical area in response to single whisker stimulation (whisker D1, 5 deflections delivered at 5 Hz) is plotted along the z-axis, with higher intrin-sic signal activity levels plotted upwards on the z-axis. Prior to plotting, data are processed with a Gaussian filter (half-width 7) to remove high frequency spatial noise.

illustrated in Figure 8.1, strong global signal captured by imaging when using a large-scale stimulus delivery is observed everywhere over a large cortical area, even in regions where suprathreshold neuronal activity are not expected based on extensive findings from single-unit recordings. For example, if a stimulus con-sisted of a vertical grating pattern, suprathreshold neuronal activation should be present only in localized cortical areas corresponding to functional columns that prefer vertical or near-vertical grating stimulation, which is in stark contrast to the strong intrinsic signal present over the *entire* imaged part of the visual cortex even including localized areas known to prefer the opposite orientation grating (hor-izontal). Similar observations have also been made for the rat barrel cortex. As illustrated in Figure 8.3, a very large cortical area ($\sim$15 mm^2) is observed to con-tain strong global signal in response to single whisker stimulation, an area that is much larger than the area expected to exhibit a suprathreshold neuronal response and is in fact about two orders of magnitude larger than the underlying layer IV barrel ($\sim$0.15 mm^2).

What, then, is the relationship between the global signal as captured with imag-ing and neuronal activation? That strong global signal is present even at cortical locations containing *no* suprathreshold neuronal response introduces the possibility that the global signal is mainly, if not exclusively, correlated with subthreshold (synaptic) neuronal response (for yet another possibility, namely hemodynamic spread or "overspill," see below). Such a suggestion would be congruent with direct intracellular electrophysiology findings describing a large subthreshold activation area within the cat primary visual cortex following small visual stimulation (Bringuier et al., 1999) and within the rat auditory cortex following pure tone stimulation (Kaur et al., 2004). We have also reported recently on a very large subthreshold activation area as measured by an eight-electrode array in the barrel cortex of the rat. The results of our local field potential (LFP) based recordings suggest that a very large, symmetrical subthreshold area of about 38 mm^2 can be evoked following single whisker stimulation, so large that it could deeply invade other primary sensory areas (Frostig et al., 2008). Our LFP finding fits well with our previous imaging findings demonstrating large symmetrical activation of the cortex following single whisker activation (Figure 8.3), and is further supported by the discovery of underlying long-range horizontal projection that could support such spread (Frostig et al., 2008). Additional findings to support the connection between the global signal and subthreshold activation were revealed using other imaging methods. By utilizing, for example, an optical imaging technique based on voltage-sensitive dyes (or VSD) that is sensitive mostly, if not exclusively, to subthreshold neuronal activity (Grinvald and Hildesheim, 2004; Berger et al., 2007), very large activation areas in response to point stimulation have been repeatedly observed for both the primary somatosensory cortex (Ferezou et al., 2007) and visual cortex (Grinvald et al., 1994; Roland et al., 2006; Sharon et al., 2007). The study by Ferezou et al. (2007) performed in the mouse somatosensory and motor cortices is especially instructive because it was performed in the awake mouse to study passive or active single whisker stimulation. The authors clearly demonstrate that roughly 40 ms following passive or active single whisker stimulation the entire hemisphere surface exhibits clear activation that originated and spread from the stimulated whisker representation areas into both the remaining somatosensory and motor cortices. Taken together, we therefore propose, based on the above studies, that the global signal spreading symmetrically away from peak activation over a large cortical area represents the *most fundamental* and therefore the *most common* type of cortical activation.

Surprisingly, a relationship to subthreshold neuronal activation may not be limited just to the global signal. Using ISOI and post-imaging verification with single unit recordings, while exploiting the mapping signal to visualize the point-spread of orientation columns, researchers (Das and Gilbert, 1995; Toth et al., 1996) observed that orientation columns were present even in cortical regions with no suprathreshold neuronal response. In addition, using the same differential techniques as ISOI to map orientation columns, VSD-based optical imaging has been successful in using mapping signals to visualize orientation columns even though this technique is sensitive to subthreshold neuronal response (Shoham et al., 1999). By simultaneously combining ISOI signals, suprathreshold neuronal responses, and subthreshold (LFPs) neuronal responses from the cat visual cortex (Niessing et al., 2005), it was demonstrated that fluctuations in the mapping signal used for imaging orientation columns are tightly correlated with subthreshold but only weakly correlated with suprathreshold neuronal response. Furthermore, the correlation between the mapping signal and subthreshold neuronal response became even tighter in the LFP high gamma oscillation band (52–90 Hz), implying that the synchrony of subthreshold responses appears to be an important contributing factor for the correlation between subthreshold neuronal activation and the ISOI mapping signal. Together, this set of findings supports an underlying subthreshold neuronal response even for the mapping signal captured with imaging.

Because subthreshold activation likely correlates with captured imaging signals (both global and mapping), we are compelled to revisit the question: what is the relationship between global/mapping intrinsic signals and *suprathreshold* activation? We would like to submit two possibilities, both entertaining the potential for suprathreshold activation to play a secondary role to that of subthreshold activation: (1) when present, suprathreshold responses contribute to the imaging signal by strengthening the amplitude of subthreshold-based imaging signals; or (2) suprathreshold responses do not contribute at all to imaging signals and the correlation observed thus far between suprathreshold responses and imaging signals is more fortuitous then actual because evoked suprathreshold activation is, in and of itself, highly correlated with evoked subthreshold activation. Interest in these possibilities was already raised in the past regarding the origin of fMRI signal (see Logothetis (2008) for a recent review, but see Mukamel et al. (2005)), and further research aiming at elucidating the origins of such signals by direct attempts to decouple suprathreshold from subthreshold activation have led to further support for the role of subthreshold activation (Mathiesen et al., 1998; Viswanathan and Freeman, 2007; Rauch et al., 2008).

A correlation between subthreshold activation and intrinsic signals captured with imaging also compels us to revisit a popular notion known as "hemodynamic overspill." As discussed previously, a point stimulus leads to an intrinsic imaging signal that spreads over a cortical area that is much larger than that containing the suprathreshold neuronal response; see Figure 8.3 for a three-dimensional example of the large intrinsic signal area evoked by single whisker stimulation. Also discussed previously, hemodynamic sources underlie intrinsic signals captured with imaging. Thus, the large area of intrinsic signal in response to a point stimulus is typically interpreted to signify that a spatial mismatch exists between the smaller area of suprathreshold activation and the much larger (i.e. "overspill", "overcompensation") area of associated hemodynamic processes (especially blood flow to the activated area) evoked by the point stimulus. However, as discussed above, the point stimulus evokes not only a small area of suprathreshold activation but also a much larger area of subthreshold activation. In considering the correlation of intrinsic imaging signal with subthreshold activation, coupled with the understanding that such evoked subthreshold activation also needs a supply of oxygen, we would like to suggest that the popular analogy for hemodynamic overspill, "the entire garden is watered for the sake of one thirsty flower" (Malonek and Grinvald, 1996), could be modified to "most, if not all, of the garden is thirsty". The idea that there may be a much smaller overspill area of hemodynamic processes than previously estimated, or none whatsoever, attests to the need for further research on the exact spatial correspondence between subthreshold activation and intrinsic signals, primarily because it could change the typical interpretation of hemodynamic-based functional imaging methods. It could lead, for example, to a change in the way functional imaging data are quantified, namely by encouraging acceptance for using lower and less conservative thresholds, because the assumption that larger areas of activation are just overspill rather than real activation does not seem valid anymore. Lowering the threshold, in turn, would therefore lead to an increase in the size of an activation area typically quantified and, in turn, increase of overlap with other activated areas; in short the brain may look functionally less modular. A clear example that demonstrates the major impact of lowering the threshold on quantifying the size of a whisker functional representation is provided in Figure 8.5 below.

In the discussion on the relationship between intrinsic signals and neuronal activity, one now also has to take into account the recently emerging and unexpected contribution from glial cells, as suggested by findings of two research groups. Using simultaneous ISOI imaging and detailed pharmacological studies that led to decoupling of neuronal from glial activation in the olfactory bulb of the rat (Gurden et al., 2006), the authors showed that following sensory stimulation, intrinsic signals are related to uptake of presynaptic glutamate release from the synapse by glial cells known as astrocytes. Using a combination of two-photon microscopy, calcium indicators, and ISOI, these surprising findings were recently corroborated and expanded in the visual cortex of ferrets (Schummers et al., 2008). These studies demonstrate that astrocytes form a critical pathway through which neuronal activity is linked to intrinsic signals. It appears that astrocytes, like neurons, can exhibit stimulus-evoked activity properties, and evoked intrinsic signals are highly sensitive to astrocyte activation. Taken together, results obtained from both the olfactory bulb and visual cortex strongly support the idea that astrocytes constitute an important underlying source of intrinsic signals.

To summarize, the relationship between neuronal activation and intrinsic signals, once thought to be just a simple relationship between the mapping signal and the underlying suprathreshold activation, requires further elucidation, but in recent years the emphasis has shifted to intrinsic signals (including both the global and mapping components) being correlated primarily with subthreshold activation that now also includes glial cells. Because a better understanding of the changing concepts regarding the relationship between intrinsic signals, neuronal and glia activation is crucial for the proper interpretation of findings obtained by hemodynamic-based methods such as ISOI and fMRI, there is a clear need for research to address these topics further. To assist in this research pursuit, the next section provides a targeted discussion of what we have found specifically using the global signals captured with ISOI (referred to simply as intrinsic signals hereafter) in the rat barrel cortex.

8.4 More on intrinsic signals in the rat barrel cortex

8.4.1 Stimulus-evoked intrinsic signals

Up to this point only one phase of the intrinsic signal has been discussed. From the first discovery of intrinsic signals, the signal has been known to be biphasic (Grinvald et al., 1986; Frostig et al., 1990), but it was the early phase, known as the "initial dip", that was typically used for mapping, including all the examples discussed so far in the present chapter. Note that the amplitude of the initial dip signal is quite small, typically between $1-10 \times 10^{-4}$ of the total amount of light that is reflected from the cortex (Figure 8.4). Nevertheless, by simple manipulations of the intrinsic signal data (Figure 8.5), clear images of an evoked initial dip are obtained with very high spatial resolution ($\sim 50\,\mu m$) from large areas of the cortex.

However, we have recently demonstrated that the intrinsic signal is reliably triphasic. Figure 8.6 contains a representative stimulus-evoked intrinsic signal obtained from a cortex illuminated with red light (630 nm) in response to stimulation of a single whisker. Intrinsic signals evoked by a short-lasting (1 s) stimulus delivery are reliably triphasic, spanning >10 s after stimulus onset and, to borrow terminology from the fMRI field, consist of an initial dip below baseline, followed by the overshoot above baseline and then the undershoot below baseline (Figure 8.6B). High spatial resolution images can be generated from the evoked triphasic signal to map the spatiotemporal profile of each signal phase (Figure 8.6A). To characterize a given signal phase comprehensively, various parameters are available for assessment pertaining to the presence/absence of an evoked signal area (e.g. onset, offset, total duration) and maximum evoked area (e.g. areal extent, peak magnitude, peak location) (see Figure 8.6B). Figure 8.7 summarizes the high spatial resolution mapping of the triphasic signal across 60 rats and Figure 8.8 summarizes in detail how the three signal phases compare to each other. Relative to stimulus onset, the evoked initial dip area typically appears

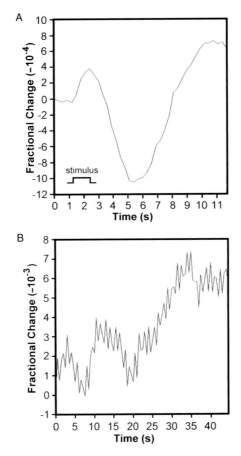

Figure 8.4 Temporal profile of evoked and spontaneous ISOI intrinsic signal obtained from the rat barrel cortex. Data were collected in either 300 ms (A) or 500 ms (B) frames and plotted on the *y*-axis in fractional change units relative to the first collected frame. By convention, decreasing light reflectance is plotted as upgoing. A. Stimulus-evoked intrinsic signal from the left cortex of a rat. Intrinsic signal was sampled over a $0.16\,mm^2$ cortical area in 300 ms frames after averaging 32 trials. On the *x*-axis, the timepoint of 0 contains intrinsic signal collected during 0–299 ms, with timepoints 1.2–2.4 s containing intrinsic signal collected during the 1.2 s stimulation of whisker C2 at 5 Hz. As compared to the slow changes in the spontaneous intrinsic signal shown in B, the amplitude of the faster changes in the stimulus-evoked cortical intrinsic signal was smaller by an order of magnitude. B. Spontaneous cortical intrinsic signal from the right barrel cortex of a rat. Intrinsic signal was sampled over a $0.16\,mm^2$ cortical area in 500 ms frames during a single 45.5 s trial. On the *x*-axis, the timepoint of 0 contains intrinsic signal collected during 0–499 ms while the timepoint of 5 s contains intrinsic signal collected during 5000–5499 ms, and so forth. The magnitude of changes in intrinsic signal remained similar when the frame duration was increased to 5 s and the trial interval increased to 455 s (data not shown). (Modified with permission from Chen-Bee et al., 1996.)

within 0.5 s (area-onset time), peaks at 1–1.5 s (area-max time), disappears by 2–2.5 s (area-offset time), and thus lasts 1.5–2.5 s (area-duration; Figure 8.8C), followed by the overshoot area (2–3 s area-onset, 3.5–4.5 s area-max, 6–9 s area-offset, 3.5–5 s area-duration) and the undershoot area (6–7.5 s area-onset, 7.5–9.5 s area-max, 10–12 s area-offset, 3–7.5 s area-duration). The overshoot signal phase is largest in spatial extent (Figure 8.8A), is strongest in peak magnitude (Figure 8.8B), and lasts the longest (Figure 8.8C), followed by the undershoot and then the initial dip, with poor co-localization of peak activity between the three signal phases (Figure 8.8D). For every signal phase, only a few parameters appear correlated, such as larger spatial extents with stronger peak magnitudes or longer total duration of evoked area presence with later offset of evoked area presence (see Table 2 of Chen-Bee et al., 2007). Correlations between signal phases are even more sparse (e.g. no correlation in peak magnitude or spatial extent between any two phases; see Table 3 of Chen-Bee et al., 2007), demonstrating the large degree of independence between the different signal phases with respect to their attributes. Such independence impedes use of the various attributes of the initial dip to predict the attributes of the subsequent overshoot and undershoot signal phases.

It remains to be determined how such a long lasting signal with three distinct phases is dependent on the wavelength of illumination and/or stimulus delivery duration used. While a similar triphasic signal is present with the use of orange illumination (Chen-Bee et al., 2007), it is possible that the characteristics of the detected signal can differ between illumination wavelengths, especially if the wavelengths differ substantially in what underlying sources contribute predominantly to the intrinsic signals. For example, in following the evoked intrinsic signals over an extended time period, it would be interesting to compare the detected signal between green, red, and infrared illumination. Duration of stimulus delivery is also likely to affect the characteristics of the detected signal; indeed, using spectroscopic optical imaging specifically to follow the hemodynamic signals Hbr, HbO2, or total hemoglobin over an extended time period, the detected signals are observed to differ substantially depending on whether the stimulus duration is brief (2 s) or long (16 s) (Berwick et al., 2008).

As with ISOI, BOLD fMRI also measures hemodynamic related changes and interestingly it too detects an evoked signal that is triphasic and exhibits similar attributes to those of ISOI (Logothetis and Pfeuffer, 2004) across various data collection conditions, suggesting that a triphasic signal with attributes as described here may to some extent transcend factors such as the technique used (albeit one based on underlying hemodynamic signal sources), stimulus parameters, cortical region, quantification method, etc. Section 8.3 is devoted to discussing the relationship of neuronal activity to the initial dip phase of the triphasic signal; involvement of reoccurring evoked neuronal activity, suprathreshold or subthreshold, is ruled out as underlying the ISOI undershoot (Figure 8.9).

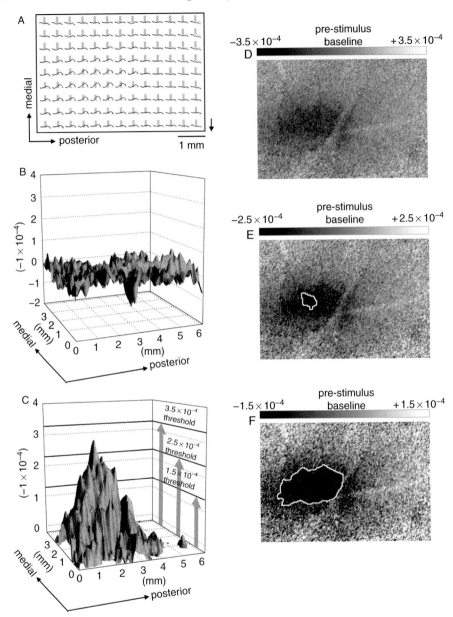

Figure 8.5 Areal extent quantification with the use of absolute thresholds combined with the prestimulus baseline. A. Array of intrinsic signals evoked by stimulation of a single whisker E2 after collection of 128 trials (downward pointing scale bar, fractional change of 1×10^{-3}). Each plot corresponds to the 4.5 s time course as averaged for the underlying 0.46 mm $\times$ 0.46 mm area, with the time epoch of 0.5 up to 1.5 s poststimulus onset highlighted in gray. Note the region of evoked intrinsic signals is located in the left center of the total imaged area. B. Determination of prestimulus baseline. Data collected from 1.0 up to 0.5 s prior to stimulus onset are converted to ratio values relative to data collected

8.4.2 Biological noise

As mentioned previously, there is no longer a need to control for heart rate and respiration artifacts when imaging rats. The absence of heart rate artifacts is presumably due to the faster heart rate of rats (relative to cats and monkeys) in combination with the lower temporal resolution (e.g. 500 ms frames) of typical ISOI experiments. At higher temporal resolutions, it is likely that heart rate artifacts will be a concern even when rats are imaged. The reason respiration artifacts are absent when imaging in rats versus cats or monkeys is less clear; it may be due to the opportunity specifically in rats to image through the thinned skull, thereby providing dampening of brain movement during breathing, and/or in general greater brain movement during breathing in cats and monkeys compared with rats prior to any surgical preparation.

Two types of biological noise are of particular interest when imaging intrinsic signals, both of whose magnitudes are much larger than that of stimulus-evoked intrinsic signals: (1) global, spontaneous fluctuations, and (2) local contributions overlying surface blood vessels. Spontaneous fluctuations in intrinsic signals can be ten times greater and occur on a slower time scale (oscillations of ~0.05–0.1 Hz or one complete cycle every 10–20 s) compared to the stimulus-evoked intrinsic signals (Figure 8.4). Because stimulus delivery evokes only a small change in

---

Caption for Figure 8.5 (cont.)

from 0.5 up to 0 s prior to stimulus onset, with the x- and y-axes indicating cortical location and the z-axis indicating strength of prestimulus intrinsic signals. The median ratio value (0.21×10^{-4} for this example) is used as a measure of the average prestimulus activity over the entire imaged area. Ratio values are filtered (Gaussian half-width 5) prior to three-dimensional plotting. C. Setting absolute thresholds when used in combination with the prestimulus baseline. Data collected 0.5 up to 1.5 s poststimulus onset are converted to ratio values relative to prestimulus data such that the processed data may be thought of as a "mountain of evoked activity," with the x- and y-axes indicating cortical location and the z-axis indicating strength of poststimulus intrinsic signals. Ratio values are filtered (Gaussian half-width 5) prior to three-dimensional plotting. Thresholds are set at absolute increments away from the prestimulus baseline (0.21×10^{-4} as indicated by z-axis minimum). Three arbitrary increments are illustrated here: 1.5, 2.5, and 3.5×10^{-4} away from the prestimulus baseline. D–F. Visualizing the quantified area of evoked intrinsic signals using absolute thresholds combined with the poststimulus baseline. An 8-bit, linear grayscale mapping function is applied to the ratio values so that the quantified area (enclosed by a white border) is visualized as a black patch within the total imaged area for each of the three absolute thresholds. Note that an area was not quantified with the highest (3.5×10^{-4}) threshold (D) as the peak ratio value for this example is 3×10^{-4}. Orientation and horizontal scale bar in A also apply to D–F. (Modified with permission from Chen-Bee et al., 2000.)

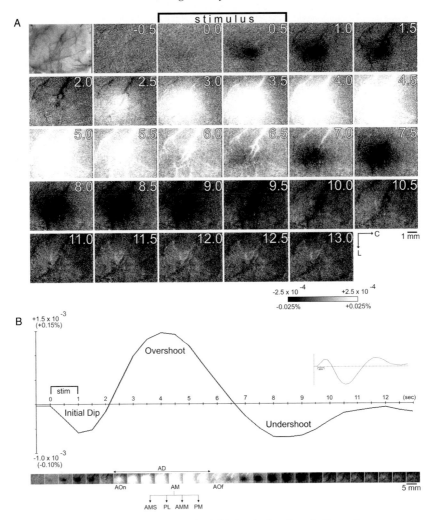

Figure 8.6 Visualization of evoked ISOI intrinsic signal. Data from a representative rat are provided to illustrate that high spatial resolution images of intrinsic signal activity can be obtained for an extended time epoch after stimulus onset (up to 13.5 s; 500 ms frames). A. A photograph is provided of the vasculature present on the surface of the imaged cortical region. Images of poststimulus data are created by first converting data into fractional change values relative to the 500 ms frame collected immediately prior to stimulus onset on a pixel-by-pixel basis before applying an 8-bit linear grayscale to the processed data such that middle gray is no change relative to prestimulus data, darker gray is larger undershoots, lighter gray is larger overshoots, and darkest (black) and lightest (white) are thresholded at $\pm 2.5 \times 10^{-4}$ or $\pm 0.025\%$. Prestimulus data can be visualized by creating an image in a similar manner, where the 500 ms frame immediately preceding stimulus onset (-0.5 s time point) is converted relative to the -1.0 s time point. Stimulus bar indicates when the 1s stimulus delivery occurred; 1 mm scale bar and neuroaxis apply to all images. B. Line plot of the fractional change values is obtained from a single

signal on top of the large spontaneous intrinsic signal fluctuations, the successful imaging of stimulus-evoked intrinsic signals requires that these spontaneous fluctuations are somehow averaged out. As they are not time locked to stimulus delivery, spontaneous intrinsic signal fluctuations can be minimized by averaging a set of stimulation trials. The number of imaging trials needed (32–128 trials) for sufficient capturing of stimulus-evoked intrinsic signals (not just the initial dip but all three signal phases) is comparable to that of single-unit recording experiments. Any residual presence of spontaneous fluctuations can then be addressed at the level of data analysis. To understand these spontaneous fluctuations better, we have collected and analyzed control trials in the same manner as stimulation trials and found that their presence can be substantial in magnitude and areal extent despite the averaging across many trials (Figure 8.10). Fortunately, these spontaneous fluctuations are non-specific in both the temporal and spatial domains, making them distinguishable from stimulus-evoked intrinsic signal. Spontaneous fluctuations have also become a topic of major interest in the fMRI field (Fox and Raichle, 2007).

Contributions from surface blood vessels within the imaged cortical region can also be time locked to stimulus delivery (Grinvald et al., 1986; Chen-Bee et al., 1996; McLoughlin and Blasdel, 1998; Sheth et al., 2004; Vanzetta et al., 2005) and thus the averaging of stimulation trials is not effective in minimizing them. The degree of vessel contributions can depend on the illumination wavelength used (e.g. Grinvald et al., 1986; Sheth et al., 2004; Vanzetta et al., 2005), but vessel contributions typically follow a slower time course compared to stimulus-evoked intrinsic signal (Chen-Bee et al., 1996; Sheth et al., 2004; Vanzetta et al., 2005). Thus, vessel contributions can be minimized by limiting analysis to only the data collected soon after stimulus onset (see Figure 8.11 for an example of data collected less than 1.5 s after stimulus onset).

Caption for Figure 8.6 (cont.)
binned pixel located centrally within the imaged cortex, with the inset containing the same data points plotted upside down as is traditionally done in ISOI studies. The same images seen in A were reduced in size and redisplayed below the line plot for easier comparison along the temporal domain (reduced images are also provided in Figures 8.2, 8.4, and 8.7). 5 mm scale bar applies to all images; the neuroaxis remains the same. Note that stimulation of a single whisker for 1 s evokes a triphasic intrinsic signal consisting of an initial dip (dark patch) followed by an overshoot (bright patch) and then an undershoot (dark patch). Each intrinsic signal phase can be comprehensively characterized by assessing such spatiotemporal parameters as: AOn area-onset time; AM area-max time; AOf area-offset time; AD area-duration; AMS area-max-size; PL peak-location; AMM area-max-magnitude; PM peak-max time. (Modified with permission from Chen-Bee et al., 2007.)

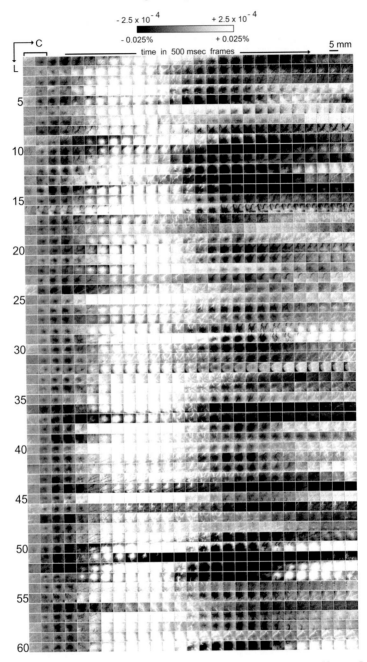

Figure 8.7 Visualization of the evoked ISOI intrinsic signal across 60 rats. Images of evoked intrinsic signal in 500 ms frames are provided through 13.5 s after onset of 1 s stimulus delivery to a single whisker (one row per rat). A total of 64 stimulation trials were collected per rat for the first 24 rats, whereas 128 trials were collected for the remaining 36 rats. The stimulus bar in the upper left corner indicates that the 1-s stimulus delivery occurs during the first two frames and

8.4.3 Additional considerations

Other considerations have arisen during our recent characterization of the three ISOI signal phases (Chen-Bee et al., 2007), where research interest is no longer restricted to just the initial dip. Although discussed within the context of characterizing three signal phases, these considerations are also relevant for other imaging studies where a comparison between different signal phases can be exchanged for research questions such as comparing between different treatment conditions, before versus after a manipulation, etc.

When trying to characterize quantitatively and compare between the three phases of the ISOI signal, one consideration is what criteria to employ in deciding which poststimulus time epoch to use per phase for quantification. As illustrated in Figures 8.5 and 8.11, a set of criteria is used to identify the typical time epoch used for the quantification of the ISOI initial dip, which took into consideration such aspects as time of signal onset and peak relative to stimulus onset as well as total signal duration. In the case where the three ISOI signal phases are quantified (or three different treatment groups, before versus after a manipulation, etc.), it is more challenging to identify one set of criteria that can be readily applied to all the three phases because these phases differ substantially in their times to signal onset and peak as well as their total durations. A relatively unambiguous option is to analyze the data collected within a short time epoch (e.g. single 500 ms frame) containing the maximum evoked area for each of the three signal phases, respectively, for quantification of such parameters as the areal size of the maximum evoked area, peak magnitude within the maximum evoked area, and peak location within the maximum evoked area (Chen-Bee et al., 2007). If it is necessary to redefine the time epoch to be used for quantifying the ISOI initial dip, then it should be noted that there are implications for analysis related to peak location. It may be that the

---

Caption for Figure 8.7 (cont.)
applies to all rows; grayscale $\pm 2.5 \times 10^{-4}$, 5 mm scale bar, and neuroaxis apply to all images. Across all rats, the evoked intrinsic signal is reliably triphasic, consisting of an initial dip (first prolonged presence of a black activity area) followed by an overshoot (first prolonged white activity area) and then an undershoot (second prolonged black activity area). While some variability existed across rats, consistent differences between the three signal phases were apparent including time after stimulus onset when an activity area first appeared (area-onset time or AOn), total duration of activity area presence (area-duration or AD), and spatial extent size of the maximum activity area (area-max-size or AMS). Note that by the end of the 13.5 s poststimulus time epoch, an additional overshoot and/or undershoot was observed in the majority of rats. (Modified with permission from Chen-Bee et al., 2007.)

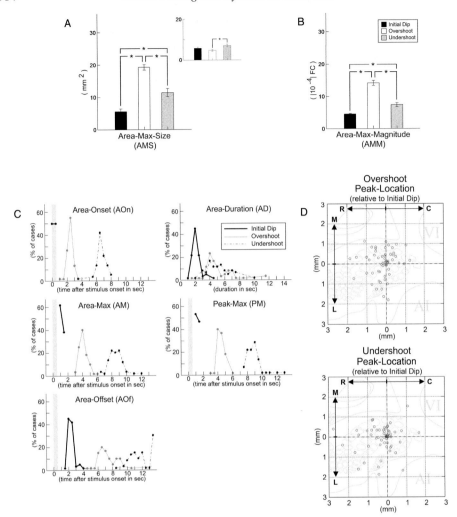

Figure 8.8 Comparison between the three ISOI signal phases. A, B. Interphase comparison of area-max-size (AMS) and area-max-magnitude (AMM). The spatial extent size of the maximum activity area (AMS) and the magnitude of peak activity within the quantified area (AMM) were assessed for all three signal phases. Means and SEs of the quantified data are provided for 60 rats. Statistically significant differences between all possible pairs of signal phases exist for both AMS and AMM (indicated by asterisks). Inset in A: the maximum activity area as quantified using a normalized threshold of 50% AMM. In striking contrast, note that the smallest maximum activity area is actually observed for the overshoot signal phase. C. Temporal profile of the evoked activity area per signal phase. Regarding the entire period that an evoked activity area is present, the following temporal characteristics were assessed per signal phase: AOn area-onset time; AM area-max time; AOf area-offset time; AD area-duration; PM peak-max time. Occurrence of the 1s stimulus delivery to a single whisker is indicated in all but the fourth panel with a vertical gray bar. While some variability existed across rats, note that the initial dip, overshoot, and

location of peak activity can vary depending on which time epoch is used for quantification. At the very least, for the initial dip, we have explicitly confirmed that the location of peak activity is comparable when using the standard time epoch of 0.5 up to 1.5 s after stimulus onset versus the 500 ms frame containing the maximum evoked area (see Chen-Bee et al., 2007 for more details).

With regard to quantifying the size of the evoked area, another consideration is the set of absolute thresholds available for use that can be applied successfully to all three phases. In this particular situation, a large difference in amplitude exists between the three signal phases (strongest phase is approximately three times that of the weakest phase). Hence, it is likely that only a narrow range of thresholds can be applied simultaneously to the three signal phases, where they are neither too high for the successful quantification of the weakest signal phase nor too low for the strongest signal phase. Interestingly, the use of a normalized threshold such as 50% peak magnitude may not be a viable option: we found that thresholding at 50% peak magnitude led to results where the smallest area was obtained for the signal

---

Caption for Figure 8.8 (cont.)
undershoot phase of the evoked intrinsic signal exhibited stereotypical time points with respect to when the area-onset, area-max, and area-offset occurred, as well as when peak-max occurred. Also, the total duration that the evoked activity area was present (area-duration) was about twice as long for the overshoot and undershoot as compared to the initial dip, and that area-max and peak-max occurred at similar time points. D. Interphase comparison of peak activity location. Interphase comparison of peak-location is meaningful especially relative to the initial dip given that the initial dip peak response to stimulation used here (single whisker C2 stimulation) has previously been shown to localize above the appropriate anatomical location (whisker C2 anatomical representation in layer IV of the barrel cortex). In order to summarize the spatial registry between the initial dip and overshoot peak-locations across all 60 rats, the overshoot peak-location was first converted to relative coordinates with respect to those of the initial dip such that a relative coordinate of (0, 0) would indicate a perfect match in peak location for the two phases. Thus, a single rat can be plotted with a single point and a total of 60 rats can be summarized in a scatterplot. To appreciate the anatomical significance of the variability in the plotted relative coordinates, the scatterplot is superimposed on an appropriately scaled schematic of tangential cortical layer IV cytochrome oxidase labeling that includes portions of primary somatosensory, visual (VI), and auditory (AI) cortex, with the relative origin centered above whisker C2 anatomical representation. Note that the overshoot peak-location did not co-localize well with that of the initial dip – only 22% of the cases exhibited a margin of error that would still be considered to lie successfully above whisker C2 anatomical representation. The overall findings were the same when comparing peak-locations for the initial dip versus the undershoot signal phase. Lastly, only a slight improvement in co-localization of peak activity was observed between the overshoot and undershoot (data not shown). (Modified with permission from Chen-Bee et al., 2007.)

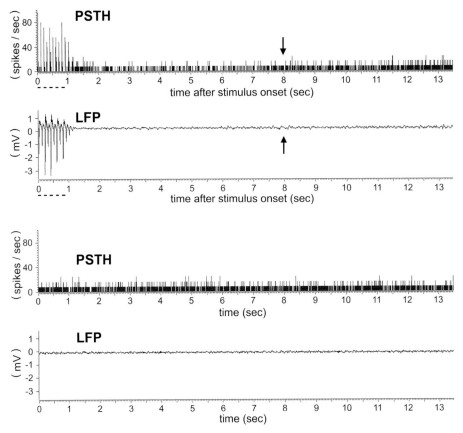

Figure 8.9 Suprathreshold and subthreshold neuronal activity recorded through
13.5 s after onset of 1s stimulus. Top: Data from the supragranular layer at the
initial dip peak-location of a representative rat are provided to illustrate that
suprathreshold (PSTH) and subthreshold (LFP) neuronal activity, recorded simultaneously from the same electrode, was followed for many seconds after stimulus
delivery. Stimulus bars indicate the 1 s delivery of 5 Hz whisker C2 stimulation,
and arrows indicate the approximate time after stimulus onset when area-max
is achieved for the ISOI undershoot phase. Other than the obvious round of
evoked suprathreshold and subthreshold neuronal activity occurring during stimulus delivery, note the lack of a second round of increased activity for both the
PSTH and LFP for the interval between stimulus offset and up to 13.5 s after stimulus onset, including within a few seconds prior to the undershoot area-max time
point. Bottom: control data from the same recording location in the same rat are
provided to illustrate that no occurrences of spontaneous activity were observed
with similar magnitudes to those of the evoked activity for either the suprathreshold (PSTH) or subthreshold (LFP) recordings. (Modified with permission from
Chen-Bee et al., 2007.)

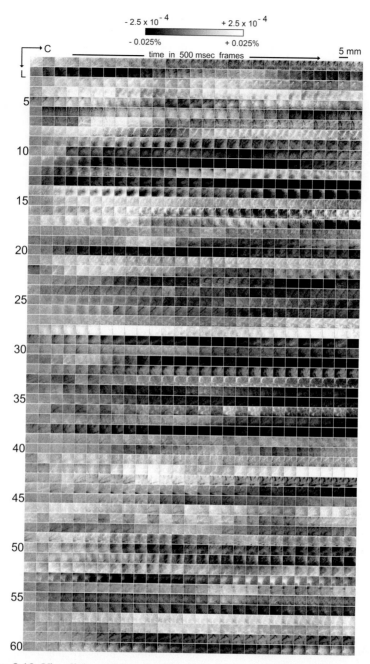

Figure 8.10 Visualizing 13.5 s of ISOI control (no stimulation) data across 60 rats. Images of ISOI intrinsic signal in 500 ms frames are provided for control trials (one row per rat) that were randomly interlaced with stimulation trials. A total of 64 control trials was collected per rat for the first 24 rats, whereas 128 trials were collected for the remaining 36 rats. Grayscale, 5 mm scale bar, and neuroaxis apply to all images. Overshoot and/or undershoot fluctuations in intrinsic signal area were observed in control trials that were collected randomly with stimulation trials, although these fluctuations did not exhibit stereotypical characteristics such as time of area-onset, area-max, or area-offset. (Modified with permission from Chen-Bee et al., 2007.)

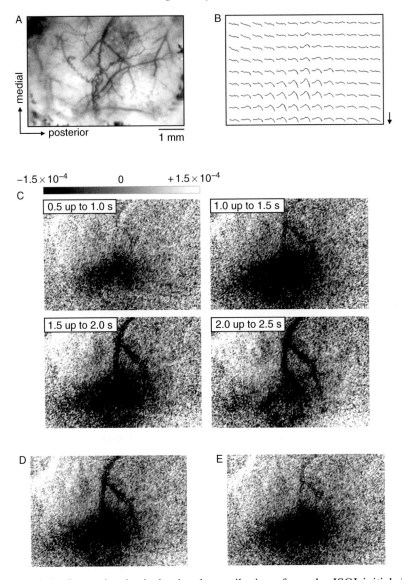

Figure 8.11 Separating intrinsic signal contributions from the ISOI initial dip phase versus those overlying surface blood vessels. A. The imaged area within the left cortex of an adult rat as imaged through a thinned skull. Dark streaks correspond to large blood vessels found on the cortical surface or dura mater. Orientation and scale bar apply to all panels. B. Array of intrinsic signals evoked by 5 Hz stimulation of whisker D1 for 1 s from a collection of 128 trials. Increasing intrinsic signal (i.e. decreasing light reflectance) is plotted as upgoing (downward pointing scale bar 1×10^{-3}), with each plot corresponding to the 4.5 s time course as averaged for the underlying 0.46 mm × 0.46 mm area. The region exhibiting an increase in the initial dip phase of the intrinsic signals overlying the cortical tissue is located in the lower center of the total imaged area, with a concurrent but slower increase in intrinsic signals overlying

phase with the strongest peak amplitude (overshoot phase; see Figure 8.8A inset), results which are in sharp contrast to the known large magnitude of the overshoot phase for both ISOI and fMRI signals (Logothetis and Pfeuffer, 2004). Presumably, the conflicting results obtained when thresholding with the 50% peak magnitude are due to the large difference in amplitude between the three ISOI signal phases (for a more detailed description of the effects on areal extent quantification when using a threshold normalized to peak magnitude, refer to Chen-Bee et al., (2000)).

Another consideration when quantifying the size of the evoked area is the usefulness of processing prestimulus data as a means to extrapolate a trend in baseline activity. It has proved useful when quantifying the ISOI initial dip, a signal phase whose peak magnitude occurs within 2 s after stimulus onset. However, it is probably not capable of extrapolating baseline activity correctly over a course of several to many seconds, as would be necessary in the case of quantifying the ISOI overshoot and undershoot phases whose peak magnitudes occur approximately 4 and 8, respectively, after stimulus onset. Indeed, as illustrated in Figure 8.10, large magnitude fluctuations in intrinsic signal activity are observed within 10+s despite the averaging across many trials. When there is no acceptable alternative method

---

Caption for Figure 8.11 (cont.)
nearby large surface blood vessels (upper center). C. Images of evoked intrinsic signals in 500 ms frames from the same data presented in B. Each image was generated after dividing a given poststimulus frame (indicated at top left of each image; 0 s stimulus onset) by a frame collected immediately prior to stimulus onset and applying an 8-bit, linear grayscale map to the processed data so that increased intrinsic signal greater than -1.5×10^{-4} is mapped to a grayscale value of black. Evoked ISOI initial dip intrinsic signal overlying the cortical tissue (black patch in lower center) is present starting 0.5 s poststimulus onset and remains elevated through 2.0 s poststimulus onset before diminishing in strength. In contrast, evoked intrinsic signals overlying large surface blood vessels (black streaks in upper center) follows a slower time course, with minimal activity present 1.0 s poststimulus onset that increases and remains elevated past 2.5 s poststimulus onset. D, E. Visualization of evoked initial dip intrinsic signals collected either 0.5 up to 2.5 s (D) or 0.5 up to 1.5 s (E) poststimulus onset. The resultant image for 0.5 up to 2.5 s (D) or 0.5 up to 1.5 s (E) poststimulus onset is equivalent to averaging all four frames or the first two frames shown in C, respectively (each frame has been divided by the frame collected immediately prior to stimulus onset). When data processing includes the time epoch of 0.5 up to 2.5 s poststimulus onset (D), evoked intrinsic signals overlying both the cortical tissue (black patch in lower center) and large surface blood vessels (black streaks in upper center) are visualized. However, data processing of the shorter post-stimulus time epoch (E) generates an image of evoked initial dip intrinsic signals overlying the cortical tissue with limited presence of evoked intrinsic signal overlying the large surface blood vessels. Grayscale bar applies to all figures (C–E). (Modified with permission from Chen-Bee et al., 2000.)

to extrapolate baseline activity levels, an absolute value of zero can be used for baseline purposes. In recent years, we have exclusively used the absolute value of zero for baseline purposes even for imaging experiments when only the ISOI initial dip is quantified.

Lastly, an important consideration for ISOI as well as other functional imaging methods is how to define a reliable criterion for choosing an appropriate activity threshold level for areal extent quantification of evoked activity areas. This is a major concern as choosing different activity thresholds can result in dramatic differences in the final quantification of the same activity area (Figure 8.5). However defined, at the very least the criterion should help identify a threshold that will reliably reflect the underlying area of activated neurons. So far, current thresholding methods (including ours as described above) are arbitrary and therefore not explicitly validated with some known correspondence with a specific level of neuronal activity. Functional imaging in general would benefit from a calibration process of functional images based on simultaneous imaging and neuronal recordings where the appropriate threshold could be identified based on maximal correspondence between the imaging-based versus neuronal-based activity area.

8.4.4 Imaging cortical plasticity

Plasticity in the cortex can occur at many scales: synaptic, neuronal, neuronal circuits, columnar organization, and functional representations of peripheral organs (see Xerri (2008) for a recent review). Because it excels in high spatial resolution functional imaging of columnar systems and functional representations of peripheral organs (e.g. whisker), ISOI is ideally suited for studying plasticity at these levels, and is typically applied for quantitative assessment of changes in such maps before and after a manipulation. The ability to image functional organization and plasticity through the skull adds an important advantage to the study of plasticity as it allows data collection that is non-invasive to the underlying cortical tissue. In addition, such a non-invasive approach (to the brain) is especially advantageous in the study of plasticity because each animal can be imaged repeatedly over time, before, during and after a manipulation (chronic imaging) to the imaged cortical system and therefore every animal can serve as its own control, a powerful experimental design for plasticity research. Figure 8.12 shows an example of a surprising plasticity that was revealed by ISOI using repeated imaging on the same animal. It should be noted that Bonhoeffer and colleagues have published impressive studies on the use of chronic imaging to study visual cortex plasticity in the developing and adult mouse (Hofer et al., 2006; Keck et al., 2008). Also, researchers working with primates have developed ingenious solutions, such as the use of artificial dura, to be able to record chronically from the primate brain, and have achieved impressive

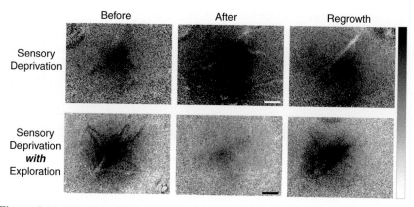

Figure 8.12 Plasticity of the spared whisker's functional representation is bidirectional depending on whisker use, and reversible upon restoration of normal sensory input. Unfiltered ratio images are provided from a sensory-deprived animal that remained in its home cage (top row) or was given an opportunity for behavioral assessment (bottom row), either before deprivation ("Before"), after 28 days of deprivation ("After"), and after 28 days of whisker regrowth ("Regrowth"). Ratio values are converted to grayscale values in which the prestimulus baseline is shown as gray and the black and white values on the grayscale bar are set to a decrease or increase of $\pm 2.5 \times 10^{-4}$ from baseline values respectively. Scale bar is 1 mm and applies to all images. Note that the two representative rats differed in the direction of plasticity of their whisker representations after sensory deprivation but were similar in that their whisker representations are nearly restored to predeprivation levels following regrowth of the whiskers. (Modified with permission from Polley et al., 1999.)

long-term chronic imaging durations (a year or more) using such solutions (Slovin et al., 2002; Chen et al., 2005).

While a main goal of the present chapter is to emphasize the advantages offered by ISOI, we are also appreciative of the advantages offered by the more traditional techniques used to study plasticity. In fact, there are several types of questions regarding cortical plasticity that can only be answered using traditional techniques such as single-unit recording techniques. Whenever possible we attempt to apply as many techniques to the same animal when pursuing the characterization of plasticity and its underlying mechanisms. We feel that the use of an imaging technique such as ISOI is an optimal first step, as it provides a clear image of the "macro" or population vantage point of cortical activity and can follow cortical plasticity within the same animal. The functional images obtained with ISOI can then be used to guide the precise placement of microelectrodes for neuronal recordings, micropipettes for injection of neuronal tracers and/or iontophoresis of drugs and microdialysis probes at "hot spots" of activity, areas of reduced activity, and areas of no activity. Recently, two-photon microscopy and ISOI have become a popular powerful combination. Consequently, the investigator can benefit from using the optical activity imaging map to ask more targeted questions about the physiology, anatomy and/or pharmacology underlying cortical plasticity. Thus, the optical map of activity can be seen as a central component of a multidisciplinary approach to the study of cortical structure and function and their relationships and their plasticity.

8.5 Current trends and future directions

Increasingly sophisticated techniques are continuously being developed to improve ISOI, including those aimed at reducing biological noise embedded in the data or aimed at improving data analysis. Examples include the use of wavelet analysis (Bathellier et al., 2007), Fourier optical imaging (Kalatsky and Stryker, 2003; Kalatsky, 2009), the use of principal component analysis (Carmona et al., 1995; Gabbay et al., 2000) together with blind separation of sources (Schiessl et al., 2000, 2008; Stetter et al., 2000), independent component analysis (Reidl et al., 2007, Siegel et al., 2007) and local similarity minimization (Fekete et al., 2009). Each of these techniques has its own advantages and limitations and their continuous refinement is promising. ISOI excels in its two-dimensional resolution but not its depth resolution and could therefore benefit from an improved ability to resolve signals from different depths of the cortex. The first successful steps towards realization of this promising direction have already been taken (Hillman et al., 2007). Another promising direction lies in the potential development of a multipurpose imaging system that combines the advantages of ISOI with the advantages offered by other imaging techniques. Combining the high-resolution spatial maps obtained with ISOI, for example, with the high temporal resolution offered by dye-based optical imaging (e.g. voltage-sensitive dyes, calcium indicators), and the detailed images of neuronal structure and function provided with multi-photon imaging, will allow researchers to address questions regarding the integrative operation of the cortex on multiple spatial and temporal scales, obtained from the same cortical area. The key advantage of a common platform is the precise ability to co-register results obtained by the different imaging methods and therefore gain insights that are difficult to see using only one technique per animal.

Finally, a most desirable direction for mapping functional organization and plasticity of the cortex is to be able to use ISOI in a non-invasive fashion in freely behaving animals. This will enable imaging of cortical activity on a continuous basis without the need for anesthesia. Ideally, a camera chip is placed directly on an animal's thinned skull and transmits optical data to the host computer by telemetry, but flexible light-guides that connect to a camera and therefore do not restrain the animal's movement could also serve this purpose. As prices of the CCD and

CMOS chips drop and the trend of miniaturization of electronic circuits that can support such imaging continues, the ability for long-term real-time imaging of cortical function and plasticity within the same freely behaving animal may be realized in the near future, as it has already been realized for the case of voltage-sensitive dye optical imaging (Ferezou et al., 2006, 2009).

Acknowledgments

We thank our previous collaborators Susan Masino, Michael Kwon, Neal Prakash, Jonathan Bakin, Yehuda Dory, Barbara Brett-Green, Silke Penschuck, Daniel Polley, Eugene Kvasnak and Ying Xiong for their original contributions. Supported by NIH grants NINDS NS-34519, NINDS NS-39760, NINDS NS-43165, NINDS NS-48350, NINDS NS-055832, and NSF IBN-9507636.

References

Bathellier, B., Van De Ville, D., Blu, T., Unser, M. and Carleton, A. (2007). Wavelet-based multi-resolution statistics for optical imaging signals: Application to automated detection of odour activated glomeruli in the mouse olfactory bulb. *NeuroImage*, **34**, 1020–1035.

Berger, T., Borgdorff, A., Crochet, S., Neubauer F. B., Lefort, S., Fauvet, B., Ferezou, I., Carleton, A., Luscher, H. R. and Petersen, C. C. (2007). Combined voltage and calcium epifluorescence imaging in vitro and in vivo reveals subthreshold and suprathreshold dynamics of mouse barrel cortex. *J. Neurophysio.*, **97**, 3751–3762.

Berwick, J., Johnston, D., Jones, M., Martindale, J., Martin, C., Kennerley, A. J., Redgrave, P., and Mayhew, J. E. (2008). Fine detail of neurovascular coupling revealed by spatiotemporal analysis of the hemodynamic response to single whisker stimulation in rat barrel cortex. *J. Neurophysio.*, **99**, 787–798.

Bringuier, V., Chavane, F., Glaeser, L., and Fregnac, Y. (1999). Horizontal propagation of visual activity in the synaptic integration field of area 17 neurons. *Science*, **283**, 695–699.

Carmona, R. A., Hwang, W. L. and Frostig, R. D. (1995). Wavelet analysis for brain-function imaging. *IEEE Trans. Med. Imaging*, **14**, 556–564.

Chen-Bee, C. H., Kwon, M. C., Masino, S. A. and Frostig, R. D. (1996). Areal extent quantification of functional representations using intrinsic signal optical imaging. *J. Neurosci. Methods*, **68**, 27–37.

Chen-Bee, C. H., Polley, D. B., Brett-Green, B., Prakash, N., Kwon, M. C. and Frostig, R. D. (2000). Visualizing and quantifying evoked cortical activity assessed with intrinsic signal imaging. *J. Neurosci. Methods*, **97**, 157–173.

Chen-Bee, C. H., Agoncillo, T., Xiong, Y. and Frostig, R. D. (2007). The triphasic intrinsic signal: implications for functional imaging. *J. Neurosci.*, **27**, 572–4586.

Chen, L. M., Friedman, R. M. and Roe, A. W. (2005). Optical imaging of SI topography in anesthetized and awake squirrel monkeys. *J. Neurosci.*, **25**, 7648–7659.

Das, A. and Gilbert, C. D. (1995). Long-range horizontal connections and their role in cortical reorganization revealed by optical recording of cat primary visual cortex [see comments]. *Nature*, **375**, 780–784.

Fekete, T., Omer, D. B., Naaman, S. and Grinvald, A. (2009). Removal of spatial biological artifacts in functional maps by local similarity minimization. *J. Neurosci. Methods*, **178**, 31–39.

Feldman, D. E. and Brecht, M. (2005). Map plasticity in somatosensory cortex. *Science*, **310**, 810–815.

Ferezou, I., Bolea, S. and Petersen, C. C. (2006). Visualizing the cortical representation of whisker touch: voltage-sensitive dye imaging in freely moving mice. *Neuron*, **50**, 617–629.

Ferezou, I., Haiss, F., Gentet, L. J., Aronoff, R., Weber, B. and Petersen, C. C. (2007). Spatiotemporal dynamics of cortical sensorimotor integration in behaving mice. *Neuron*, **56**, 907–923.

Ferezou I., Matyas, F. and Petersen, C. C. H. (2009). Imaging the brain in action – real-time voltage-sensitive dye imaging of sensorimotor cortex of awake behaving mice. In: R. D. Frostig (editor), *In Vivo Optical Imaging of Brain Function* (2nd edition). Boca Raton, FL: CRC Press.

Fox, M. D. and Raichle, M. E. (2007). Spontaneous fluctuations in brain activity observed with functional magnetic resonance imaging. *Nature Revi.*, **8**, 700–711.

Frostig, R. D., Lieke, E. E., Ts'o, D. Y. and Grinvald, A. (1990). Cortical functional architecture and local coupling between neuronal activity and the microcirculation revealed by in vivo high-resolution optical imaging of intrinsic signals. *Proc. Nati. Acad. Sci. USA*, **87**, 6082–6086.

Frostig, R. D., Xiong, Y., Chen-Bee, C. H., Kvasnak, E. and Stehberg, J. (2008). Large-scale organization of rat sensorimotor cortex based on a motif of large activation spreads. *J. Neurosci.*, **28**, 13274–13284.

Gabbay, M., Brennan, C., Kaplan, E. and Sirovich, L. (2000). A principal components-based method for the detection of neuronal activity maps: application to optical imaging. *NeuroImage*, **11**, 313–325.

Grinvald, A. and Hildesheim, R. (2004). VSDI: a new era in functional imaging of cortical dynamics. *Nature Revi*, **5**, 874–885.

Grinvald, A., Lieke, E., Frostig, R. D., Gilbert, C. D. and Wiesel, T. N. (1986). Functional architecture of cortex revealed by optical imaging of intrinsic signals. *Nature*, **324**, 361–364.

Grinvald, A., Lieke, E. E., Frostig, R. D. and Hildesheim, R. (1994). Cortical point-spread function and long-range lateral interactions revealed by real-time optical imaging of macaque monkey primary visual cortex. *J. Neurosci.*, **14**, 2545–2568.

Gurden, H., Uchida, N. and Mainen, Z. F. (2006). Sensory-evoked intrinsic optical signals in the olfactory bulb are coupled to glutamate release and uptake. *Neuron*, **52**, 335–345.

Hillman, E. M., Devor, A., Bouchard, M. B., Dunn, A. K., Krauss, G. W., Skoch, J., Bacskai, B. J., Dale, A. M. and Boas, D. A. (2007). Depth-resolved optical imaging and microscopy of vascular compartment dynamics during somatosensory stimulation. *NeuroImage*, **35**, 89–104.

Hofer, S. B., Mrsic-Flogel, T. D., Bonhoeffer, T. and Hubener, M. (2006). Prior experience enhances plasticity in adult visual cortex. *Nature neurosci.*, **9**, 127–132.

Husson, R. T. and Issa, N. P. (2009). Functional imaging with mitochondrial flavoprotein autofluorescence: theory, practice, and applications. In: R. D. Frostig (editor), *In Vivo Optical Imaging of Brain Function* (2nd edition). Boca Raton, FL: CRC Press.

Kalatsky, V. A. (2009). Fourier approach to optical imaging. In: R. D. Frostig (editor), *In Vivo Optical Imaging of Brain Function* (2nd edition). Boca Raton, FL: CRC Press.

Kalatsky, V. A. and Stryker, M. P. (2003). New paradigm for optical imaging: temporally encoded maps of intrinsic signal. *Neuron*, **38**, 529–545.

Kaur, S., Lazar, R. and Metherate, R. (2004). Intracortical pathways determine breadth of subthreshold frequency receptive fields in primary auditory cortex. *J. Neurophysio.*, **91**, 2551–2567.

Keck, T., Mrsic-Flogel, T. D., Vaz Afonso, M., Eysel, U. T., Bonhoeffer, T. and Hubener, M. (2008). Massive restructuring of neuronal circuits during functional reorganization of adult visual cortex. *Nature Neurosci.*, **11**, 1162–1167.

Logothetis, N. K. (2008). What we can do and what we cannot do with fMRI. *Nature*, **453**, 869–878.

Logothetis, N. K. and Pfeuffer, J. (2004). On the nature of the BOLD fMRI contrast mechanism. *Magn. Reson. Imaging*, **22**, 1517–1531.

Malonek, D. and Grinvald, A. (1996). Interactions between electrical activity and cortical microcirculation revealed by imaging spectroscopy: implications for functional brain mapping. *Science*, **272**, 551–554.

Masino, S. A., Kwon, M. C., Dory, Y. and Frostig, R. D. (1993). Characterization of functional organization within rat barrel cortex using intrinsic signal optical imaging through a thinned skull. *Proc. Nati. Acad. Sci. USA*, **90**, 9998–10002.

Mathiesen, C., Caesar, K., Akgoren, N. and Lauritzen, M. (1998). Modification of activity-dependent increases of cerebral blood flow by excitatory synaptic activity and spikes in rat cerebellar cortex. *J. Physiol.*, **512** (2), 555–566.

Mcloughlin, N. P. and Blasdel, G. G. (1998). Wavelength-dependent differences between optically determined functions maps from macaque striate cortex. *NeuroImage*, **7**, 326–336.

Merzenich, M. M., Nelson, R. J., Stryker, M. P., Cynader, M. S., Schoppmann, A. and Zook, J. M. (1984). Somatosensory cortical map changes following digit amputation in adult monkeys. *J. Comp. Neurol.*, **224**, 591–605.

Mukamel, R., Gelbard, H., Arieli, A., Hasson, U., Fried, I. and Malach, R. (2005). Coupling between neuronal firing, field potentials, and FMRI in human auditory cortex. *Science*, **309**, 951–954.

Nicolelis, M. A. and Ribeiro, S. (2002). Multielectrode recordings: the next steps. *Curr. Opinion Neurobiol.*, **12**, 602–606.

Niessing, J., Ebisch, B., Schmidt, K. E., Niessing, M., Singer W. and Galuske, R. A. (2005). Hemodynamic signals correlate tightly with synchronized gamma oscillations. *Science*, **309**, 948–951.

Petersen, C. C. (2007). The functional organization of the barrel cortex. *Neuron*, **56**, 339–355.

Prakash, N., Vanderhaeghen, P., Cohen-Cory, S., Frisen, J., Flanagan, J. G. and Frostig, R. D. (2000). Malformation of the functional organization of somatosensory cortex in adult ephrin-A5 knock-out mice revealed by in vivo functional imaging. *J. Neurosci.*, **20**, 5841–5847.

Rauch, A., Rainer, G. and Logothetis, N. K. (2008). The effect of a serotonin-induced dissociation between spiking and perisynaptic activity on BOLD functional MRI. *Proc. Nati. Acad. Sci. USA*, **105**, 6759–6764.

Rector, D. M., Yao, X., Harper, R. M. and George, J. S. (2009). In-vivo observations of rapid scattered light changes associated with neurophysiological activity. In: R. D. Frostig (editor), *In Vivo Optical Imaging of Brain Function* (2nd edition). Boca Raton, FL: CRC Press.

Reidl, J., Starke, J., Omer, D. B., Grinvald, A. and Spors, H. (2007). Independent component analysis of high-resolution imaging data identifies distinct functional domains. *NeuroImage*, **34**, 94–108.

Roland, P. E., Hanazawa, A., Undeman, C., Eriksson, D., Tompa, T., Nakamura, H., Valentiniene, S. and Ahmed, B. (2006). Cortical feedback depolarization waves: a mechanism of top-down influence on early visual areas. *Proc. Nati. Acad. Sci. USA,* **103**, 12586–12591.

Schiessl, I., Stetter, M., Mayhew, J. E., McLoughlin, N., Lund, J. S. and Obermayer, K. (2000). Blind signal separation from optical imaging recordings with extended spatial decorrelation. *IEEE Trans. Biomed. Eng.,* **47**, 573–577.

Schiessl, I., Wang, W. and McLoughlin, N. (2008). Independent components of the haemodynamic response in intrinsic optical imaging. *NeuroImage,* **39**, 634–646.

Schummers, J., Yu, H. and Sur, M. (2008). Tuned responses of astrocytes and their influence on hemodynamic signals in the visual cortex. *Science,* **320**, 1638–1643.

Sharon, D., Jancke, D., Chavane, F., Na'aman, S. and Grinvald, A. (2007). Cortical response field dynamics in cat visual cortex. *Cereb. Cortex,* **17**, 2866–2877.

Sheth, S. A., Nemoto, M., Guiou, M., Walker, M., Pouratian, N., Hageman, N. and Toga, A. W. (2004). Columnar specificity of microvascular oxygenation and volume responses: implications for functional brain mapping. *J. Neurosci.,* **24**, 634–641.

Shibuki, K., Hishida, R., Tohmi, M., Takahashi, K., Kitaura, H. and Kubota, Y. (2009). Flavoprotein fluorescence imaging of experience-dependent cortical plasticity in rodents. In: R. D. Frostig (editor), *In Vivo Optical Imaging of Brain Function* (2nd edition). Boca Raton, FL: CRC Press.

Shoham, D., Glaser, D. E., Arieli, A., Kenet, T., Wijnbergen, C., Toledo, Y., Hildesheim, R. and Grinvald, A. (1999). Imaging cortical dynamics at high spatial and temporal resolution with novel blue voltage-sensitive dyes. *Neuron,* **24**, 791–802.

Siegel, R. M., Duann, J. R., Jung, T. P. and Sejnowski, T. (2007). Spatiotemporal dynamics of the functional architecture for gain fields in inferior parietal lobule of behaving monkey. *Cereb. Cortex,* **17**, 378–390.

Slovin, H., Arieli, A., Hildesheim, R. and Grinvald, A. (2002). Long-term voltage-sensitive dye imaging reveals cortical dynamics in behaving monkeys. *J. Neurophysio.,* **88**, 3421–3438.

Stetter, M., Schiessl, I., Otto, T., Sengpiel, F., Hubener, M., Bonhoeffer, T. and Obermayer, K. (2000). Principal component analysis and blind separation of sources for optical imaging of intrinsic signals. *NeuroImage,* **11**, 482–490.

Toth, L. J., Rao, S. C., Kim, D. S., Somers, D. and Sur, M. (1996). Subthreshold facilitation and suppression in primary visual cortex revealed by intrinsic signal imaging. *Proc. Nati. Acad. Sci. USA,* **93**, 9869–9874.

Ts'o, D. Y., Frostig, R. D., Lieke, E. E. and Grinvald, A. (1990). Functional organization of primate visual cortex revealed by high resolution optical imaging. *Science,* **249**, 417–420.

Vanzetta, I., Hildesheim, R. and Grinvald, A. (2005). Compartment-resolved imaging of activity-dependent dynamics of cortical blood volume and oximetry. *J. Neurosci.,* **25**, 2233–2244.

Vanzetta, I. and Grinvald, A. (2008). Coupling between neuronal activity and microcirculation: implications for functional brain imaging. *HFSP J.,* **2**, 79–98.

Viswanathan, A. and Freeman, R. D. (2007). Neurometabolic coupling in cerebral cortex reflects synaptic more than spiking activity. *Nature neurosci.,* **10**, 1308–1312.

Xerri, C. (2008). Imprinting of idyosyncratic experience in cortical sensory maps: Neural substrates of representational remodeling and correlative perceptual changes. *Behavi. Brain Res.,* **192**, 26–41.

9

Voltage-sensitive dye imaging

S. CHEMLA AND F. CHAVANE

9.1 Introduction

Modern neuroimaging and computational neuroscience are two recent neuro-science disciplines that are very important for understanding brain mechanisms. Optical imaging gives the opportunity of observing the brain in activity at the level of large populations of neurons with high resolution. Many types of optical imaging techniques exist, but only two are usually used in vivo (see Grinvald et al., 1999, for a detailed review): the first is based on intrinsic optical signals and records brain activity indirectly, the second is based on voltage-sensitive dyes (VSDs) and reports postsynaptic neuronal activation in real time. In this review, we focus on the second technique, aiming at a better understanding of the origin of the optical signal. Extensive reviews of VSDI have been published elsewhere (Roland, 2002; Grinvald and Hildesheim, 2004).

This amazing technique is based on complex interaction with the system which is not yet fully understood. Indeed, the recorded signal (VSD signal) originates from a large amount of intermingled neuronal and glial membrane components and it seems difficult to isolate the contributions from the different components. Combined intracellular recording with VSDI has demonstrated a linear correspondence between the VSD signal and membrane potential of an individual neuron, but so far no studies have focused on what exactly the VSD signal actually measures when applied to a cortical population in vivo.

Experimental approaches are not really feasible because the available methodologies do not offer the possibility to inspect simultaneously all the components that may contribute to the signal. Therefore, this review suggests modeling as the appropriate solution to determine the origin of the VSD signal. Only a few models of the VSD signal exist which help to explain the optical signal in terms of functional organization and dynamics of a population neural network. A closer interaction between VSDI experimentalists and modelers is desirable.

Handbook of Neural Activity Measurement, ed. Romain Brette and Alain Destexhe. Published by Cambridge University Press. © Cambridge University Press 2012.

In the first part of this review, we present the general principle of VSDI and its main applications. The second part makes explicit the multi-component origin of the VSD signal. Finally, we show the benefit of brain activity modeling for VSD signal analysis.

9.2 Voltage-sensitive dye imaging: basics

9.2.1 Principle

Many techniques aim at functionally recording or visualizing the activity of individual neurons or groups of neurons. Each technique has its own advantages and each provides different information about brain structure and function. However, our understanding of these brain structures is often limited by the spatial and temporal resolution of the techniques. Therefore, the combination of two or more of these techniques is an emergent solution to identify brain activity more precisely. Figure 9.1 shows a three-dimensional classification of a representative subset of recording techniques, according to their spatial and temporal resolutions but also taking into account the spatial scale (field of view) they cover.

The techniques usually used for studying neuronal activity directly consist of inserting one or several electrodes in a specific region of the brain, in order to

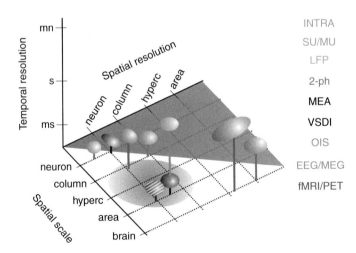

Figure 9.1 (See plate section for color version.) Imaging and non-imaging techniques are classified according to their resolutions, both spatial and temporal, and their spatial scale (INTRA intracellular recording, SU/MU single-unit/multi-unit recording, LFP local field potential, 2-ph two-photon imaging, MEA multi-electrode array, VSDI voltage-sensitive dye imaging, OIS optical imaging of intrinsic signals, EEG/MEG electroencephalography/magnetoencephalography, fMRI/PET functional magnetic resonance imaging/positron emission tomography). The mesoscopic scale is represented by the oval shaded area. (Modified from Chemla and Chavane, 2010b).

measure the individual activity of one or several neurons (single or multi-unit recording). These techniques have thus a very high spatial and temporal resolution, but they have the disadvantage that they do not allow study of the behavior of neurons as interconnected networks and are subject to the so-called sampling problem. Conversely, fMRI/PET and EEG/MEG provide non-invasive methods to map the entire brain but with lower spatial resolution and in particular very poor temporal resolution. For recording a large population of neurons simultaneously, the method of choice is optical imaging techniques. Optical imaging of intrinsic signals (OIS) has a high spatial resolution and can cover the spatial scale of the cortical area but unfortunately has a poor temporal resolution. Two-photon microscopy has excellent spatial resolution. However, it measures slow signals (calcium signals) and only reaches the columnar spatial scale. MEA provides an excellent alternative, with recordings of many neurons or LFP at multiple sites over a larger area. However, it does not allow the recording of neuronal activity uniformly within this area, and each recorded position is subject to the sampling problem. Finally, VSDI offers the possibility of visualizing, in real time, the cortical activity of large neuronal populations (10 mm or more diameter) with continuous and high spatial resolution (down to 20–50 μm) and high temporal resolution (down to the millisecond). With such resolutions, VSDI appears to be the best technique to study the dynamics of cortical processing at mesoscopic population level.

This technique is also called "extrinsic optical imaging" because it relies on the use of external dyes which are sensitive to voltage (Cohen et al., 1974; Ross et al., 1977; Waggoner and Grinvald, 1977; Gupta et al., 1981). After opening the skull and the dura mater of the animal, the dye molecules are applied on the surface of the cortex (Figure 9.2). They bind to the external surface of the membranes of all cells with a gradient from cortical surface to white matter, as illustrated in Figure 9.2A (Kleinfeld and Delaney, 1996; Petersen et al., 2003a; Lippert et al., 2007). Bound dyes do not interrupt the cell's normal function and act as molecular transducers that transform linearly changes in membrane potential into fluorescent optical signals. More precisely, once excited with the appropriate wavelength (Figure 9.2B), dyes emit instantaneously an amount of fluorescent light that is a function of changes in membrane potential, thus providing excellent temporal resolution for imaging of neuronal activity (Figure 9.2C). The fluorescent signal is proportional to the membrane area of all stained elements under each measuring pixel.

"All elements" means all neuronal cells present in the cortex but also all non-neuronal cells, such as glial cells (see Section 9.3.1 for more details). Moreover, neuronal cells include excitatory cells and inhibitory cells, whose morphology and intrinsic properties are quite different (see Salin and Bullier, 1995, for a review of the different types of neurons and connections in the visual cortex). Lastly, each cell has various compartments, including dendrites, somata and axons, each of

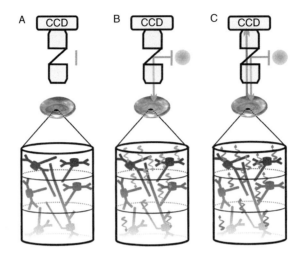

Figure 9.2 (See plate section for color version.) The principle of VSDI in three steps. The imaging chamber allows direct access to the primary visual cortex V1 represented as a patch of cortex with its six layers. A. The dye, applied on the surface of the cortex, penetrates through the cortical layers of V1. B. All neuronal and non-neuronal cells are now stained with the dye and when the cortex is illuminated, the dye molecules act as molecular tranducers that transform changes in membrane potential into optical signals. C. The fluorescent signal (red arrow) is recorded by a CCD camera. (Modified from Chemla and Chavane, 2010b).

them carrying signals with different temporal properties (spike versus postsynaptic potentials). The measured signal thus combines all these components, which are all likely to be stained in the same manner, the dye concentration only being dependent on cortical depth.

The fluorescent signal is then recorded by the camera of the optical video imaging device and displayed as dynamic sequences on a computer. The submillisecond temporal resolution is obtained using an ultrasensitive charge-coupled device (CCD) camera, whereas the spatial resolution is only limited by optical scattering of the emitted fluorescence (Orbach and Cohen, 1983).

9.2.2 Applications

9.2.2.1 General history

The first optical imaging experiments were carried out at the single neuron level, from cultured cells (Tasaki et al., 1968), and from various invertebrate preparations such as ganglia of the leech (Salzberg et al., 1973), or the giant axon of the squid (Davila et al., 1973). A crucial step in VSDI experiments has been the successful application of VSD at the population level (see Table 9.1 for a summary).

Table 9.1 *Non-exhaustive list of publications related to VSDI, classified by experimental conditions (either in vitro or in vivo) and by species (modified from Chemla and Chavane, 2010b).*

Conditions	Species	Related publications	Structure	Dye	λ_{exc} (nm)
In vitro Invertebrate preparations, cultured cells or brain slices	Invertebrate (squid, skate, snail, leech)	Tasaki et al. (1968), Davila et al. (1973), Salzberg et al. (1973), Woolum and Strumwasser (1978), Gupta et al. (1981), Konnerth et al. (1987), Cinelli and Salzberg (1990), Antic and Zecevic (1995), Zochowski et al. (2000)	Giant neurons	Styryl JPW1114 optimized for intracellular applications	540
			Axons Cerebellar parallel fibres	JPW1114 (fluorescence) Pyrazo-oxonol RH482, RH155 (absorption)	520
	Goldfish	Manis and Freeman (1988)	Optic tectum	Styryl RH414 (fluorescence)	540
	Salamander	Orbach and Cohen (1983), Cinelli and Salzberg (1992)	Olfactory bulb	Merocyanine XVII optimized for absorption measurements (Ross et al., 1977; Gupta et al., 1981), RH414, RH155	

Table 9.1 (cont.).

Conditions	Species	Related publications	Structure	Dye	λ_{exc} (nm)
	Rodent	Grinvald et al. (1982), Bolz et al. (1992), Albowitz and Kuhnt (1993), Yuste et al. (1997), Antic et al. (1999), Petersen and Sakmann (2001), Contreras and Llinas (2001), Laaris and Keller (2002), Jin et al. (2002), Kubota et al. (2006), Berger et al. (2007), Carlson and Coulter (2008), Kee et al. (2009)	Visual cortex	Fluorochrome Di-4-ANEPPS, RH414, Styryl RH795 (fluorescence)	500, 540
			Barrel cortex	JPW2038, RH155, RH482, NK3630, JPW1114, RH414, RH795	
			Auditory cortex	RH795 for fluorescence, Oxonol NK3630 for absorption	520, 705
			Hippocampus	WW401	520
	Ferret	Nelson and Katz (1995), Tucker and Katz (2003a), Tucker and Katz (2003b)	Visual cortex	RH461 (fluorescence)	590

In vivo Anesthetized or awake		References	Preparation	Dye	Wavelength
Frog		Grinvald et al. (1984)	Visual cortex	Styryl RH414	520
Salamander		Orbach and Cohen (1983), Kauer (1988)	Olfactory bulb	styryl RH160 and RH414 optimized for fluorescence measurements (Grinvald et al., 1982)	510, 540
Rodent		Orbach et al. (1985), Orbach and Van Essen (1993), Petersen et al. (2003a, 2003b), Derdikman et al. (2003), Civillico and Contreras (2006), Ferezou et al. (2006), Berger et al. (2007), Lippert et al. (2007), Xu et al. (2007), Takagaki et al. (2008), Brown et al. (2009)	Barrel cortex	RH795, Oxonol RH1691, RH1692 and RH1838 optimized for in vivo fluorescent measurements (Shoham et al., 1999; Spors and Grinvald, 2002)	540, 630
Ferret		Roland et al. (2006), Ahmed et al. (2008)	Visual cortex, Visual cortex	RH1691, RH1838, RH795, RH1691	630, 530, 630

Table 9.1 (cont.).

Conditions	Species	Related publications	Structure	Dye	λ_{exc} (nm)
	Cat	Arieli et al. (1995), Sterkin et al. (1998), Shoham et al. (1999), Sharon and Grinvald (2002), Jancke et al. (2002), Sharon et al. (2004), Benucci et al. (2007), Benucci et al. (2007)	Visual cortex (area 17/18)	RH795, RH1692	530–540, 630
	Monkey	Grinvald et al. (1994), Shoham et al. (1999), Slovin et al. (2002), Seidemann et al. (2002), Reynaud et al. (2007), Yang et al. (2007)	Visual cortex (V1/V2)	RH1691, RH1692, RH1838	630
			FEF	RH1691	630

This was first accomplished in vitro on brain slices, mainly of rodent and ferret. Population experiments were devoted to optical recording from the hippocampus (Grinvald et al., 1982), the visual cortex (Bolz et al., 1992; Albowitz and Kuhnt, 1993; Nelson and Katz, 1995; Yuste et al., 1997; Contreras and Llinas, 2001; Tucker and Katz, 2003a, 2003b), the somatosensory cortex (Yuste et al., 1997; Antic et al., 1999; Contreras and Llinas, 2001; Petersen and Sakmann, 2001; Jin et al., 2002; Laaris and Keller, 2002; Berger et al., 2007) and from the auditory cortex (Jin et al., 2002; Kubota et al., 2006).

The salamander, largely used in vitro (Orbach and Cohen, 1983; Cinelli and Salzberg, 1992), was the first species also used in vivo for the study of the olfactory system using VSDI (Orbach and Cohen, 1983), followed by the frog for the visual system (Grinvald et al., 1984), and the rodent for the somatosensory system. Indeed, initial in vivo studies of the somatosensory cortex were carried out using anesthetized rodents (Orbach et al., 1985). More recently, VSDI in freely moving mice has also been performed with success (Ferezou et al., 2006).

Rodent and ferret were also used for studying the visual cortex in vivo (Roland et al., 2006; Lippert et al., 2007; Xu et al., 2007; Ahmed et al., 2008). However, the main VSDI experiments on visual modality were conducted on two other mammalian species: cat and monkey (Grinvald et al., 1994; Arieli et al., 1995; Sterkin et al., 1998; Shoham et al., 1999; Seidemann et al., 2002; Sharon and Grinvald, 2002; Slovin et al., 2002; Jancke et al., 2004; Benucci et al., 2007; Reynaud et al., 2007; Sharon et al., 2007; Yang et al., 2007). Experiments on anesthetized cats are very attractive for mapping and studying the primary visual cortex, whereas monkey experiments can also be combined with behavioral measures.

Importantly, evidence has been gained from single cell to population recording in vitro and in vivo that the VSD signal is always linearly correlated with the depolarization dynamics of single cells (Ross et al., 1977; Contreras and Llinas, 2001; Petersen and Sakmann, 2001; Petersen et al., 2003a). Furthermore, VSD is continuously evolving as a result of the development of novel dyes since the end of the twentieth century (Shoham et al., 1999; Grinvald and Hildesheim, 2004; Kee et al., 2009).

9.2.2.2 Cortical cartography

In contrast to most functional brain imaging, optical imaging can provide helpful information on brain mapping at the mesoscopic scale, i.e. cartographic maps within a cortical area. Two-dimensional cartographic maps were first determined for single conditions using autoradiography techniques ([14C]2-deoxyglucose or [3H]proline for example), showing the arrangement of ocular dominance stripes (Wiesel et al., 1974; Kennedy et al., 1976), orientation maps (Hubel et al.,

1977; LeVay et al., 1985), and also retinotopy (Tootell et al., 1982). However, optical imaging based on intrinsic signals (OIS) soon became the predominant technique for cortical mapping since it allows cartography in response to more than one condition (Blasdel and Salama, 1986; Grinvald et al., 1986, 1991; Ts'o et al., 1990; Bonhoeffer and Grinvald, 1991; Hubener et al., 1997; Rubin and Katz, 1999). Similar to OIS, VSDI also allows the construction of high-resolution functional maps: direct comparison between the two imaging techniques (Grinvald et al., 1999; Shoham et al., 1999; Slovin et al., 2002) indeed confirmed the high spatial resolution of VSDI methodology for mapping the functional architecture of the visual cortex. The dynamics of orientation and ocular-dominance maps (Grinvald et al., 1999; Shoham et al., 1999; Sharon and Grinvald, 2002; Slovin et al., 2002) were hence revealed. However, although it is possible to do this brain mapping using VSDI, it does not take advantage of the available dynamic measurement: spatial maps were found to appear at once over the full cortical area imaged.

9.2.2.3 Dynamics of cortical processing

The main benefit of the VSDI technique is indeed the possibility to record in real time the development of activity over a large cortical area. With VSDI, neuroscientists have the possibility of extracting information complementary to that determined from electrophysiological studies and low resolution (either temporal or spatial) imaging techniques (see Figure 9.1) by visualization at high spatial resolution of cortical network temporal dynamics at mesoscopic level. For instance, VSDI allows the investigation of how a sensory stimulus is represented dynamically on the cortical surface in space and time (Grinvald et al., 1984, 1994; Arieli et al., 1996; Civillico and Contreras, 2006). More precisely, the spatiotemporal dynamics of the response to simple stimuli, for example local drifting-oriented gratings or single whisker stimulation, have been visualized using VSDI on in vivo preparations (Ćat, Sharon et al., 2007; Ŕodent, Petersen et al., 2003a, 2003b). Complex stimuli, for example line motion or apparent motion illusions, have also been achieved using VSDI in the visual cortex of cats (Jancke et al., 2004) and ferrets (Ahmed et al., 2008), revealing fundamental principles of cortical processing in vivo. Nowadays, rapid and precise dynamic functional maps can even be obtained from behaving animals, as shown by Seidemann et al. (2002), Slovin et al. (2002) and Yang et al. (2007) for behaving monkeys, and by Ferezou et al. (2006) for freely moving mice.

9.2.2.4 Functional connectivity

Another application of VSDI is the possibility of studying the functional connectivity of neuronal populations. Yuste et al. (1997), for example, investigated the connectivity diagram of rat visual cortex using VSDI. Vertical and horizontal connections were detected. More generally, intracortical and intercortical interactions, occurring during sensory processing (especially visual), have been explored using VSDI, both in vitro and in vivo. Functional connections have been mapped using VSDI in vitro in the rat visual cortex (Bolz et al., 1992; Carlson and Coulter, 2008), in the guinea pig visual cortex (Albowitz and Kuhnt, 1993) and in the ferret visual cortex (Nelson and Katz, 1995; Tucker and Katz, 2003a, 2003b), providing not only functional but also anatomical and physiological information on the local network. For example, Tucker and Katz (2003a) investigated with VSDI how neurons in layer 2/3 of the ferret visual cortex integrate convergent horizontal connections.

Orbach and Van Essen (1993) used VSDI in the visual system of the rat in vivo to map striate and extrastriate pathways. Feedforward propagating waves from V1 to other cortical areas, and feedback waves from V2 to V1 have been reported recently by Xu et al. (2007), using VSDI. In addition, feedback depolarization waves (from areas 21 and 19 toward areas 18 and 17) were studied extensively by Roland et al. (2006) in ferrets after staining the visual cortex with VSD.

9.2.3 Conclusion

VSDI is unique in the sense that no other functional brain imaging techniques can report directly the spatiotemporal dynamics of the mesoscopic activity of neuronal populations (Figure 9.1). VSDI studies have revealed spatiotemporal patterns of activity occurring in different parts of the CNS, in vitro and in vivo, on several preparations or animal species. Table 9.1 lists most articles presenting experimental results using VSDI techniques. The publications are first classified by the condition of the experiment, either in vitro or in vivo, and then by the experimental preparation or animal species. Additional information about dyes is available in the last column (see Ebner and Chen, 1995, for a compilation of the commonly used dyes and their properties).

9.3 On the origin of the VSD signal

9.3.1 Glial cells

Glial cells have long been neglected, in particular because they do not carry action potentials. However, glial cells have important functions (see Cameron and Rakic, 1991, for a review) and can theoretically contribute to the VSD signal. Glial cells are known as the "supporting cells" of the CNS and are estimated to outnumber neurons by as much as 50 to 1. However, their role in information representation or processing remains unresolved. Indeed, in vitro studies have shown increasing evidence for an active role of astrocytes in brain function. However, little is known about the behavior of astrocytes in vivo.

When interpreting the VSD signal, we face two conflicting viewpoints. Konnerth and Orkand (1986), Lev-Ram and Grinvald (1986), Konnerth et al. (1987, 1988) and Manis and Freeman (1988) showed that the optical signal has two components: a "fast" followed by a "slow" signal. The latter was revealed by successive staining with different dyes (e.g. RH482 and RH155), since each dye can stain different neuronal membranes preferentially. The authors then presented evidence that this slow signal has a glial origin. However, Kelly and Van Essen (1974) showed that the glial responses are weak (depolarizations of only 1 to 7 mV in response to visual stimuli) and have a time scale of seconds. A recent paper by Schummers et al. (2008) shows that the astrocyte response is delayed 3 to 4 seconds from stimulus onset, which is a very slow temporal response compared to the neuron response. Generally, in VSDI, only the first 1000 ms are considered, since intrinsic activity can contaminate the signal after this time (Shoham et al., 1999, Figure 9).

We understand here that the controversy over glial contribution is linked directly to the dye used (Ebner and Chen, 1995), and the time course of the optical signal generated. Thus, glial activity is very unlikely to participate significantly in the VSD signal (when considering recent fast dyes), since the amplitude of the glial response is weak and its time course is very slow.

9.3.2 Excitatory versus inhibitory cells

In the neocortex, excitatory neurons represent about 80% of the cortical cells, and inhibitory neurons represent about 20% of the cortical cells (Douglas and Martin, 1990). Thus, it is tempting to say that the VSD signal mainly reflects the activity of excitatory neurons.

However, the VSD signal is proportional to changes in membrane potential. Thus, both excitatory and inhibitory neurons contribute positively to the VSD signal and it is hard to tease apart contributions from excitatory and inhibitory cells. An additional level of complexity arises from the fact that inhibition operates generally in a shunting "silent" mode (Borg-Graham et al., 1998). In this mode, inhibition suppresses synaptic excitation without hyperpolarizing the membrane potential. To conclude, estimating the contribution of inhibitory cells to the VSD signal is not trivial and would obviously benefit from modeling studies.

9.3.3 Somas versus axons versus dendrites

Neurons are also made up of various functional compartments, whose surfaces and electrical activities are different.

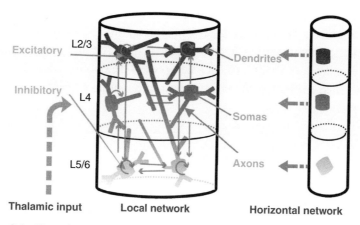

Figure 9.3 (See plate section for color version.) Contributions of the optical signal. Once neurons are stained by the VSD, every neuronal membrane contributes to the resulting fluorescent signal, but from where and in what proportion? Answering these four questions could clarify the origins of optical signals: (1) which cells? (2) which parts of the cell? (3) which layers? (4) which presynaptic origins? (Modified from Chemla and Chavane, 2010b).

1. The dendrites integrate presynaptic action potential (AP) information from other cells. The electrical activity is mainly synaptic, however, back-propagating AP can be recorded in the dendrites (see Waters et al., 2005, for a review). The dendritic surface area of mammalian neurons has been estimated by Sholl (1955), Aitken (1955), and Young (1958) to be ten to twenty times larger than the cell body surface area, and to represent 90% of the total neuronal cell membrane (Eberwine, 2001).
2. The soma receives convergent postsynaptic activity from the dendrite but also conveys spiking activity.
3. The axon is the initiation site of spikes (Stuart et al., 1997) and carries spiking signals to the axon terminal. In contrast with dendrites, the surface area of axons represents 1% of the total neuronal cell surface (Eberwine, 2001).

Regarding the difference in membrane area of the various neuronal components and the nature of the signal, it is generally accepted that the optical signal, in a given pixel, originates mostly from the dendrites of cortical cells, and therefore mainly reflects dendritic postsynaptic activity (Orbach et al., 1985; Grinvald and Hildesheim, 2004). Extensive comparisons between intracellular recordings from a single neuron and VSDI also showed that the optical signal correlates closely with changes in synaptic membrane potential (Contreras and Llinas, 2001; Petersen et al., 2003a). However, no real quantitative analysis has been performed to date and it is more correct to state that the optical signal is multi-component since the VSD signal reflects the summed changes in intracellular membrane potential of all the neuronal compartments at a given cortical site. The aim is then to determine what is the exact contribution of each component. Important questions could then be tackled, such as what is the contribution of dendritic activity, and can spiking activity be neglected?

9.3.4 Superficial versus deep layers

The neocortex is made up of six horizontal layers over approximately 2 mm depth, segregated principally by cell types and neuronal connections. Layer II mostly contains small pyramidal neurons that make strong connections with large pyramidal neurons of layer V (Thomson and Morris, 2002).

Improved dyes, when placed on the surface of the exposed cortex, can reach a depth of about 400 to 800 μm from the cortical surface, which mainly corresponds to superficial layers (Grinvald et al., 1999; Petersen et al., 2003a). Furthermore, measures of the distribution of dye fluorescence intensity in rat visual and barrel cortex confirm that the optical signal mostly originates from superficial layers I–III (Ferezou et al., 2006; Lippert et al., 2007). Note that Lippert et al. (2007) used a special staining procedure, i.e. keeping the dura mater intact, but dried.

However, this is not taking into account the fact that the activity in superficial layers arises in part from neurons whose soma is located in deep layers with dendritic arborization reaching upper layers. Therefore, the exact contribution of neurons belonging to the different cortical layers to the global VSD signal is not obvious.

9.3.5 Thalamic versus horizontal connections

The signal is the result of postsynaptic integration of presynaptic activity from a wide variety of anatomical origins: thalamic afference, local recurrent circuits, horizontal inputs or feedback pathway. For instance, in response to a local stimulation, slow propagating waves can be recorded (Grinvald et al., 1994; Jancke et al., 2004; Roland et al., 2006; Benucci et al., 2007; Xu et al., 2007) from well beyond the expected retinotopic representation limit. A highly probable candidate for such a slow spread of activity is the horizontal intracortical network (Bringuier et al., 1999). For a given pixel, it would thus be of tremendous interest to know what are the relative contributions of all the synaptic input sources contributing to the global signal, i.e. feedforward, recurrent, horizontal or feedback inputs. Dedicated models could help tease apart these various contributions.

9.3.6 Conclusion

Figure 9.3 summarizes the four main questions we have introduced.

1. What are the contributions of the various neurons and neuronal components to the optical signal?
2. What is the ratio between spiking and synaptic activity?
3. What are the respective contributions of cells from deep versus superficial layers?
4. What is the origin of the synaptic input? More precisely, what are the respective contributions of thalamic, local and long-range inputs?

To answer these questions, we propose that computational models that reproduce and/or analyze VSD signals can provide invaluable information. We will now review the different existing models of VSD signals.

9.4 Models of VSDI signals

In this section, we review the different models that have been developed to reproduce and analyze the VSD signal. These models were developed at the mesoscopic scale and were successful in reproducing known VSD dynamics of cortical maps, in vitro and in vivo patterns of activity. However, none of them could give precise insights about the source of the VSD signal (see Section 9.3) which is actually submesoscopic in scale. Therefore, we propose a list of specifications that a model adapted for this purpose, i.e. detailed analysis of the origin of the VSD signal, must satisfy. Following these requirements, an "intermediate" biophysical model of a cortical column has been proposed and is detailed in the final section.

9.4.1 The scale of the model

The choice of the scale of the model depends on what exactly the model is designed for. In the following paragraphs, we will show that several authors have proposed the *mesoscopic scale* as the best scale for analyzing and reproducing the population VSD signal. In neuroscience, this scale lies between the microscopic scale of neurons and the macroscopic scale of brain activity. The mesoscopic scale is thus the scale at which the behavior of large assemblies of neurons is observed with VSD, without entering into microscopic considerations, mathematically unwieldy. Instead, it is a size which allows the manipulation of average values, thus requiring only simple computations. Mean-field theory is successful in describing the activity of interacting populations of neurons (Faugeras et al., 2009).

Since the pioneering work of Mountcastle (1957), we know that the cerebral cortex has a columnar organization. In cat primary somatosensory cortex, Mountcastle found that neurons sharing common receptive field properties were arranged vertically in columns, crossing the six layers of the cortical tissue. In the 1960s and 1970s, Hubel and Wiesel followed Mountcastle's discoveries by showing that ocular dominance and orientations are organized in a columnar manner in cat and monkey visual cortex (Hubel and Wiesel, 1962, 1965, 1977). Today, the notion of cortical column is controversial since the original concept – a discrete structure spanning the cellular layers of the cortex, which contains cells responsive to only a single modality – is expanding, year after year, discovery after discovery, to embrace a variety of different structures, principles and names. A "column" now refers to cells in any vertical cluster that share the same tuning for any given receptive field attribute (see Horton and Adams, 2005, for a detailed review of the cortical column concept). A novel and useful concept is to propose that each definition of cortical column depends of its type (anatomical or functional) and its spatial scale, as detailed in Table 9.2. A minicolumn or a microcolumn is an anatomical column of about one hundred neurons, since its spatial scale is about 40 μm. A macrocolumn is an anatomofunctional column, but at a larger scale, from 0.4 to 1 mm. Orientation and ocular dominance columns, as well as barrels, blobs and stripes are classified as functional columns with spatial scale between 200 and 400 μm, containing several minicolumns. A hypercolumn then represents a larger functional unit containing a full set of values for any given functional parameter. Its spatial scale can be up to 750 μm and it contains about 10^4 neurons.

Following this concept, a cortical column, i.e. the elementary processing unit in the brain, can be viewed as a mesoscopic unit, which is smaller than a macroscopic unit such as the brain area and larger than a microscopic unit such as a neuron. It is therefore quite natural to use the mesoscopic scale to study the VSD signal since it is a population signal which reveals the columnar organization of the cortex (Section 9.2). Several models have been built at this scale (La Rota, 2003; Miikkulainen et al., 2005; Rangan et al., 2005; Grimbert et al., 2007; El Boustani and Destexhe, 2009; Symes and Wennekers, 2009), allowing reproduction of the dynamics of the VSD signal, i.e. time course and spatial extent. Those models are reviewed in the next section.

An alternative point of view is to choose a much finer scale allowing the construction of a more detailed biophysical model in order to estimate quantitatively the exact contribution of the VSD signal (excitation verses inhibition, parts of the neuron, layer participation, ...). In optical imaging, the visual scale studied, which is about 50 μm, corresponds to one pixel. It is still a population activity since it represents about 200 neurons, but because the scale is relatively small, we will call it

Table 9.2 *The different types of cortical columns (modified from Chemla and Chavane, 2010b).*

Type	Anatomical	Anatomofunctional	Functional		OI pixel
Name	Microcolumn, minicolumn	Macrocolumn	Ocular dominance, orientation, blobs, barrels, stripes	Hypercolumn	Optical column
Spatial scale (μm)	40	400–1000	200–400	600–1500	50
Number of neurons	80–100	10 000–60 000	2500–10 000	20 000–140 000	150–200

the "intermediate mesoscopic scale." This model, detailed in Chemla and Chavane (2010a), is reviewed in the last subsection.

9.4.2 A review of mesoscopic VSDI Models

9.4.2.1 A LISSOM model to account for dynamic maps

The LISSOM (laterally interconnected synergetically self-organizing map) family of models was developped by Bednar, Choe, Miikkulainen and Sirosh, at the University of Texas (Sirosh and Miikkulainen, 1994; Miikkulainen et al., 2005), as models of the visual cortex at a neural column level. It is based on the self-organizing maps (SOM) algorithm (from Kohonen, 2001) used to visualize and interpret large high-dimensional data sets. When extended, the LISSOM neural network models take into account lateral interactions (excitatory and inhibitory connections), allowing reproduction of the pinwheel organization of the primary visual cortex map, such as orientation, motion direction selectivity and ocular dominance maps.

Sit and Miikkulainen used a LISSOM model to represent V1 and tried to show how the activity of such a computational model of V1 can be related to the VSD signal (Sit and Miikkulainen, 2007). Indeed, with an extented LISSOM model including propagation delays in the cortical connections, they showed that the orientation tuning curve and the response dynamics of the model were similar to those measured with VSDI.

The model is a couple of two layers of neural units that represent the retina and V1. In V1, neural units account for a whole vertical column of cells. They receive input from the retina and also from neighbor columns (short-range lateral excitatory and long-range lateral inhibitory connections). Thus, the neuronal activity of unit $\mathbf{r}$ in V1 can be written as:

$$A(\mathbf{r}, t) = \sigma\left(V(\mathbf{r}, t)\right),$$
$$V(\mathbf{r}, t) = \sum_{\rho} \gamma_{\rho} \sum_{\mathbf{r}'} W_{\rho, \mathbf{r}, \mathbf{r}'} A(\mathbf{r}', t - d(\mathbf{r}, \mathbf{r}')) + \sum_{s} \chi_s R_{s, \mathbf{r}}, \tag{9.1}$$

where σ is a sigmoid activation function and the two terms are respectively the weighted sum of the lateral activations and the input activation from the retina. $W_{\rho, \mathbf{r}, \mathbf{r}'}$ and $R_{s, \mathbf{r}}$ are respectively the synaptic weight matrices of lateral and retinal connections, and $d(\mathbf{r}, \mathbf{r}')$ is the delay function between unit $\mathbf{r}$ and unit $\mathbf{r}'$. This is thus a scalar model of neural activity.

Then, the VSD signal is computed by looking only at the subthreshold activity $V(\mathbf{r}, t)$, given by the weighted sum of presynaptic activity. To simplify, the authors extended the LISSOM model with delayed lateral connections to compute the VSD signal from subthreshold signals. This is thus a scalar linear model of the VSD signals built on convolutions.

This model, based on Hebbian self-organizing mechanisms, is simple and efficient in replicating the detailed development of the primary visual cortex. It is thus very useful in the study of VSDI functional maps. However, this model is not specific enough to answer the previously asked questions (see Section 9.3.6).

9.4.2.2 Mean field models to inspect neural network dynamics

A mean field model to inspect upper layer dynamics Symes and Wennekers (2009) have recently developed a mean field computational model representing a patch of layer 2/3 visual cortex. The model represents a small patch of approximately 3 mm^2, using two arrays of cells (51 x 51 grid), one excitatory and one inhibitory. Each cell location is denoted by a spatial position $\mathbf{r} = (x, y)$ and the membrane potential at this location, representing the average local activity of a small population of excitatory and inhibitory neurons, is governed by:

$$\tau \dot{V}(\mathbf{r}, t) = -(V(\mathbf{r}, t) - V_{rest}) - g_E (V(\mathbf{r}, t) - V_E) - g_I ((V(\mathbf{r}, t) - V_I) \quad (9.2)$$

where V_{rest} set to -70 mV is the resting potential of the cell, τ is its passive membrane time constant (10.4 ms and 7.6 ms for respectively excitatory and inhibitory cells), g_E and g_I represent respectively excitatory (AMPA) and inhibitory (GABA$_A$) synaptic inputs to the cell, and V_E, V_I their associated reversal potentials (0 mV and -70 mV respectively).

The model includes both local and long-range lateral connectivity, from anatomical data (Kisvarday et al., 1997). The connectivity strength within layer 2/3 is thus represented by a Gaussian function:

$$\phi(d) = e^{-d^2/2\sigma^2} \quad (9.3)$$

where d represents the distance between cells. Different values of σ are used for excitatory ($\sigma = 300$ μm) and inhibitory ($\sigma = 200$ μm) connections, according to the data of Kisvarday et al. (1997) suggesting different extents of lateral excitatory and inhibitory connections.

The voltage-sensitive dye signal S generated by a model cell at location $\mathbf{r}$ is then computed by the following formula:

$$\tau \dot{S}(\mathbf{r}) = -S(\mathbf{r}) + K(\mathbf{r}) \quad (9.4)$$

where $K(\mathbf{r})$ is the sum of excitatory and inhibitory synaptic inputs (see Equation (9.2)) to the model cell located at $\mathbf{r}$, whilst $\tau = 5$ ms. Explicitly separating the relative contribution of both excitatory and inhibitory cells to the VSD signal, the final relation obtained is

$$VSD(\mathbf{r}) = p \, S_E(\mathbf{r}) + (1 - p) \, S_I(\mathbf{r}) \quad (9.5)$$

where S_E and S_I are the excitatory and inhibitory VSD signals defined previously by Equation (9.4), and p set to 0.8 represents the ratio between excitatory and inhibitory combined surface areas of dendritic trees.

This model was constructed specifically to investigate the mechanisms that underlie specific patterns of spatiotemporal activity observed in vitro in layer 2/3 of the ferret primary visual cortex (Tucker and Katz, 2003b). However, the authors only did a qualitative comparison between the model and experimental data, mainly because of the uncertainties related to the origin of the VSD signal.

A mean field model to study local cortical microcircuit or a "pixel" in VSDI
A recent development by El Boustani and Destexhe (2009) also considered a mean field approach based on spiking neuron models (Amit and Brunel, 1997; van Vreeswijk and Sompolinsky, 1998; Brunel and Hakim, 1999), but also taking into account the variance of the activity, which must be considered in order to describe the activity of some in vivo states (e.g. balance states) where the mean is irrelevant (i.e. null). More precisely, they proposed a master equation formalism appropriate for a second-order meanfield description of network activity in irregular states. The corresponding Markovian transition function for $i \in \{1, ..., N\}$ populations is given by

$$P(m_i^\tau | \{m_k^{\tau-1}\}_{k=1}^N) = \sqrt{\frac{N_i}{2\pi v_i(1-v_i)}} \exp\left(-\frac{N_i(m_i^\tau - v_i)^2}{2v_i(1-v_i)}\right) \tag{9.6}$$

where m_k^τ is the mean activity of population k at time τ, and N_k is the number of neurons in this population. The transfer function $v_i(\{m_k^{\tau-1}\}_{k=1}^N)$ gives the mean activity of the population i, knowing the previous state of all populations. These functions also depend on the coupling between every population so they used a maximum likelihood estimation and a simplified transfer function to find these couplings from the activity.

Their model performs well when considering the net activity computed over excitatory and inhibitory neurons within a "pixel." So using this framework, the authors showed that this could be applied to one pixel in VSDI, from which a complete transition matrix can be extracted and mapped to this model.

9.4.2.3 Neural field models to reproduce correlates of illusory motion

Grimbert et al. (2007, 2008) as well as Markounikau et al. (2010) proposed neural fields as a suitable mesoscopic model of cortical areas, combined with VSD. Neural fields are continuous networks of interacting neural masses, describing the dynamics of the cortical tissue at the population level (Wilson and Cowan, 1972, 1973). They could thus be applied to solve the direct problem of the VSD signal, providing the right parameters.

More precisely, Grimbert et al. (2007) showed that neural fields can easily integrate biological knowledge of the cortical structure, especially horizontal and vertical connectivity patterns. Hence, they proposed a biophysical formula to compute the VSD signal in terms of the activity of a field.

The classical neural field model equation is used, written either in terms of membrane potential or in terms of activity of the different neural masses present in a cortical column. For example, if $\mathbf{r}$ represents one spatial position of the spatial domain defining the area, then the underlying cortical column is described, at time t, by either a vector $\mathbf{V}(\mathbf{r}, t)$ or $\mathbf{A}(\mathbf{r}, t)$:

$$\dot{\mathbf{V}}(\mathbf{r}, t) = -\mathbf{L}\,\mathbf{V}(\mathbf{r}, t) + \int_{\Omega} \mathbf{W}(\mathbf{r}, \mathbf{r}')\,\mathbf{S}(\mathbf{V}(\mathbf{r}', t))\,d\mathbf{r}' + \mathbf{I}_{ext}(\mathbf{r}, t), \tag{9.7}$$

and

$$\dot{\mathbf{A}}(\mathbf{r}, t) = -\mathbf{L}\,\mathbf{A}(\mathbf{r}, t) + \mathbf{S}\left(\int_{\Omega} \mathbf{W}(\mathbf{r}, \mathbf{r}')\,\mathbf{A}(\mathbf{r}', t))\,d\mathbf{r}' + \mathbf{I}_{ext}(\mathbf{r}, t)\right). \tag{9.8}$$

Here, $\mathbf{V}(\mathbf{r}, t)$ contains the average soma membrane potentials of the different neural masses present in the column (the dimension of the vector then represents the number of neuronal types considered in every column). $\mathbf{A}(\mathbf{r}, t)$ contains the average activities of the masses. For example, A_i is the potential quantity of postsynaptic potential induced by mass i on the dendrites of all its postsynaptic partners. The actual quantity depends on the strength and sign (excitatory or inhibitory) of the projections (see Grimbert et al., 2007, 2008; Faugeras et al., 2008, for more details of the model's equations). The model includes horizontal intercolumnar connections and also vertical intracolumnar connections between neural masses. The latter gives this model an advantage compared to the previous model, since the vertical connectivity was not taken into account in the extended LISSOM model. Furthermore, extracortical connectivity is not made explicit here, although it was taken into account by Grimbert et al. (2007).

Hence, based on this biophysical formalism (and especially the activity-based model, which is more adapted than the voltage-based model), the authors propose a formula involving the variables and parameters of a neural field model to compute the VSD signal:

$$\mathrm{OI}(\mathbf{r}, t) = \sum_{j=1}^{N} \int_{\Omega} \tilde{w}_j(\mathbf{r}, \mathbf{r}')\,A_j(\mathbf{r}', t)\,d\mathbf{r}', \tag{9.9}$$

where $\tilde{w}_j(\mathbf{r}, \mathbf{r}')$ contains all the biophysical parameters accounting for a cortical area structure stained by a voltage-sensitive dye, i.e. the different layers, the number of neurons, the number of dye molecules per membrane surface unit, the attenuation coefficient of light and also the horizontal and vertical distribution patterns of intracortical and intercortical connectivities.

This formula is the result of many decompositions of the total optical signal, from layer level to cellular membrane level, where the signal is simply proportional to the membrane potential.

Better than the Lissom model for our considerations, this large-scale model reproduces the spatiotemporal interactions of a cortical area in response to complex stimuli, for example line motion illusion, and allows, on average, to answer some of the previous questions at the mesoscopic scale (see paragraph Section 9.3.6). However, improvements in parameter tuning are still needed.

More recently, Markounikau et al. (2010) confirmed that a one-dimensional two-layer neural field model can simulate the main VSD results on the line-motion paradigm, i.e. non-linear characteristics of cortical activity. Compared to the previous models, this model clearly separates the activity of the excitatory layer from the activity of the inhibitory layer, resulting in a system of two integrodifferential equations of the form of Equation (9.7). The same sigmoidal functions are used to relate membrane depolarization to firing rate, while synaptic connectivities between the two layers are simply represented by Gaussian functions. In contrast, Grimbert et al. (2007) used a central local patch surrounded by six satellite patches to represent accurately the patchy projections made by pyramidal cells from layers II/III and V. With this model, Markounikau et al. (2010) proposed that inhibition participates as much as excitation in shaping the observed VSDI dynamics in response to the line-motion paradigm.

9.4.2.4 Conductance-based IAF neuronal network model to reproduce correlates of illusory motion

Another large-scale computational model of the primary visual cortex has been proposed by Rangan et al. (2005). The model is a two-dimensional patch of cortex, containing about 10^6 neurons with preferred orientation, 80% of which are excitatory and 20% of which are inhibitory. The dynamics of a single cell i is described by a single-compartment, conductance-based, exponential integrate-and-fire equation (see Geisler et al., 2005, for more details of this neuron model). The derivation of this equation gives the membrane potential of neuron i of spatial position $\mathbf{r}_i$:

$$V(\mathbf{r}_i, t) = \frac{g^L\, V^L + (g_i^A(t) + g_i^N(t))\, V^E + g_i^G(t)\, V^I}{g^L + g_i^A(t) + g_i^N(t) + g_i^G(t)} \qquad (9.10)$$

where g^L, g_i^A, g_i^N and g_i^G are respectively the leak, AMPA, NMDA and GABA conductances, and V^L, V^E and V^I are respectively the leak, excitatory and inhibitory reversal potentials.

The authors then use $V(\mathbf{r}, t)$ to represent the VSD signal, i.e. the subthreshold dendritic activity in the superficial layers of the cortex. Poisson processes are used to simulate inputs from the thalamus and background noise.

Like the previous model (Grimbert et al., 2007), this model reproduces the spatiotemporal activity patterns of V1, as revealed by VSDI, in response to complex stimuli, for example the line motion illusion. However, in comparison with Grimbert et al. (2007), no laminar structure is taken into account.

9.4.2.5 A linear model to inspect the origin of the VSD signal

With the same scale of analysis, La Rota (2003) presented an interesting linear model to study the neural sources of the mesoscopic VSD signal. The author chose a compromise between a detailed and a "black-box" model of the signal, by taking into account the important properties of the VSD signal and also the artifacts linked directly to its measurement, in a mesoscopic, linear and additive model. The VSD signal of a cortical area can then be modeled by an intrinsic and an extrinsic component:

$$\mathrm{OI}(t) = A(t) + \rho(t), \tag{9.11}$$

where $A(t)$ represents the activity of the intrinsic component of the optical signal (i.e. the synaptic activity of the cortical area observed) and $\rho(t)$ represents all the noise and artifacts due to the measurement (e.g. hemodynamic artifact, cardiovascular and respiratory movements, instrumental noise, etc.). In this model, inputs from the thalamus are considered as background noise and thus enter in the ρ component.

The model is interesting because it takes into account both the intrinsic and the extrinsic variability of the VSD signal. The latter is neglected in the other models presented.

9.4.3 A submesoscopic model to study the VSD signal

9.4.3.1 The submesoscopic sources of the VSD signal

Although the aforementioned mesoscopic models reproduce accurately the VSD signal dynamics, none were specific enough to determine the different contributions to the optical signal. This is not surprising since they were not built for that purpose. As the sources of the VSD signal are actually submesoscopic, there is a need to develop a model at this spatial scale. The idea is thus to choose a scale adjusted to the methodological constraints of optical imaging. To quantify the contributions from excitatory verses inhibitory neurons, such a model should take into account several types of neurons, or at least one type of excitatory neuron and one type of inhibitory neuron, with realistic behaviors. To determine whether the VSD signal mainly reflects postsynaptic activity, the modeled neurons must have at least three compartments, i.e. soma, axon and dendrite. Furthermore, since it is important to inspect the contribution of the different layers to the signal, neurons in deep layers should have more than one dendrite, actually at least one per layer, in order to differentiate neurons whose dendrites reach superficial layers.

In optical imaging, a pixel size is about 50 μm. We thus decided to consider a cortical column of 50 μm diameter. Using the novel concept of cortical column previously proposed, it is then easy to introduce a new distinction of cortical column into Table 9.2 (last column), with spatial scale 50 μm, corresponding to one pixel of optical imaging. Therefore, this "optical" column lies between the anatomical and the functional columns. Because the scale is relatively small, compared with those used in the seven models previously presented, we will call this an "intermediate mesoscopic scale." This cortical column must be embedded into a larger network, in order to be realistic.

Following all these requirements, a detailed biophysically inspired model of a cortical column, embedded into an artificial hypercolumn (in the case of V1) has been proposed (Chemla and Chavane, 2010a). This model is detailed in the following section and was built to solve the direct VSD problem, i.e. generate a VSD signal, given the neural substrate parameters and activities.

9.4.3.2 Specifications of the proposed biophysical cortical column model

Given the spatial scale of 50 μm, the number of neurons, evaluated from Binzegger et al. (2004), is about 200. We then consider a class of models based on a cortical microcircuit (see Raizada and Grossberg, 2003; Douglas and Martin, 2004; Haeusler and Maass, 2007, for more details on this concept), with synaptic connections made only between six specific populations of neurons: two populations (excitatory and inhibitory) for three main layers (2/3, 4, 5/6).

Each neuron is represented by a reduced compartmental description (see Bush and Sejnowski, 1993, for more details on the reduction method) with a conductance-based Hodgkin–Huxley neuron model (see Hodgkin and Huxley, 1952) in the soma and the axon. Thus, the dynamics of single cells are described by the following equation:

$$C_m \frac{dV}{dt} = I_{ext} - \sum_i Gi(V)(V - V^i) \qquad (9.12)$$

where V is the membrane potential, I_{ext} is an external current injected into the neuron, C_m is the membrane capacitance, and where three types of current are represented: leak, potassium and sodium conductances or respectively G_L, G_K and G_{Na}. G_L is independent of V and determines the passive properties of the cells near resting potential. The sodium and potassium conductances are responsible for the spike generation. Furthermore, a slow potassium conductance was included in the dynamics of the excitatory population to reproduce the observed adaptation of the spike trains emitted by these neurons (see Nowak et al., 2003). This feature seems to be absent in inhibitory neurons, as taken into account in this model.

Only passive dendrites were considered. Each neuron is represented with seven to nine compartments. The link between compartments can then be described by Equation (9.13) (Hines and Carnevale, 1997):

$$C_j \frac{dV_j}{dt} + I_{ion_j} = \sum_k \frac{V_k - V_j}{R_{jk}} \tag{9.13}$$

where V_j is the membrane potential in compartment j, I_{ion_j} is the net transmembrane ionic current in compartment j, C_j is the membrane capacitance of compartment j and R_{jk} is the axial resistance between the centers of compartment j and the adjacent compartment k.

Synaptic inputs are modeled as conductance changes. Excitatory AMPA synapses are converging on the soma and dendrites of each neuron, whereas inhibitory GABA synapses are only converging on the soma of each neuron (Salin and Bullier, 1995). The number of synapses involved in the projections between these different neuronal types, including the afferent from the LGN, were recalculated for a 50 μm cortical column, based on Binzegger et al. (2004) for the considered layers, while latencies were introduced for each connection following Thomson and Lamy (2007).

Input signals from the thalamus into the neocortex layer IV were simulated by applying random spike trains to each neuron in layer IV and random latency was introduced for each input connection to simulate the temporal properties of the geniculocortical pathway. Then we increased the frequency of the spike trains in order to represent stimulus contrast and see how the model transforms an increasing input, i.e. the contrast response function (see Albrecht and Hamilton, 1982). At this point, the column is isolated. As a step further, the conditions relative to a larger network were reproduced as follows. First, "background noise" was introduced in each neuron of the column. Typically, noise can be introduced in the form of stochastic fluctutation of a current or an ionic conductance. The stochastic model of Destexhe et al. (2001), containing two fluctuating conductances, was used here, allowing us to simulate synaptic background activity similar to in vivo measurements, for a large network. Second, lateral connections between two neighboring columns were reproduced by introducing another set of random spike train inputs whose frequency, synaptic delays and synaptic weights were adapted for fitting experimental data. Figure 9.4A shows a schematic diagram of the model, with thalamic input, background activity and lateral interactions.

A Cortical column model

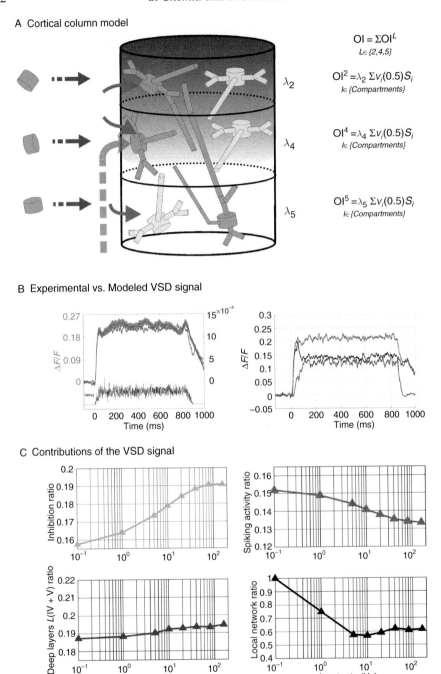

$$OI = \Sigma OI^L$$
$$L \in \{2,4,5\}$$

$$OI^2 = \lambda_2 \, \Sigma v_i(0.5)S_i$$
$$I \in \{Compartments\}$$

$$OI^4 = \lambda_4 \, \Sigma v_i(0.5)S_i$$
$$I \in \{Compartments\}$$

$$OI^5 = \lambda_5 \, \Sigma v_i(0.5)S_i$$
$$I \in \{Compartments\}$$

B Experimental vs. Modeled VSD signal

C Contributions of the VSD signal

Figure 9.4 (See plate section for color version.) Biophysical model implementation. A. Model representation with thalamic input (red), background activity (green) and lateral connections (blue). The model offers the possibility of computing the VSD signal with a linear formula (right panel) taking into account the dye concentration (λ parameter), and the membrane potential (V_i) at each

9.4.3.3 VSD signal computation

The VSD signal is simulated using a linear integration over the membrane surface of all neuronal components (Grinvald and Hildesheim, 2004). Here, the use of a compartmental model has a real interest. Indeed, the computation of the VSD signal, for a given layer L, is then given by a linear formula:

$$OI^L = \lambda^L \sum_{i \in \{Compartments\}} V_i\, S_i \tag{9.14}$$

where S_i and V_i are respectively the surface and the membrane potential of the ith compartment and λ^L represents the fluorescence gradient or the illumination intensity of the dye in layer L.

Thus, this model takes into account soma, axon and dendrite influences, introduces three-dimensional geometrical properties (e.g. dendrites of large pyramidal neurons in layer 5 can reach superficial layers) and a fluorescence gradient depending on depth. According to Lippert et al. (2007) and Petersen et al. (2003a), $\lambda^2 = 0.95$, $\lambda^4 = 0.05$ and $\lambda^5 = 0$. Then, the total optical imaging signal is given by the following formula:

$$OI = \sum_{L \in \{Layers\}} OI^L. \tag{9.15}$$

Following this framework, the VSD signal is simulated in response to known stimuli (Figure 9.4B, left panel) and compared to experimental results (see Chemla and Chavane, 2010a, for more details).

Thanks to its compartmental construction, this model can predict the different contributions of the VSD signal. It thus gives the possibility of answering quantitatively the relative contributions to the VSD signal of excitation versus inhibition, spiking versus synaptic activity, superficial versus deep layers and local versus horizontal inputs. This model confirms that the VSD signal mainly reflects dendritic activity (77%) of excitatory neurons (83%) in superficial layers (80%). However, these numbers change when the level of input activity is increased (Chemla and

---

Caption for Figure 9.4 (cont.)
compartment. B. *Left*: Temporal evolution of the VSD signal in response to an input of 800 ms and for a given input strength (80% contrast for the experimental and 300 Hz for the model). Comparison between modeled (red) and experimental (gray, mean ± sem) responses. *Right*: Model prediction of the relative contributions of local (black line) and lateral (blue line) inputs to the total VSD signal (red line). C. The different contributions of the VSD signal: inhibition vs. excitation, axonic vs. dendritic activity, deep vs. superficial layers and local vs. global activity. (Modified from Chemla and Chavane, 2010a).

Chavane, 2010a). At a high level of activity, inhibitory cells, spiking activity and deep layers become non-negligible (Figure 9.4C, left and top right panels), and should be taken into account in the computation of the VSD signal. Furthermore, it allows us to quantify that the proportion of activity arising from the local connectivity also changes with input rate: from 100% at low input to 50% at high input frequency (Figure 9.4C, bottom right panel), the rest of the VSD global signal coming from inputs from lateral connectivity that contribute more and more with increasing input rate (Figure 9.4B, right panel). Finally, it predicts that only 60% of the signal comes from the expected combined dendritic activity of excitatory neurons in superficial layers.

9.5 Conclusion

VSDI offers an unprecedented solution for recording large neural networks with high spatial and temporal resolution. However, as we have discussed here, the recorded optical signal is multi-component and its origins and their relative contributions to the signal are still unresolved.

In this review, we suggest that computational models are an appropriate solution for a better understanding of the VSDI signal. We reported seven existing models that reproduce and analyze the VSD signal on various dynamical aspects, giving insights on what are the network mechanisms at the origin of the actual experimental observations. However, none of these models was constructed to provide information about the origin of the signal. We propose that biophysical cortical column models at a submesoscopic scale are suitable for inspecting the biological sources of the VSD signal. Recently, a wide variety of models devoted to understanding, replicating and studying VSDI has emerged. We think that this enthusiasm is crucial and needs to be pursued in parallel with the use of this multi-component signal. Hopefully, future studies will succeed in intermingling both approaches and will open new perspectives, for example giving direct access to all the subcomponents constituting the VSD signal.

Acknowledgements

The authors are grateful to Thierry Vieville, from the CORTEX Lab., INRIA, Sophia-Antipolis, for his active participation in the model development and his help in writing this review. They also thank Francois Grimbert, at Northwestern University in the Department of Neurobiology and Physiology, for helpful discussions on the subject. This work was partially supported by the EC IP project FP6015879, FACETS & the MACCAC ARC projects.

References

Ahmed, B., Hanazawa, A., Undeman, C., Eriksson, D., Valentiniene, S. and Roland, P. E. (2008). Cortical dynamics subserving visual apparent motion. *Cereb. Cortex*, **18** (12), 2796–2810.

Aitken, J. T. (1955). Observations on the larger anterior horn cells in the lumbar region of the cat's spinal cord. *J. Anat.*, **89**, 571.

Albowitz, B. and Kuhnt, U. (1993). The contribution of intracortical connections to horizontal spread of activity in the neocortex as revealed by voltage sensitive dyes and a fast optical recording method. *Eur. J. Neurosci.*, **5** (10), 1349–1359.

Albrecht, D. and Hamilton, D. (1982). Striate cortex of monkey and cat: contrast response function. *Neurophysiol.*, **48** (1), 217–233.

Amit, D. and Brunel, N. (1997). Model of global spontaneous activity and local structured delay activity during delay periods in the cerebral cortex. *Cereb. Cortex*, **7**, 237–252.

Antic, S. and Zecevic, D. (1995). Optical signals from neurons with internally applied voltage-sensitive dyes. *J. Neurosci.*, **15** (2), 1392–1405.

Antic, S., Major, G. and Zecevic, D. (1999). Fast optical recordings of membrane potential changes from dendrites of pyramidal neurons. *J. Neurophysiol.*, **82** (3), 1615–1621.

Arieli, A., Shoham, D., Hildesheim, R. and Grinvald, A. (1995). Coherent spatiotemporal patterns of ongoing activity revealed by real-time optical imaging coupled with single-unit recording in the cat visual cortex. *J. Neurophysiol.*, **73** (5), 2072–2093.

Arieli, A., Sterkin, A., Grinvald, A. and Aerster, A. (1996). Dynamics of ongoing activity: explanation of the large variability in evoked cortical responses. *Science*, **273** (5283), 1868–1871.

Benucci, A., Robert, A. F. and Carandini, M. (2007). Standing waves and traveling waves distinguish two circuits in visual cortex. *Neuron*, **55** (1), 103–117.

Berger, T., Borgdorff, A., Crochet, S., Neubauer, F. B., Lefort, S., Fauvet, B., Fere-zou, I., Carleton, A., Luscher, H. R. and Petersen, C. C. (2007). Combined voltage and calcium epifluorescence imaging in vitro and in vivo reveals subthreshold and suprathreshold dynamics of mouse barrel cortex. *J. Neurophysiol.*, **97** (5), 3751–3762.

Binzegger, T., Douglas, R. and Martin, K. (2004). A quantitative map of the circuit of cat primary visual cortex. *Neurosci.*, **24** (39), 8441–8453.

Blasdel, G. G. and Salama, G. (1986). Voltage-sensitive dyes reveal a modular organization in monkey striate cortex. *Nature*, **321** (6070), 579–585.

Bolz, J., Novak, N. and Staiger, V. (1992). Formation of specific afferent connections in organotypic slice cultures from rat visual cortex cocultured with lateral geniculate nucleus. *J. Neurosci.*, **12** (8), 3054–3070.

Bonhoeffer, T. and Grinvald, A. (1991). Iso-orientation domains in cat visual cortex are arranged in pinwheel-like patterns. *Nature*, **353** (6343), 429–431.

Borg-Graham, L., Monier, C. and Fregnac, Y. (1998). Visual input evokes transient and strong shunting inhibition in visual cortical neurons. *Nature*, **393**, 369–373.

Bringuier, V., Chavane, F., Glaeser, L. and Fregnac, Y. (1999). Horizontal propagation of visual activity in the synaptic integration field of area 17 neurons. *Science*, **283** (5402), 695–699.

Brown, C. E., Aminoltejari, K., Erb, H., Winship, I. R. and Murphy, T. H. (2009). In vivo voltage-sensitive dye imaging in adult mice reveals that somatosensory maps lost to stroke are replaced over weeks by new structural and functional circuits with prolonged modes of activation within both the peri-infarct zone and distant sites. *J. Neurosci.*, **29** (6), 1719–1734.

Brunel, N. and Hakim, V. (1999). Fast global oscillations in networks of integrate-and-fire neurons with low firing rates. *Neural Comput.*, **11** (7), 1621–1671.

Bush, P. and Sejnowski, T., (1993). Reduced compartmental models of neocortical pyramidal cells. *Neurosci. Methods*, **46**, 159–166.

Cameron, R. S. and Rakic, P. (1991). Glial cell lineage in the cerebral cortex: a review and synthesis. *Glia*, **4**, 124–137.

Carlson, G. C. and Coulter, D. A. (2008). In vitro functional imaging in brain slices using fast voltage-sensitive dye imaging combined with whole-cell patch recording. *Nature Protocols*, **3** (2), 249–255.

Chemla, S. and Chavane, F. (2010a). A biophysical cortical column model to study the multi-component origin of the vsd signal. *NeuroImage*, **53** (2), 420–438.

Chemla, S. and Chavane, F. (2010b). Voltage-sensitive dye imaging: technique review and models. *J. Physiol. Paris*, **104** (1–2), 40–50.

Cinelli, A. R. and Salzberg, B. M. (1990). Multiple site optical recording of transmembrane voltage (msortv), single-unit recordings, and evoked field potentials from the olfactory bulb of skate (*Raja erinacea*). *J. Neurophysiol.*, **64** (6), 1767–1790.

Cinelli, A. R. and Salzberg, B. M. (1992). Dendritic origin of late events in optical recordings from salamander olfactory bulb. *J. Neurophysiol.*, **68** (3), 786–806.

Civillico, E. F. and Contreras, D. (2006). Integration of evoked responses in supra-granular cortex studied with optical recordings in vivo. *J. Membr. Biol.*, **96** (1), 336–351.

Cohen, L., Salzberg, B. M., Davila, H. V., Ross, W. N., Landowne, D., Waggoner, A. S. and Wang, C. H. (1974). Changes in axon fluorescence during activity: molecular probes of membrane potential. *J. Membr. Biol.*, **19** (1), 1–36.

Contreras, D. and Llinas, R. (2001). Voltage-sensitive dye imaging of neocortical spatiotemporal dynamics to afferent activation frequency. *J. Neurosci.*, **21** (23), 9403–9413.

Davila, H. V., Salzberg, B. M., Cohen, L. B. and Waggoner, A. S. (1973). A large change in axon fluorescence that provides a promising method for measuring membrane potential. *Nature (London) New Biol.*, **241** (109), 159–160.

Derdikman, D., Hildesheim, R., Ahissar, E., Arieli, A. and Grinvald, A. (2003). Imaging spatiotemporal dynamics of surround inhibition in the barrels somatosensory cortex. *J. Neurosci.*, **23** (8), 3100–3105.

Destexhe, A., Rudolph, M., Fellous, J. and Sejnowski, T. (2001). Fluctuating synaptic conductances recreate in-vivo-like activity in neocortical neurons. *Neuroscience*, **107**, 13–24.

Douglas, R. J. and Martin, K. A. C. (1990). Neocortex. In: G Shepeherd (editor), *Synaptic Organization of the Brain*. New York: Oxford University Press, pp. 220–248.

Douglas, R. and Martin, K. A. C. (2004). Neuronal circuit of the neocortex. *Ann. Rev. Neurosci.*, **27**, 419.

Eberwine, J. (2001). Molecular biological of axons: a turning point... *Neuron*, **32** (6), 959–960.

Ebner, T. J. and Chen, G. (1995). Use of voltage-sensitive dyes and optical recordings in the central nervous system. *Prog. Neurobiol.*, **46** (5), 463–506.

El Boustani, S. and Destexhe, A. (2009). A master equation formalism for macroscopic modeling of asynchronous irregular activity states. *Neural Comput.*, **21** (1), 46–100.

Faugeras, O., Grimbert, F. and Slotine, J.-J. (2008). Abolute stability and complete synchronization in a class of neural fields models. *SIAM J. Appl. Math.*, **61** (1), 205–250.

Faugeras, O., Touboul, J. and Cessac, B. (2009). A constructive mean-field analysis of multi-population neural networks with random synaptic weights and stochastic inputs. *Front. Comput. Neurosci.*, **3** (1), doi:10.3389/neuro.10.001.2009.

Ferezou, I., Bolea, S. and Petersen, C. C. H. (2006). Visualizing the cortical representation of whisker touch: voltage-sensitive dye imaging in freely moving mice. *Neuron*, **50**, 617–629.

Geisler, C., Brunel, N. and Wang, X.-J. (2005). Contributions of intrinsic membrane dynamics to fast network oscillations with irregular neuronal discharges. *J. Neurophysiol.*, **94**, 4344–4361.

Grimbert, F., Faugeras, O. and Chavane, F. (2007). From neural fields to VSD optical imaging. In: *Sixteenth Annual Computational Neuroscience Meeting, CNS*. www.cnsorg.org.

Grimbert, F., Faugeras, O. and Chavane, F. (2008). Neural field model of VSD optical imaging signals. In: Areadne08. www.areadne.org/

Grinvald, A. and Hildesheim, R. (2004). VSDI: a new era in functional imaging of cortical dynamics. *Nature*, **5**, 874–885.

Grinvald, A., Manker, A. and Segal, M. (1982). Visualization of the spread of electrical activity in rat hippocampal slices by voltage-sensitive optical probes. *J. Physiol.*, **333**, 269–291.

Grinvald, A., Anglister, L., Freeman, J. A., Hildesheim, R. and Manker, A. (1984). Real time optical imaging of naturally evoked electrical activity in the intact frog brain. *Nature*, **308** (5962), 848–850.

Grinvald, A., Lieke, E., Frostig, R. D., Gilbert, C. D. and Wiesel, T. N. (1986). Functional architecture of cortex revealed by optical imaging of intrinsic signals. *Nature*, **324** (6095), 361–364.

Grinvald, A., Frostig, R. D., Siegel, R. M. and Bartfeld, E. (1991). High-resolution optical imaging of functional brain architecture in the awake monkey. *Proc. Natl. Acad. Sci. USA*, **88** (24), 11559–11563.

Grinvald, A., Lieke, E., Frostig, R. D. and Hildesheim, R. (1994). Cortical point-spread function and long-range lateral interactions revealed by real-time optical imaging of macaque monkey primary visual cortex. *J. Neurosci.*, **14**, 2545–2568.

Grinvald, A., Shoham, D., Shmuel, A., Glaser, D., Vanzetta, I., Shtoyerman, E., Slovin, H. and Arieli, A. (1999). In-vivo optical imaging of cortical architecture and dynamics. In: U. Windhorst and H. Johansson (editors), *Modern Techniques in Neuroscience Research*. New York: Springer, pp. 893–969.

Gupta, R. G., Salzberg, B. M., Grinvald, A., Cohen, L., Kamino, K., Boyle, M. B., Waggoner, S. and Wang, C. H. (1981). Improvements in optical methods for measuring rapid changes in membrane potential. *J. Membr. Biol.*, **58** (2), 123–137.

Haeusler, S. and Maass, W. (2007). A statistical analysis of information-processing properties of lamina-specific cortical microcircuits models. *Cereb. Cortex*, **17**, 149–162.

Hines, M. and Carnevale, N. (1997). The neuron simulation environment. *Neural Comput.*, **9**, 1179–1209.

Hodgkin, A. and Huxley, A. (1952). A quantitative description of membrane current and its application to conduction and excitation in nerve. *J. Physiol.*, **117**, 500–544.

Horton, J. and Adams, D. (2005). The cortical column: a structure without a function. *Philos. Trans. R. Soc. London B. Ser.*, **360** (1456), 837–862.

Hubel, D. and Wiesel, T. (1962). Receptive fields, binocular interaction and functional architecture in the cat visual cortex. *J. Physiol.*, **160**, 106–154.

Hubel, D. and Wiesel, T. (1965). Receptive fields and functional architecture in two nonstriate visual areas (18 and 19) of the cat. *J. Neurophysiol.*, **28**, 229–289.

Hubel, D. and Wiesel, T. (1977). Functional architecture of macaque monkey. *Proc. R. Soc. London, Ser. B*, **198**, 1–59.

Hubel, D. H., Wiesel, T. N. and Stryker, M. P. (1977). Orientation columns in macaque monkey visual cortex demonstrated by the 2-deoxyglucose autoradiographic technique. *Nature*, **269** (5626), 328–330.

Hubener, M., Shoham, D., Grinvald, A. and Bonhoeffer, T. (1997). Spatial relationships among three columnar systems in cat area 17. *J. Neurosci.*, **17** (23), 9270–9284.

Jancke, D., Chavane, F., Naaman, S. and Grinvald, A. (2004). Imaging cortical correlates of illusion in early visual cortex. *Nature*, **428**, 423–426.

Jin, W., Zhang, R. and Wu, J. (2002). Voltage-sensitive dye imaging of population neuronal activity in cortical tissue. *J. Neurosci. Methods*, **115**, 13–27.

Kauer, J. S. (1988). Real-time imaging of evoked activity in local circuits of the salamander olfactory bulb. *Nature*, **331** (6152), 166–168.

Kee, M. Z., Wuskell, J. P., Loew, L. M., Augustine, G. J. and Sekino, Y. (2009). Imaging activity of neuronal populations with new long-wavelength voltage-sensitive dyes. *Brain Cell. Biol.*, **36** (5–6), 157–172.

Kelly, J. P. and Van Essen, D. C. (1974). Cell structure and function in the visual cortex of the cat. *J. Physiol.*, **238**, 515–547.

Kennedy, C., Des Rosiers, M., Sakurada, O., Shinohara, M., Reivich, M., Jehle, J. and Sokoloff, L. (1976). Metabolic mapping of the primary visual system of the monkey by means of the autoradiographic (14c)deoxyglucose technique. *Proc. Natl. Acad. Sci. USA*, **73** (11), 4230–4234.

Kisvarday, Z., Toth, E., Rausch, M. and Eysel, U. (1997). Orientation-specific relationship between populations of excitatory and inhibitory lateral connections in the visual cortex of the cat. *Cereb. Cortex*, **7** (7), 605–618.

Kleinfeld, D. and Delaney, K. (1996). Distributed representation of vibrissa movement in the upper layers of somatosensory cortex revealed with voltage-sensitive dyes. *J. Comp. Neurol.*, **375**, 89–108.

Kohonen, T. (2001). *Self-Organizing Maps* (3rd edition). New York: Springer.

Konnerth, A. and Orkand, R. K. (1986). Voltage-sensitive dyes measure potential changes in axons and glia of the frog optic nerve. *Neurosci. Lett.*, **66** (1), 49–54.

Konnerth, A., Obaid, A. L. and Salzberg, B. M. (1987). Optical recording of electrical activity from parallel fibres and other cell types in skate cerebellar slices in vitro. *J. Physiol.*, **393**, 681–702.

Konnerth, A., Orkand, P. M. and Orkand, R. K. (1988). Optical recording of electrical activity from axons and glia of frog optic nerve: potentiometric dye responses and morphometrics. *Glia*, **1** (3), 225–232.

Kubota, M., Hosokawa, Y. and Horikawa, J. (2006). Layer-specific short-term dynamics in network activity in the cerebral cortex. *NeuroReport*, **17** (11), 1107–1110.

La Rota, C. (2003). Analyse de l'activité électrique multi-ties du cortex auditif chez le cobaye. *Ph.D. thesis*, Université Joseph Fourier, Grenoble I.

Laaris, N. and Keller, A. (2002). Functional independence of layer IV barrels. *J. Neurophysiol.*, **87** (2), 1028–1034.

Lev-Ram, V. and Grinvald, A. (1986). Ca^{2+}- and K^+-dependent communication between central nervous system myelinated axons and oligodendrocytes revealed by voltage-sensitive dyes. *Proc. Natl. Acad. Sci. USA*, **83** (17), 6651–6655.

LeVay, S., Connolly, M., Houde, J. and Van Essen, D. (1985). The complete pattern of ocular dominance stripes in the striate cortex and visual field of the macaque monkey. *J. Neurosci.*, **5**, 486–501.

Lippert, M. T., Takagaki, K., Xu, W., Huang, X. and Wu, J. Y. (2007). Methods for voltage-sensitive dye imaging of rat cortical activity with high signal-to-noise ratio. *Neurophysiol.*, **98**, 502–512.

Manis, P. B. and Freeman, J. A. (1988). Fluorescence recordings of electrical activity in goldfish optic tectum in vitro. *J. Neurosci.*, **8** (2), 383–394.

Markounikau, V., Igel, C., Grinvald, A. and Jancke, D. (2010). A dynamic neural field model of mesoscopic cortical activity captured with voltage-sensitive dye imaging. *PLoS Comput. Biol.*, **6** (9), 1–14.

Miikkulainen, R., Bednar, J. A., Choe, Y. and Sirosh, J. (2005). *Computational Maps in the Visual Cortex.* Berlin: Springer.

Mountcastle, V. (1957). Modality and topographic properties of single neurons of cat's somatosensory cortex. *Neurophysiol.*, **20**, 408–434.

Nelson, D. A. and Katz, L. C. (1995). Emergence of functional circuits in ferret visual cortex visualized by optical imaging. *Neuron*, **15** (1), 23–34.

Nowak, L. G., Azouz, R., Sanchez-Vives, M. V., Gray, C. and McCormick, D. (2003). Electrophysiological classes of cat primary visual cortical neurons in vivo as revealed by quantitative analyses. *J. Neurophysiol.*, **89**, 1541–1566.

Orbach, H. S. and Cohen, L. B. (1983). Optical monitoring of activity from many areas of the in vitro and in vivo salamander olfactory bulb: a new method for studying functional organization in the vertebrate central nervous system. *J. Neurosci.*, **3**, 2251–2262.

Orbach, H. S. and Van Essen, D. C. (1993). In vivo tracing of pathways and spatio-temporal activity patterns in rat visual cortex using voltage sensitive dyes. *Exp. Brain Res.*, **94** (3), 371–392.

Orbach, H. S., Cohen, L. B. and Grinvald, A. (1985). Optical mapping of electrical activity in rat somatosensory and visual cortex. *J. Neurosci.*, **5**, 1886–1895.

Petersen, C. and Sakmann, B. (2001). Functional independent columns of rat somatosensory barrel cortex revealed with voltage-sensitive dye imaging. *J. Neurosci.*, **21** (21), 8435–8446.

Petersen, C., Grinvald, A. and Sakmann, B. (2003a). Spatiotemporal dynamics of sensory responses in layer 2/3 of rat barrel cortex measured in vivo by voltage-sensitive dye imaging combined with whole-cell voltage recordings and neuron reconstructions. *J. Neurosci.*, **23** (3), 1298–1309.

Petersen, C., Hahn, T., Mehta, M., Grinvald, A. and Sakmann, B. (2003b). Interaction of sensory responses with spontaneous depolarization in layer 2/3 barrel cortex. *Proc. Natl. Acad. Sci. USA*, **100** (23), 13638–13643.

Raizada, R. and Grossberg, S. (2003). Towards a theory of the laminar architecture of the cerebral cortex: computational clues from the visual system. *Cereb. Cortex*, **13**, 100–113.

Rangan, A. V., Cai, D. and McLaughlin, D. W. (2005). Modeling the spatiotemporal cortical activity associated with the line-motion illusion in primary visual cortex. *Proc. Natl. Acad. Sci. USA*, **102** (52), 18793–18800.

Reynaud, A., Barthelemy, F., Masson, G. and Chavane, F. (2007). Input-ouput transformation in the visuo-oculomotor network. *Arch. Ital. Biol.*, **145**, 251–262.

Roland, P. E. (2002). Dynamic depolarization fields in the cerebral cortex. *Trends Neurosci.*, **25**, 183–190.

Roland, P. E., Hanazawa, A., Undeman, C., Eriksson, D., Tompa, T., Nakamura, H., Valentiniene, S. and Ahmed, B. (2006). Cortical feedback depolarization waves: a mechanism of top-down influence on early visual areas. *Proc. Natl. Acad. Sci. USA*, **103** (33), 12586–12591.

Ross, W. N., Salzberg, B. M., Cohen, L. B., Grinvald, A., Davila, H. V., Waggoner, A. S. and Wang, C. H. (1977). Changes in absorption, fluorescence, dichroism, and birefringence in stained giant axons: optical measurements of membrane potential. *J. Membr. Biol.*, **33** (1–2), 141–183.

Rubin, B. D. and Katz, L. C. (1999). Optical imaging of odorant representations in the mammalian olfactory bulb. *Neuron*, **23** (3), 499–511.

Salin, P. and Bullier, J. (1995). Corticocortical connections in the visual system: structure and function. *Psychol. Bull.*, **75**, 107–154.

Salzberg, B. M., Davila, H. V. and Cohen, L. B. (1973). Optical recording of impulses in individual neurons of an invertebrate central nervous system. *Nature*, **246**, 508–509.

Schummers, J., Yu, H. and Sur, M. (2008). Tuned responses of astrocytes and their influence on hemodynamic signals in the visual cortex. *Science*, **320**, 1638–1643.

Seidemann, E., Arieli, A., Grinvald, A. and Slovin, H. (2002). Dynamics of depolarization and hyperpolarization in the frontal cortex and saccade goal. *Science*, **295** (5556), 862–865.

Sharon, D. and Grinvald, A. (2002). Dynamics and constancy in cortical spatiotemporal patterns of orientation processing. *Science*, **295** (5554), 512–515.

Sharon, D., Jancke, D., Chavane, F., Na'aman, S. and Grinvald, A. (2007). Cortical response field dynamics in cat visual cortex. *Cereb. Cortex*, **17** (12), 2866–2877.

Shoham, D., Glaser, D., Arieli, A., Kenet, T., Wijnbergeb, C., Toledo, Y., Hildesheim, R. and Grinvald, A. (1999). Imaging cortical dynamics at high spatial and temporal resolution with novel blue voltage-sensitive dyes. *Neuron*, **24** (4), 791–802.

Sholl, D. A. 1955. The organization of the visual cortex in the cat. *J. Anat.*, **89**, 33–46.

Sirosh, J. and Miikkulainen, R. (1994). Cooperative self-organization of afferent and lateral connections in cortical maps. *Biol. Cybernet.*, **71**, 66–78.

Sit, Y. F. and Miikkulainen, R. (2007). A computational model of the signals in optical imaging with voltage-sensitive dyes. *Neurocomputing*, **70** (10–12), 1853–1857.

Slovin, H., Arieli, A., Hildesheim, R. and Grinvald, A. (2002). Long-term voltage-sensitive dye imaging reveals cortical dynamics in behaving monkeys. *J. Neurophysiol.*, **88** (6), 3421–3438.

Spors, H. and Grinvald, A. (2002). Spatio-temporal dynamics of odor representations in the mammalian olfacgtory bulb. *Neuron*, **34** (2), 301–315.

Sterkin, A., Lampl, I., Ferster, D., Grinvald, A. and Arieli, A. (1998). Realtime optical imaging in cat visual cortex exhibits high similarity to intracellular activity. *Neurosci. Lett.*, **51**, S41.

Stuart, G., Spruston, N., Sakmann, B. and Hausser, M. (1997). Action potential initiation and backpropagation in neurons of the mammalian CNS. *Trends Neurosci.*, **20** (3), 125–131.

Symes, A. and Wennekers, T. (2009). Spatiotemporal dynamics in the cortical microcircuit: a modelling study of primary visual cortex layer 2/3. *Neural Networks*, **22** (8), 1079–1092.

Takagaki, K., Lippert, M. T., Dann, B., Wanger, T. and Ohl, F. W. (2008). Normalization of voltage-sensitive dye signal with functional activity measures. *PloS one*, **3** (12), 1–12.

Tasaki, I., Watanabe, A. and Carnay, L. (1968). Changes in fluorescence, turbidity, and bireference associated with nerve excitation. *Proc. Natl. Acad. Sci. USA*, **61**, 883–888.

Thomson, A. and Lamy, C. (2007). Functional maps of neocortical local circuitry. *Front. Neurosci.*, **1** (1), 19–42.

Thomson, A. and Morris, O. (2002). Selectivity in the inter-laminar connections made by neocortical neurones. *J. Neurocytol.*, **31** (3-5), 239–246.

Tootell, R., Silverman, M., Switked, E. and De Valois, R. (1982). Deoxyglucose analysis of retinotopic organization in primate striate cortex. *Science*, **218** (4575), 902–904.

Ts'o, D. Y., Frostig, R. D., Lieke, E. E. and Grinvald, A. (1990). Functional organization of primate visual cortex revealed by high resolution optical imaging. *Science*, **249** (4967), 417–420.

Tsunoda, K., Yamane, Y., Nishizaki, M. and Tanifuji, M. (2001). Complex objects are represented in macaque inferotemporal cortex by the combination of feature columns. *Nature Neurosci.*, **4** (8), 832–838.

Tucker, T. R. and Katz, L. C. (2003a). Recruitment of local inhibitory networks by horizontal connections in layer 2/3 of ferret visual cortex. *J. Neurophysiol.*, **89** (1), 501–512.

Tucker, T. R. and Katz, L. C. (2003b). Spatiotemporal patterns of excitation and inhibition evoked by the horizontal network in layer 2/3 of ferret visual cortex. *J. Neurophysiol.*, **89** (1), 488–500.

van Vreeswijk, C. and Sompolinsky, H. (1998). Chaotic balanced state in a model of cortical circuits. *Neural Comput.*, **10** (6), 1321–1371.

Waggoner, A. S. and Grinvald, A. (1977). Mechanisms of rapid optical changes of potential sensitive dyes. *Ann. NY Acad. Sci.*, **30** (303), 217–241.

Waters, J., Schaefer, A. and Sakmann, B. (2005). Backpropagating action potentials in neurones: measurement, mechanisms and potential functions. *Prog. Biophys. Mol. Biol.*, **87** (1), 145–170.

Wiesel, T. N., Hubel, D. H. and Lam, D. M. (1974). Autoradiographic demonstration of ocular-dominance columns in the monkey striate cortex by means of transneuronal transport. *Brain Res.*, **79** (2), 273–279.

Wilson, H. and Cowan, J. (1972). Excitatory and inhibitory interactions in localized populations of model neurons. *Biophys. J.*, **12**, 1–24.

Wilson, H. and Cowan, J. (1973). A mathematical theory of the functional dynamics of cortical and thalamic nervous tissue. *Biol. Cybernet.*, **13** (2), 55–80.

Woolum, J. C. and Strumwasser, F. (1978). Membrane-potential-sensitive dyes for optical monitoring of activity in aplysia neurons. *J. Neurobiol.*, **9** (3), 185–193.

Xu, W., Huang, X., Takgaki, K. and Wu, J. (2007). Compression and reflection of visually evoked cortical waves. *Neuron*, **55** (1), 119–129.

Yang, Z., Heeger, D. J. and Seidemann, E. (2007). Rapid and precise retinotopic mapping of the visual cortex obtained by voltage-sensitive dye imaging in the behaving monkey. *J. Neurophysiol.*, **98** (2), 1002–1014.

Young, J. Z. (1958). Anatomical considerations. *EEG Clin. Neurophysiol.*, **10**, 9–11.

Yuste, R., Tank, D. K. and Kleinfeld, D. (1997). Functional study of the rat cortical microcircuitry with voltage-sensitive dye imaging of neocortical slices. *Cereb. Cortex*, **7** (6), 546–558.

Zochowski, M., Wachowiak, M., Falk, C. X., Cohen, L. B., Lam, Y. W., Antic, S. and Zecevic, D. (2000). Imaging membrane potential with voltage-sensitive dyes. *Biol. Bull.*, **198** (1), 749–762.

10

Calcium imaging

FRITJOF HELMCHEN

Over the past 30 years calcium-sensitive fluorescent dyes have emerged as powerful tools for optical imaging of cell function. Calcium ions subserve a variety of essential functions in all cell types. For example, changes in intracellular free calcium concentration ($[Ca^{2+}]_i$) underlie fundamental cellular processes such as muscle contraction, cell division, exocytosis, and synaptic plasticity. Most of these processes rely on the steep gradient of calcium ion concentration that is actively maintained across the plasma membrane. Moreover, cells store calcium ions in intracellular organelles, enabling them to release a surge of Ca^{2+} into the cytosol where and when needed. Calcium ions act through molecular binding to various Ca^{2+}-binding proteins, inducing conformational changes and thereby activating or modulating protein function. The development of optical reporters of calcium concentration has opened great opportunities to read out $[Ca^{2+}]_i$ directly as a crucial intracellular messenger signal. A major application of calcium indicators is the quantitative study of a specific calcium-dependent process X, for example, neurotransmitter release, with the goal to reveal the function $X = X([Ca^{2+}]_i)$. However, this is not the only type of application. Because neuronal excitation in the form of receptor activation or generation of action potentials typically is linked to calcium influx, calcium indicators are also used to reveal neural activation patterns, either within the dendritic tree of individual cells or within cell populations.

Today, a large palette of fluorescent calcium indicators is available. All of them act through binding of Ca^{2+}. In the simplest case the fluorescence intensity depends on $[Ca^{2+}]_i$ but there are several other fluorescence parameters that may change and that can be read out. In order to understand fully the action of calcium ions within cells we need to understand the spatiotemporal dynamics of the variable $[Ca^{2+}]_i(x, t)$ and the complex interactions of Ca^{2+} with Ca^{2+}-binding partners. In particular, we need to consider the calcium indicator itself as one of the interacting reaction partners. For simplicity we will use the notation $[Ca](x, t)$ for the

Handbook of Neural Activity Measurement, ed. Romain Brette and Alain Destexhe. Published by Cambridge University Press. © Cambridge University Press 2012.

intracellular free calcium concentration, where the vector $x = (x, y, z)$ refers to the three-dimensional spatial coordinates and t is the temporal variable. It may seem a daunting task to keep track of calcium dynamics throughout a cell but fortunately such detail often is not necessary and various useful mathematical models can be derived based on reasonable simplifications. A fundamental aspect that the experimenter should be aware of is the fact that calcium indicators inevitably perturb intracellular calcium dynamics (to a lesser or greater degree). This notion means that indicators interfere with the variable they are supposed to report, which may limit the information that can be gathered. On the flipside, knowledge about how calcium indicators influence intracellular calcium dynamics can be exploited by the experimenter as we will discuss in this chapter. Being familiar with the principles of how calcium indicators work and with the mathematical description of calcium dynamics thus is essential for the design and interpretation of calcium imaging experiments.

The chapter starts with a brief introduction of the different types of fluorescent calcium indicators. Subsequently, the fundamental processes involved in intracellular calcium dynamics and their mathematical descriptions are introduced. To help the reader a list of key parameters used in the mathematical description is provided in Table 10.1. Following a treatment of the Ca^{2+}-dependence of fluorescence we then discuss simplified models of calcium dynamics and their applications to reveal specific aspects of neural dynamics. The chapter concludes with a brief comparison with other methods and a discussion of future perspectives.

10.1 Fluorescent calcium indicators

The first optical $[Ca^{2+}]_i$ measurements were achieved in the 1960s and 1970s using the bioluminescent protein Aequorin (e.g. Ridgway and Ashley, 1967) or metal-lochromic arsenazo dyes that change their light absorbance depending on $[Ca^{2+}]_i$ (e.g. Brown et al., 1975). Although these indicators are still sometimes used we focus here entirely on fluorescent calcium indicators, the prevailing indicator form today. Fluorescence readout is beneficial because even at low indicator concentrations it allows for high-contrast labeling against a low background (Tsien, 1989). Two different types of fluorescent calcium indicators exist (Figure 10.1): (1) small-molecule indicators (SMIs), which are synthetic organic dyes that have been developed since the beginning of the 1980s (Tsien, 1980); and (2) calcium-sensitive fluorescent proteins, the first one of which was reported in 1997 (Miyawaki et al., 1997). The latter type often is referred to as the group of genetically-encoded calcium indicators (GECIs). It is beyond the scope of this chapter to provide a complete overview of the various indicator types and the history of their development (for detailed reviews see Tsien, 1989; Miyawaki, 2003; Garaschuk et al., 2007;

Table 10.1 *Key parameters of intracellular calcium dynamics and fluorescent calcium indicators.*

Parameter	Definition	Typical range/unit
$[Ca^{2+}]_i$ or $[Ca]$	intracellular free Ca^{2+} concentration, in general space and time dependent	$50\,nM$–$100\,\mu M$
$[Ca^{2+}]_{rest}$	resting intracellular free Ca^{2+} concentration	50–$150\,nM$
k_{on}	association rate constant of Ca^{2+} binding	10^6–$10^9\,M^{-1}\,s^{-1}$
k_{off}	dissociation rate constant of Ca^{2+} binding	1–$10^3\,s^{-1}$
K_d	dissociation constant of a Ca^{2+} binding molecule	$100\,nM$–$0.1\,mM$
κ_B	Ca^{2+} binding ratio, "strength of buffering"	10–1000 for endogenous buffers
κ_m	Ca^{2+} binding ratio of a mobile buffer	
κ_f	Ca^{2+} binding ratio of an immobile ("fixed") buffer	
j_{in}	concentration change due to Ca^{2+} influx	$M\,s^{-1}$
j_{out}	concentration change due to Ca^{2+} extrusion	$M\,s^{-1}$
j_{leak}	leakage term to maintain $[Ca^{2+}]_{rest}$	$M\,s^{-1}$
F	Faraday's constant	$96485\,C\,mol^{-1}$
Q_{Ca}	"calcium charge," related to total number of calcium ions entering a cell during influx	C
v_{max}	maximum velocity of a Ca^{2+} extrusion pump	$mol\,m^{-2}\,s^{-1}$
γ	linear extrusion rate of a simplified extrusion mechanism	100–$2000\,s^{-1}$
D_{Ca}	diffusion constant of free calcium ions	$220\,\mu m^2\,s^{-1}$
$D_{B,i}$	diffusion constant of Ca^{2+} binding molecule species i	$\mu m^2\,s^{-1}$
D_{app}	apparent diffusion constant in the presence of Ca^{2+} binding molecules	$\mu m^2\,s^{-1}$
ε	molar dye extinction coefficient	$20\,000$–$100\,000\,M^{-1}\,cm^{-1}$
Q_F	quantum yield	between 0 and 1
Q_D	quantum efficiency of a photodetector	0–100%, typically between 10–40%
R_f	dynamic range of an indicator	
τ_F	fluorescence lifetime of an indicator	1–$10\,ns$
Φ_D	collection efficiency of the detector system	0–100%, typically far less than 100%
$(K_{eff}, R_{min}, R_{max})$	triplet of calibration parameters for ratiometric measurements	
$(K_d, \Delta F/F_{max}, [Ca^{2+}]_{rest})$	triplet of calibration parameters for single-wavelength measurements	
(K_d, R_f, F_{max})	alternative triplet of calibration parameters for single-wavelength measurements	
$(K_{app}, \tau_{min}, \tau_{max})$	triplet of calibration parameters for fluorescence lifetime measurements	

Mank and Griesbeck, 2008). Nonetheless, it is useful to illustrate briefly how these indicator molecules work before we start the mathematical treatment.

10.1.1 Small-molecule indicators

The first type of calcium indicators are small organic molecules. A major step in their development was the modification of the well-known Ca^{2+}-chelating molecule EGTA to BAPTA, a Ca^{2+} chelator with fast binding kinetics (Tsien, 1980). Many of the SMIs are designed by covalently attaching a fluorophore to the BAPTA-derived Ca^{2+} chelating part. As an example, Oregon Green BAPTA-1 is shown in Figure 10.1A. Binding of Ca^{2+} to the octacoordinate pocket of the chelator leads to a reconfiguration of the conjugated electron system, which then induces changes in the fluorescence properties. The exact nature as well as the strength of the fluorescence changes can be tuned by the choice of the

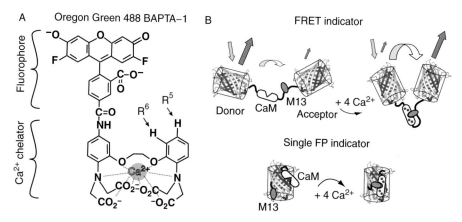

Figure 10.1 Examples of the two major classes of fluorescent calcium indicators. A. Oregon Green BAPTA-1 is a small-molecule indicator consisting of a Ca^{2+}-chelating part and a fluorophore part. Variants of Oregon Green BAPTA indicators differ in their substituents at key positions: Oregon Green BAPTA-5N contains a nitro group at position R^5 and Oregon Green BAPTA-6F a contains fluorine atom at position R^6. B. Genetically encoded calcium indicators are further subdivided into two classes. Top: FRET-based indicators rely on the Ca^{2+}-dependent change in distance between two fluorescent proteins, mediated by a Ca^{2+}-binding linker (here Calmodulin which binds four calcium ions and interacts with the M13 peptide). One fluorescent protein acts as donor, the other as acceptor for FRET. FRET efficiency increases when the two proteins are brought closer together. Bottom: Ca^{2+}-binding domains are incorporated into single circularly permutated fluorescent proteins. In these single-protein indicators Ca^{2+}-binding leads to a conformational change that is sensed by the chromophore inside the fluorescent protein barrel structure, causing a change in fluorescence intensity.

substituents at critical intramolecular positions, so that a multitude of indicator variants can be created. With the exception of some indicators that have been targeted to the membrane by attaching lipophilic side chains, most of the SMIs are water soluble and readily diffuse throughout the cytosol. In their salt form, indicators dissociate in solution. As charged molecules they are not membrane permeable and therefore need to be loaded into cells with special techniques. Direct physical methods include electroporation and filling via intracellular recording pipettes. An elegant alternative method is to load SMIs into cells using their acetomethoxy(AM)-conjugated form (Tsien, 1981). In this chemically modified form the carboxy-groups are esterified and thus turned into uncharged side groups. Consequently, the AM-form of indicators is membrane permeable. Once inside the cytosol, however, the ester-groups are cleaved by endogenous esterases, so that the original indicator molecule is recovered but is now trapped inside the cell. AM-ester loading of calcium has been used for many years to label cells in cell culture and in brain slices. Only recently, bolus injection of AM-indicators directly into neural tissue has enabled labeling of large cell populations in the intact brain (Stosiek et al., 2003). This breakthrough has opened new opportunities for optical recording of neural network activity in vivo (Garaschuk et al., 2006; Göbel and Helmchen, 2007a; Grewe and Helmchen, 2009).

10.1.2 Genetically encoded calcium indicators

The second group of fluorescent calcium indicators is the still rapidly growing group of GECIs. Similar to SMIs, a large variety of GECIs is available. Two major classes can be distinguished (Figure 10.1B). In the first class, GECIs are designed by coupling two fluorescent proteins (e.g. cyan fluorescent protein, CFP, and yellow fluorescent protein, YFP) via a Ca^{2+}-binding linker. Due to their enforced close proximity the two proteins can exchange energy via fluorescence resonance energy transfer (FRET) (Stryer, 1978). One protein acts as acceptor and the other as donor molecule. Ca^{2+}-binding to the linker induces a conformational change that brings the two proteins closer together. Because of the strong distance dependence of FRET, this leads to a change in the fluorescence emissions of the two proteins that can be read out ratiometrically (see below). The first constructs of this kind used the Ca^{2+}-dependent interaction of calmodulin (CaM) with the M13 protein domain to translate $[Ca^{2+}]_i$ changes into FRET changes (Miyawaki et al., 1997). Meanwhile many variants of such "chameleons" have been generated (Nagai et al., 2004; Palmer et al., 2006). In addition, other FRET-based designs have been introduced using the Troponin C protein as Ca^{2+}-dependent linker molecule (Heim and Griesbeck, 2004; Garaschuk et al., 2007).

The second major class of GECIs comprises single-protein indicators rather than tandem pairs of proteins (Figure 10.1B). Example proteins are Inverse Pericam or members of the GCaMP family (Tian et al., 2009). In these indicators, the protein has been modified by insertion of a Ca^{2+}-binding domain such that Ca^{2+}-binding leads to a conformational change that either increases or decreases the fluorescence yield of the chromophore (Baird et al., 1999; Nakai et al., 2001). Compared to SMIs the general advantages of GECIs are the possibilities of long-term expression and of targeting them to specific subtypes of neurons or subcellular locations. Many of the initial problems have now been overcome and one can expect rapid expansion of the application of GECIs for functional measurements in the upcoming years.

10.2 Intracellular calcium dynamics

This section gives a general description of the spatiotemporal dynamics of $[Ca^{2+}]_i$ within a cell before we will treat Ca^{2+}-sensitive fluorescence changes of calcium indicators in the next section. For the main dynamic processes involved in intracellular calcium handling (binding, influx, extrusion, and diffusion) we introduce the key parameters and generic mathematical formulations. First considered individually, we then combine these aspects in a general set of differential equations that describes intracellular calcium dynamics.

10.2.1 Calcium binding

Within the cytosol, calcium ions bind to a multitude of endogenous proteins. In addition, Ca^{2+}-binding is the fundamental process by which calcium indicators work. In the case of Oregon Green BAPTA-1 Ca^{2+}-binding occurs at an octa-coordinate binding site of the chelator part (Figure 10.1A). This indicator is an example of one-to-one binding of Ca^{2+} to a molecule. In many proteins, including GECIs, the Ca^{2+}-binding sites are so-called EF-hand domains (Mank and Griesbeck, 2008; Schwaller, 2009), of which several may be present in the protein. For example, calmodulin is made up of four EF-hand domains. These domains are not independent in their structural rearrangements upon Ca^{2+}-binding, resulting in cooperativity of the binding processes. We first consider independent binding before we briefly discuss cooperative binding.

Independent calcium binding In the simplest case calcium ions are bound either individually or, if multiple calcium ions can bind to the molecule, binding events occur independent from each other. In this case we only need to consider the simple binding scheme

$$Ca^{2+} + B \quad \overset{k_{on}}{\underset{k_{off}}{\rightleftharpoons}} \quad CaB \qquad (10.1)$$

where B denotes the binding site and CaB the calcium-bound complex. B thus stands for an intracellular protein or an exogenous Ca^{2+}-binding molecule that has been artificially introduced into the cell, for example, an indicator dye. Binding occurs with an association rate k_{on} (unit $mM^{-1}s^{-1}$) and the complex CaB dissociates with a rate k_{off} (unit s^{-1}). The temporal dynamics of this process is described by the following differential equations:

$$\frac{\partial [Ca]}{\partial t} = -k_{on} [Ca] [B] + k_{off} [CaB]$$

$$\frac{\partial [B]}{\partial t} = -k_{on} [Ca] [B] + k_{off} [CaB]$$

$$\frac{\partial [CaB]}{\partial t} = -\frac{\partial [B]}{\partial t} = k_{on} [Ca] [B] - k_{off} [CaB]$$

$$[Ca]_T = [Ca] + [CaB]$$

$$[B]_T = [B] + [CaB]. \qquad (10.2)$$

Here, we have introduced the total concentrations $[Ca]_T$ and $[B]_T$ as the sum of the free and bound forms. $[Ca]_T$ is conserved in the absence of Ca^{2+} influx into or extrusion from the cytosol. $[B]_T$ is conserved unless changes in expression levels of Ca^{2+}-binding proteins occur or exogenous Ca^{2+} buffers are artificially added.

In many cases one is mainly interested in the steady-state concentrations when equilibrium has been reached. In this case the concentration changes on the left hand side are zero. From Equations (10.2) the law of mass action follows, in which the affinity of B for binding Ca^{2+} is described by the dissociation constant

$$K_d = \frac{k_{off}}{k_{on}} = \frac{[Ca] [B]}{[CaB]} = \frac{[Ca] ([B]_T - [CaB])}{[CaB]}. \qquad (10.3)$$

The dissociation constant K_d has the intuitive meaning that it is equal to the $[Ca^{2+}]_i$ level, at which 50% of the binding sites have bound Ca^{2+}. This is obvious by rearranging Equation (10.3) to yield the so-called saturation curve of the binding species B:

$$S = \frac{[CaB]}{[B]_T} = \frac{[Ca]}{[Ca] + K_d}. \qquad (10.4)$$

This saturation curve is plotted in Figure 10.2A. For limited ranges of $[Ca^{2+}]_i$ this curve can be piecewise approximated by a linear function. Note that there is an inverse relationship between the expressions "affinity" and "dissociation constant" with a low K_d value implying a high affinity and vice versa. This notion is

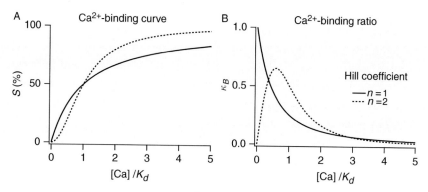

Figure 10.2 Saturation curve and Ca^{2+}-binding ratio for Ca^{2+}-binding molecules showing either independent binding (Hill coefficient $n = 1$) or cooperative binding ($n = 2$). A. Percentage of saturation as a function of calcium concentration level normalized to the dissociation constant. B. Ca^{2+}-binding ratio κ_B as a function of normalized calcium concentration level.

important for working with calcium indicators because the affinity of the indicator should match the expected range of calcium concentrations "seen" by the indicator. Typically, indicators are only sensitive in a narrow window of the entire range of possible intracellular $[Ca^{2+}]_i$ levels that spans from resting concentrations of below 10^{-4} mM to high concentrations around 0.1 mM, which for example can occur locally at sites of calcium influx. If the affinity of the indicator is too low (K_d too high) it will not be sensitive in the low concentration range; if the affinity is too high the indicator will be nearly saturated. Thus, depending on the experimental aims, the most appropriate indicator has to be chosen from the available spectrum of dyes with widely different affinities. High affinity calcium indicators have dissociation constants in the submicromolar range (e.g. K_d of OGB-1 is about 200–300 nM), while K_d values reach around 50 μM for low affinity indicators.

A second important aspect is that the Ca^{2+}-binding efficiency depends on the $[Ca^{2+}]_i$ level. This is intuitively clear as the probability of Ca^{2+}-binding will decrease once a certain number of ions are already bound. In other words, fewer and fewer binding sites will be available until – at relatively high $[Ca^{2+}]_i$ levels – most binding sites are occupied and this particular species of Ca^{2+}-binding molecules becomes saturated. Formally this dependency is expressed by the so-called "Ca^{2+}-binding ratio" κ_B, which is obtained by differentiating Equation (10.4):

$$\kappa_B = \frac{\partial[CaB]}{\partial[Ca]} = \frac{[B]_T K_d}{([Ca] + K_d)^2}. \tag{10.5}$$

The Ca^{2+}-binding ratio scales with the total concentration of the binding species. For independent calcium binding, κ_B has a maximal value of $[B]_T/K_d$ at zero

$[Ca^{2+}]_i$ and decreases monotonically for increasing $[Ca^{2+}]_i$ levels (Figure 10.2B). In the literature the Ca^{2+}-binding ratio is also often termed "buffering capacity" or "buffering strength." As we will see further below, the concept of Ca^{2+}-binding ratio is very helpful in describing key aspects of calcium buffers and indicators. In general, it is desirable to know (or at least to have a good estimate of) the Ca^{2+}-binding ratios of both the endogenous calcium buffers and the fluorescent calcium indicator that is used for the measurement.

Cooperative calcium binding While independent Ca^{2+}-binding is a good model for small-molecule calcium indicators, it does not adequately describe the situation for many protein binding reactions. The reason is that Ca^{2+}-binding proteins often contain multiple, structurally connected binding sites that are no longer independent. More generally, we therefore have to consider cooperative binding as it is known from the classic case of O_2-binding to hemoglobin.

A general description of cooperative binding is fairly complex. For our purposes it is sufficient to incorporate cooperativity using the Hill equation, which is an empirical description of the equilibrium case. In this simplified view, binding of multiple calcium ions is described using the Hill coefficient n and an apparent dissociation constant K_A

$$K_A = \frac{[Ca]^n \, [B]}{[Ca^n B]} = \frac{[Ca]^n \, ([B]_T - [Ca^n B])}{[Ca^n B]}. \tag{10.6}$$

Note that the Hill coefficient is not equal to the number of binding sites but rather is an empirical parameter that describes the overall effect of cooperative binding. The saturation curve in this case takes a sigmoidal shape according to the equation

$$S = \frac{[Ca^n B]}{[B]_T} = \frac{[Ca]^n}{[Ca]^n + K_A} = \frac{[Ca]^n}{[Ca]^n + K_d^n}, \tag{10.7}$$

where K_d again is the calcium concentration at half-maximal occupancy (Figure 10.2A). Hill coefficients for GECIs are in the range of 0.7–3.8 (Mank and Griesbeck, 2008). From Equation (10.7) we can also derive the Ca^{2+}-binding ratio for the case of cooperative binding:

$$\kappa_B = \frac{\partial \, [Ca^n B]}{\partial \, [Ca]} = [B]_T \, \frac{n \, [Ca]^{n-1} \, K_d^n}{([Ca]^n + K_d^n)^2}. \tag{10.8}$$

This relationship is plotted in Figure 10.2B for $n = 2$, showing a non-monotonic $[Ca^{2+}]_i$ dependence of κ_B with a maximum at some intermediate concentration level. This highly non-linear behavior makes the interpretation of calcium measurements with GECIs more difficult so that the binding characteristics of the specific indicator need to be taken into account.

Often Ca^{2+}-binding is referred to as "buffering" and many proteins are classified as "calcium buffers." While "binding" refers to the physical process itself, "buffering" has a functional meaning and refers to the ability of a molecular species to maintain the $[Ca^{2+}]_i$ level within a certain concentration range by dampening changes in free calcium concentration following calcium influx. For a number of proteins, calcium buffering appears to be the main function, although additional Ca^{2+}-sensing functions may be unknown and await to be discovered (Schwaller, 2009).

10.2.2 Calcium influx

The steady-state equilibrium of Ca^{2+} binding and unbinding in the cytosol is continually perturbed by Ca^{2+} influx from various sources (Figure 10.3). For example, neural excitation leads to Ca^{2+} flux across the plasma membrane through either voltage-gated calcium channels or calcium-permeable receptor channels. Moreover, Ca^{2+} may be released from intracellular stores, including the endoplasmic reticulum and mitochondria. In general, we denote the spatially and temporally varying influx as $j_{in}(\mathbf{x}, t)$. In the absence of extrusion mechanisms, the spatiotemporal integral of j_{in} equals the total calcium charge Q_{Ca} that enters the cell during a

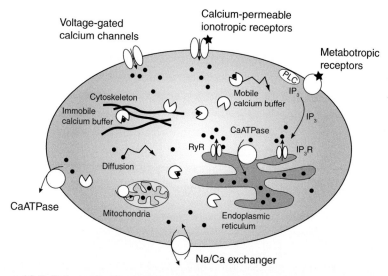

Figure 10.3 Schematic illustration of various processes that contribute to intracellular calcium dynamics. Several pathways exist for entry of calcium ions into the cytosol (black dots), including influx through voltage- or ligand-gated channels and release from intracellular organelles. Intracellularly, calcium ions diffuse and bind to various mobile or immobile Ca^{2+}-binding proteins. Several extrusion mechanisms remove calcium from the cytosol. PLC phospholipase C, RyR ryanodine receptor, IP_3R inositol-tri-phosphate receptor.

given time period. This calcium load will change the total concentration of calcium ions (free or bound), which is given by

$$[Ca]_T = [Ca] + \sum_i [CaB_i]. \tag{10.9}$$

The index i in B_i runs over all Ca^{2+}-binding molecular species that are present. While the spatiotemporal pattern of influx in general is highly complex, approximate expressions can be used for j_{in} in particular situations. For example, if calcium influx is brief compared to extrusion times and if it can be assumed to occur rather homogenously throughout the cell compartment, then j_{in} can be approximated by a pulse-like, instantaneous influx, which can be mathematically expressed by the Kronecker delta function

$$j_{in} = \Delta [Ca]_T \ \delta(t - t_p) = \frac{Q_{Ca}}{2\,F\,V} \ \delta(t - t_p), \tag{10.10}$$

where t_p is the point in time when influx occurs, F is Faraday's constant, and V is the volume of the cell compartment under consideration. This approximation is particularly useful for describing action-potential evoked Ca^{2+} influx, because voltage-gated calcium channels open for less than a millisecond during the repolarizing phase of the action potential (Borst and Helmchen, 1998). Ca^{2+} extrusion in the soma, dendrites or axons is typically at least an order of magnitude slower. For a train of n action potentials, Equation (10.10) can be extended to a sum of delta functions with spike times at time points t_k with $k = 1, \ldots, n$. If Ca^{2+} influx does not occur in a pulse-like way but rather is prolonged, for example, during excitatory postsynaptic currents or during release from intracellular stores, the time course of j_{in} needs to be modeled with other suitable analytical functions.

How is the extra calcium load following Ca^{2+} influx distributed among the Ca^{2+}-binding partners present? After an initial non-equilibrium phase that is governed by binding kinetics and diffusional exchange, a new steady state will be reached. The steady-state concentration *changes* of all partners (expressed with the greek "delta" Δ) depend on each other according to:

$$\Delta [Ca]_T = \Delta [Ca] + \sum_i \Delta [CaB_i] = \Delta [Ca] + \sum_i \kappa_{B,i} \Delta [Ca] = \Delta [Ca] \left(1 + \sum_i \kappa_{B,i}\right). \tag{10.11}$$

Here, we approximated concentration changes of Ca^{2+}-bound molecules with the help of their Ca^{2+}-binding ratios (assuming relatively small changes). We conclude that the change in free calcium concentration is determined by the total change in concentration divided by one plus the sum of all Ca^{2+}-binding ratios:

$$\Delta [Ca] = \frac{\Delta [Ca]_T}{(1 + \sum_i \kappa_{B,i})} = \frac{Q_{Ca}}{2\,F\,V} \frac{1}{(1 + \sum_i \kappa_{B,i})}. \tag{10.12}$$

As we will discuss in more detail further below, this equation in particular implies that addition of a calcium indicator to the system may significantly affect changes in free calcium concentration. The magnitude of this effect will, however, also depend critically on the concentrations and properties of all other Ca^{2+}-binding molecules present.

10.2.3 Calcium extrusion

Calcium extrusion is vital for the cell because low $[Ca^{2+}]_i$ levels need to be re-established following surges of Ca^{2+} influx in order to keep the cell alive. Cells maintain a very low resting free calcium concentration through effective calcium pump mechanisms (Figure 10.3). Calcium ions are either sequestered into intra-cellular organelles or extruded via the plasma membrane until a resting $[Ca^{2+}]_i$ level of about 30–100 nM has been re-established (Helmchen et al., 1996; Maravall et al., 2000). The requirement to keep $[Ca^{2+}]_i$ levels low may have arisen early in evolution because calcium compounds are prone to precipitation. The result-ing steep concentration gradient across the membrane then presumably provided opportunities to use Ca^{2+} as a fast and effective signaling ion.

In general terms we denote the spatiotemporal calcium extrusion with $j_{out}(\mathbf{x}, t)$. A common formalization is the description as Ca^{2+}-dependent, saturable mecha-nism with first-order Michaelis–Menten kinetics:

$$j_{out}([Ca]) = v_{max}\frac{A}{V}\frac{[Ca]}{[Ca] + K_m}, \tag{10.13}$$

where v_{max} is the maximal efflux rate per unit area of cell membrane, A is the com-partment surface, and K_m is the concentration at which extrusion is half-maximal. This equation does not account for the fact that $[Ca^{2+}]_i$ is not zero at rest. Under resting conditions a steady state exists between some leakage Ca^{2+} influx and the extrusion of Ca^{2+}, resulting in a zero net flux. Resting conditions are most easily incorporated in the mathematical description of calcium dynamics by including a constant leakage term that just balances the pump mechanisms at rest:

$$j_{leak} = -v_{max}\frac{A}{V}\frac{[Ca]_{rest}}{[Ca]_{rest} + K_m}. \tag{10.14}$$

If the pumps operate well below saturation ($[Ca^{2+}]_i \ll K_m$) the net extrusion can be further simplified by linearization of Equations (10.13) and (10.14) using a single clearance rate γ

$$j_{out}([Ca]) = \gamma \; \Delta[Ca] = \gamma([Ca] - [Ca]_{rest}). \tag{10.15}$$

Because many possible calcium extrusion pathways exist, characterization of cal-cium dynamics in a particular cell type or cell compartment should first aim to identify the predominant pathway and then examine whether saturation is reached during the experimental protocol because distinct extrusion mechanisms may dominate at low and high $[Ca^{2+}]_i$ levels. Such investigation, for example using pharmacological tools, can provide a good starting point for adequate modeling of calcium extrusion. As a complicating issue, slow Ca^{2+}-binding mechanisms sometimes operate on a time scale similar to extrusion, which makes it difficult to distinguish between these mechanisms which both represent calcium "sinks" (see below). While extruded Ca^{2+} is lost, slowly bound Ca^{2+} remains in the cytosol and can still shift to other Ca^{2+}-binding partners and exert functional control.

10.2.4 Calcium diffusion

Another important aspect is the spatial distribution and mobility of calcium ions as well as of Ca^{2+}-binding molecules. While some proteins are anchored to the cytoskeleton in strategic places and thus exhibit low mobility, others can diffuse freely throughout the cytosol. As mentioned above, many calcium indicators show a rather high mobility and can fill the entire cell by diffusion. For freely diffusible substances, the concentration is homogenous throughout a given cell compartment at steady state. Localized Ca^{2+} influx and efflux will cause short-lived concentration gradients, which then equilibrate by diffusion. Diffusion is a probabilistic process due to the thermal agitation of molecules. For any diffusible substance S the temporal change of the concentration $[S](\mathbf{x}, t)$ at a position x is given by the difference of influx and efflux in the local volume element, which both depend on the local concentration gradient. For the one-dimensional case the resulting balance equation is the *diffusion equation*

$$\frac{\partial [S](x, t)}{\partial t} = D_S \frac{\partial^2 [S](x, t)}{\partial x^2} \tag{10.16}$$

where D_S is the diffusion coefficient. A characteristic feature of diffusional spread is that the mean displacement $\langle x \rangle$ of the diffusing particles is proportional to the square-root of the time period Δt

$$\langle x \rangle \propto \sqrt{D_S \, \Delta t}. \tag{10.17}$$

For example, a substance with a diffusion coefficient of $100 \, \mu m^2 \, s^{-1}$ spreads from a point source about $0.3 \, \mu m$ in $1 \, ms$, $1 \, \mu m$ in $10 \, ms$, $3 \, \mu m$ in $100 \, ms$, and $10 \, \mu m$ in $1 \, s$. In cells with elaborate morphologies such as neurons, this square-root-law of diffusion has the important consequence that intracellular chemical signals can be effectively compartmentalized so that their action is restricted to a certain spatial range. For the three-dimensional case we rewrite Equation (10.16) in vector notation. For the diffusion of free calcium ions we write for example

$$\frac{\partial\,[Ca]\,(\mathbf{x},t)}{\partial t} = D_{Ca}\,\nabla(\nabla\,[Ca]\,(\mathbf{x},t)), \tag{10.18}$$

where ∇ denotes the gradient operator. We avoid the Laplace operator $\Delta f = \nabla(\nabla f)$ here because of the potential confusion with Δ, the "delta" sign used to express signal changes.

For large molecules (molecular weight $M > 1000$) the diffusion coefficient depends on the molecule radius and hence on the inverse of the cubic root of M. Due to the formation of hydration shells in aqueous solution, ions have relatively small diffusion coefficients. For free calcium ions D_{Ca} is about $600\,\mu m^2/s^{-1}$ in water but is reduced to $D_{Ca} \approx 220\,\mu m^2\,s^{-1}$ in the cytosol (Allbritton et al., 1992). In general, diffusion coefficients are two to three times lower inside cells compared to in water because of the higher viscosity (Woolf and Greer, 1994). If multiple diffusible molecules are present, each will spread according to its diffusion coefficient. As a consequence the diffusional spread of calcium ions can be either promoted or slowed down in the presence of Ca^{2+}-binding molecules, depending on their mobility (see below). In particular, loading a cell with calcium indicator may affect not only Ca^{2+}-buffering but also the overall diffusional spread of calcium ions.

10.2.5 General formulation of calcium dynamics

We can now give a general description of intracellular calcium dynamics that incorporates all the different aspects of Ca^{2+}-binding, influx, efflux, and diffusion. Assuming the presence of multiple Ca^{2+}-binding molecules B_i we obtain the following set of *reaction-diffusion equations*

$$\frac{\partial\,[Ca]}{\partial t} = D_{Ca}\nabla(\nabla\,[Ca]) - [Ca]\sum_i k_{on,i}\,[B_i] + \sum_i k_{off,i}\,[CaB_i]$$

$$+\,j_{in} - j_{out} + j_{leak}$$

$$\frac{\partial\,[B_i]}{\partial t} = D_{B,i}\nabla(\nabla\,[B_i]) - k_{on,i}\,[Ca]\cdot[B_i] + k_{off,i}\,[CaB_i]$$

$$\frac{\partial\,[CaB_i]}{\partial t} = D_{CaB,i}\nabla(\nabla\,[CaB_i]) + k_{on,i}\,[Ca]\cdot[B_i] - k_{off,i}\,[CaB_i]$$

$$[B_i]_T = [B_i] + [CaB_i]$$

$$[Ca]_T = [Ca] + \sum_i [CaB_i]. \tag{10.19}$$

For each species of Ca^{2+}-binding molecules we have to write three equations analogous to the middle three equations. Note that this formulation can be simplified under the assumptions that (1) the total concentration of mobile binding partner $[B_i]_T$ is spatially uniform and (2) the diffusion coefficients of the free and the bound forms are equal ($D_{B,i} = D_{CaB,i}$) (Wagner and Keizer, 1994). Under these conditions the partial derivatives of $[B_i]$ and $[CaB_i]$ are equal with opposite signs. For immobile binding partners, the diffusive terms vanish but the total concentration typically will be spatially non-uniform. Equations (10.19) illustrate that calcium dynamics is governed by a fairly complex coupled system of partial differential equations, which in general is difficult to solve analytically. This notion highlights the increased complexity of signaling when intracellular binding sites are present at a high density. It also indicates that it might not be intuitively clear how such a complex signaling system is perturbed by the addition of a calcium indicator. In general, Equations (10.19) can be solved numerically by discretizing them in both time and space (see for example DeSchutter and Smolen, 1998; Markram et al., 1998). Several modeling environments for simulating neuronal dynamics, for example NEURON (Carnevale and Hines, 2009) and GENESIS (Bower and Beeman, 1998), support the simulation of calcium dynamics within cell compartments of various geometries and within entire compartmental models of neurons. Because our knowledge of the identity, subcellular distribution and kinetic properties of endogenous Ca^{2+}-binding proteins in most cases is limited, realistic high-resolution simulations, attempting to take into account many details, are still difficult. On the other hand, for many applications such detail is not needed. Depending on the spatial scale of interest, one can focus on a few relevant aspects of calcium dynamics and thereby derive helpful analytical approximations, which will be further eluded on in Section 10.4.

After this introduction to the basic mechanisms governing intracellular calcium dynamics, the next section will treat the different types of fluorescence changes of calcium indicators, how they are utilized for calcium imaging, and how they can be related back to the underlying $[Ca^{2+}]_i$ changes.

10.3 Calcium-dependent fluorescence properties

Fluorescent calcium indicators translate Ca^{2+}-binding into a measurable change in fluorescence that can be read out in an optical setup. In principle, Ca^{2+}-binding may affect various fluorescence properties, including absorption or emission spectra, overall fluorescence yield, and fluorescence lifetime. Moreover, FRET indicators are designed to show large changes in FRET efficiency because of the Ca^{2+}-sensitive spacing between the two coupled fluorophores. Which of the induced changes are most suitable for reporting $[Ca^{2+}]_i$ changes depends on the specific indicator molecule as well as on the complexity of the imaging setup. In any case, the observed fluorescence changes need to be quantified, normalized, and – if the absolute $[Ca^{2+}]_i$ values are relevant – calibrated in terms of absolute concentration levels.

10.3.1 Fluorescence intensity

The foremost readout mode of calcium indicators is a change in fluorescence intensity. Figure 10.4 illustrates a few typical cases schematically. Calcium indicator dyes can be excited by absorption of either a single or multiple photons. For example, two-photon excitation involves near-simultaneous absorption of two lower energy photons, a non-linear excitation mode that is highly suitable for deep imaging in biological tissues (Helmchen and Denk, 2005). Possible spectral effects of Ca^{2+}-binding include scaling of the emission spectrum and spectral shifts of either

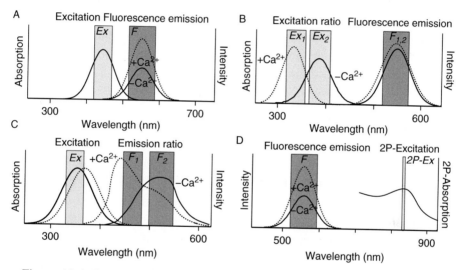

Figure 10.4 Common modes for measuring fluorescence changes of calcium indicators. A. Intensity measurement for indicators that scale their fluorescence spectrum upon Ca^{2+}-binding, showing either an increase (as illustrated here) or a decrease in fluorescence. The two boxes schematize spectral windows for excitation (*Ex*) and fluorescence collection (*F*), respectively. Windows can be set for example using band-pass optical filters. B. Fluorescence excitation ratioing can be performed with indicators that show a strong Ca^{2+}-dependent shift of their absorption spectrum. Typically, alternating excitation in two spectral windows is used in this case. C. Fluorescence emission ratioing is possible with indicators that show a strong shift of their emission spectrum. Here, emitted fluorescence is measured separately in two spectral windows. D. Two-photon excitation of calcium indicators with a pulsed near-infrared laser (bandwidth about 10 nm). In this case emitted fluorescence is of shorter wavelength than the excitation light. The examples shown here approximate the situation for Calcium Green and Oregon Green BAPTA dyes (A), fura-2 (B), indo-1 (C), and two-photon excitation of Calcium Green or Oregon Green BAPTA dyes (D).

the excitation or the emission spectrum. In Figure 10.4A a calcium indicator is single-photon excited in the visible wavelength range and fluorescence emission is collected in a separate, longer wavelength band (spectral windows are set for instance with bandpass optical filters in the excitation and detection pathways). In the case shown here, fluorescence intensity F is indicated to increase with increasing $[Ca^{2+}]_i$ but there also exist indicators that show a decrease in fluorescence instead. In general, the absorption and emission spectra of indicators at various $[Ca^{2+}]_i$ are well characterized in vitro and are available from the suppliers; one should keep in mind, however, that spectral properties depend on ionic strength, pH and viscosity and therefore may deviate from standard conditions in the cytosolic environment.

How does the fluorescence signal F relate to the $[Ca^{2+}]_i$ level? To understand this relationship we first need to clarify how fluorescence signals are generated and detected. Let us consider a cellular compartment of volume V loaded with a generic fluorescent dye X at concentration $[X]$. For simplicity we assume illumination of a flat sheet with area A and thickness l. If the excitation intensity is I_0 then the thin sheet will absorb light according to the Beer–Lambert law

$$I_{abs} = I_0 \, (1 - 10^{-\varepsilon l \, [X]}) \approx I_0 \ln(10) \, \varepsilon l \, [X], \qquad (10.20)$$

where ε is the molar extinction coefficient of the dye. Extinction coefficients of calcium indicators are typically in the range $20\,000$–$100\,000\,M^{-1}\,cm^{-1}$. The approximation of a linear relationship between I_{abs} and $[X]$ is only valid if $[X] < (\ln(10)\varepsilon l)^{-1}$. Equation (10.20) thus provides an upper bound to the useful dye concentration range. For measurements on small cells with $10\mu m$ diameter, dye concentration should be well below 5–$20\,mM$. Another constraint is that Ca^{2+}-binding ratios become substantial for Ca^{2+} indicators at relatively low concentration, in particular for high affinity indicators. This extra Ca^{2+}-buffer load potentially severely alters intracellular calcium dynamics, and high affinity indicators are therefore mostly applied at micromolar concentrations.

The excited molecules within V will emit fluorescence photons isotropically with a quantum yield Q_F. The quantum yield is defined as the ratio of the number of photons emitted to the number of photons absorbed. Only a certain fraction of the emitted photons (Φ_D) is collected by the photodetection system and these photons are detected with a limited likelihood given by the detector's quantum efficiency Q_D. Taking all these factors into account, we can express the fluorescence readout signal as

$$F = Q_D \Phi_D Q_F \, I_{abs} = Q_D \Phi_D Q_F \, I_0 \ln(10) \, \varepsilon l \, [X] = S \cdot [X]. \qquad (10.21)$$

Here, all dye- and setup-specific factors have been merged into a single proportionality constant S. For a calcium indicator we need to consider both the free and the Ca^{2+}-bound form at concentrations [B] and [CaB], respectively. They are treated as two different molecular species because they differ with respect to quantum yield and absorption properties and therefore contribute to F with different factors S_f and S_b, respectively:

$$
\begin{aligned}
F &= S_f \, [B] + S_b \, [CaB] \\
&= F_{min} + (S_b - S_f) \, [CaB] \\
&= F_{max} - (S_b - S_f) \, [B] \, .
\end{aligned}
\tag{10.22}
$$

By rearranging this equation we have introduced the minimal and maximal fluorescence values $F_{min} = S_f[B]_T$ and $F_{max} = S_b[B]_T$ (assuming an increase in fluorescence upon Ca^{2+} binding). The ratio of these two parameters $R_f = F_{max}/F_{min}$ determines the maximal possible fluorescence change and is referred to as the *dynamic range* of the indicator. Note that the dynamic range depends on dye properties and also on the experimental setup, for example, the spectral windows used for excitation and fluorescence detection.

In the intermediate range, between F_{min} and F_{max}, the measured fluorescence intensity depends on the relative amounts of Ca^{2+}-bound and unbound indicator, which are determined by the indicator's affinity according to Equation (10.3) or (10.7). For 1:1 complexation we can convert F to $[Ca^{2+}]_i$ by expanding the fraction in Equation (10.3) and then substituting Equations (10.22):

$$
[Ca] = K_d \frac{[CaB] \, (S_b - S_f)}{[B] \, (S_b - S_f)} = K_d \frac{F - F_{min}}{F_{max} - F} .
\tag{10.23}
$$

This equation is the basic conversion equation for relating the measured fluorescence intensity to the intracellular calcium concentration. However, although useful for cuvette measurements (e.g. measurements of cell suspensions or calibration solutions), Equation (10.23) turns out to be impractical for imaging experiments because the optical path length, total dye concentration, and illumination intensity – and consequently F_{min} and F_{max} – usually vary over the field of view. A calibration of these parameters for each imaging pixel is not feasible. To overcome this problem various ratioing approaches are commonly employed which we describe in the next sections.

10.3.2 Relative fluorescence change $\Delta F/F$

The simplest way to account for spatial inhomogeneities is to normalize the time-dependent fluorescence change $F(t)$ to the initial baseline fluorescence

F_0 (Figure 10.5A), which yields the relative percentage change in fluorescence intensity $\Delta F/F$ (read "delta F over F"):

$$\Delta F/F = (F - F_0)/F_0. \tag{10.24}$$

Equation (10.24) as written here assumes that background fluorescence values are already subtracted from both F and F_0 beforehand. Optical components, the bathing solution and endogenous fluorophores all add background to the indicator fluorescence. Thus, the readout values from the photodetector always contain a background component that must be subtracted ($F = F_{observed} - F_{background}$). If background subtraction is omitted, $\Delta F/F$ calculated from Equation (10.24) will

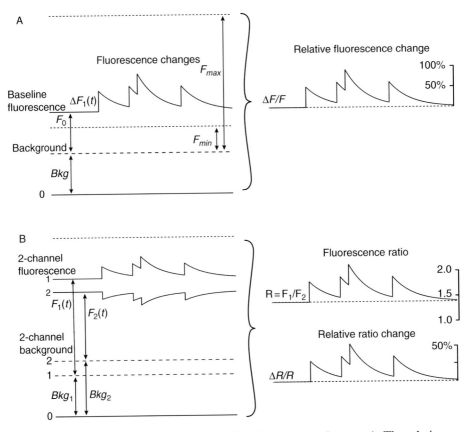

Figure 10.5 Standard methods to normalize fluorescence changes. A. The relative fluorescence change $\Delta F/F$ is calculated by dividing the background-corrected fluorescence trace by the background-corrected baseline fluorescence. $\Delta F/F$ is typically expressed as percentage change. B. For ratiometric measurements the ratio R of the background-corrected fluorescence intensities for the two channels is calculated. Note that the background needs to be determined separately for each channel. Often the change in ratio is expressed as percentage change $\Delta R/R$.

be too small, underestimating the true percentage changes. Background correction may seem trivial but actually often is not straightforward. For example, bulk-loading with AM-ester calcium indicators leads to a widespread, rather diffuse staining, for which it is difficult to obtain a good estimate of the true fluorescence background. A novel method to estimate background values, based solely on the dynamics of pixel intensities within a region of interest (ROI) can help to solve this problem (Chen et al., 2006).

How does $\Delta F / F$ relate to absolute $[Ca^{2+}]_i$ levels? Obviously, $\Delta F / F = 0$ corresponds to the resting $[Ca^{2+}]_i$ level. From Equation (10.23) one can derive the following conversion equation (Lev-Ram et al., 1992)

$$[Ca] = \frac{[Ca]_{rest} + K_d \frac{(\Delta F/F)}{(\Delta F/F)_{max}}}{(1 - \frac{(\Delta F/F)}{(\Delta F/F)_{max}})} \tag{10.25}$$

where $(\Delta F/F)_{max}$ denotes the maximal change upon dye saturation, which can be estimated for example using strong stimulation at the end of the experiment. The major drawback of Equation (10.25) and related single-wavelength equations (Vranesic and Knöpfel, 1991; Neher and Augustine, 1992; Wang et al., 1995) is that $[Ca^{2+}]_{rest}$ cannot be inferred from the fluorescence signal but needs to be determined in an independent manner, for example using a dual-wavelength ratiometric measurement (see below). If this is not possible, reasonable values of $[Ca^{2+}]_{rest}$ can be assumed (50–100 nM) given that the healthiness of cells can be assessed independently, for example based on morphological criteria or electrophysiological recording. Finally, for small fluorescence changes far from indicator saturation it may be sufficient to approximate changes in $[Ca^{2+}]_i$ by linearization of Equation (10.25)

$$\Delta [Ca] = \frac{K_d}{(\Delta F/F)_{max}} (\Delta F/F) \quad (\Delta F/F \ll (\Delta F/F)_{max}). \tag{10.26}$$

Note that this equation provides an estimate of the *change* in calcium concentration but not of its absolute value. It assumes that $[Ca^{2+}]_{rest}$ has the same value as for the $(\Delta F/F)_{max}$ calibration measurements.

To circumvent the necessity for an independent measurement of $[Ca^{2+}]_{rest}$ and to obtain estimates of absolute concentrations rather than concentration changes, an alternative single-wavelength approach has been introduced that is based on a different rearrangement of Equation (10.23) (Maravall et al., 2000):

$$[Ca] = K_d \frac{F/F_{max} - 1/R_f}{1 - F/F_{max}}. \tag{10.27}$$

Here, the idea is that the ratio of actual fluorescence F to saturating fluorescence F_{max} directly reflects the $[Ca^{2+}]_i$ level if the dynamic range R_f is known. While

R_f can be determined beforehand for a particular experimental setup, the maximal fluorescence F_{max} is best determined during or at the end of the experiment using strong stimulation to induce a indicator-saturating Ca^{2+} load.

10.3.3 Fluorescence ratio

Some indicators exhibit wavelength shifts in either the excitation or emission spectrum upon Ca^{2+} binding (Figure 10.4B and C). These spectral shifts can be exploited for $[Ca^{2+}]_i$ measurements because in this case the ratio $R = F_1/F_2$ of the fluorescence intensities measured in two different spectral windows (either for excitation or for emission detection) is $[Ca^{2+}]_i$ dependent. According to Equation (10.22) the intensities F_1 and F_2 are given by

$$F_1 = S_{f1}\,[B] + S_{b1}\,[CaB]$$
$$F_2 = S_{f2}\,[B] + S_{b2}\,[CaB]\,. \tag{10.28}$$

Using these two equations and the law of mass action one arrives at the standard equation for ratiometric measurements (Grynkiewicz et al., 1985)

$$[Ca] = K_{eff}\,\frac{R - R_{min}}{R_{max} - R} \tag{10.29}$$

where $R_{min} = (S_{f1}/S_{f2})$ and $R_{max} = (S_{b1}/S_{b2})$ denote the ratios at zero and saturating $[Ca^{2+}]_i$ level, respectively, and $K_{eff} = K_d\,(S_{f2}/S_{b2})$ is an effective binding constant. This ratiometric approach normalizes for inhomogeneities of dye concentration, optical path length, and illumination intensity. As in the case of $\Delta F/F$, background subtraction in each channel at the corresponding excitation or emission wavelengths is essential before taking the ratio (Figure 10.5B). Notably, the ratiometric method is not limited to indicators that show spectral shifts but can also be extended to mixtures of non-ratiometric dyes that result in Ca^{2+}-sensitive fluorescence ratios (Lipp and Niggli, 1993; Oheim et al., 1998).

10.3.4 Fluorescence lifetime

Changes in fluorescence lifetime can also be used to quantify Ca^{2+}-binding of the indicator (Lakowicz et al., 1992). Following an excitation pulse the decay in fluorescence intensity reflects the lifetime of the excited state of the dye molecules (Figure 10.6A). Microscopy techniques that inherently use a pulsed laser source for excitation such as two-photon laser scanning microscopy thus may easily be adapted for lifetime measurements (Wilms et al., 2006). In the simplest case, the fluorescence decay is described by a single exponential curve with a fluorescence lifetime constant τ_L, which is typically in the nanosecond range.

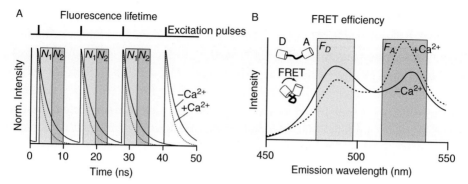

Figure 10.6 Other Ca^{2+}-dependent fluorescence properties. A. Fluorescence lifetime. Some indicators show a Ca^{2+}-dependent change in fluorescence lifetime,
which is here depicted as exponential decay following each brief excitation pulse.
Lifetime changes can be quantified for example by measuring the ratio of photon counts in two temporal windows (N_1, N_2) following each excitation pulse.
B. FRET changes. FRET indicators utilize the Ca^{2+}-dependent change in FRET
efficiency between the donor (D) and the acceptor (A) protein domain. A distance decrease between donor and acceptor leads to a decrease in detected donor
fluorescence and an increase in acceptor fluorescence.

Several calcium indicators, including Fura-2, Fluo-3 and Oregon Green BAPTA-
1, respond to Ca^{2+} binding with either increases or decreases in fluorescence
lifetime.

Fluorescence lifetime changes can be measured using time-gated photon detection following a brief exciting laser pulse. The ratio of the numbers of photons N_1
and N_2, detected in two time windows following the excitation pulse, provides an
effective fluorescence lifetime $\tau_{eff} = \Delta t / \log(N_1/N_2)$ where Δt is the width of
the windows (Figure 10.6A). Both the free and the Ca^{2+}-bound form of the indicator contribute to the effective fluorescence lifetime, which relates to $[Ca^{2+}]_i$ via a
equation similar to the one derived for ratiometric measurements:

$$[Ca] = K_{app} \frac{\tau_{eff} - \tau_{min}}{\tau_{max} - \tau_{eff}}. \tag{10.30}$$

Here, τ_{min} and τ_{max} denote the lifetimes of the bound and free indicator forms,
respectively, and K_{app} denotes an apparent dissociation constant. Alternatively,
lifetime changes can be inferred from time-correlated single-photon counting with
a pulsed laser source (Wilms et al., 2006) or from the phase shifts and amplitude modulation generated by a modulated light source (Lakowicz et al., 1992).
Importantly, lifetimes are independent of dye concentration, optical path length,
and illumination intensity and thus avoid the problem associated with a calibration
based on absolute fluorescence intensities.

10.3.5 FRET efficiency

For FRET-based calcium indicators it is straightforward to adopt a ratiometric detection scheme (Figure 10.6B). Ca^{2+}-binding to the linker molecule leads to a conformational change of the entire protein such that the two interacting fluorescent proteins are brought closer together. As a result FRET efficiency increases, i.e. more energy is transferred directly from the donor protein (or protein domain) to the acceptor protein. Macroscopically this change is detectable as a reduction in donor fluorescence intensity and an increase in acceptor fluorescence. The Ca^{2+}-dependent FRET change can thus be quantified by taking the ratio R of the intensities measured in two spectral windows that reflect donor and acceptor fluorescence changes, respectively (Palmer and Tsien, 2006). Ratio changes can then be expressed as percentage changes $\Delta R/R$ relative to the baseline ratio R_0, similar to the case of $\Delta F/F$. It is also possible to use a pair of fluorescent proteins with similar emission spectra and then read out changes in FRET efficiency via fluorescence lifetime measurements (Harpur et al., 2001).

10.3.6 Calibration of calcium indicators

For some research questions it is sufficient to measure uncalibrated changes in fluorescence. For example, Ca^{2+} indicators are often employed as indirect markers of cellular or dendritic electrical excitation, which typically induces rather stereotyped Ca^{2+} influx through the activation of voltage-gated calcium channels. We will return to this type of measurement further below when we discuss how temporal patterns of action potential firing can be reconstructed from calcium imaging experiments. For this type of measurement the detection of events is the main goal and the absolute value of $[Ca^{2+}]_i$ is of secondary interest. It is nevertheless important to choose a suitable indicator with an appropriate K_d value for reporting the expected calcium transients. Even without thorough calibration, a rough estimate of the absolute changes in $[Ca^{2+}]_i$ can be obtained based on reasonable assumptions of K_d, $[Ca^{2+}]_{rest}$ and $\Delta F/F_{max}$.

Many other applications, however, do require an accurate calibration of the measured fluorescence changes in terms of absolute free Ca^{2+} concentration. For example, calibration is essential in experiments that aim to quantify the $[Ca^{2+}]_i$-dependence of specific cellular processes such as exocytosis or synaptic plasticity. The calibration procedure consists of determining the parameter sets in the above equations, $(K_{eff}, R_{min}, R_{max})$ or $(K_d, \Delta F/F_{max}, [Ca^{2+}]_{rest})$ or (K_d, R_f, F_{max}) or $(K_{app}, \tau_{min}, \tau_{max})$. For this purpose measurements with at least three calibration solutions that contain different known $[Ca^{2+}]_i$ levels are required. Note that calibration is specific to the experimental setup used and therefore cannot be transported to other setups. Calibration measurements can be performed either "in vitro" on Ca^{2+}-buffered solutions (e.g. in small capillaries or in droplets in oil on a cover glass) or "in vivo," i.e. in living cells. A mixture of in vitro and in vivo approaches is also possible. Wherever possible, an in vivo calibration is preferable because Ca^{2+}-binding properties depend on solution properties such as its viscosity that cannot be exactly mimicked in vitro. While it is rather easy to prepare solutions that either contain no free calcium ions (through strong buffering) or saturating levels of calcium, the calibration solution with an intermediate level of $[Ca^{2+}]_i$ is most critical. The problem is that the calibration solution itself is based on a Ca^{2+} buffer (e.g. EGTA or BAPTA) and thus requires assumptions about the K_d-value of this reference buffer, which may also depend on ionic strength, pH, temperature etc. (for further reference on calibration procedures see Groden et al., 1991; Tsien and Pozzan, 1989; Neher, 2005; Palmer and Tsien, 2006; Helmchen, 2011).

Having learned about the functioning of fluorescent calcium indicators in this section, we now return to the application of simplified models of calcium dynamics to extract relevant information about neural dynamics on various spatiotemporal scales.

10.4 Simplified models of calcium dynamics

10.4.1 Calcium microdomain model

For many cellular signaling processes the exact route of calcium entry is important. The reason is that calcium-conducting channels are often strategically placed so that the activation of Ca^{2+}-binding proteins is locally restricted. Locally, $[Ca^{2+}]_i$ increases are much higher than the average ("bulk") $[Ca^{2+}]_i$ change that remains after initial concentration gradients have equilibrated by diffusion. As a consequence, the same calcium influx can exert multiple functions, activating low affinity binding proteins with subcellular specificity on the one hand and regulating distinct calcium-dependent processes with higher affinity in a more widespread fashion (Neher and Sakaba, 2008). On the molecular scale, opening of calcium-permeable ion channels causes a steep calcium concentration gradient in the immediate vicinity of the channel pores, producing "microdomains" of high $[Ca^{2+}]_i$ that develop beneath the plasma membrane (Figure 10.7A). Microdomains extend over only a few hundred nanometers with peak concentrations $>100\,\mu M$ (Neher, 1998b; Berridge, 2006) making them difficult to investigate with calcium imaging techniques (but see Llinas et al., 1992; DiGregorio et al., 1999). Calcium domains develop and collapse rapidly within a few tens or hundreds of microseconds following channel opening and closure, respectively (Roberts, 1994). At short distances the steady-state profile as a function of the radial distance r is given by

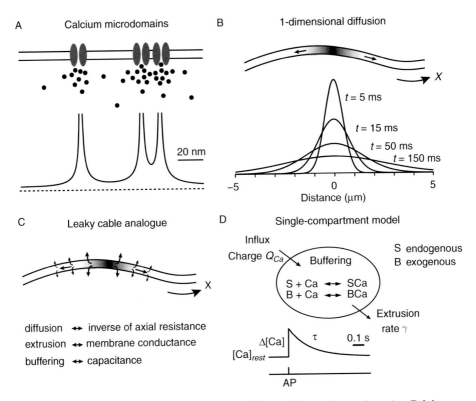

Figure 10.7 Various models of intracellular calcium dynamics. A. Calcium microdomains form beneath the mouth of open Ca^{2+}-permeable channel pores. B. Ca^{2+} redistribution in cellular processes can be modeled as one-dimensional buffered calcium diffusion. The traces show the temporal spread of Gaussian-shaped calcium concentration profiles following a localized Ca^{2+} influx. C. If Ca^{2+} extrusion mechanisms are included one obtains a leaky "chemical" cable with the main parameters bearing analogies to a passive electrical cable. D. If diffusion can be neglected the structure of interest can be considered a well-mixed compartment. In this single-compartment model, a perturbation by a brief calcium influx (e.g. during an action potential, AP) causes an exponentially shaped elementary calcium transient.

$$\Delta [Ca] (r) = \frac{i_{Ca}}{4\pi F D_{Ca} r} \exp(-r/\lambda_D) \qquad (10.31)$$

where i_{Ca} is the Ca^{2+} current of a single channel and λ_D is the length constant (Neher, 1986). How does the presence of Ca^{2+} buffers shape this spatial profile? While immobile Ca^{2+} buffers have no influence on the steady-state spatial profile of $[Ca^{2+}]_i$, because they bind and unbind Ca^{2+} locally at equal rates, mobile buffers have a chance to capture calcium ions near the channel, carry them away and then be replenished by free buffer molecules. Therefore they can effectively narrow the $[Ca^{2+}]_i$ profile near the channel mouth and thus restrict binding to nearby targets

(Figure 10.7A). This effect depends critically on the speed with which mobile buffer molecules capture calcium ions. Accordingly the length constant is given by (Naraghi and Neher, 1997; Neher, 1998a)

$$\lambda_D = \sqrt{D_{Ca}\,\tau_D} = \sqrt{D_{Ca}/(k_{on}\,[B])} \tag{10.32}$$

where τ_D is the mean time until a calcium ion is captured by a buffer molecule and k_{on} and $[B]$ refer to the association rate and concentration of a mobile buffer. Note that on the molecular scale binding kinetics rather than equilibrium properties of the buffer are important. As λ_D depends on the association rate, mobile buffers with either slow (e.g. EGTA) or fast (e.g. BAPTA) kinetics are differentially effective and therefore can be used to estimate the distance between intracellular targets and calcium channels (Neher, 1998a).

10.4.2 Buffered calcium diffusion

On a next higher spatial level one aims to describe Ca^{2+} redistribution in cell processes such as dendrites and axons. We consider the simple case of calcium diffusion in the presence of Ca^{2+} buffers that are either immobile or mobile. We consider only longitudinal diffusion along the axis of the process, assuming relatively thin processes. Using

$$\partial\,[CaB_i]/\partial t = \kappa_{B,i}\,\partial\,[Ca]/\partial t \tag{10.33}$$

we can express the sum of the partial derivatives of $[Ca]$ and all $[CaB_i]$ according to Equation (10.19) as

$$\frac{\partial\,[Ca]}{\partial t} + \sum_i \frac{\partial\,[CaB_i]}{\partial t} = \left(1 + \sum_i \kappa_{B,i}\right) \frac{\partial\,[Ca]}{\partial t}$$

$$= D_{Ca}\frac{\partial^2\,[Ca]}{\partial x^2} + \sum_i D_{B,i}\frac{\partial^2\,[CaB_i]}{\partial x^2}. \tag{10.34}$$

Our assumptions imply that κ_B is spatially uniform so that we can write

$$\frac{\partial^2\,[CaB_i]}{\partial x^2} = \frac{\partial}{\partial x}\kappa_{B,i}\frac{\partial\,[Ca]}{\partial x} = \kappa_{B,i}\frac{\partial^2\,[Ca]}{\partial x^2} \tag{10.35}$$

and further simplify Equation (10.34) to

$$\left(1 + \sum_i \kappa_{B,i}\right)\frac{\partial\,[Ca]}{\partial t} = \left(D_{Ca} + \sum_i \kappa_{B,i}\,D_{B,i}\right)\frac{\partial^2\,[Ca]}{\partial x^2}. \tag{10.36}$$

Note that on the left hand side all Ca^{2+}-binding molecules contribute, while on the right hand side only mobile buffers ($D_{B,i} \neq 0$) are relevant. The equation thus in general reduces to a simple diffusion equation

$$\frac{\partial [Ca]}{\partial t} = D_{app} \frac{\partial^2 [Ca]}{\partial x^2} \tag{10.37}$$

with an apparent diffusion coefficient of (Gabso et al., 1997; Irving et al., 1990; Neher, 1998a):

$$D_{app} = \frac{D_{Ca} + \sum_i \kappa_{B,i} D_{B,i}}{(1 + \sum_i \kappa_{B,i})} = \frac{D_{Ca} + \sum_j \kappa_{m,j} D_{m,j}}{(1 + \sum_j \kappa_{m,j} + \sum_k \kappa_{f,k})}. \tag{10.38}$$

Here, we highlighted the distinct roles of mobile versus immobile buffers by splitting the sums in mobile (m) and fixed (f) molecules. If only immobile Ca^{2+}-binding molecules are present, the apparent diffusion constant is reduced, i.e. calcium diffusion is effectively slowed down. Intuitively this is clear, because fixed buffers hold calcium ions in place and thus hinder their diffusional spread. If both mobile and fixed buffers are present the situation is more complex and the outcome depends critically on both the Ca^{2+}-binding ratios and the mobility of the Ca^{2+}-binding molecules. For example, if the diffusion constant of a mobile buffer is larger than the apparent diffusion coefficient in the presence of a fixed buffer $(D_{Ca}/(1 + \kappa_f))$, the addition of this mobile buffer even speeds up diffusional exchange because $D_{app} = (D_{Ca} + \kappa_m D_m)/(1 + \kappa_f + \kappa_m)$ is increased. The explanation is that a highly mobile buffer captures the substance near the source and facilitates diffusional spread by transporting it in "piggyback" fashion. In summary, intracellular buffers can both slow down and facilitate the diffusional spread of a substance depending on their concentration, affinity and mobility. This leaves room for speculation that the spatial extent of dendritic calcium compartments may not be stationary but many be dynamically regulated depending on the expression level of Ca^{2+}-binding proteins.

For the one-dimensional case considered here the well-known solution for Equation (10.37) is a spreading Gaussian shaped concentration profile (Gabso et al., 1997)

$$Ca[x, t] = \frac{A_0}{\sqrt{4\pi D_{app} t}} \exp\left(-\frac{(x - x_0)^2}{4 D_{app} t}\right) \tag{10.39}$$

where A_0 indicates the strength of the pulse-like injection at time zero and position x_0 (Figure 10.7B).

10.4.3 Cable-equation analog

Most cell compartments contain effective calcium extrusion mechanisms. This is illustrated in Figure 10.7C showing a "leaky" membrane that will extrude Ca^{2+} along the cell process. If we include a simple linear removal mechanism in Equation 10.34 – with both mobile and fixed buffers present – and rewrite the equation for concentration changes from resting level (assuming spatial homogeneity at rest), we obtain the *linearized reaction-diffusion equation*

$$\left(1 + \sum_i \kappa_{m,i} + \sum_j \kappa_{f,j}\right) \frac{\partial \Delta [Ca]}{\partial t}$$

$$= \left(D_{Ca} + \sum_j \kappa_{m,j} D_{m,j}\right) \frac{\partial^2 \Delta [Ca]}{\partial x^2} - \gamma \Delta [Ca]. \qquad (10.40)$$

This equation is equivalent to the cable equation used to describe electrotonic spread along a cable, enabling a useful analogy between chemical and electrical signaling (Kasai and Petersen, 1994; Zador and Koch, 1994). The diffusion coefficient relates to the intracellular resistivity, the removal rate corresponds to the membrane conductance, and buffers act similar to a capacitance (for more details see Koch (1998)). In the idealized case of an infinite cylinder the chemical equivalents of space and time constants can be defined based on cable theory. The *chemical length constant* or diffusion length is given by

$$\lambda_{ch} = \sqrt{D_{app} \tau_{ch}} = \sqrt{\frac{D_{Ca} + \sum_i \kappa_{m,i} D_{m,i}}{\gamma}}. \qquad (10.41)$$

with the time constant defined as

$$\tau_{ch} = \frac{1 + \sum_i \kappa_{m,i} + \sum_j \kappa_{f,j}}{\gamma}. \qquad (10.42)$$

Thus, the diffusion length is relatively small for a large removal rate, which could help to restrict spatially the effect of Ca^{2+} influx. Indeed, synaptic activation of aspiny dendrites of neocortical interneurons has been reported to cause localized calcium microdomains ($<1\,\mu m$) along the dendrites, which were dependent on fast Ca^{2+} influx through AMPA receptors and effective local extrusion mechanisms (Goldberg et al., 2003). On a note aside, an analogous mathematical description also holds for other intracellular signaling molecules, such as cAMP or IP_3, which however typically have longer diffusion lengths of tens of micrometers or more (Allbritton et al., 1992; Kasai and Petersen, 1994).

10.4.4 Single-compartment model

The above simplifications are useful to describe the spatiotemporal redistribution of Ca^{2+} in subcellular compartments and cell processes. Further simplification is possible if Ca^{2+} influx occurs rather homogeneously throughout a cell compartment and diffusional equilibration is faster than Ca^{2+} extrusion. In this case diffusion can be neglected altogether (Figure 10.7D). Nearly homogeneous Ca^{2+} influx occurs, for example, when voltage-dependent Ca^{2+} channels are activated during an action potential that spreads rapidly and with little attenuation throughout the neuron. Diffusional equilibration in axonal and dendritic segments will occur within milliseconds. Even in large cell bodies initial gradients typically disappear relatively fast compared to extrusion (within less than 300 ms for a 40-μm diameter neuron; Hernandez-Cruz et al., 1990). As a first approximation we can therefore assume a stepwise homogenous increase in $[Ca^{2+}]_i$ in the initial phase. Neglecting the diffusion term in Equation (10.40) and adding repetitive Ca^{2+} influx during a train of action potentials, we obtain

$$\left(1 + \sum_i \kappa_{B,i}\right) \frac{d\,\Delta\,[Ca]}{dt} = \frac{Q_{Ca}}{2FV} \sum_k \delta(t - t_k) - \gamma\,\Delta\,[Ca], \tag{10.43}$$

where Q_{Ca} is the Ca^{2+} charge per action potential. This balancing equation simply states that the change in total calcium concentration (left hand side, assuming constant $[Ca^{2+}]_{rest}$) equals influx minus extrusion (right hand side):

$$\frac{d[Ca]_T}{dt} = \frac{d\Delta[Ca]_T}{dt} = j_{in} + j_{out}. \tag{10.44}$$

Because the variable $\Delta[Ca]$ now depends only on the time variable, we have written Equation (10.44) as an *ordinary* differential equation. Assuming constant Ca^{2+}-binding ratios, the analytical solution of Equation (10.44) for a single Ca^{2+} influx is a sharp rise of $[Ca^{2+}]_i$ at time point t_0 with amplitude

$$A = \frac{Q_{Ca}/(2FV)}{(1 + \sum_i \kappa_{B,i})} \tag{10.45}$$

followed by an exponential decay with time constant

$$\tau = \frac{(1 + \sum_i \kappa_{B,i})}{\gamma}. \tag{10.46}$$

The general solution for a train of brief Ca^{2+} injections, for example as they occur during a neural spike train, is given by

$$\Delta\,[Ca]\,(t) = \sum_k A\,\exp\left(-\frac{(t - t_k)}{\tau}\right) \theta(t - t_k) \tag{10.47}$$

where θ denotes the Heaviside step function. Notably, this result is equivalent to the convolution of the impulse train with an impulse response function given by a stepwise increase in amplitude A and an exponential decay with time constant τ.

For practical applications, the single-compartment model often assumes two distinct pools of rapid Ca^{2+} buffers: one pool S that represents endogenous buffers and another pool B that represents the exogenously added calcium indicator. Equations (10.45) and (10.46) can thus be rewritten as (Neher and Augustine, 1992; Regehr et al., 1994; Helmchen et al., 1996)

$$A = \frac{Q_{Ca}/(2FV)}{(1 + \kappa_S + \kappa_B)} \qquad (10.48)$$

and

$$\tau = \frac{(1 + \kappa_S + \kappa_B)}{\gamma}. \qquad (10.49)$$

These equations are the simplest formulation of the competition for Ca^{2+} between endogenous buffers and calcium indicator. Both A and τ depend on the Ca^{2+}-binding ratios but it is the relative amount of added buffering capacity (not its absolute value) that determines how strongly the calcium transient shape will be affected by the addition of indicator dye. Note also that the product $A\tau$ which is equal to the integral of the $[Ca^{2+}]_i$ transient (the area "underneath" the exponential curve), does not depend on Ca^{2+} binding ratios. Figure 10.8A summarizes how changes in Ca^{2+} influx, total Ca^{2+}-binding ratio and extrusion rate affect amplitude and decay time of an action potential evoked calcium transient (Sabatini and Regehr, 1995; Helmchen and Tank, 2011). While changes in influx or

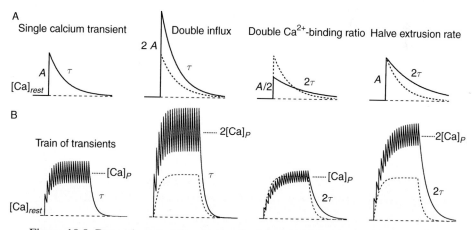

Figure 10.8 Dependence of the shape of calcium transients on the main parameters of the single-compartment model. A. Effects of doubling calcium influx, total Ca^{2+}-binding ratio, and extrusion rate on the single action potential evoked calcium transient. B. Effect of the same manipulations on the $[Ca^{2+}]_i$ accumulation during a train of action potentials at constant frequency. Note that the mean plateau level of $[Ca^{2+}]_i$ that is reached during steady state does not depend on the Ca^{2+}-binding ratio.

extrusion rate specifically alter A and τ, respectively, both parameters depend on the Ca^{2+}-binding ratio.

Because action potential evoked calcium transients are relatively slow, with typical decay times of tens to hundreds of milliseconds and even longer for cell somata, individual calcium transients summate during trains of action potentials. For rapid bursts of action potentials, Ca^{2+} influx adds up so that the calcium transient amplitude scales approximately with the number of spikes in a short burst. If spikes occur at a lower but constant frequency we can derive an analytical expression for the buildup of $[Ca^{2+}]_i$. Consider a train of stimuli starting at time $t = 0$ and with a time interval of Δt (frequency $f = 1/\Delta t$). The $[Ca^{2+}]_i$ level above resting level immediately before the $(n + 1)$th stimulus is given by a geometric progression (Regehr et al., 1994)

$$\Delta[Ca](n\Delta t) = A \sum_{i=1}^{n} \exp(-(i\Delta t)/\tau) = A \frac{1 - \exp(-(n\Delta t)/\tau)}{\exp(\Delta t/\tau) - 1}. \quad (10.50)$$

For stimulation frequencies $f < 1/(2\tau)$ there is little summation with individually spaced calcium transients, at higher frequencies calcium transients summate and $[Ca^{2+}]_i$ exponentially reaches a steady state, fluctuating around a plateau level with influx and extrusion balancing each other (Figure 10.8B). The rise time and decay of the $[Ca^{2+}]_i$ buildup in this case are governed by the time constant τ. A simple calculation reveals that the mean plateau level P that is reached at steady state is proportional to the spike frequency (Tank et al., 1995; Helmchen et al., 1996):

$$P = A\tau f \propto f \quad (10.51)$$

with the integral ($A\tau$) as proportionality constant. The plateau level is therefore independent of Ca^{2+}-binding ratios, which means that Ca^{2+} buffers affect the transient dynamics of $[Ca^{2+}]_i$ but not its steady-state level. The relationship between mean $[Ca^{2+}]_i$ level and the frequency of the underlying calcium influx events also indicates that intracellular calcium concentration encodes spike rate on this relatively slow time scale. Via this mechanism intracellular signaling pathways could be modulated in a spike rate dependent manner. We conclude that the single-compartment model provides a useful framework for describing the temporal pattern and the summation of elementary calcium transients.

10.4.5 Non-linear calcium dynamics

In the last sections we have introduced simplified models of intracellular calcium dynamics that are suitable to describe fluorescence measurements under various conditions. These models rely on linear approximations of the relevant processes and often assume chemical equilibrium in addition. In this section we discuss several limitations of this approach, which are important to realize in order to understand how calcium dynamics and fluorescence measurements are affected when the assumption of linearity breaks down.

The first issue is that the approximation of constant Ca^{2+}-binding ratios is only valid for $[Ca^{2+}]_i$ changes which are small compared to the dissociation constant (Figure 10.2B). For larger changes and in particular when buffering molecules start to saturate, Ca^{2+}-binding ratios change substantially. At sufficiently high $[Ca^{2+}]_i$ levels the effective Ca^{2+}-binding ratio decreases and as a consequence $\Delta[Ca]$ induced by a certain amount of Ca^{2+} influx tends to become larger (Tank et al., 1995). Because of the overall reduced Ca^{2+}-binding ratio, decay times will also tend to become shorter. Further complexity arises from cooperative Ca^{2+}-binding due to multiple binding sites on one protein. This is relevant not only for the activation of certain intracellular enzymes, for example the Ca^{2+}-calmodulin complex (Ghosh and Greenberg, 1995), but also for the interpretation of fluorescence signals from GECIs, which typically have Hill coefficients different from unity. Indeed, non-linear response properties have been reported for various GECIs at low $[Ca^{2+}]_i$ levels, exhibiting either supralinear responses with little sensitivity for single action potentials (Pologruto et al., 2004) or single-spike sensitivity but rapid saturation with bursts of action potentials (Wallace et al., 2008).

A second important issue is saturation of Ca^{2+} extrusion mechanisms, which will lead to an opposite effect on decay time constant compared to buffer saturation, that is a prolongation rather than shortening of decay times at very high $[Ca^{2+}]_i$ levels. In addition, pump saturation is expected to alter the linear frequency dependence of the steady-state plateau level during long stimulus trains. Since buffer and pump saturation in most situations are likely to occur in parallel, the net effect may be difficult to predict. Moreover, cells typically contain a mixture of different Ca^{2+} extrusion mechanisms with distinct affinities (e.g. plasma membrane Ca^{2+}-ATPase, Na^+-Ca^{2+}-exchanger, and ATPases in intracellular organelles), which eventually makes detailed modeling of such a complex Ca^{2+} removal system necessary.

Another assumption in the above models (except for the calcium microdomain model) is that at any location within the cytosol, chemical equilibration is reached virtually instantaneously. However, the kinetics of Ca^{2+}-binding is not infinitely fast and thus cannot always be neglected, especially not when Ca^{2+}-binding proteins with very different on-rates are mixed together. For example, the addition of slow Ca^{2+}-binding molecules such as EGTA was found to have little effect on the amplitude of action potential evoked calcium transients but a marked acceleration of the initial decay (Atluri and Regehr, 1996; Markram et al., 1998). In contrast, addition of BAPTA at similar concentrations reduced amplitudes and prolonged decay times as expected from the extra Ca^{2+}-binding ratio. The differential effect of EGTA and BAPTA can only be explained by their different Ca^{2+}-binding rates. Initially, calcium ions escape binding to EGTA because of its slow on-rate. Only in a second phase do calcium ions shift to EGTA molecules until these eventually have received their equilibrium share. As a result, calcium transients measured with calcium indicators, which typically belong to the pool of fast buffers, show a multi-exponential decay time course. The initial decay is governed by slow Ca^{2+} binding to EGTA acting as an additional sink, yielding an approximate fast time constant (Atluri and Regehr, 1996)

$$\tau_{fast} = \frac{(1 + \sum_i \kappa_i)}{\gamma + k_{EGTA}^{on}[\text{EGTA}]}. \tag{10.52}$$

Here, the sum runs over all rapid buffers. The slow second decay phase is described by the equilibrium expression, now including the Ca^{2+}-binding ratio of EGTA:

$$\tau_{slow} = \frac{(1 + \kappa_{EGTA} + \sum_i \kappa_i)}{\gamma}. \tag{10.53}$$

This example illustrates how important kinetic effects can be, especially when multiple molecules with different binding kinetics (and possibly differential strategic localizations) are present (Markram et al., 1998). Proteins with slow binding kinetics may be bypassed altogether in the case of brief and localized Ca^{2+} influx while they can still be activated significantly during long, sustained influx. Hence, the exact spatiotemporal pattern of Ca^{2+} binding (and thus the pattern of target activation) can be a complicated function of the properties of all Ca^{2+} binding partners, which in the end needs to be explored in numerical simulations.

In summary, this section has shown that starting from a full description of intracellular calcium dynamics, certain aspects of the dynamics can be emphasized and molded in simplified models depending on the specific scientific questions of interest. Similarly, the application of calcium indicators can be tuned to uncover various aspects of neural function as will be shown in the next section.

10.5 Application modes

Calcium indicators are applied in various ways and for many purposes. In this section we discuss a few principal modes of application, which use the insights from the above theoretical considerations. One common topic is that, having realized that calcium indicators inevitably affect intracellular calcium dynamics, we can use these perturbations to our advantage for a systematic and quantitative characterization of certain aspects of calcium dynamics.

10.5.1 How to estimate unperturbed calcium dynamics

Because of the interference of calcium indicators with intracellular calcium handling it is not a trivial task to determine the unperturbed temporal dynamics of $[Ca^{2+}]_i$ as it occurs under physiological conditions. The concentration of indicator cannot be lowered infinitely in order to minimize perturbation effects because the signal to noise ratio of the fluorescence measurement will become too low. Moreover, there is no general rule on which absolute dye concentration is sufficiently low because the amount of perturbation also depends on cell properties, in particular the endogenous Ca^{2+}-binding ratio. A simple method to test whether addition of indicator severely affects the $[Ca^{2+}]_i$ measurements is to perform measurements at twice the concentration. If amplitudes and decay time constant change little, this would show that the indicator is interfering little at the original concentration. In extension of this approach, a systematic variation of the indicator concentration can be performed, from which unperturbed amplitude and decay time constants can be determined via extrapolation to zero indicator concentration (Figure 10.9). Such an approach is particularly helpful when investigating reproducible, stimulus evoked calcium influx such as during an action potential (Neher and Augustine, 1992; Helmchen et al., 1996, Helmchen et al., 1997). As an alternative, one can compare several indicators with very different Ca^{2+}-binding affinities under comparable conditions. At the same concentration, a low affinity indicator (e.g. MagFura-2) adds much less Ca^{2+}-binding ratio to the cell than its high affinity analog and therefore perturbs the cytosolic calcium dynamics much less. On the other hand, even though the fluorescence yield might be large, the evoked fluorescence changes will tend to be very small ($<1\%$) so that averaging of many trials may be necessary to resolve small calcium transients clearly (Helmchen et al., 1997).

10.5.2 How to estimate the endogenous calcium binding ratio

The systematic variation of calcium indicator concentration also enables the estimate of the endogenous Ca^{2+}-binding ratio κ_S and the clearance rate γ. For this purpose we assume the single-compartment model and plot the decay time constant of stereotype calcium transient as a function of exogenous Ca^{2+}-binding ratio κ_B (Figure 10.9A). According to Equation (10.49) this relationship is described by a linear function

$$\tau = a_1 \kappa_B + a_0 \tag{10.54}$$

with

$$a_1 = 1/\gamma$$
$$a_0 = (1 + \kappa_S)/\gamma. \tag{10.55}$$

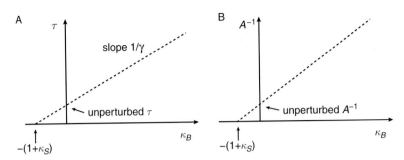

Figure 10.9 Estimate of unperturbed calcium dynamics and endogenous Ca^{2+}-binding ratio by systematically varying the indicator concentration. A. Fitting a linear relationship to a plot of decay time constant versus exogenous Ca^{2+}-binding ratio κ_B reveals the unperturbed time constant and the endogenous Ca^{2+}-binding ratio κ_S from the y- and x-axis interception, respectively. B. Similarly, a plot of the inverse of the calcium transient amplitude A versus κ_B reveals the unperturbed amplitude and κ_S from the y- and x-axis interception, respectively.

Thus, the inverse of the measured slope (a_1) provides an estimate of the clearance rate and the negative x-axis intercept ($-a_0/a_1$) provides an estimate of the endogenous buffer capacity κ_S (Neher and Augustine, 1992; Helmchen et al., 1996). Similarly, the unperturbed amplitude of the calcium transient can be estimated from a plot of the inverse of the amplitude A^{-1} versus κ_B by extrapolation (Figure 10.9B).

In practice, mainly two methods have been employed to vary the calcium indicator concentration systematically. First, whole-cell patch clamp recordings can be obtained with pipette solutions containing known indicator concentrations (alternatively, injections through intracellular sharp microelectrodes can be used but then concentrations have to be inferred indirectly from the fluorescence intensity values after loading (Tank et al., 1995)). With sufficiently low access resistance and after some waiting period one can assume diffusional equilibration between the pipette and the cell compartment. The plots in Figure 10.9 can then be built up from measurements on many cells with different indicator concentrations (Neher and Augustine, 1992; Helmchen et al., 1996; Maravall et al., 2000; Sabatini et al., 2002). A second more efficient method is to monitor the effect of adding calcium indicator during the dye loading process in individual experiments (Neher and Augustine, 1992; Helmchen et al., 1996; Sabatini et al., 2002). This approach avoids cell-to-cell variability. The actual indicator concentrations at the various measurement time points are back-calculated from the final fluorescence intensity of the cell, which is presumed to reflect complete loading with a concentration equal to the pipette concentration. With this approach κ_S can thus be estimated from single experiments. The endogenous Ca^{2+}-binding ratio κ_S has been determined for a variety of cell types and subcellular compartments and typically yielded values between 50 and 1000 (for a summary see Helmchen and Tank, 2011).

10.5.3 How to quantify total calcium fluxes

An estimation of unperturbed calcium dynamics is important for revealing $[Ca^{2+}]_i$ changes under physiological conditions. A different goal that requires a different approach is to quantify the total Ca^{2+} flux (Schneggenburger et al., 1993; Neher, 1995). For this purpose one can load excess amounts of calcium indicator into the cellular compartment, so that the indicator molecules outcompete any other endogenous Ca^{2+} buffers and capture virtually all calcium ions. As a result the change in absolute fluorescence intensity under these "overload" conditions directly reflects the total calcium charge that has entered the cytosol. In a sense the situation is analogous to the different modes of electrical recording (Neher, 1995): while the "minimal indicator concentration" approach corresponds to "voltage recording" (or "current clamp" mode), with the aim of minimally perturbing the observable (membrane potential and $[Ca^{2+}]_i$, respectively), "indicator overload" corresponds to "voltage clamp" mode, where one variable is forced artificially to a constant value in order to gain valuable information about another variable. In the case of dye overload, $[Ca^{2+}]_i$ is effectively clamped to a low concentration while any extra Ca^{2+} influx is reported through a fluorescence change.

Mathematically we can treat the indicator overload situation as follows. From Equations (10.12) and (10.22) we see that under these conditions the evoked fluorescence change is directly proportional to the total calcium charge Q_{Ca} that has entered the cell

$$\Delta F = (S_b - S_f)\Delta[CaB] \cong (S_b - S_f)\Delta[Ca]_T = \frac{(S_b - S_f)}{2FV}Q_{Ca}. \qquad (10.56)$$

The proportionality constant often is defined as f_{max} (Schneggenburger et al., 1993)

$$\Delta F = f_{max}\, Q_{Ca}. \qquad (10.57)$$

If f_{max} is known one thus has a direct readout of the integral of the Ca^{2+} influx that has occurred during a certain time period. In order to use this equation the experimental setup needs to be calibrated, however, by measuring fluorescence changes that are induced by known calcium fluxes. This can be achieved, for example, by simultaneous electrical recordings of pure calcium currents (Schneggenburger et al., 1993; Helmchen et al., 1997). It should be noted, however, that ΔF is measured at one wavelength (non-ratiometric) and that S_b and S_f depend on features of the imaging setup, in particular on the illumination power, which may change over time due to lamp aging. Therefore, regular measurements on some fluorescence standard (e.g. fluorescent beads) should be used to normalize for changes in illumination power over the course of experiments. The overload method has been applied for instance to quantify fractional Ca^{2+} currents through ligand-gated ion channels (Schneggenburger et al., 1993; Garaschuk et al., 1996; Bollmann et al., 1998) and the total calcium influx during an action potential (Helmchen et al., 1997; Bollmann et al., 1998).

10.5.4 How to characterize calcium-dependent processes

Ca^{2+}-dependent intracellular signaling pathways underlie many fundamental cell functions. To understand fully the molecular mechanisms involved we need to characterize the Ca^{2+} sensitivity of the relevant Ca^{2+}-sensing proteins in the living cell. For cellular processes, which essentially "see" the average "global" $[Ca^{2+}]_i$ changes, such as for example short-term facilitation or the recruitment of synaptic vesicle pools (Neher and Sakaba, 2008), the required measurements can be made with a well-calibrated calcium imaging setup and do not require particularly high spatial resolution. However, many other processes sense the local rather than global $[Ca^{2+}]_i$ changes and it is therefore difficult if not impossible to apply calcium imaging techniques for quantitative measurements on this scale. Calcium sensors are often anchored to protein complexes and placed strategically in the vicinity of calcium sources. For example, neurotransmitter release is triggered by localized calcium influx through voltage-gated calcium channels positioned very close to the presynaptic protein complexes triggering vesicle fusion with the membrane (Neher and Sakaba, 2008). Precise optical measurements of the absolute changes in Ca^{2+} concentration reached within calcium microdomains are beyond what optical imaging technology currently can achieve. Perhaps novel nanoresolution light microscopy techniques (Hell, 2007) in the future will enable functional calcium measurements with a spatial resolution below 100 nm, but even then the effective Ca^{2+} concentration at the proteins of interest will be hard to determine. Analysis of local $[Ca^{2+}]_i$ values is further hindered by the complicated effects that the indicator molecules have on intracellular calcium dynamics on this molecular scale.

A method to circumvent these problems is flash photolysis of caged calcium compounds (Neher and Zucker, 1993). In this approach, a brief flash of light causes release of Ca^{2+} from the caged compound by rapidly changing its affinity for Ca^{2+}-binding. The goal is to increase the average cytosolic Ca^{2+} concentration uniformly in a step-like fashion and to quantify the absolute $[Ca^{2+}]_i$ level reached using a simultaneous measurement with a fluorescent calcium indicator. At the same time the resulting action of Ca^{2+}-binding (e.g. induction of a postsynaptic current) is read out and quantified. Because flash photolysis leads to a spatially homogenous elevation of $[Ca^{2+}]_i$ it circumvents the problem of highly localized calcium domains. More details on the principles and application of the calcium uncaging technique, in particular its use for quantifying the calcium dependence of neurotransmitter release at central synapses, can be found in recent reviews (Schneggenburger and Neher, 2005; Neher and Sakaba, 2008).

10.5.5 How to reconstruct neural spike trains

The prospect of optical measurement of the spiking activity in large neuronal networks has fascinated researchers for many years. While membrane potential imaging using voltage-sensitive dyes still suffers from low signal to noise ratios and difficulties in achieving single-trial measurements in intact tissue, calcium imaging offers a promising alternative by providing an indirect measure of action potential firing. For a given cell type, each action potential elicits a stereotype Ca^{2+} influx in the soma (and dendrites), which induces an elementary calcium transient. We neglect here subtle frequency-dependent changes in Ca^{2+} influx and we also neglect potential other sources of calcium, for example, release from intracellular stores. The total amount of Ca^{2+} influx depends on action potential shape as well as on the density of voltage-gated calcium channels and therefore may vary between different cell types. The duration of the evoked calcium transients is typically much longer than the Ca^{2+} influx itself, with the activity of extrusion mechanisms and the total Ca^{2+}-binding ratio as main determinants of the decay time course. The main issue is that for specific experimental conditions each cell type displays a characteristic "impulse response," given by the single action potential evoked calcium transient $C_1(t)$. For a train of n action potentials the time course of the resulting $[Ca^{2+}]_i$ change $C_n(t)$ is a filtered version of the action potential train A_n or – mathematically more precise – it is the convolution of the spike train A_n with the impulse response $C_1(t)$ (Figure 10.10)

$$C_n(t) = C_1(t) * A_n(t). \tag{10.58}$$

For an action potential train with spike times t_k ($k = 1, \ldots, n$) and assuming a single-exponentially decaying elementary calcium transient, this equation is equivalent to the expression that was given in Equation (10.47). Note that in general more complex elementary calcium transients with several decay components or a slow onset (filtered for example by dye kinetics or by the time it takes Ca^{2+} to equilibrate by diffusion throughout a large cell compartment) may need to be assumed.

In practice, fluorescence measurements using fluorescent calcium indicators are always confounded by experimental noise (Figure 10.10). In more general terms we thus have to write

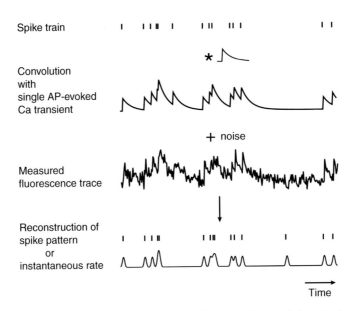

Figure 10.10 Reconstruction of neuronal spike trains from calcium indicator fluorescence measurements. Assuming a strict relationship between action potentials
and evoked calcium influx, the fluorescence time course reflects the convolution
of the spike train with the single action potential evoked calcium transient as
"impulse response". Depending on the level of experimental noise, various methods can be used to reconstruct more-or-less accurately the spike train or the time
course of instantaneous spike rate. Note a missed spike and the occurrence of a
false-positive spike in the schematic example shown here.

$$C_n(t) = C_1(t) * A_n(t) + N(t). \tag{10.59}$$

The noise term $N(t)$ may originate from various sources, including detector noise
and photon shot noise. In order to reveal the dynamic pattern of action potential
firing in an entire neuronal population, we face the challenge of reconstructing
(or "deconvolving") the action potential train in each cell and of finding the best
estimate for the spike pattern underlying the observed noisy fluorescence traces.
The temporal precision of such a reconstruction is clearly limited by the acquisition rate of the imaging system and in most cases is far from reaching millisecond
accuracy. For many purposes it may be sufficient, however, to reconstruct the temporal dynamics as instantaneous spike rate, which is a smooth function "blurred"
over a certain temporal window.

   Several algorithms have been proposed for the reconstruction of action potential
patterns from fluorescence traces (see for example Smetters et al., 1999; Ramdya
et al., 2006; Yaksi and Friedrich, 2006; Kerr et al., 2007; Moreaux and Laurent, 2007; Holekamp et al., 2008; Greenberg et al., 2008; Sasaki et al., 2008;
Vogelstein et al., 2009, 2010; Grewe et al., 2010). Here we only introduce a

basic approach using the Wiener filter as deconvolution filter (Press et al., 1988). The task is to find an optimal solution for $A_n(t)$ given the noisy $C_n(t)$. Rewriting Equation (10.59) in Fourier space transforms the convolution operation into a multiplication and yields

$$\hat{C}_n(\omega) = \hat{C}_1(\omega) \cdot \hat{A}_n(\omega) + \hat{N}(\omega). \tag{10.60}$$

In the presence of "white noise" the optimal filter for retrieving $A_n(\omega)$ in the Fourier domain is (Press et al., 1988)

$$\hat{A}_n(\omega) = \hat{H}(\omega) \, \hat{C}_n(\omega) \tag{10.61}$$

where $H(\omega)$ is the Wiener filter given by

$$\hat{H}(\omega) = \frac{\hat{C}_1(\omega)}{\left|\hat{C}_1(\omega)\right|^2 + \left|\hat{N}(\omega)\right|^2} = \frac{\hat{C}_1(\omega)}{\left|\hat{C}_1(\omega)\right|^2 (1 + \alpha(\omega))} \tag{10.62}$$

with $\alpha = |N(\omega)|^2/|C_1(\omega)|^2$. Inverse fourier transformation of $A_n(\omega)$ then yields an estimate for the original spike train $A_n(t)$. $N(\omega)$ may be estimated from the noise spectrum in the absence of neural activity and $C_1(\omega)$ from the assumed model of elementary calcium transient. Alternatively the parameter α can be fine tuned empirically. Special pre-processing steps, including non-linear filters (Yaksi and Friedrich, 2006), may further increase the accuracy of the reconstruction. The Wiener filter is clearly not the best solution to this problem and can be outperformed by more sophisticated model-based approaches (Greenberg et al., 2008; Vogelstein et al., 2009, 2010), which in principle can also deal with non-linear effects such as dye saturation. With new methods for high-speed two-photon calcium imaging (Rochefort et al., 2009; Grewe et al., 2010; Rothschild et al., 2010; Cheng et al., 2011) there are also new options for spike train reconstruction with improved temporal precision. For example, using a novel "peeling" algorithm that reconstructs complex spike trains based on iterative subtraction of a template calcium transient waveform, neural spike times could be determined with near-millisecond precision (Grewe et al., 2010). An important caveat is that a one-to-one relationship between action potential firing and somatic calcium transient does not hold for all cell types; for example somatic calcium transients have been reported in granule cells of olfactory bulb upon subthreshold activation (Egger, 2007; Lin et al., 2007). For several reasons it is thus important to verify the performance and validity of a reconstruction algorithm using simultaneous electrical and optical recordings, also allowing the quantification of success rate and rate of false positives/negatives (Kerr et al., 2005; Sasaki et al., 2008; Grewe et al., 2010). Further progress in calcium indicator performance and imaging techniques in the near future is likely to drive optimization of the methods for optical reconstruction of neural spike trains from calcium imaging data, especially under in vivo conditions.

10.6 Comparison with other techniques

With the large toolbox of fluorescent calcium indicator now at hand and with the continual progress of in vivo labeling techniques, calcium imaging currently is the prevailing optical technique for studying the function of cell compartments, individual cells, and populations of cells. As a fluorescence technique that is easily implemented in high-resolution microscopes, calcium imaging has the key advantage of providing cellular and subcellular resolution. It thus complements other non-optical techniques, such as extracellular recordings, and optical techniques for functional imaging that lack cellular resolution, such as fMRI or intrinsic optical signal imaging.

Calcium imaging can provide quantitative data about the fundamental role of Ca^{2+} in various physiological processes and in different cell types. In neural tissue calcium imaging is a versatile tool for dissecting principles of neural excitation from the synaptic to the neural network level. As an imaging technique, calcium measurements are complementary to electrophysiological recordings in several ways. Imaging data can be obtained simultaneously from many recording sites, revealing spatiotemporal dynamics, and even very thin dendritic processes and synaptic structures can be investigated that are hardly accessible to recording pipettes. On the other hand the temporal resolution of calcium imaging typically is traded off against the spatial extent of sampling and in most applications is poor compared to electrophysiological recordings. Different from voltage-sensitive dye signals, calcium indicator signals are indirectly related to membrane potential changes. However, because action potential induced Ca^{2+} influx appears to be a quite universal neuronal feature and because VSD imaging still suffers from low signal to noise ratios, in vivo calcium imaging from neuronal populations is currently the most promising approach for investigating network dynamics in local neuronal microcircuits.

Another advantage of calcium imaging is that it can probe activity patterns not only in neurons but additionally in glial cells that also show $[Ca^{2+}]_i$ changes upon their activation. For example, astrocytes in the neocortex (Hirase et al., 2004; Nimmerjahn et al., 2004; Wang et al., 2006) as well as Bergmann glia in the cerebellum (Hoogland et al., 2009; Nimmerjahn et al., 2009) display spontaneous as well as sensory- or behavior-related calcium signals. The ability to monitor activation patterns simultaneously in the intermingled local networks of neurons and glial cells opens particularly interesting opportunities to study neuro–glial interactions.

I like to emphasize that many of the principles of indicator fluorescence measurements that were treated in this chapter are generally applicable and might be equally well used for interpreting fluorescence measurements with other small-molecule ion indicators or genetically encoded sensors (e.g. for Na^+, Cl^-, cAMP, etc.).

10.7 Future perspectives

Several aspects of calcium indicator measurements and their interpretation are likely to gain more interest in the near future. First, more information is needed about the differences in calcium dynamics in the various cell types, for example, the many subtypes of GABAergic neurons found in the cortical microcircuits. This will be important in order to achieve a comprehensive reconstruction of activity patterns in local populations taking all their diversity into account. Second, non-linear features of calcium indicators, such as saturation, will increasingly be considered to understand the relationship between electrical excitation, calcium concentration changes, and fluorescence signals more completely. This is particularly true for newest generation GECIs, for which the binding curves, conformational changes, and interactions with other proteins can be complex. Third, consideration of kinetic properties of calcium indicators will become essential not only when analyzing calcium dynamics on the very fine spatial scale, in synaptic and dendritic structures, but also when starting to model the multiple parallel effects of calcium influx on the cytosolic protein networks. Finally, the combination of electrical recordings – be it juxtacellular or multi-electrode extracellular recordings – will remain a key technique for verifying the interpretation of calcium measurements in terms of neuronal spiking.

The principles of calcium imaging are rather independent of the microscope setup but the interpretation of measured calcium dynamics clearly depends on specific parameters of the imaging procedure such as sampling frequency, spatial resolution, and sensitivity. Implementations of calcium imaging may crucially depend on the specific imaging technology, which is obvious in the case of two-photon calcium imaging because it is the only method that allows cellular imaging several hundred micrometers deep inside neural tissue (Helmchen and Denk, 2005; Kerr and Denk, 2008). Therefore, further advances in microscopy and labeling techniques may open entire new fields of application for calcium imaging. One promising direction is the development of three-dimensional laser scanning technologies, which enable comprehensive volumetric measurements from local cell populations and thus new insights into the fundamentals of neural network dynamics (Göbel and Helmchen, 2007b; Göbel et al., 2007). The same holds true for imaging at higher speed, which is essential to capture rapid signal flow through neural networks (Grewe et al., 2010). Moreover, the ongoing improvement and further development of GECIs (Mank et al., 2008; Wallace et al., 2008, Lütcke et al., 2010) and their great potential for combination with other methods such as viral tracing techniques (Granstedt et al., 2009), can be foreseen to revolutionize the analysis of neural circuits in the neurosciences. Soon it will be possible to monitor neural network activity repeatedly over days to weeks from the same cell population, which will open tremendous opportunities to study the formation, plasticity, and reconfiguration of neural circuits.

After forty years of calcium imaging we are still experiencing only the beginning of the use of this powerful method for the study of cellular function. Most of the principles of fluorescence measurements that were covered in this chapter will remain valid so that we hope that young researchers have gained a solid foundation for exploring new research fields.

Acknowledgments

Work in the author's department was supported by research grants from the Swiss National Science Foundation (grants 3100A0-114624 and 310030-127091), the EU-FP7 program (projects 200873, 223524, 243914, 269921), and the Swiss SystemsX.ch initiative (Neurochoice project).

References

Allbritton, N. L., Meyer, T. and Stryer, L. (1992). Range of messenger action of calcium ion and inositol 1,4,5-trisphosphate. *Science*, **258**, 1812–1815.

Atluri, P. P. and Regehr, W. G. (1996). Determinants of the time course of facilitation at the granule cell to Purkinje cell synapse. *J. Neurosci.*, **16**, 5661–5671.

Baird, G. S., Zacharias, D. A. and Tsien, R. Y. (1999). Circular permutation and receptor insertion within green fluorescent proteins. *Proc. Natl. Acad. Sci. USA*, **96** etc., 11241–11246.

Berridge, M. J. (2006). Calcium microdomains: organization and function. *Cell Calcium*, **40** etc., 405–412.

Bollmann, J. H., Helmchen, F., Borst, J. G. and Sakmann, B. (1998). Postsynaptic Ca^{2+} influx mediated by three different pathways during synaptic transmission at a calyx-type synapse. *J. Neurosci.*, **18**, 10409–10419.

Borst, J. G. and Helmchen, F. (1998). Calcium influx during an action potential. *Methods Enzymol.*, **293**, 352–371.

Bower, J. M. and Beeman, D. (1998). *The Book of GENESIS: Exploring Realistic Neural Models with the GEneral NEural SImulation System* (2nd edition). New York: Springer Verlag.

Brown, J. E., Cohen, L. B., De Weer, P., Pinto, L. H., Ross, W. N. and Salzberg, B. M. (1975). Rapid changes in intracellular free calcium concentration. Detection by metallochromic indicator dyes in squid giant axon. *Biophys. J.*, **15**, 1155–1160.

Carnevale, N. T. and Hines, M. L. (2009). *The NEURON Book*. Cambridge: Cambridge University Press.

Chen, T. W., Lin, B. J., Brunner, E. and Schild, D. (2006). In situ background estimation in quantitative fluorescence imaging. *Biophys. J.*, **90**, 2534–2547.

Cheng, A., Goncalves, J. T., Golshani, P., Arisaka, K. and Portera-Cailliau, C. (2011). Simultaneous two-photon calcium imaging at different depths with spatiotemporal multiplexing. *Nature Methods*, **8**, 139–142.

DeSchutter, E. and Smolen, P. (1998). Calcium dynamics in large neuronal models. In: C. Koch and I. Segev (editors), *Methods in Neuronal Modeling*. Cambridge, MA: MIT Press, pp. 211–250.

DiGregorio, D. A., Peskoff, A. and Vergara, J. L. (1999). Measurement of action potential-induced presynaptic calcium domains at a cultured neuromuscular junction. *J. Neurosci.*, **19**, 7846–7859.

Egger, V. (2007). Imaging the activity of neuronal populations: when spikes don't flash and flashes don't spike. *J. Physiol.*, **582**(1), 7.

Gabso, M., Neher, E. and Spira, M. E. (1997). Low mobility of the Ca^{2+} buffers in axons of cultured Aplysia neurons. *Neuron*, **18**, 473–481.

Garaschuk, O., Schneggenburger, R., Schirra, C., Tempia, F. and Konnerth, A. (1996). Fractional Ca^{2+} currents through somatic and dendritic glutamate receptor channels of rat hippocampal CA1 pyramidal neurones. *J. Physiol. (London)*, **491**, 757–772.

Garaschuk, O., Milos, R. I., Grienberger, C., Marandi, N., Adelsberger, H. and Konnerth, A. (2006). Optical monitoring of brain function in vivo: from neurons to networks. *Pflügers Arch.*, **453**, 385–396.

Garaschuk, O., Griesbeck, O. and Konnerth, A. (2007). Troponin C-based biosensors: a new family of genetically encoded indicators for in vivo calcium imaging in the nervous system. *Cell Calcium*, **42** (4–5), 351–361.

Ghosh, A. and Greenberg, M. E. (1995). Calcium signaling in neurons: molecular mechanisms and cellular consequences. *Science*, **268**, 239–247.

Göbel, W. and Helmchen, F. (2007a). In vivo calcium imaging of neural network function. *Physiology (Bethesda)*, **22**, 358–365.

Göbel, W. and Helmchen, F. (2007b). New angles on neuronal dendrites in vivo. *J. Neurophysiol.*, **98**, 3770–3779.

Göbel, W., Kampa, B. M. and Helmchen, F. (2007). Imaging cellular network dynamics in three dimensions using fast 3D laser scanning. *Nature Methods*, **4**, 73–79.

Goldberg, J. H., Tamas, G., Aronov, D. and Yuste, R. (2003). Calcium microdomains in aspiny dendrites. *Neuron*, **40**, 807–821.

Granstedt, A. E., Szpara, M. L., Kuhn, B., Wang, S. S. and Enquist, L. W. (2009). Fluorescence-based monitoring of in vivo neural activity using a circuit-tracing pseudorabies virus. *PLoS One*, **4**, e6923.

Greenberg, D. S., Houweling, A. R. and Kerr, J. N. (2008). Population imaging of ongoing neuronal activity in the visual cortex of awake rats. *Nature Neurosci.*, **11** (7), 749–751.

Grewe, B. F. and Helmchen, F. (2009). Optical probing of neuronal ensemble activity. *Curr. Opinion Neurobiol.*, **19**, 520–529.

Grewe, B. F., Langer, D., Kasper, H., Kampa, B. M. and Helmchen, F. (2010). High-speed in vivo calcium imaging reveals spike trains in neuronal networks with near-millisecond precision. *Nature Methods*, **7**, 399–405.

Groden, D. L., Guan, Z. and Stokes, B. T. (1991). Determination of Fura-2 dissociation constants following adjustment of the apparent Ca-EGTA association constant for temperature and ionic strength. *Cell Calcium*, **12**, 279–287.

Grynkiewicz, G., Poenie, M. and Tsien, R. Y. (1985). A new generation of Ca^{2+} indicators with greatly improved fluorescence properties. *J. Biol. Chem.*, **260**, 3440–3450.

Harpur, A. G., Wouters, F. S. and Bastiaens, P. I. (2001). Imaging FRET between spectrally similar GFP molecules in single cells. *Nature Biotechnol.*, **19**, 167–169.

Heim, N. and Griesbeck, O. (2004). Genetically encoded indicators of cellular calcium dynamics based on troponin C and green fluorescent protein. *J. Biol. Chem.*, **279**, 14280–14286.

Hell, S. W. (2007). Far-field optical nanoscopy. *Science*, **316**, 1153–1158.

Helmchen, F. (2011). Calibration of fluorescent calcium indicators. In: *R. Yuste (editor), Imaging: A Laboratory Manual.* Cold Spring Harbor, NY: Cold Spring Harbor Laboratory Press.

Helmchen, F. and Denk, W. (2005). Deep tissue two-photon microscopy. *Nature Methods*, **2**, 932–940.

Helmchen, F. and Tank, D. W. (2011). A single-compartment model of calcium dynamics in nerve terminals and dendrites. In: F. Helmchen and A. Konnerth (editors), *Imaging in Neuroscience: A Laboratory Manual.* Cold Spring Harbor, NY: Cold Spring Harbor Laboratory Press.

Helmchen, F., Imoto, K. and Sakmann, B. (1996). Ca^{2+} buffering and action potential-evoked Ca^{2+} signaling in dendrites of pyramidal neurons. *Biophys. J.*, **70**, 1069–1081.

Helmchen, F., Borst, J. G. and Sakmann, B. (1997). Calcium dynamics associated with a single action potential in a CNS presynaptic terminal. *Biophys. J.*, **72**, 1458–1471.

Hernandez-Cruz, A., Sala, F. and Adams, P. R. (1990). Subcellular calcium transients visualized by confocal microscopy in a voltage-clamped vertebrate neuron. *Science*, **247**, 858–862.

Hirase, H., Qian, L., Bartho, P. and Buzsaki, G. (2004). Calcium dynamics of cortical astrocytic networks in vivo. *PLoS Biol.*, **2**, E96.

Holekamp, T. F., Turaga, D. and Holy, T. E. (2008). Fast three-dimensional fluorescence imaging of activity in neural populations by objective-coupled planar illumination microscopy. *Neuron*, **57**, 661–672.

Hoogland, T. M., Kuhn, B., Göbel, W., Huang, W., Nakai, J., Helmchen, F., Flint, J. and Wang, S. S. (2009). Radially expanding transglial calcium waves in the intact cerebellum. *Proc. Natl. Acad. Sci. USA*, **106**, 3496–3501.

Irving, M., Maylie, J., Sizto, N. L. and Chandler, W. K. (1990). Intracellular diffusion in the presence of mobile buffers. Application to proton movement in muscle. *Biophys. J.*, **57**, 717–721.

Kasai, H. and Petersen, O. H. (1994). Spatial dynamics of second messengers: IP_3 and cAMP as long-range and associative messengers. *Trends Neurosci.*, **17**, 95–101.

Kerr, J. N. and Denk, W. (2008). Imaging in vivo: watching the brain in action. *Nature Rev. Neurosci.*, **9**, 195–205.

Kerr, J. N. D., Greenberg, D. and Helmchen, F. (2005). Imaging input and output of neocortical networks in vivo. *Proc. Natl. Acad. Sci. USA*, **102**, 14063–14068.

Kerr, J. N., de Kock, C. P., Greenberg, D. S., Bruno, R. M., Sakmann, B. and Helmchen, F. (2007). Spatial organization of neuronal population responses in layer 2/3 of rat barrel cortex. *J. Neurosci.*, **27**, 13316–13328.

Koch, C. (1998). *Biophysics of Computation.* Oxford: Oxford University Press.

Lakowicz, J. R., Szmacinski, H., Nowaczyk, K. and Johnson, M. L. (1992). Fluorescence lifetime imaging of calcium using Quin-2. *Cell Calcium*, **13**, 131–147.

Lev-Ram, V., Miyakawa, H., Lasser-Ross, N. and Ross, W. N. (1992). Calcium transients in cerebellar Purkinje neurons evoked by intracellular stimulation. *J. Neurophysiol.*, **68**, 1167–1177.

Lin, B. J., Chen, T. W. and Schild, D. (2007). Cell type-specific relationships between spiking and $[Ca^{2+}]_i$ in neurons of the Xenopus tadpole olfactory bulb. *J. Physiol.*, **582** (1), 163–175.

Lipp, P. and Niggli, E. (1993). Ratiometric confocal Ca^{2+}-measurements with visible wavelength indicators in isolated cardiac myocytes. *Cell Calcium*, **14**, 359–372.

Llinas, R., Sugimori, M. and Silver, R. B. (1992). Microdomains of high calcium concentration in a presynaptic terminal. *Science*, **256**, 677–679.

Lütcke, H., Murayama, M., Hahn, T., Margolis, D. J., Astori, S., Meyer zum Alten Borgloh, S., Göbel, W., Yang, Y., Tang, W., Kügler, S., Sprengel, R., Nagai, T., Miyawaki, A., Larkum, M. E., Helmchen, F. and Hasan, M.T. (2010). Optical recording of neuronal activity with a genetically-encoded calcium indicator in anesthetized and freely moving mice. *Front. Neural Circ.*, **4**, 9.

Mank, M. and Griesbeck, O. (2008). Genetically encoded calcium indicators. *Chem. Rev.*, **108**, 1550–1564.

Mank, M., Santos, A. F., Direnberger, S., Mrsic-Flogel, T. D., Hofer, S. B., Stein, V., Hendel, T., Reiff, D. F., Levelt, C., Borst, A. et al. (2008). A genetically encoded calcium indicator for chronic in vivo two-photon imaging. *Nature Methods*, **5**, 805–811.

Maravall, M., Mainen, Z. F., Sabatini, B. L. and Svoboda, K. (2000). Estimating intracellular calcium concentrations and buffering without wavelength ratioing. *Biophys. J.*, **78**, 2655–2667.

Markram, H., Roth, A. and Helmchen, F. (1998). Competitive calcium binding: implications for dendritic calcium signaling. *J. Comput. Neurosci.*, **5**, 331–348.

Miyawaki, A. (2003). Fluorescence imaging of physiological activity in complex systems using GFP-based probes. *Curr. Opinion Neurobiol.*, **13**, 591–596.

Miyawaki, A., Llopis, J., Heim, R., McCaffery, J. M., Adams, J. A., Ikura, M. and Tsien, R. Y. (1997). Fluorescent indicators for Ca^{2+} based on green fluorescent proteins and calmodulin. *Nature*, **388**, 882–887.

Moreaux, L. and Laurent, G. (2007). Estimating firing rates from calcium signals in locust projection neurons in vivo. *Front. Neural Circ.*, **1**, 2.

Nagai, T., Yamada, S., Tominaga, T., Ichikawa, M. and Miyawaki, A. (2004). Expanded dynamic range of fluorescent indicators for Ca^{2+} by circularly permuted yellow fluorescent proteins. *Proc. Natl. Acad. Sci. USA*, **101**, 10554–10559.

Nakai, J., Ohkura, M. and Imoto, K. (2001). A high signal-to-noise Ca^{2+} probe composed of a single green fluorescent protein. *Nature Biotechnol.*, **19**, 137–141.

Naraghi, M. and Neher, E. (1997). Linearized buffered Ca^{2+} diffusion in microdomains and its implications for calculation of $[Ca^{2+}]$ at the mouth of a calcium channel. *J. Neurosci.*, **17**, 6961–6973.

Neher, E. (1986). Concentration profiles of intracellular calcium in the presence of a diffusible chelator. In: U. Heinemann, M. Klee, E. Neher and W. Singer (editors), Series 14, Calcium Electrogenesis and Neuronal Functioning. Berlin: Springer, pp. 80–96.

Neher, E. (1995). The use of fura-2 for estimating Ca buffers and Ca fluxes. *Neuropharmacology*, **34**, 1423–1442.

Neher, E. (1998a). Usefulness and limitations of linear approximations to the understanding of Ca^{2+} signals. *Cell Calcium*, **24**, 345–357.

Neher, E. (1998b). Vesicle pools and Ca^{2+} microdomains: new tools for understanding their roles in neurotransmitter release. *Neuron*, **20**, 389–399.

Neher, E. (2005). Some quantitative aspects of calcium fluorimetry. In: R. Yuste and A. Konnerth (editors), *Imaging in Neuroscience and Development: A Laboratory Manual*. Cold Spring Harbor: Cold Spring Harbor Laboratory Press, pp. 245–252.

Neher, E. and Augustine, G. J. (1992). Calcium gradients and buffers in bovine chromaffin cells. *J. Physiol. (London)*, **450**, 273–301.

Neher, E. and Sakaba, T. (2008). Multiple roles of calcium ions in the regulation of neurotransmitter release. *Neuron*, **59**, 861–872.

Neher, E. and Zucker, R. S. (1993). Multiple calcium-dependent processes related to secretion in bovine chromaffin cells. *Neuron*, **10**, 21–30.

Nimmerjahn, A., Kirchhoff, F., Kerr, J. N. and Helmchen, F. (2004). Sulforhodamine 101 as a specific marker of astroglia in the neocortex in vivo. *Nature Methods*, **1**, 31–37.

Nimmerjahn, A., Mukamel, E. A. and Schnitzer, M. J. (2009). Motor behavior activates Bergmann glial networks. *Neuron*, **62**, 400–412.

Oheim, M., Naraghi, M., Muller, T. H. and Neher, E. (1998). Two dye two wavelength excitation calcium imaging: results from bovine adrenal chromaffin cells. *Cell Calcium*, **24**, 71–84.

Palmer, A. E. and Tsien, R. Y. (2006). Measuring calcium signaling using genetically targetable fluorescent indicators. *Nature Protocols*, **1**, 1057–1065.

Palmer, A. E., Giacomello, M., Kortemme, T., Hires, S. A., Lev-Ram, V., Baker, D. and Tsien, R. Y. (2006). Ca^{2+} indicators based on computationally redesigned calmodulin-peptide pairs. *Chem. Biol.*, **13**, 521–530.

Pologruto, T. A., Yasuda, R. and Svoboda, K. (2004). Monitoring neural activity and $[Ca^{2+}]$ with genetically encoded Ca^{2+} indicators. *J. Neurosci.*, **24**, 9572–9579.

Press, W. H., Flannery, B. P., Teukolsky, S. A. and Vetterling, W. T. (1988). *Numerical Recipes in C*. Cambridge: Cambridge University Press.

Ramdya, P., Reiter, B. and Engert, F. (2006). Reverse correlation of rapid calcium signals in the zebrafish optic tectum in vivo. *J. Neurosci. Methods*, **157**, 230–237.

Regehr, W. G., Delaney, K. R. and Tank, D. W. (1994). The role of presynaptic calcium in short-term enhancement at the hippocampal mossy fiber synapse. *J. Neurosci.*, **14**, 523–537.

Ridgway, E. B., and Ashley, C. C. (1967). Calcium transients in single muscle fibers. *Biochem. Biophys. Res. Commun.*, **29**, 229–234.

Roberts, W. M. (1994). Localization of calcium signals by a mobile calcium buffer in frog saccular hair cells. *J. Neurosci.*, **14**, 3246–3262.

Rochefort, N. L., Garaschuk, O., Milos, R. I., Narushima, M., Marandi, N., Pichler, B., Kovalchuk, Y. and Konnerth, A. (2009). Sparsification of neuronal activity in the visual cortex at eye-opening. *Proc. Natl. Acad. Sci. USA*, **106**, 15049–15054.

Rothschild, G., Nelken, I. and Mizrahi, A. (2010). Functional organization and population dynamics in the mouse primary auditory cortex. *Nature Neurosci.*, **13**, 353–360.

Sabatini, B. L. and Regehr, W. G. (1995). Detecting changes in calcium influx which contribute to synaptic modulation in mammalian brain slice. *Neuropharmacology*, **34**, 1453–1467.

Sabatini, B. L., Oertner, T. G. and Svoboda, K. (2002). The life-cycle of Ca^{2+} ions in dendritic spines. *Neuron*, **33**, 439–452.

Sasaki, T., Takahashi, N., Matsuki, N. and Ikegaya, Y. (2008). Fast and accurate detection of action potentials from somatic calcium fluctuations. *J. Neurophysiol.*, **100**, 1668–1676.

Schneggenburger, R. and Neher, E. (2005). Presynaptic calcium and control of vesicle fusion. *Curr. Opinion Neurobiol.*, **15**, 266–274.

Schneggenburger, R., Zhou, Z., Konnerth, A. and Neher, E. (1993). Fractional contribution of calcium to the cation current through glutamate receptor channels. *Neuron*, **11**, 133–143.

Schwaller, B. (2009). The continuing disappearance of "pure" Ca^{2+} buffers. *Cell Mol. Life Sci.*, **66**, 275–300.

Smetters, D., Majewska, A. and Yuste, R. (1999). Detecting action potentials in neuronal populations with calcium imaging. *Methods*, **18**, 215–221.

Stosiek, C., Garaschuk, O., Holthoff, K. and Konnerth, A. (2003). In vivo two-photon calcium imaging of neuronal networks. *Proc. Acad. Sci. USA*, **100**, 7319–7324.

Stryer, L. (1978). Fluorescence energy transfer as a spectroscopic ruler. *Annu. Rev. Biochem.*, **47**, 819–846.

Tank, D. W., Regehr, W. G. and Delaney, K. R. (1995). A quantitative analysis of presynaptic calcium dynamics that contribute to short-term enhancement. *J. Neurosci.*, **15**, 7940–7952.

Tian, L., Hires, S.A., Mao, T., Huber, D., Chiappe, M. E., Chalasani, S. H., Petreanu, L., Akerboom, J., McKinney, S. A., Schreiter, E. R. et al. (2009). Imaging neural activity in worms, flies and mice with improved GCaMP calcium indicators. *Nature Methods*, **6**, 875–881.

Tsien, R. Y. (1980). New calcium indicators and buffers with high selectivity against magnesium and protons: design, synthesis, and properties of prototype structures. *Biochemistry*, **19**, 2396–2404.

Tsien, R. Y. (1981). A non-disruptive technique for loading calcium buffers and indicators into cells. *Nature*, **290**, 527–528.

Tsien, R. Y. (1989). Fluorescent probes of cell signaling. *Annu. Rev. Neurosci.*, **12**, 227–253.

Tsien, R. and Pozzan, T. (1989). Measurement of cytosolic free Ca^{2+} with quin2. *Methods Enzymol.*, **172**, 230–262.

Vogelstein, J. T., Watson, B. O., Packer, A. M., Yuste, R., Jedynak, B. and Paninski, L. (2009). Spike inference from calcium imaging using sequential Monte Carlo methods. *Biophys. J.*, **97**, 636–655.

Vogelstein, J. T., Packer, A. M., Machado, T. A., Sippy, T., Babadi, B., Yuste, R. and Paninski, L. (2010). Fast nonnegative deconvolution for spike train inference from population calcium imaging. *J. Neurophysiol.*, **6**, 3691–3704.

Vranesic, I. and Knöpfel, T. (1991). Calculation of calcium dynamics from single wavelength fura-2 fluorescence recordings. *Pflugers Arch.*, **418**, 184–189.

Wagner, J., and Keizer, J. (1994). Effects of rapid buffers on Ca^{2+} diffusion and Ca^{2+} oscillations. *Biophys. J.*, **67**, 447–456.

Wallace, D. J., Zum Alten Borgloh, S. M., Astori, S., Yang, Y., Bausen, M., Kugler, S., Palmer, A. E., Tsien, R. Y., Sprengel, R., Kerr, J. N. et al. (2008). Single-spike detection in vitro and in vivo with a genetic Ca^{2+} sensor. *Nature Methods*, **5**, 797–804.

Wang, S. S., Alousi, A. A. and Thompson, S. H. (1995). The lifetime of inositol 1,4,5-trisphosphate in single cells. *J. Gen. Physiol.*, **105**, 149–171.

Wang, X., Lou, N., Xu, Q., Tian, G. F., Peng, W. G., Han, X., Kang, J., Takano, T. and Nedergaard, M. (2006). Astrocytic Ca^{2+} signaling evoked by sensory stimulation in vivo. *Nature Neurosci.*, **9**, 816–823.

Wilms, C. D., Schmidt, H. and Eilers, J. (2006). Quantitative two-photon Ca^{2+} imaging via fluorescence lifetime analysis. *Cell Calcium*, **40**, 73–79.

Woolf, T. B. and Greer, C. A. (1994). Local communication within dendritic spines: models of second messenger diffusion in granule cell spines of the mammalian olfactory bulb. *Synapse*, **17**, 247–267.

Yaksi, E. and Friedrich, R. W. (2006). Reconstruction of firing rate changes across neuronal populations by temporally deconvolved Ca^{2+} imaging. *Nature Methods*, **3**, 377–383.

Zador, A. and Koch, C. (1994). Linearized models of calcium dynamics: formal equivalence to the cable equation. *J. Neurosci.*, **14**, 4705–4715.

11

Functional magnetic resonance imaging

ANDREAS BARTELS, JOZIEN GOENSE AND NIKOS
LOGOTHETIS

11.1 Introduction

Functional magnetic resonance imaging (fMRI) allows the non-invasive measurement of neural activity nearly everywhere in the brain. The structural predecessor, MRI, was invented in the early 1970s (Lauterbur, 1973) and has been used clinically since the mid-1980s to provide high-resolution structural images of body parts, including rapid successions of images for example of the beating heart. However, it was the advent of blood oxygenation level dependent (BOLD) functional imaging developed first by Ogawa et al. (1990) that made the method crucial especially for the human neurosciences, leading to a vast expansion of both the method of fMRI as well as the field of human neurosciences. fMRI is now a mainstay of neuroscience research and by far the most widespread method for investigations of neural function in the human brain as it is entirely harmless, relatively easy to use, and the data are relatively straightforward to analyze. It is therefore no surprise that fMRI has provided a wealth of information about the functional organization of the human brain. While many publications initially confirmed knowledge derived from invasive animal experiments or from clinical studies, it is now frequently fMRI that opens up a new field of investigation that is then later followed up by invasive methods.

It is important to note that fMRI does not measure electrical or neurochemical activity directly. Physically, it relies on decay time-constants of water protons, which are affected by brain tissue and the concentration of deoxyhemoglobin. Biologically, fMRI relies on a combination of changes in blood deoxygenation and blood flow that are induced by neural activity. That neural activation increases blood flow to the activated brain region was first observed by Mosso in the late 1800s (see Iadecola (2004) for a review). Belliveau et al. (1990) used an exogenous paramagnetic contrast agent to measure changes in blood volume upon activation, but it was Ogawa et al. (1990) who showed that deoxyhemoglobin acts as

Handbook of Neural Activity Measurement, ed. Romain Brette and Alain Destexhe. Published by Cambridge University Press. © Cambridge University Press 2012.

an endogenous contrast agent, laying the basis for most of today's completely non-invasive fMRI measurements. Compared to other functional imaging methods, fMRI offers much higher spatial resolution and it is entirely non-invasive, which has allowed us to improve the mapping of the functional parcelation in human brain.

Because of its obvious link to neural activity, fMRI activation is typically interpreted without much question as a marker of neural function, which many scientists link intuitively to neural spiking activity. There is no doubt that fMRI measures neural function, but how this measurement relates to the various different types of neural activity is still a matter of investigation. If one plans to interpret fMRI signals in terms of other known signals (such as spiking activity, synaptic activity, neuromodulation, etc.), a deeper understanding of their physiological link to fMRI signals is crucial. For example, it is not known whether the different neural processes mentioned above are all represented equally in the fMRI signal, or whether some are overrepresented or underrepresented. This is still relatively unknown territory that many different groups are currently trying to address with multidisciplinary approaches.

This understanding is also crucial in order to relate the neurophysiological work performed for over 50 years in non-human primates and other animals, often painstakingly mapping the functional properties of isolated neurons. And although this has offered a wealth of knowledge about brain function, for obvious reasons a direct extrapolation to human brain function is difficult.

Therefore there has been an increasing amount of work in recent years aimed at elucidating how the neural signals that are measured invasively with intracortical microelectrodes in animals relate to the non-invasive measures of brain function that can be researched in humans, like fMRI, electroencephalography (EEG) and magnetoencephalography (MEG). One way to achieve has been by combining invasive methods with fMRI in monkeys or other animals, either simultaneously (Logothetis et al., 2001; Goense and Logothetis, 2008) or consecutively (Disbrow et al., 2000; Smith et al., 2002; Lipton et al., 2006; Tsao et al., 2006; Lu et al., 2007; Maandag et al., 2007). Neurophysiological recording in humans also offers opportunities to study directly the relationship between neural signals and fMRI (Mukamel et al., 2005; Privman et al., 2007). However, most invasive neurophysiology in humans is conducted in a surgical or presurgical context and is usually limited to certain areas such as seizure-prone areas like the temporal lobe (Engel et al., 2005; Mukamel et al., 2005; Privman et al., 2007). Another approach to a better understanding and interpretation of the human fMRI studies is to perform comparative fMRI in humans and monkeys, which allows us to look at homology questions and provides a better basis for extrapolating animal work to human work (Nakahara et al., 2002; VanDuffel et al., 2002; Tsao et al., 2003;

Denys et al., 2004; Koyama et al., 2004; Sawamura et al., 2005; Tsao et al., 2008; Rajimehr et al., 2009).

As both electrophysiological and imaging methods have their strengths and limitations and measure only particular aspects of brain function, the integrative approach allows us to obtain a more complete picture. This is expected to improve our interpretation of both methods. The most obvious limitation of intracortical microelectrodes is their highly restricted spatial sampling and, depending on the electrodes used, a bias towards large pyramidal neurons; for EEG it is the limited spatial resolution of its reconstructed sources and for fMRI it is the limited temporal resolution, which also stems from the fact that it is a hemodynamic signal and does not measure the neural events themselves.

Since the blood oxygen level dependent (BOLD) signal has a time-constant in the range of seconds, fMRI cannot resolve the temporal sequences of fast processes constituting neural processing, such as those resolved by invasive or non-invasive measurements of electrical or magnetic activity.

Nevertheless, fMRI can be used to measure a broad variety of neural events, even if they are very brief, since most have metabolic consequences that are reflected in the BOLD signal. The exact mechanisms linking distinct types of neural activity to the BOLD signal are nevertheless still a matter of debate.

The spatial resolution of fMRI depends on the biological spread of the vascular response as well as on imaging parameters, magnetic field strength and the construction of the coil.

In typical human whole-brain imaging experiments carried out at 3 T magnetic field strength, the spatial resolution usually ranges between 2 and 4 mm, with a temporal sampling rate between 1 and 4 s. For special purposes small custom-made coils allow resolutions of a few hundred micrometers, and single-slice acquisitions lasting a few dozen milliseconds. Current 3 T human fMRI scanners are extremely easy to operate, so that after a few hours of training any investigator can collect their own data. Also, unlike EEG recordings, fMRI datacollection is relatively fast, in that a full experiment can be completed within one hour including setup time.

This combination of factors may explain the extraordinary number of fMRI-based publications, which has continued to rise in an exponential fashion since its invention. A search in the ISI Web of Science publication database revealed for the year 2010 about 4500 publications for fMRI, about 5000 publications for EEG, 650 for MEG and 3200 for PET. Figure 11.1 shows the publications per year for these non-invasive brain imaging methods since 1990. fMRI has the steepest exponential ascent, likely to overtake all other methods in terms of published papers per year within coming years, with more than 12 publications per day in 2010.

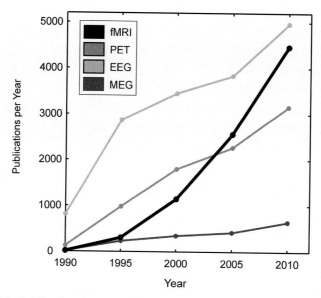

Figure 11.1 Publications per year in the fields of fMRI, PET, EEG and MEG. Data were retrieved from the ISI Web of Science using the topic keywords "fMRI" or "functional MRI" or "functional magnetic resonance imaging" for fMRI, "positron emission tomography" for PET, "EEG" or "evoked potential" or "electroenceph*" for EEG, and "MEG" or "magnetoenceph*" for MEG, searching for articles only.

In this chapter we aim to provide an overview of the very basic concepts of fMRI. We will first provide an extremely brief explanation of the underlying physics, continue with properties of the BOLD signal and some anatomical properties of the vascularization underlying it, and then briefly review the main analysis techniques and assumptions underlying them in common fMRI analyses. We will then cover in more detail perhaps one of the least understood yet most important questions regarding the method, namely the relation between the BOLD signal and the underlying physiological processes. For this, we will focus mainly on signals recorded with microelectrodes, i.e. single- and multi-unit activity (SUA and MUA) (i.e. spikes) and local field potentials (LFP). The increases in local neural activity upon stimulus presentation and the concomitant increased energy demands of neurotransmission and spiking, leads to an increase in blood flow to the activated area, which ultimately drives the BOLD response. The work by Logothetis et al. (2001) showed that the BOLD signal is not as well correlated to single-unit activity, but correlates better with the LFP, a more comprehensive signal that includes membrane potential fluctuations, oscillations, postsynaptic and presynaptic events. This raises questions about whether all aspects of neural activity drive the BOLD response equally, and if not, which ones are more important: for instance the input versus the output from an area, or inhibition versus excitation, stimulus-driven or neuromodulatory activity. Answers to such questions will directly affect our interpretation of fMRI results and thus help us to understand results obtained with fMRI. We will discuss which neural events are thought to drive the hemodynamic response, but to get some insight into the coupling between neural activity and BOLD, we will also discuss neurovascular coupling and the specificity of the BOLD response, as these issues have direct bearing on our understanding of the coupling between BOLD and neurophysiological signals.

11.2 Physical basis of the fMRI signal

Here we provide only a brief overview of the basic physical principles underlying MRI and fMRI, as the primary focus of this chapter is the relation of different phys-iological signals to the BOLD signal. There are several excellent web-resources as well as textbooks providing outstanding and detailed introductions to MR physics (see e.g. Haacke et al., 1999; Liang and Lauterbur, 1999; Huettel et al., 2008; Buxton, 2009; Hornak, www.cis.rit.edu/htbooks/mri).

Atomic nuclei possess a quantum physical property called spin. When it is non-zero, such as in protons (^{1}H) of water or fat in body tissue that have spin-1/2, it provides them with a magnetic moment that can be visualized as a spinning magnet. When placed in a strong magnetic field ($\mathbf{B}_0$), the spin will precess around an axis along the external field, with the so-called Larmor frequency. The Larmor frequency is specific to the nucleus and depends on the field strength of $\mathbf{B}_0$. The alignment of the spin can be parallel or antiparallel (energetically less favorable) to the field, with the two states being separated by a small energy band. At a field strength of 1.5 T there is only a small excess of the number of parallel spins, in the order of 1:10 000 000, leading to a net polarization of the spins along the z-axis, which in turn introduces a net magnetization along the z-axis. The application of a radiofrequency (RF) field of amplitude $\mathbf{B}_1$ rotating synchronously with the precessing spins can null the netpolarization in the z-axis, (with a so-called 90° RF excitation, Figure 11.2A) and tilt the net magnetization by 90° into the xy-plane, i.e. the longitudinal magnetization is converted to transverse magnetization (see Figure 11.2A). This happens only when the carrier frequency of the RF pulse is equal to the Larmor frequency, hence the term magnetic resonance (note that this was formerly known as nuclear magnetic resonance, which was abandoned in clinical and human imaging environments because of unfavorable connotations).

In addition to changing the spin-polarization, the application of $\mathbf{B}_1$ also intro-duces phase coherence among the spins (Figure 11.2B). The transverse component of the in-phase rotation of the spins leads to a magnetic oscillation in the xy-plane (the so-called free induction decay, FID) that can be measured by an RF coil. In

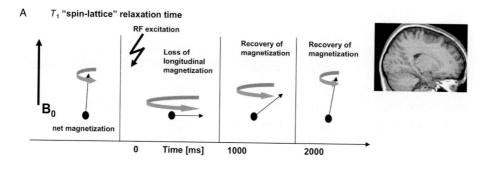

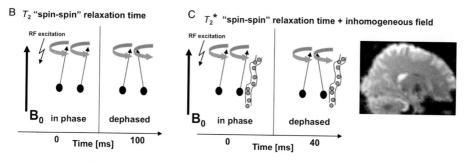

Figure 11.2 Illustration of spin properties and T_1, T_2 and T_2^* relaxation time constants A. T_1 relaxation is the time constant for net magnetization to recover after the application of an RF field (at time 0) that tilted the net magnetization away from the z-axis into the xy-plane. Note that T_1 denotes solely the magnetization recovery along the z-axis, and that the magnetization in the xy-plane disappears much earlier due to spindephasing as described in B and C. B. T_2 relaxation is the time constant for the precession of the spins to dephase due to random spin–spin interactions after an RF pulse equalized their phases (at time 0). C. T_2^* relaxation also describes spin dephasing, but due to a combination of both spin–spin interactions and inhomogeneities of the magnetic field, such as induced by blood vessels passing through tissue.

fact, this is the key signal measured in MRI and fMRI experiments. After the RF field is turned off, the spins recover to their original positions. This recovery is governed by several relaxation time-constants T that follow exponential functions. T_1 denotes the time constant for the longitudinal relaxation of the spins back to their equilibrium state (aligned with the magnetic field $\mathbf{B_0}$). This so-called spin-lattice relaxation is influenced by the proton density, and thus differs between different types of tissue. At 1.5 T, it takes on the following approximate values: cerebral gray matter. 920 ms, white matter 780 ms, muscle 880 ms, liver 490 ms, blood 1350 ms, fat 240 ms. Differences in T_1 between tissues provide contrast, and are therefore an important measure used in many clinical applications that visualize tissues and defects therein (see the T_1-weighted contrast image of a healthy human brain in Figure 11.2A). T_2 is the time-constant for the dephasing of the rotation of the spins, the so-called spin–spin relaxation that happens as a result of quantum mechanical spin–spin interactions (Figure 11.2B). Importantly, the T_2 relaxation is the time-constant pertaining to the irreversible component of spin–spin dephasing, as this is a random process. However, the most important contributor to contrast in fMRI applications is T_2^*. This is the time-constant pertaining to both the above irreversible spin–spin component of dephasing and the dephasing due to inhomogeneities of the magnetic field – for example due to the vast amount of small-diameter blood vessels innervating the tissue that carry paramagnetic deoxyhemoglobin and thus alter the local magnetic field (Figure 11.2C). T_2^* is always shorter than T_2. Dephasing due to field inhomogeneities is not random (as long as the field inhomogeneities are constant and spatially stationary), and can be reversed by applying a so-called re-focusing pulse that is 180°, in a so-called spin-echo experiment. This way, the irreversible T_2 component can be measured independently from T_2^*. T_2 is usually around 5–20 times shorter than T_1 and is more tissue dependent.

A key aspect of magnetic resonance imaging is its ability to obtain a spatial map of signals non-invasively. Spatial coding and the various imaging sequences employed to obtain images of different contrasts are highly complex and still provide opportunities for innovation and improvements. The basic principle of most of the commonly used sequences is fairly simple: during application of the RF field (used to tilt the net magnetization and to 0-phase the spins) a "slice selection" magnetic gradient is turned on (in addition and parallel to $\mathbf{B}_0$). The consequence of this is that the RF pulse will now only excite one plane of tissue that matches the Larmor frequency at a particular magnetic field strength. A second "phase-encoding" magnetic gradient orthogonal to the first is then applied for a brief duration after the RF field has been applied. Its function is to introduce a controlled dephasing of spins, such that spins in each line of the activated slice differ in their phase. A third "read-out" or "frequency-encoding" gradient orthogonal to the previous two is applied during the acquisition of the data, such that distinct lines of spins have a different Larmor frequency. The three gradients thus allow the resolution of three-dimensional tissue, with slices having to be acquired sequentially.

In functional imaging, the repetition time (T_R) denotes the time intervals at which the same tissue slice is excited by an RF pulse. The echo time (T_E) denotes the time that passes between the excitation pulse and the time of measuring RF signals. The flip-angle (F_A) indicates the angle by which the net magnetization was tilted away from its alignment with $\mathbf{B}_0$. These parameters determine together with many others the weighting towards different types of contrasts that are optimal for different types of applications.

$\mathbf{B}_0$ is typically produced using a superconducting coil that is cooled by liquid helium and nitrogen. This field is always on, even when the scanner is not in use.

The gradient coils for three-dimensional a spatial encoding are only turned on during scanning, and need to switch fields at a rapid rate. The mechanical stress this exerts on the mechanics of their fixpoints, on the scanner bore, and on other parts leads to the extremely loud noise emitted during scanning, requiring the subjects to wear ear-protection. Finally, a limitation especially of fMRI is field-homogeneity. Since the subject's body and head induce field changes inside the scanner bore, these are compensated for (to make B_0 homogeneous) usually prior to initiation of the main scan using additional coils located around the head between the magnet and the gradients (this process is called "shimming"). Very sharp transitions of air-tissue boundaries, and especially deviations from a sphere (such as the ear-canals or the sinuses near the orbitofrontal cortex) cannot well be compensated for using standard setups. The resulting field-inhomogeneities prevent signal acquisition (due to rapid dephasing, short T_2^* and deviant Larmor frequencies) in sequences that do not use refocusing pulses. Areas of signal loss and distortion resulting from the above reasons are referred to as "susceptibility artifacts." This makes fMRI of some brainregions difficult or impossible, for example some temporal regions and parts of the orbitofrontal cortex. Similarly, movement during the scan can introduce severe artifacts in fMRI, as it voids the initial shimming – unless the field is reshimmed. All in all, (f)MR imaging still has many challenges and one can expect enormous improvements in the future, as it is still a very young and relatively under-explored technology.

11.3 BOLD contrast mechanism

The first researchers to use magnetic resonance techniques in order to measure changes in cerebral blood flow and oxygenation were Ogawa et al. (1990) in rats and Belliveau et al. (1990) in dogs. Belliveau et al. (1990) used an exogenous contrast agent injected into the bloodstream to measure changes in cerebral blood volume (CBV). Ogawa et al. (1990) discovered deoxyhemoglobin (dHb) as an endogenous contrast agent. By manipulating O_2 and CO_2 levels, they showed that excess deoxygenated blood leads to signal loss near vessels, which could be reversed by increasing the flow of freshly oxygenated blood. Soon after that it was shown in humans that changes in brain function lead to changes in blood oxygenation and flow that are measurable with MRI (Belliveau et al., 1991; Kwong et al., 1992; Ogawa et al., 1992). The origin of the functional BOLD contrast is actually an oversupply of freshly oxygenated blood that increases the signal in the activated areas. The increase in neural activity in a given brain area by a stimulus or performing a task triggers an increase in the flow of fresh blood to the activated area in order to meet the increased metabolic demands. The BOLD contrast is based on the concentration of deoxyhemoglobin in the blood since dHb is paramagnetic and acts as a contrast agent: an increase in its concentration decreases T_2^*. The increase in oxygen supply to the active tissue is larger than the oxygen that is used by the cells, and hence there is a relative increase in the oxyhemoglobin (Hb) concentration, and a decrease in the deoxyhemoglobin concentration which increases T_2^* and leads to a signal increase in the gradient-echo (GE) images (Kwong et al., 1992; Ogawa et al., 1992). Belliveau et al. (1990, 1991) showed functional MRI based on similar principles but using an exogenous contrast agent which is sensitive to changes in CBV.

Although the first fMRI experiments were performed in the early 1990s, what exactly the BOLD signal represents is still unclear. Questions remain not only about which neural or metabolic changes exactly trigger the BOLD signal, but also about the relative contributions of flow increases versus oxygen extraction, or which parts of the vascular tree contribute most to the BOLD signal. And given these uncertainties, the BOLD signal is not (yet) suitable for quantitative measurements.

11.3.1 Properties of the BOLD signal

The presence of paramagnetic deoxyhemoglobin in the blood results in susceptibility gradients, which are local variations in the magnetic field. These gradients exist near vessels because of their high dHb content, and their size depends on the vessel size and dHb concentration. Spins within these gradients dephase which leads to T_2^*-based signal loss. Based on theory and simulations (Kennan et al., 1994; Weisskoff et al., 1994; Boxerman et al., 1995) it has been determined that with the commonly used fMRI methods, one observes a BOLD signal arising from vessels ranging from capillaries to large veins. Large draining veins can have very strong BOLD signals, but they are downstream from the neural activation and can be quite remote from the activated area. However, the relative contribution of large and small vessels to the BOLD signal is still debated. One of the reasons why there is still debate about this is that the properties of the BOLD signal depend strongly on MR-hardware (most obviously field strength) and acquisition parameters (Yacoub et al., 2001; Duong et al., 2003; Goense and Logothetis, 2006; Jin et al., 2006) which complicates the comparison of results across laboratories or studies. fMRI in humans is usually performed at field strengths of 1.5 and 3 T, although 7 T scanners are becoming increasingly common. At low magnetic field (e.g. 1.5 T) the BOLD signal is very sensitive to large vessel signal, and this progressively decreases at higher fields (Yacoub et al., 2001). But although the vessel-fraction is decreased at high field, high-resolution studies show that there is still a substantial large vessel contribution (Figure. 11.3) (Zhao et al., 2004; Harel et al., 2006a; Jin et al., 2006). Because BOLD signals from large vessels decrease the accuracy with

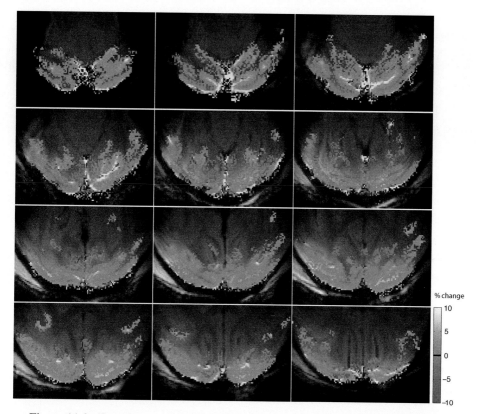

Figure 11.3 (See plate section for color version.) Functional activation acquired in the visual cortex of an anesthetized monkey at 7 T at a spatial resolution of $0.5 \times 0.5 \times 2\,\text{mm}^3$. The map shows the percentage signal change in response to a black-and-white rotating checkerboard presented to both eyes. Functional activation is seen in V1–V5. The scanner and experimental procedures are described in Logothetis et al. (1999) and Pfeuffer et al. (2004).

which functional activity can be localized, it is important to decrease the sensitivity to the large vessel signal.

Figure 11.3 shows the BOLD signal elicited by visual stimulation in a monkey visual cortex acquired at 7 T. Functional activation is seen in the entire early visual cortex (V1–V5) and at this resolution it can easily be seen that the BOLD signal occurs both at the cortical surface and in gray matter, with the strongest BOLD signals at the cortical surface and in the calcarine sulcus. This is the large vessel contribution, which can be easily distinguished at high resolution. Note that statistical maps (instead of percentage change maps) are often used to display activation, and they depict the correlation with the stimulus. The highest p-values do not usually occur at the surface, because even though vessels have larger signals they also have higher noise (Goense and Logothetis, 2006). The sensitivity to large vessels has the drawback that an area of activation identified in a functional map could in fact be a vessel remote from the area of neural activity. Also, when a blood vessel in a sulcus shows activation, at low resolution this could lead to ambiguity about the exact location of the activated tissue.

Many details of the BOLD signal are still not completely understood, like the relative contributions of venous, arterial and capillary fractions to the BOLD signal, and the relative contributions of blood flow increases and oxygen consumption (Haacke et al., 2001; Buxton et al., 2004). This is the case for the positive BOLD response, for the initial dip, the post-stimulus undershoot, as well as for fMRI signals recorded with different methods (Ugurbil et al., 2006; Yacoub et al., 2006; Kim et al., 2007; Zhao et al., 2007). A discussion of the different fMRI methods and their properties is beyond the scope of this chapter. Other MR-based functional imaging methods are used in addition to the standard BOLD approach, like spin-echo (SE) BOLD, and functional CBV and CBF measurements. We will not go into detail here and note only that the general consensus is that GE-BOLD represents mostly venous signal, which becomes more strongly weighted towards smaller venules and capillaries as the field strength increases, high-field SE-BOLD represents mostly capillary signal, the CBV signal represents smaller vessels (arteries and veins) and capillaries, and the functional CBF signal represents mostly arterioles and capillaries.

11.3.2 *Spatial resolution and specificity of fMRI*

How well functional activation is localized to the actual place of neural activation depends on the achievable fMRI resolution and the specificity of the hemodynamic signal. The achievable spatial resolution is determined by scanner hardware and the signal to noise ratio (SNR). The specificity is determined by the fMRI method that is used and how closely the hemodynamic response reflects the actual neural activity. Large vessel BOLD signal can be remote from the site of activation, but BOLD signal from capillaries can reasonably be assumed to be closely related to the neural activity in that area. Hence, the specificity is not only determined by biological factors, but also by the choice of fMRI method, hardware and sequence parameters (Ugurbil et al., 2003; Harel et al., 2006b).

With advanced MRI hardware the spatial resolution achievable for structural imaging in vivo is of the order of 200–300 μm in-plane for whole-head imaging (Ugurbil et al., 2006; Wald et al., 2006) and ∼100 μm for localized imaging (Nakada et al., 2005) in humans. In animals, resolutions of 70–100 μm have been achieved in macaques (Logothetis et al., 2002) (Figure 11.4) and a few tens of

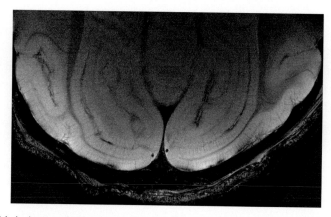

Figure 11.4 Anatomical image of macaque V1 acquired at 4.7 T. The high-resolution GE image at a resolution of $100 \times 100 \times 1000\,\mu m^3$ and volume of $0.01\,\mu l$ shows the small perpendicular intracortical veins and layer IV, which is indicated by the Gennari line. The Gennari line has a higher myelin content than the rest of the cortex. T_E 23.5 ms, T_R 2000 ms. (Adapted from Goense and Logothetis, 2010, with permission.)

micrometers in rodents. In principle, fMRI can also be done at very high resolutions, although it is constrained by the relatively low contrast to noise ratio (CNR) of the functional activation and the limited amount of time available for the acquisition of each image. Because of its speed, echo planar imaging (EPI) (Mansfield, 1977) is typically used for fMRI (it can collect one image per excitation or repetition time T_R). However, EPI requires high-performance hardware and the limitations of the gradients often limit the maximally achievable resolution. Despite this, by using segmented EPI or parallel imaging, high-resolution fMRI can be done, and in monkeys functional maps at $125\,\mu m$ in-plane resolution have been shown (Logothetis et al., 2002) while in rats, maps with $50–100\,\mu m$ have been demonstrated (Silva and Koretsky, 2002; Xu et al., 2003). In human fMRI studies, the typical resolution is $\sim 3 \times 3 \times 3\,mm^3$, and currently the highest resolution achieved is about $0.5\,mm^2$ in-plane (Pfeuffer et al., 2002; Yacoub et al., 2005). The magnitude of the functional activation in gray matter is only a few percent and although image SNR decreases for smaller voxels, the functional signal tends actually to increase as the voxel size decreases, due to a decrease in partial volume effects. At a few hundred micrometers or less, the fMRI resolution can be higher than the point spread function (PSF) of the activation. Hence, the theoretically achievable spatial resolution is probably limited by the spatial extent of the neural signals and the hemodynamic regulation, even though this issue is still not fully resolved.

At high resolution specificity becomes important for the ability to visualize structures like cortical columns or layers. But also at low resolution increasing the specificity of the functional signal is important to eliminate the effect of draining veins and increase the accuracy of the mapping. Specificity depends on (1) the anatomy of the capillary bed, (2) the fMRI method used, and (3) neurovascular coupling and the spatial scale of blood flow regulation.

11.3.2.1 Anatomy of the cortical vascular system

Since the fMRI signal depends on blood flow, its specificity ultimately depends on the anatomy of the vascular bed and the spatial scale of blood flow regulation. Figure 11.5 shows a corrosion cast of the cortical vasculature in macaque V1. The cortical blood supply is characterized by large arteries that run along the surface of the cortex, branch into smaller pial arteries and finally into intracortical arteries that enter the cortical gray matter perpendicular to the surface. The intracortical vessels eventually branch into capillaries, and blood is collected in intracortical

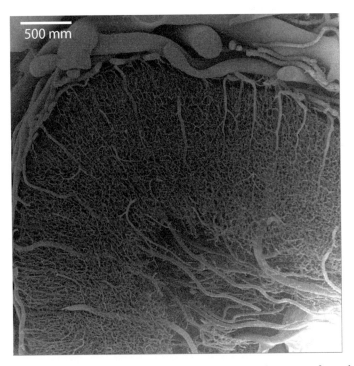

Figure 11.5 (See plate section for color version.) Corrosion cast of monkey V1 showing the microvasculature of the cortex (from Weber et al., 2008). Arteries and large draining are located in the pial layer. Penetrating arterioles and venules can be seen in addition to the capillary network. Intracortical arteries are shown in red and intracortical veins in blue. The vascular density in white matter is lower than in gray matter. (Adapted from Weber et al., 2008, with permission.)

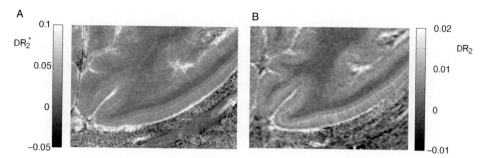

Figure 11.6 High-resolution steady-state CBV image of macaque V1 acquired at 4.7 T showing the relaxivity changes induced by MION-injection. CBV is higher in gray matter than in white matter and it is high in the pial layers where the large vessels are located. In addition, differences in CBV can be observed within the cortex with layer IV having a higher blood volume. Intracortical vessels can also be seen. GE, resolution $100 \times 100 \times 2000\,\mu m^3$, T_E 20 ms, T_R 2000 ms, 8 mg/kg MION. (Adapted from Goense and Logothetis, 2010.)

veins, which form larger veins on the surface of the cortex (Duvernoy et al., 1981). The distance between capillaries at any place in the gray matter of the cortex is about $40\,\mu m$ (Weber et al., 2008). Gray matter is more highly vascularized than white matter, and there are differences in vascular density between different cortical areas; areas with higher vascularization, like primary sensory cortex, typically also show higher fMRI responses.

The differences in CBV in vivo are shown in Figure 11.6. The steady-state CBV image shows the change in relaxivity DR_2* induced upon injection of the intravascular iron-based contrast agent MION (monocrystalline iron oxide nanocolloid) (Dennie et al., 1998; Wu et al., 2004). The figure shows the higher blood volume in gray matter than in white matter and high blood volume at the surface of the cortex. There are also laminar differences in capillary density within the cortex (Lauwers et al., 2008; Weber et al., 2008) which are reflected in differences in perfusion. Figure 11.6 shows these differences in macaque V1 showing that the middle cortical layers have higher blood volume (Goense et al., 2007).

11.3.2.2 Regulation of cortical blood flow

Because of the high energy use of the brain, it needs a constant supply of nutrients and oxygen and it is sensitive to perfusion changes. To protect the brain from injury CBF is tightly regulated at multiple levels (Faraci and Heistad, 1998; Iadecola, 2004; Hamel, 2006). An important aspect of cerebrovascular regulation is its autoregulation whereby changes in perfusion induce changes in vascular resistance, and its function is to maintain CBF within the normal range despite changes in systemic blood pressure.

Cerebral blood flow is coupled to neural metabolism, meaning that changes in neural activity produce concomitant changes in blood flow. There are multiple mechanisms whereby changes in neural activity and metabolism lead to a change in CBF. Factors affecting perfusion are the partial pressure of CO_2 and oxygen, the pH, or tissue concentrations of metabolites (Faraci and Heistad, 1998). There are also multiple signaling molecules and pathways involved in the neurovascular coupling, for instance, NO, prostaglandins, etc. (Iadecola, 2004; Hamel, 2006).

The specificity of the BOLD response to localized neural activation depends on the spatial scale of the blood flow regulation. Changes in blood flow are mediated by dilation and constriction of arteries and arterioles, while venules and veins have no smooth muscle and are mostly kept open. The neurovascular response can be quite localized as shown in optical imaging and two-photon microscopy experiments. For instance, with optical imaging ocular dominance columns (ODCs) and blobs, which are of the order of a few $100 \, \mu m$, have been shown in macaque V1 (Ts'o et al., 1990). Chaigneau et al. (2003) showed that blood flow in the rat olfactory bulb in vivo is regulated at the level of individual glomeruli and with two-photon microscopy it has been shown that blood flow is regulated at the level of individual arterioles or even capillaries. Neurons and associated astrocytes form the so-called neurovascular units, and astrocytes are thought to play an important role in mediating the blood flow response (Schummers et al., 2008). Constriction and dilation of individual arterioles were shown to be mediated by astrocytes (Zonta et al., 2003; Mulligan and Macvicar, 2004; Metea and Newman, 2006; Takano et al., 2006). Dilation and constriction of capillaries by specialized structures called pericytes has been shown in the retina (Peppiatt and Attwell, 2004) and pericytes may have a similar function in the brain, although this is still unclear (Hirase et al., 2004).

11.3.2.3 Specificity of different fMRI methods

Based on the above, the expectation is that the scale of the hemodynamic regulation is currently not the limiting factor for most fMRI applications except possibly for ultra-high-resolution fMRI. The specificity of the fMRI signal further depends on the fMRI method that is used, hardware (for instance field strength) and acquisition parameters. Different fMRI methods have different specificity (Kennan et al., 1994; Weisskoff et al., 1994; Boxerman et al., 1995). We already mentioned that GE-BOLD, although sensitive to both vessels and capillaries, is dominated by large vessels at low field, and the contribution of the capillary, signal increases at higher field. The GE signal is sensitive to phase dispersal near large vessels, i.e. water-protons in voxels near large vessels exhibit a range of phases due to the susceptibility gradients, and phase dispersal in a voxel causes signal within a voxel to cancel out (called static dephasing). SE-BOLD, CBV and CBF methods are less sensitive to large vessels. SE is not sensitive to static dephasing because the accumulated phase dispersal is refocused by the 180° pulse. However, near capillaries a dynamic effect is dominant which gives rise to the SE-BOLD signal, that is, spins that move within the field gradients accrue phase dispersal and the change in phase due to this movement cannot be refocused. This is called dynamic averaging, characterized by the "apparent T_2" or T_2' (Ugurbil et al., 2000). In addition, there is a T_2-effect which arises from intravascular protons. At fields up to 3 T this is the dominant contributor to the SE-BOLD signal (Duong et al., 2003), but at higher field the T_2-effect becomes less important, because of the short T_2 of blood at high field.

Differences in specificity between the different methods can easily be seen in high-resolution fMRI. At spatial resolutions of less than 1 mm partial volume effects make it difficult to differentiate clearly between the signals originating from vessels or from within gray matter. Figure 11.5 shows that large vessels are located on the surface of the cortex and vessels within gray matter are small. In high-resolution fMRI it becomes obvious that the largest GE-BOLD functional changes occur at the surface and near large vessels (Figure 11.7). The largest SE-BOLD functional changes occur within gray matter, approximately in layer IV, and not much functional activation is seen near vessels, illustrating the improved specificity of SE-BOLD over GE-BOLD.

The specificity of the different fMRI methods can also be demonstrated by visualizing cortical columns, like ocular dominance columns in humans ($\sim$1 mm in diameter (Adams et al., 2007)) or orientation columns in cats ($\sim$1 mm in diameter (Lowel et al., 1987)). Columnar resolution was successfully demonstrated in humans and cats using GE-EPI (Menon and Goodyear, 1999; Dechent and Frahm, 2000; Cheng et al., 2001; Moon et al., 2007; Yacoub et al., 2008). When GE-BOLD is used, typically subtraction paradigms are needed that subtract out the signals common to both stimuli, and thus remove the non-specific vessel signals. Alternatively vessel signals are thresholded to ameliorate the predominance of vessel signal in GE-BOLD (Cheng et al., 2001; Logothetis et al., 2002; Moon et al., 2007). The drawback of subtraction paradigms is that orthogonal stimuli are needed which are often not available. With SE-fMRI ocular dominance columns have been shown in humans (Yacoub et al., 2007, 2008) and using CBV and CBF methods single condition maps of orientation columns were demonstrated in cats (Duong et al., 2001; Zhao et al., 2005). Methods that have high specificity also allow functional mapping of laminar differences within the cortex, as demonstrated in V1 of cats and monkeys (Zappe et al., 2008; Goense et al., 2007; Harel et al., 2006a; Zhao et al., 2006).

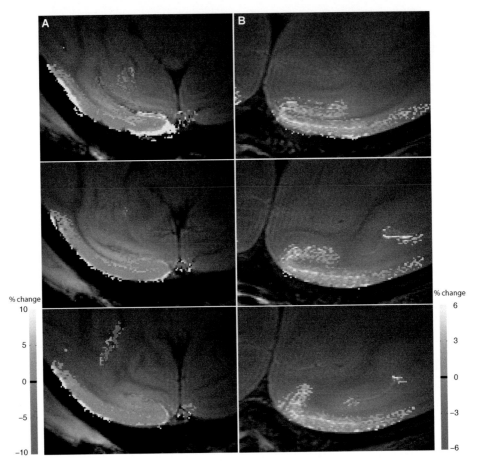

Figure 11.7 (See plate section for color version.) Specificity of GE- and SE-BOLD fMRI acquired at 4.7 T. The GE-BOLD map at an in-plane resolution of $333 \times 333\,\mu m^2$ (A) shows highest activity at the cortical surface and near vessels in the sulcus (Goense et al., 2007; Logothetis, 2008), whereas the SE-BOLD map in-plane resolution of $250 \times 180\,\mu m^2$ shows a BOLD signal that is better confined to the gray matter (B). Laminar structure can be seen in the SE-fMRI signal. Partial volume effects in the slice direction were not detrimental due to the anatomy of monkey V1 and by positioning the slices perpendicular to the cortical surface.

11.4 Analysis of fMRI signals

11.4.1 Overview

fMRI signal analysis is relatively simple: compared to electrophysiological signals, the BOLD signal has an extremely slow time-constant, making many complicated analysis techniques developed for fast time-varying signals unnecessary. Spike-sorting, spike timing analyses, frequency-separated power analyses, phase-power coupling, spectrograms, Hilbert transforms – all unnecessary for fMRI. Despite this comparable simplicity, it turns out that fMRI time-series analysis – done correctly – is extremely complex, even for relatively "simple" analyses. Since statisticians have paid very close attention to fMRI data-analysis, and since most fMRI laboratories use highly standardized analysis protocols and software, from a statistical perspective the standard fMRI analyses are carried out at the highest possible standards. Also, the statistical awareness of researchers is extremely high in the field of fMRI, not least because data collection is relatively simple and education can focus on paradigms and analysis. It is important to point this out to counter the occasional prejudices among people outside the field. It is therefore hardly the technique nor the state of analysis techniques that is to blame for some of the less convincing fMRI studies, but most often – as in most fields of science – the researchers themselves, their experimental designs, their interpretation of the data and, importantly, the grant agencies enforcing the policy of "publish or perish" in terms of numbers rather than quality. This obviously affects all fields of science.

There are several highly accessible books describing the principles and problems, theory and practice concerning the analysis of fMRI data (Friston et al., 2007). In particular we recommend the freely available chapters written by some of the foremost experts in fMRI data analysis (www.fil.ion.ucl.ac.uk/spm/doc/books/hbf2/). Here we provide a highly summarized overview of problems and common practices in fMRI data analysis, as employed by many of the standard analysis packages that are available for free, such as SPM, FSL, AFNI, Neurolens, or packages that are available commercially such as BrainVoyager or ANALYZE.

11.4.2 Properties of the data

In fMRI data, several problems have to be dealt with. One is the vast number of (nearly) independent measurements over space. With a typical resolution of $3 \times 3 \times 3 \, mm^3$ an imaging experiment yields about 50 000 voxels in a human brain, each with for example 600 time points collected at 2 s spacing.

There are three obvious properties of these data and each requires attention during statistical analysis.

First, the time courses contain auto-correlations (in signal and noise) due to the slow time-constant of the BOLD signal, due to slow scanner drifts and due to subject-induced drifts (getting tired, etc.). Therefore, a time point collected at t_2 is not independent of a time-point collected at time t_1, thus either requiring a reduction of the degrees of freedom in subsequent statistical tests or requiring a removal of the auto-correlation in the time series. Many packages, among the widespread and freely available matlab-toolbox "SPM," estimate and remove auto-correlations using a so-called AR(1) model, after applying a high-pass filter to the data (Woolrich et al., 2001; Worsley, 2005).

Second, there is also a spatial dependence of the signal: limited k-space sampling (Gibbs ringing), deviations from exact Cartesian sampling due to limitations in gradient rise times, and imperfect RF pulses for slice selection leave some overlap between voxels, but also the neural or vascular response (of signal or of noise) may spread over several voxels. Thus, not every voxel provides an independent measurement. Several packages estimate the spatial smoothness of the data in a spatially resolved way and smooth data additionally using Gaussian kernels to allow for a correct estimation of the actual number of spatially independent "resels" (Kiebel et al., 1999; Worsley, 2005).

Third, the vast number of statistical tests (e.g. performed on the time course of each voxel in a typical voxel-wise analysis) requires a correction for multiple comparisons. Several approaches can be distinguished: first, Bonferroni correction at the single-voxel level (the strictest way of dealing with the problem). This would theoretically leave not a single-false positive across the whole brain, but would also increase the number of misses. This type of correction is dubbed "family wise error correction" (FWE-correction) which, in context of the "theory of random Gaussian fields" takes into account the number of resels rather than the larger number of voxels, thus being less strict compared to a Bonferroni correction applied to N voxels (Friston et al., 1996; Kiebel et al., 1999). Another approach is to keep control of the expected fraction of false positives by setting the statistical threshold such that for example 5% of all voxels that pass the threshold are expected to be false positives (false discovery rate, FDR correction) (Genovese et al., 2002). One additional approach is to calculate the probability of finding a certain number of neighboring voxels activated at a given threshold (of course taking into account the spatial dependence of voxels in the data) – the mere extent of an activation cluster can thus be translated into a p-value in so-called cluster-level statistics (Poline et al., 1997).

Finally, many research groups circumvent the problem of multiple comparisons by defining regions of interest (ROIs) independently from the main experiment, and subsequently performing statistical tests on the averaged signal of these ROIs – thus only making it necessary to correct for the number of ROIs (and tests performed). An in-between way is to perform tests for each voxel in a ROI, and to correct for the number of resels present in it.

Lastly, it is noteworthy that the simultaneous measurement of spatially distributed activity can be exploited to examine consistent spatio temporal relations between the voxels imaged. One of the simplest, yet perhaps easiest interpretable way of doing this is termed "functional connectivity" – describing raw BOLD signal time course correlations of a "seed"-voxel or "seed"-region with all remaining voxels in the brain (Friston et al., 1993; Biswal et al., 1995; Bartels and

Zeki, 2005a). The strength of such correlations may also be compared between two task-sets, allowing one to examine whether some areas increase their communication during task A compared to task B (a so-called psychophysiological interaction) (Friston et al., 1997). In addition, one can use Granger causality to examine which brain areas lead and which lag with respect to the seed-area, thus using temporal delays of such correlations to infer directional information flow across the cortex (Goebel et al., 2003). Other, considerably more model-based and assumption-loaded methods can be used to assess causal dependencies between a user-defined and limited set of brain regions, such as "dynamic causal modeling" (Friston et al., 2003). Alternatively, there are also entirely data driven methods that use clustering algorithms such as independent component analysis to segregate functional areas or networks from each other in a model-free approach (McKeown et al., 1998; Bartels and Zeki, 2005b). We limit ourselves to pointing out the above references for the interested reader, but focus in the following on analyses of localized functional responses.

11.4.3 Preprocessing

The above steps are the last in a series of analysis steps typically performed. First, many packages allow the user to re-sample the fMRI data across time so as to compensate for the time delays between acquisitions of different brain slices – the first and last slice are separated by T_R, which can amount to several seconds. This is usually done in Fourier space of the time series to obtain whole-brain images that are acquired at a single time point. Secondly, a rigid-body motion correction can be applied. This estimates the rotation and planar motion of the head during the scanning sequence, and re-samples the data to provide a spatially stationary sequence of brain images (this step may not be necessary when subjects use bite-bars for head-stabilization). The resulting six motion parameters (a matrix of $6 \times n$ time steps) can later be included in the time series analysis to remove residual artifacts induced by head motion or its correction. Some packages, for example SPM, even calculate the expected distortions of the B_0 field induced by the head motion and the resulting distortions of the brain images, in order to "unwarp" them. After motion correction, the brain can be "morphed" into the shape of the standard brain (an average of several hundred brains), which allows for easier comparison of results with the literature in a standard coordinate system, and for statistics across multiple subjects. This step is called "normalization." This processing step is very complex, and employs several substeps such as tissue categorization (gray matter, white matter, CSF) and probability maps for each of them prior to calculation of the appropriate warping parameters. The Montreal Neurological Institute brain template is currently a frequently used standard (referred to as

"MNI-space"). Standard brain spaces (such as the MNI-space) are often mistakenly referred to as to "Talairach"-space – mistakenly because merely the method of aligning some key anatomical features was borrowed from the brain-atlas provided by the anatomist Talairach, but not the unfortunately rather untypical brain used in his atlas (Talairach and Tournoux, 1988). Nevertheless, some care should be taken when comparing results obtained from different analysis packages, as not all use the same standard brain template, which can lead to discrepancies in the range of several millimeters or even centimeters (in the case of the true Talairach brain).

11.4.4 General linear model (GLM) statistics and design efficiency

Perhaps the most widely used approach to analyzing fMRI data employs the concept of the general linear model (GLM), adapted for fMRI (Friston et al., 1995, 2007). Note that the adaptation to fMRI concerns primarily the above-mentioned signal auto-correlation and different methods to correct for the multiple comparison problems. The above-mentioned book and downloadable chapters provide excellent reviews of the GLM approach, so we provide merely an overview here. All ANOVAs and t-tests are special instances of the GLM, and so is the time series analysis: it all comes down to the equation

$$y = X b + e.$$

The BOLD signal time series y can be explained by the weighted sum of predictors (contained in the columns of the matrix X – each predictor being a time series) plus a residual error e (also a time series). The weights b express the extent to which each predictor time–series in X contributes to y. The residual error e is the unexplained remainder – it should be random Gaussian noise without auto-correlations. The aim of the statistics engine in analysis packages is to come up with robust estimates of b and associated confidence intervals (which can be quite complex), while the user provides X (Figure 11.8).

   In a so-called block-design experiment, different stimulus classes alternate at a slow pace, for example a checkerboard stimulus is alternated with a blank screen every 30 s, so a reasonable X would contain a time series alternating for example between zeros and ones, plus a time series containing only ones to account for a mean-signal offset. To improve on this, one can convolve the square-wave predictor with the standard hemodynamic response function (hrf) (see Figure 11.8). The hrf is the canonical BOLD response to a very brief (e.g. 0.5 s) stimulus. Convolving the expected neural response (our 30 s on-off square-wave) with the hrf yields a good estimate of the expected BOLD response. In so-called event-related paradigms, stimuli are presented only briefly, allowing more different stimulus categories to be shown within a shorter time span, and perhaps more importantly event-related paradigms also remove confounds such as raised expectations or adaptation to a

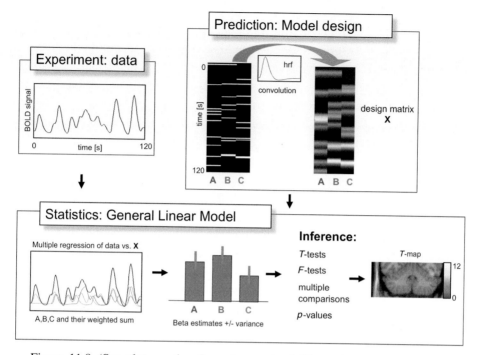

Figure 11.8 (See plate section for color version.) Illustration of the statistical parametric mapping approach. The aim is to determine whether and to what extent the BOLD signal time series of a given voxel ("data") can be explained by experimental conditions, here denoted as A, B and C. Here, each condition was presented ten times for 1 s at random intervals within a 120 s period, as shown in the left panel of the model design. The stimulus time series are convolved by the hrf to obtain the expected effect of each condition on the BOLD signal of the voxels. A subsequent GLM analysis estimates the weights by which each predictor (regressor) contributes to the data. In a last step (inference), the contributions of the different predictors (betas) are compared against zero or against each other. In a whole-brain analysis, the resulting statistical T-values are projected in a color code onto each voxel, providing a statistical spatial map.

condition that may occur during presentation of long blocks. What concerns the analysis, there is a continuum between block-designs and event-related designs. Since the experimenter cannot be certain that every brain area will have the response characteristics of the canonical hrf (or that every stimulus-type will be processed with the same delay), one can include additional predictors – for example the derivative of the hrf based predictor. This is orthogonal to the hrf predictor, and inclusion of this additional regressor will therefore not alter the beta-estimate of the hrf predictor, but simply remove variance from the residual, which in turn will improve the statistics (Hopfinger et al., 2000). If one is entirely uncertain with regards to the response to a given stimulus or task, there are powerful alternatives to the above approach. Instead of including a single regressor, one can include a whole family of cosines of different frequencies (similar to a Fourier basis set), whose onset is locked to the stimulus onsets. In order to test in which voxels a stimulus condition had any significant effect on the BOLD response, this time a simple t-test will not work, as we have multiple regressors modeling the stimulus, with unknown and likely varying weights. Here an F-test is performed, that simply tests whether the whole set of regressors explains a significant amount of variance in the data. The most extreme form of this approach employs a finite impulse response (FIR) model. Here, a separate regressor consisting of zeros except for the T_R vallies of the stimulus onsets of a given condition, and additional time-shifted regressors up to for example 16 s after stimulus offset constitute the model, allowing modeling of any waveform consistently occurring following the stimulus onsets of a given condition. However, the use of such models usually requires the use of F-tests to test for an overall significance of *any* combination of the regressors used, which can make comparisons between distinct conditions that evoke distinct time courses difficult. For this reason many studies use more simple models to account for their data.

One important aspect is the timings and order of the stimulus types presented: if two stimuli A and B alternate too fast, the BOLD signal cannot distinguish between regions responding to A and to B. If the spacing is too far, the responses can be distinguished, but weak statistics result since most of the scanning time is spent on inter-stimulus intervals. Thus, there is an optimal stimulus spacing, leading to a so-called efficient (stimulus) design. Note though that this optimum depends on what is to be optimized. One can distinguish between optimizing contrast detection power and hrf estimation efficiency. The former optimizes duration and intervals of stimulus types such that, given the BOLD signal properties, one or several contrasts (such as the comparison between conditions A and B) yield the best possible statistical power. The latter optimizes the design such that the hrf for one or several conditions can be estimated optimally. Block-designs tend to be optimal for the former, event-related ones for the latter. But apart from statistical considerations, psychological reasons often dominate the choice of the stimulus timing parameters, as mentioned above. Another important point to consider is history effects, in that good designs should ensure that a given condition is equally often preceded by all conditions, to ensure that long-lasting hemodynamic effects or neural adaptation effects of preceding conditions will not affect the results of subsequent conditions. There are several excellent papers explaining the theory and practical considerations of design efficiency in an accessible way (Buracas and Boynton, 2002; Wager and Nichols, 2003; Liu, 2004; Liu and Frank, 2004) and see also the above-mentioned chapters and http://imaging.mrc-cbu.cam.ac.uk/imaging/DesignEfficiency.

However, it should be noted that experimental designs ignoring any of the above considerations can also yield highly significant results. Figure 11.9 shows a human

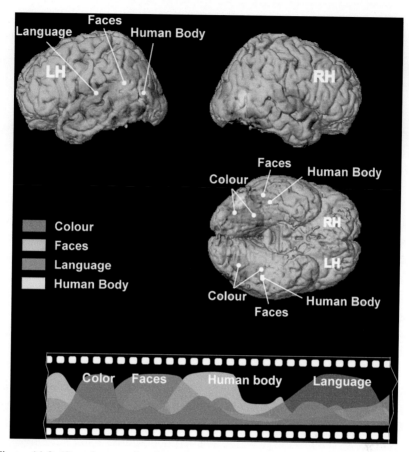

Figure 11.9 (See plate section for color version.) Activation maps obtained using a GLM analysis of volunteers' brains watching a James Bond movie. Voxels that achieved significant beta-estimates for one of the four feature regressors were color coded to identify the feature and superimposed on a surface rendering of the anatomical image of a single brain. Statistics were calculated on eight volunteers, and activity is thresholded at $p < 0.001$ using FWE correction. (Modified with permission from Bartels and Zeki, 2003.)

brain with superimposed activation related to faces, human bodies, language and color. However, instead of using controlled stimuli and experimental designs to activate the regions specialized for these features, the observers watched a James Bond movie, and the GLM design matrix was based on the occurrence of these features in the movie clip (Bartels and Zeki, 2003). While this approach worked very well, also for more fine distinctions, for example between local object motion in the movie and simulated observer motion affecting the full screen (Bartels et al., 2008a), this approach is obviously non-standard. While it provides results of high general validity (since voxels only achieve significance if their response is consistently correlated with a given feature across the huge variability of scenes in the movie) it does not allow comparisons of the sort highly controlled experiments allow for.

11.4.5 fMRI adaptation experiments

In comparison to invasive electrophysiology, a major limitation of fMRI is its spatial resolution – at least in standard human imaging setups using 3 T scanners with whole-head coils. Several methods have been developed to get around this problem at least partially.

One tool used in fMRI to circumvent the problem of resolution has relied on neural adaptation (Grill-Spector et al., 2006; Krekelberg et al., 2006). Note that in psychophysics, adaptation is highly valuable as it provides unambiguous indication of a neural interaction between distinct stimulus conditions. Electrophysiology studies adaptation in its own right. fMRI in contrast has attempted to *use* adaptation (fMRI-A) as a means to derive and map functional properties of neuronal subpopulations that are mixed within the resolution of a voxel.

The idea is to expose an observer to one stimulus (adaptation phase, e.g. to leftward motion), and then show either another stimulus (test phase, e.g. rightward motion) or the same stimulus again. The idea is that if the two different stimuli excite distinct sets of neurons, only one population of neurons is adapted by the first stimulus, and the other (unadapted) population should give a strong response in the test phase when a different stimulus is used because (1) it is not adapted and (2) it receives less inhibition from the adapted population. Thus, a stronger BOLD response after adaptation was thought to be a good indicator of two separate sets of functionally distinct neurons. An important confound in this thinking is the fact that attention can modulate the strength of responses in virtually any site of the cortex, and that adaptation experiments compare responses evoked by repetitions of the same stimulus and compare these responses to those evoked by non-repeated stimulus features. Thus, it cannot be excluded that part of the effect stems from repeated stimuli being perceived as more boring by the subject – on the other hand, this boredom may be a consequence of adaptation and reduced neural response. A carefully conducted study very nicely demonstrated that, when the attentional load was equated between repeat and non-repeat conditions (using a threshold-detection task), a previously observed adaptation result disappeared (Huk et al., 2001). This is indicative of two issues: firstly, that attentional modulation appeared to account for the whole effect previously observed, and secondly that fMRI was very sensitive to attention.

Another problem concerns the localization of the actual site of adaptation in fMRI-A. While psychophysical experiments provide clear evidence for adaptation somewhere in the brain, it is unclear whether fMRI is the best method to local-ize the site of adaptation reliably. One example stems from experiments on the processing of directional visual motion stimuli.

Elegant electrophysiological experiments demonstrated that the site of direction-selective adaptation during prolonged unidirectional stimulation cannot reside in the motion-responsive area V5, since adaptation effects are spatially specific within the receptive field (RF) of a given V5 neuron – in other words, if a motion adap-tation stimulus was presented to only part of a V5 RF, only this part of the RF would reveal adaptation. This suggests that the actual adaptation does not occur in the V5 neuron, but in an input stage with smaller RFs, most likely to be V1 (Kohn and Movshon, 2003). Despite this, most fMRI motion adaptation studies localize adaptation to V5 – a downstream site with respect to the actual site of adaptation.

Even though direction-selective neurons in both V1 (i.e. this site of adaptation) and V5 (downstream) reduce their firing rates following adaptation, the underlying reasons for this are entirely distinct and with important consequences for BOLD signals: synaptic input to V1 from subcortical sites is largely non-adapting, and V1 neurons reduce their firing rate primarily due to mechanisms residing within the neurons (such as opening of slow hyperpolarizing channels) (Sanchez-Vives et al., 2000). The converse is true for V5. Following adaptation it receives a strongly reduced synaptic input due to spike rate adaptation of its primary input area V1.

Now consider that BOLD signal may primarily reflect (peri-)synaptic activity instead of principal neuron spiking activity (see sections below). fMRI adaptation results may not only report downstream effects. Instead, fMRI may de-emphasize (if not fail to detect) the site of the actual origin of adaptation, and emphasize sites receiving input from adapted sites one or more levels downstream.

This goes beyond "inheritance" of adaptation effects when the receiving region has no or little neuronal specificity for the adapted property. In such situations, adaptation effects would still be observed in an area whose neurons were totally unselective to the direction of motion (see Figure 11.10). A recent study mea-sured spike rate adaptation to object images in the inferotemporal cortex of monkeys, intentionally using paradigms similar to those used in fMRI-A (Sawa-mura et al., 2006). Neurons that responded similarly well to images A and B nevertheless showed greater adaptation to same-image repeats (A-A, or B-B) than to different-image repeats (A-B). This demonstrates that adaptation effects are not equivalent to direct measures of neuronal specificity. If the interpretation that is commonly used in fMRI adaptation was applied to this result, one can infer three populations of neurons – one responding to both A and B, and one image-A and one image-B selective population – even though these measurements stem from a single population. The observed effect may originate from adaptation in earlier areas or in neurons providing input to the ones measured.

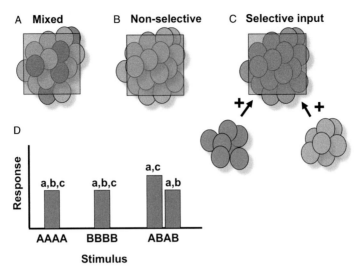

Figure 11.10 (See plate section for color version.) Ambiguities in interpreting fMRI adaptation results. Panels A–C illustrate neural functional organizations within a voxel (in blue). Direct recordings of neurons (not relying on adaptation) would differentiate stimulus-selective (A) from non-selective (B, C) neurons. The adaptation effects that fMRI-A relies on can however lead to ambiguous net effects (D). A mixed neural population (A) would adapt its spike rates more to same-stimulus repeats (AAAA or BBBB) than to stimulus alternations (ABAB). Neurovascular coupling may render the BOLD effect of this differential spike rate adaptation more apparent in the downstream site (e.g. V5/MT) whose input is affected rather than in the site where it originates (e.g. V1, Kohn and Movshon, 2003), especially when the mechanism of adaptation involves primarily intraneural spike rate adaptation of the output neurons (with potentially little BOLD signal effect) rather than network adaptation (substantial BOLD effect). Hence the ambiguous outcome of (A) illustrated in (D). Non-selective sites should not reveal differential adaptation (B) – unless they get input from selective sites (C). In this case, adaptation can render neurons or sites that are not selective to a feature to appear selective following adaptation (Tolias et al., 2005a, 2005b). (Modified from Bartels et al., 2008b.)

All in all, the presence or absence of adaptation in an area measured using fMRI therefore does not allow for the conclusive inference of either presence or absence of the neural property in question.

fMRI-A studies thus often interpret their results using prior knowledge of the underlying single-cell physiology, rather than providing new, unambiguous knowledge of underlying physiological processes. Adaptation is interesting to study in its own right, and its downstream effects may provide crucial cues as to "who talks to whom." But its use to infer and map neural population properties using fMRI appears questionable (for reviews see also Grill-Spector et al., 2006; Krekelberg et al., 2006).

11.4.6 Classifiers and high-resolution imaging

Apart from fMRI adaptation, a second tool has gained enormous popularity in the fMRI community more recently. It relies on multivariate rather than univariate statistics to extract weak signals from whole populations of voxels rather than from single voxels.

The idea is that even relatively large voxels may receive small response biases towards one feature (e.g. towards leftwards or rightwards motion) due to the underlying neural organization, such as functionally specific columns (Figure 11.11A). Since signal biases may be small and insignificant for single voxels, the approach relies on multivariate statistics (often implemented in the form of classifiers), that can detect and combine consistent trends of many data points (voxels) (Figure 11.11B) (Cox and Savoy, 2003; Haynes and Rees, 2006; Kamitani and Tong, 2006).

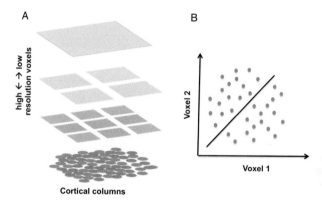

Figure 11.11 (See plate section for color version.) Effects of cortical functional organization on pattern classification and high-resolution imaging. A. Schematic diagram illustrating how a region (left) packet with feature-specific functional columns (circles, e.g. green responding to leftwards and red to rightwards motion) may lead to less feature bias in voxel responses (squares) as the voxel size increases (from bottom to top). Top level: the areal net response is not biased to red/green (e.g. left/right motion). Second level: medium high resolution may lead to voxel-wise yet subthreshold biases, allowing multivariate statistics to inform about the presence of feature-specific information in this region. Third level: high-resolution imaging may reveal the actual functional organization of the region using voxel-wise, standard univariate statistics. B. Multivariate statistics can combine weak signal biases from many voxels. The example shows the signal strength of one voxel plotted against that of one other voxel, each dot representing the response to one trial, of two stimulus classes. While the mean response to the red stimulus class may not be statistically different from that to the green stimulus class in any of the two voxels, their combined responses allow classification of the two stimulus classes.

Since classifiers rely on feature selectivity *and* a spatial inhomogeneity of feature selective responses, only their *conjunction* can lead to biased responses in segregated voxels. Therefore, the degree of successful response classification in an area can by no means be interpreted as a degree of feature specificity in it. Despite this, a successful classification result clearly indicates the presence of information about the respective features in the studied region. However, the *interpretation* of such a result requires extreme caution. One reason for this is that classifier analyses are much more sensitive to very small signal differences than the community is used to from single-voxel analyses. For example, it is conceivable that a classifier may distinguish between house or face stimuli based on BOLD signal responses in V1, even though V1 is not known to contain object selective neurons. Post hoc, one could attribute such a result for example to distinct spatial frequency properties of the two stimulus classes. Without a priori knowledge though, one may be led to interpret V1 as a "face-selective" region.

Therefore, although the presence of feature selective neurons can potentially produce a successful classification result, the argument does not necessarily work the other way round.

It is worth noting that, like fMRI-A, the use of multivariate classifiers is merely a work-around to the low spatial resolution of fMRI. In principle, extremely high-resolution imaging could replace these approaches if the voxel size can be brought down to match that of functional columns. To achieve this, the BOLD signal must be spatially sufficiently specific, and the imaging method must be able to resolve this (Logothetis, 2008). High field experiments suggest that imaging with resolutions in the hundreds of micrometer range should be possible. Indeed, such experiments have reliably mapped columns in primary visual cortex with ocular and even orientation selectivity (Cheng et al., 2001; Moon et al., 2007; Yacoub et al., 2007). Interestingly, such experiments also begin to unravel novel functional patterns, such as of domains with differential preferences for temporal frequencies (Sun et al., 2007).

11.5 Neural basis of BOLD signals

To be able to answer the question how BOLD relates to the neural signals and what aspects of the neural signals are best represented by the fMRI signals, we need to take a closer look at what we actually mean when we talk about "the neural signals." While the previous sections have addressed electrophysiological recordings and the measured signals in detail, we nevertheless briefly review the key points again here in order to explain key relations between the BOLD signal on the one hand and electrophysiological signals on the other.

Usually the assumption is that "neural signals" represent signals measured by intracortical electrodes with standard extracellular methods. And we do wish to compare the BOLD signal to invasive electrophysiological recordings because a large amount of our knowledge about neural-Function and brain function is based on these extracellular techniques. Typically single-unit or multiple unit activity is recorded, and we look at the specificity and spatial extent of these signals, for example, what types of neurons are recorded and how large is the area that is sampled. However, what is recorded by an electrode is not always fully representative of all the processes that occur in the brain because electrodes measure only a subset of the neural processes.

The signal measured by an electrode placed at a neural site is the mean extra-cellular field potential (mEFP) from the weighted sum of the electrical sinks and sources along multiple cells. Its waveform is characterized by fast action potentials superimposed on relatively slowly varying field potentials. Different signals are recorded depending on the impedance of the electrode. If a microelectrode with a small tip is placed close to the soma or axon of a neuron, then the measured mEFP directly reports the spiking of that neuron and frequently also that of its immediate neighbors. With electrodes of the order of several hundreds of kilo-ohms, single spikes or single-unit activity (SUA), multi-unit activity (MUA) and local field potentials (LFP) are recorded. Low impedance electrodes record pre-dominantly LFP, while sharp electrodes are typically used to record single units, and not much LFP is observed.

11.5.1 Single-unit and multi-unit activity

The standard technique in neurophysiological research has been the recording of single spikes. Single spike monitoring has the best possible spatial and temporal resolution and it has been and it will continue to be the method of choice when single cell properties are the subject of investigation. It provides information on the spike output of the isolated cell and its response properties, for instance its receptive field or its tuning to different stimuli. Depending on the location and impedance of the electrode, multiple neurons are often recorded simultaneously. If MUA is recorded, the spikes generated by different neurons can be sorted based on their shape. The spike shape that is recorded varies depending on the location of the electrode with respect to the neuron. However, for accurate sorting tetrodes (four-contact electrodes) or multi-contact electrodes are advantageous or often necessary, particularly when the spike shapes of the neurons are similar.

A drawback of single-unit and multi-unit recording is that they are biased towards certain cell types and sizes (Towe and Harding, 1970; Stone, 1973). Figure 11.12 shows a drawing of the cortical circuit and illustrates the variation in size and morphology of neurons within the cortical sheet. The measured spikes, however, mostly represent a very small population of large neurons, which are by and large the pyramidal cells in the cerebral cortex and Purkinje neurons in

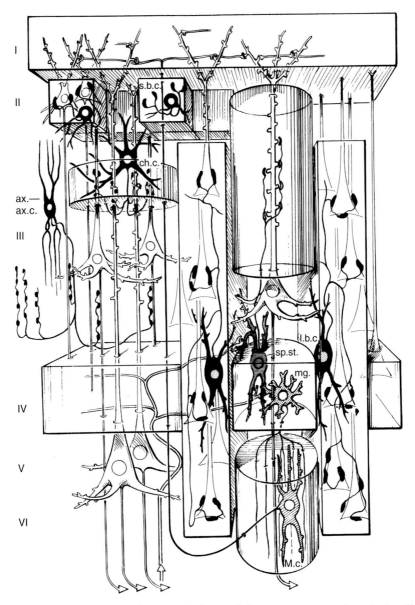

Figure 11.12 Schematic diagram of the modular arrangement of cortical cells and connections (from Szentagothai, 1978). The cortex has a columnar and well ordered parallel structure. The drawing shows the different types of neurons and connections. pyramidal cells, spiny stellate (sp.st.), small and large basket cells (s.b.c. and l.b.c.), microgliform cells (mg.), chandelier cells (ch.c.), and Martinotti cells (m.c.).

the cerebellar cortex. The magnitude of EFPs in the MUA range, for example, was shown to be a function of cell size and axon size (Gur et al., 1999). Combined physiology-histology experiments also demonstrated that the magnitude of MUA is site size specific (Buchwald and Grover, 1970) and cell size specific (Nelson, 1966), varying considerably from one brain region to another (e.g. neocortex versus hippocampus) but remaining relatively constant within a particular region. Homogeneous populations of large cells were found to occur systematically at sites of large-amplitude fast activity and vice versa (Grover and Buchwald, 1970). Similarly, the magnitude of axonal spikes is directly correlated with the size of the transmitting axon (Gasser and Grundfest, 1939; Hunt, 1951).

Recording from non-pyramidal cell types such as interneurons is often difficult because of their small size and because their response is often uncorrelated to the stimulus or to the behavioral state of the animal. Since response to a stimulus is often the criterion for successful isolation of a neuron or cluster of neurons, this can lead to a bias against certain neurons, for instance inhibitory neurons, against neuromodulatory neurons, or against neurons that have very low firing rates. This could introduce a substantial bias because there is reason to believe that neurons with very low firing rates may actually be very common in the cortex (Henze et al., 2000; Shoham et al., 2006).

11.5.2 Local field potentials

The obvious drawback of single-unit recording is that it provides information mainly on the output of the recorded neuron with no access to its subthreshold integrative processes or the associational operations taking place. To this end, the LFP is measured, to which these processes do contribute. LFPs are recorded when the impedance of the microelectrode is sufficiently low and its exposed tip is farther from the spike generating sources, so that action potentials do not dominate the signal. The electrode then monitors the totality of the potentials. LFPs are related both to integrative processes (dendritic events) and to spikes generated by several hundreds of neurons (Lorente de Nó, 1947) and they represent mostly slow events reflecting cooperative activity in neural populations. They rather reflect the input of a given cortical area as well as its local intracortical processing including the activity of excitatory and inhibitory interneurons. Based on current source density (CSD) analysis and combined field potential and intracellular recordings, Mitzdorf (1985, 1987) suggested that LFPs reflect a weighted average of synchronized dendrosomatic components of the synaptic signals of a neural population near the electrode tip. Studies of inhibitory networks in the hippocampus (Buzsaki and Chrobak, 1995; Kandel and Buzsaki, 1997; Kocsis et al., 1999) have shown that other types of slow activity, including voltage-dependent membrane oscillations (Kamondi et al., 1998) and spike afterpotentials, also contribute to the LFP (Buzsaki et al., 1988).

Another finding in studies combining EEG and intracortical recording was that, unlike MUA, the magnitude of the slow fluctuations was not correlated with cell size, but instead reflected the extent and geometry of dendrites at the recording site (Fromm and Bond, 1964; Buchwald et al., 1965; Fromm and Bond, 1967). Cells in a so-called *open field* geometrical arrangement, in which dendrites extend in one direction and somata in another produce strong dendrite-to-soma dipoles when they are activated by synchronous synaptic input. The pyramidal cells with their apical dendrites running parallel to each other and perpendicular to the pial surface (Figure 11.12) form an ideal open field arrangement, and contribute maximally to both the macroscopically measured EEG and the LFP. However, the dependency on geometry also implies that neurons that are oriented horizontally or have spherical symmetric dendritic fields (closed field arrangement) contribute less efficiently or not at all to the sum of potentials. Because of this large contributions arise from pyramidal/Purkinje neurons with interneurons contributing less.

LFP and MUA signals can be separated by filtering; a high-pass filter with cutoff of approximately 500 Hz was used to obtain MUA and a low-pass filter cutoff of ~300 Hz to obtain the LFP. The modulations in the LFP are classified in a number of specific frequency bands initially introduced in the EEG literature. EEG is subdivided into frequency bands known as *delta* (DC–4 Hz), *theta* (4–8 Hz), *alpha* (8–12 Hz), *beta* (12–24 Hz) and *gamma* (24–40/80 Hz) (Lindsley and Wicke, 1974; Steriade and Hobson, 1976; Basar, 1980; Steriade, 1991). The classification is based on the strong correlation of each band with a distinct behavioral state. An alternative band separation is based on information theory where the information carried by the different bands in the LFP or MUA range was calculated in recordings from monkeys that were viewing movies with natural images (Belitski et al., 2008). The most informative LFP frequency ranges were 1–8 Hz and 60–100 Hz. Positive signal correlations were found between LFPs (60–100 Hz) and spikes, and between frequencies within the 60–100 Hz LFP range, suggesting that the 60–100 Hz LFP range and spikes are possibly generated within the same network. LFP in the range of 20–60 Hz carried very little information about the stimulus, although they shared strong trial-to-trial correlations, indicating that they might be influenced by a common source such as diffuse neuromodulatory input.

11.5.3 Spatial extent and propagation of neural signals

The volume from which electrical signals are measured by a recording electrode depends on the properties of the electrode. The activity from each point within the volume is weighted depending on the distance from the tip of the electrode.

Single-unit activity or separable spikes are typically recorded from near the electrode, for instance, Henze et al. (2000) and Gray et al. (1995) measured spikes within 50–150 μm from a tetrode. Electrodes with exposed tips of approximately 100 μm were estimated to record from a sphere with a radius of 50–350 μm (Grover and Buchwald, 1970; Nicholson and Llinas, 1971; Legatt et al., 1980). LFP signals arise from a larger volume and the spatial extent of LFP summation can be calculated by computing the coherence of LFP as a function of interelectrode distance in experiments using simultaneous multi-electrode recordings (Juergens et al., 1996). The volume from which LFPs arise is estimated to range from 0.25–3 mm distance from the electrode tip (Juergens et al., 1999; Mitzdorf, 1987; Katzner et al., 2009; Nauhaus et al., 2009).

The electrical properties of the conductive medium through which the current travels will obviously affect the voltages that are recorded. Within the physiological frequency range a current source can be described as a simple static point source. The description of the volume conductor is further simplified by the fact that the propagation of signals is independent of their frequency. Intracortical measurements showed that the tissue impedance is actually frequency independent and isotropic, allowing the description of cortex as an Ohmic resistor (Logothetis et al., 2007). Hence, this implies that the distance over which the signals can be measured is not dependent on the frequency of the signals but is determined by the relative size of the electrical sources.

The spatial extent of the different neural signals can be estimated by coherence analysis. The coherence between two electrodes is measured as a function of the distance between the electrodes for the different neural signals (Figure 11.13). This showed that the half-maximum of the coherence-to-distance functions was 2.9 mm for the 2–8 Hz LFP range, 2.4 mm for the 8–15 Hz range, 1.9 mm for the 20–60 Hz range and 1.5 mm for MUA (Goense and Logothetis, 2008). Comparable distance ranges were found for the LFP in V1/V2 (Juergens et al., 1999) and for the similarity of object preferences encoded by MUA and LFP in macaque inferotemporal cortex (Kreiman et al., 2006). These findings suggest that the range over which the neural signals are measured in V1 are similar, and in the range of the typical voxel sizes in fMRI, and that this holds for the LFP band as well as for the MUA.

11.5.4 Combined measurements of fMRI and electrophysiology

To investigate the neural origins of the fMRI response, simultaneous fMRI and electrophysiology was performed in anesthetized (Logothetis et al., 2001; Shmuel et al., 2006; Rauch et al., 2008) and awake monkeys (Goense and Logothetis, 2008). Measurement of BOLD and neural signals is ideally done simultaneously; an important reason for this is that non-simultaneous

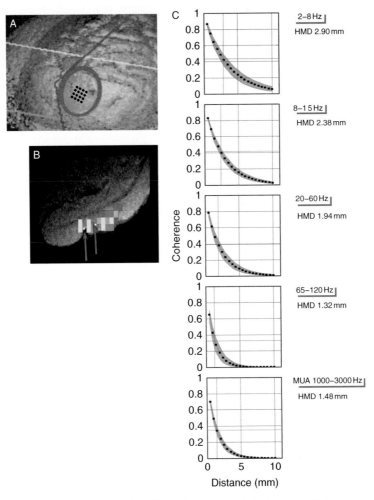

Figure 11.13 (See plate section for color version.) Coherence between signals recorded at different electrodes as a function of electrode distance for the different neural signals. A. Diagram of the 4 × 4 electrode array used for measuring the neural signals. B. Position of two electrodes (arrows) in combined MRI and physiology experiments. C. Average coherence-over-distance for the different neural signals, which shows the decrease in coherence with interelectrode distance. The gray shading shows 1–99% confidence intervals. The distance at which the coherence is halved is indicated by the red lines and shown next to the panels. The loss of coherence with distance was 1–3 mm and was comparable for all neural frequency bands. (Adapted from Goense and Logothetis, 2008.)

fMRI/electrophysiology increases experimental variability, making subtle or not so subtle effects harder to discriminate. For instance, the response of an animal to anesthesia can be different from one animal to the next or from one experiment to the next. In awake subjects also, the magnitude of the BOLD response is different for different individuals and different sessions, and variability can also arise due to behavioral factors like attention, arousal etc. Given that the analysis is based on finding trends or correlations between phenomena that often yield weak signals, like the BOLD signal, or are highly correlated, like the different neural signals, the added variance of non-simultaneous measurement can further complicate interpretation of the data. That said, to record BOLD and intracortical electrophysiological signals simultaneously some formidable hurdles need to be overcome. These are the interference of the scanner with the electrophysiological recordings and measuring equipment, and the possible degrading effect of the recording hardware on MR image quality. The scanner introduces electrical and mechanical interference with the electrophysiological signal due to the switching of the gradients and strong RF signals for excitation that interfere with the low-voltage neurophysiological recording. The recording hardware can cause artifacts in the images wherever metal is used, and conducting leads into the magnet bore can introduce noise. This is a problem because the MR signal is a low-voltage signal and hence image quality is sensitive to RF interference. In addition, all equipment that enters the bore needs to be non-magnetic, and it should function in strong magnetic fields. The latter is a problem for equipment that enters the magnet bore, and for equipment that stays outside if the magnet is not shielded. For details on the setup and experimental procedures we refer to the relevant publications (Logothetis et al., 1999; Logothetis et al., 2001; Oeltermann et al., 2007; Goense and Logothetis, 2008).

11.5.5 Neural basis of the BOLD response

In the well-known 2001 study by Logothetis et al. (2001) it was shown that the LFP is generally a better predictor of the BOLD response than the MUA. Figure 11.14 shows an example of a comparison of the time course of the BOLD signal and the neural signals in the MUA and LFP bands in an awake monkey.

The figure shows time courses of seven band-limited power signals extracted from the comprehensive neurophysiological signal following removal of gradient interference, band separation and rectification. The first three bands are known from the EEG literature, while the other bands were defined based on work by Belitski et al. (2008). There the relationship between visual information carried by different frequency bands of LFP and spikes was investigated in recordings where the monkey was viewing color movie clips. This ensured that the stimulation was diverse and stimulated all visual cortices not affected by anesthesia. Note that in Figure 11.14 the stimuli were simple geometrical shapes that optimally drive V1, and that stimuli could only be shown for short periods in sequential

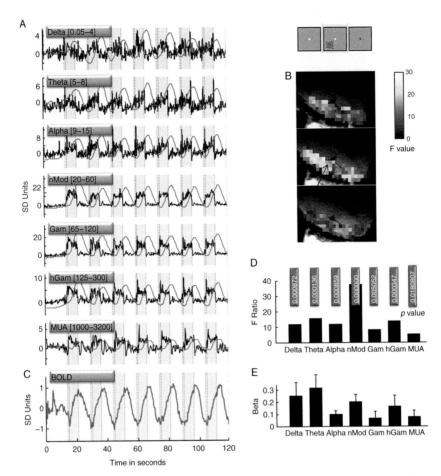

Figure 11.14 (See plate section for color version.) Dependencies between BOLD and neural signals in V1 of an awake monkey. A. Band-separated neural signals (black) in response to a 6° rotating checkerboard (inset). Green shading indicates the times the stimulus was presented. The different band-limited power signals were convolved with a theoretical HRF and used as regressors (red). The response to the stimulus was especially pronounced in the neuromodulatory and gamma bands. B. Functional activation maps superimposed on GE anatomical images. The location of the electrode is indicated by the arrow. C. The BOLD time course acquired at a temporal resolution of 250 ms shows clear modulation to the stimulus. The output of the GLM analysis and F-test yielded significant p-values for all neural bands (D) indicating that all bands contributed to the BOLD response. E. Beta values lacked dramatic differences across bands.

trials due to the limitations imposed by the behavioral task. It is therefore not surprising that the time courses of the signals in all bands are relatively similar to each other due to the presence of onset responses. Nonetheless, GLM analysis revealed a differential contribution of the different neural signals to the BOLD response. Whether a frequency band has a unique contribution to the BOLD signal was assessed by calculating the F-ratio, which showed that all frequency bands contributed significantly to the BOLD signal (Goense and Logothetis, 2008). The beta values were comparable across frequency bands, suggesting that under these stimulus conditions no single band in particular determined the BOLD response.

Studies in awake and anesthetized monkeys have shown higher correlation coefficients between LFP and the BOLD signal than between MUA and the BOLD signal (Logothetis et al., 2001; Goense and Logothetis, 2008). This implies that the overall synaptic activity or the input of an area is a stronger generator of the BOLD signal than its output. But although the correlation of the LFP to the BOLD signal is consistently higher than the correlation of the MUA to the BOLD signal, MUA is also positively correlated. High correlations of spiking with the BOLD response were also found in humans (Mukamel et al., 2005). This is not surprising given that in most cases the MUA is correlated to the LFP, and in most cases a positive correlation exists between the input and the output of a neural system. Thus, based on differences in correlation coefficients we cannot unambiguously determine which signal best drives the BOLD response. Because of the correlation between MUA and the BOLD signal, one needs to find circuits or stimulus conditions where there is dissociation to obtain more conclusive evidence, in other words, conditions where the LFP is not or only weakly correlated with the MUA.

An example of such a case in the awake monkey is shown in Figure 11.15. Dissociation of the LFP and MUA can occur when there is strong adaptation and the LFP stays elevated long after the MUA has returned to baseline (Logothetis et al., 2001). In these cases the BOLD response also stayed elevated and this shows a better correlation between BOLD and LFP than between BOLD and MUA. Similar results, better coupling of the hemodynamic response to the LFP in cases of dissociation, were also observed by other groups, for instance in V1 of cats using optical imaging (Niessing et al., 2005). Another way to induce dissociation was demonstrated by Viswanathan and Freeman (2007) who used high temporal frequency stimuli that did not elicit spikes in V1 and they observed that LFP activity elicits changes in tissue oxygen in the absence of spiking.

Dissociation of LFP and MUA can also be induced by intracortical injection of serotonin or a 5-HT1A serotonin receptor agonist which abolished spiking of the output neurons that are typically recorded in the MUA (Logothetis, 2003; Rauch

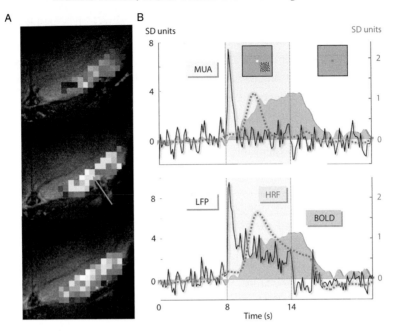

Figure 11.15 (See plate section for color version.) Dissociation between the MUA and BOLD response. A. Functional activation maps in response to a 6° peripheral rotating checkerboard stimulus (inset in B). The arrow indicates the location of the electrode. B. The average MUA, LFP in the 20–60 Hz range and BOLD time courses show that while the neuromodulatory component of the LFP stayed elevated for the duration of the stimulus, the MUA rapidly returned to baseline after a transient onset response. The prolonged time course of the BOLD response suggests a more sustained driving mechanism of the BOLD response as opposed to the transient MUA signal. The dotted line shows the regressor, i.e. the neural signal convolved with the theoretical hrf, which indicates that the MUA-derived regressor cannot capture the sustained part of the BOLD response. (Adapted from Goense and Logothetis, 2008.)

et al., 2008). The LFP on the other hand did not change substantially and similarly the BOLD response was not changed. Such cases indicate that the signals recorded in the LFP and representing the input in an area, are more likely driving the BOLD signal than the MUA signal that mostly represents the output of large pyramidal neurons.

Other cases of dissociation of neural activity and metabolism or blood flow have been observed in structures where the anatomical and functional properties of the neural circuit allow a clear segregation between input and output. For instance, Nudo and Masterton (1986) observed that inhibitory synaptic activity increased 2-DG labeling in the lateral superior olive of the cat although postsynaptic spiking was suppressed. In the cerebellum, Mathiesen et al. (1998) observed that when the parallel fibre system was stimulated, this inhibits Purkinje cell firing but the

CBF response remained. The important point to be taken from these studies is that BOLD signal can be expected not to be limited to report excitatory activity, but also reports inhibitory activity.

Given the similar volumes from which the neural signals are recorded (see above) the stronger coupling of LFP to BOLD (Logothetis et al., 2001; Goense and Logothetis, 2008) cannot be explained by differences in spatial summation of the neural signals. Also the spatial area from which the neural signals are sampled provides justification for the use of fMRI resolutions of 1–2 mm to determine the correlation between MR and neural signals.

11.5.6 LFP, spikes, metabolism and blood flow

In the previous sections we showed that the BOLD response is better correlated with LFP than with MUA, implying that the BOLD response is better correlated with the input and local processing of the neurons in an area than with the output of the large pyramidal cells. Note that "synaptic input" does not necessarily imply input from another cortical or subcortical area, but it also includes local (intrinsic) connections. Similarly to the positive BOLD response, negative BOLD responses were also associated with decreases in LFP and MUA (Shmuel et al., 2006).

The energy demands of neurotransmission and spiking determine the blood flow to an activated area. Ultimately, the function of perfusion is the supply of nutrients and O_2, removal of waste products and removal of heat. This raises questions about which processes have the highest metabolic demands (synaptic processes versus spiking) and whether the neurovascular response is driven by the processes that have the highest metabolic demands or is there some kind of anticipatory or feedforward process.

The brain consumes 20% of the total energy (Sokoloff, 1960) and it is oxidative phosphorylation that feeds the brain. Most of the energy (50–80%) is used for neural signaling, while cellular maintenance processes use a minor (5–15%) fraction of the energy budget (Ames, 2000; Shulman et al., 2004; Raichle and Mintun, 2006; Riera et al., 2008). Because the sodium–potassium pump or Na^+/K^+-ATPase maintains the transmembrane electrochemical gradients that are the driving force for most signaling processes, the main energy consumer in the brain is the Na^+/K^+-ATPase (Sokoloff, 1999). The pump depends on oxidative phosphorylation for its energy needs (Erecinska and Silver, 1994) and high concentrations of Na^+/K^+-ATPase are co-localized with high concentrations of cytochrome oxidase (CytOx) (Wong-Riley, 1989; Hevner et al., 1992). Cytochrome oxidase is the enzyme in the electron transport chain that catalyzes the reduction of O_2 to H_2O and is a marker for oxidative metabolism. The vascularization of the cortex also reflects its energy use: areas that have high energy needs are more densly vascularized. For instance, in V1 energy use is highest in layer IV and it is higher in blobs than in interblobs. This is evidenced by higher CytOx-reactivity (Wong-Riley, 1989), higher ^{14}C deoxyglucose (2-$[^{14}C]$-DG) uptake (Kennedy et al., 1976) and is reflected in the higher vascularization of these areas (Weber et al., 2008).

At the cellular level, we can also ask which signaling processes use the most energy, for instance, spiking, synaptic transmission, neurotransmitter recycling, or the maintenance of transmembrane gradients (Ames, 2000; Attwell and Laughlin, 2001). It is generally believed that synaptic transmission and associated processes use more energy (Ames, 2000; Shulman et al., 2004) although the relative contribution of spiking is still debated (Attwell and Laughlin, 2001; Lennie, 2003). The higher presence of cytochrome oxidase and mitochondria in dendrites (especially in postsynaptic areas) compared to cell bodies and axons indicates that these are the more metabolically active sites (Wong-Riley, 1989). Similarly, Schwartz et al. (1979) also found that nerve terminals have a higher uptake of radiolabeled glucose than areas with cell bodies.

Despite these energy needs, however, the cerebral metabolic rate of oxygen consumption ($CMRO_2$) measured with PET showed little increase under visual stimulation while blood flow and glucose consumption did increase, indicating an uncoupling between blood flow and oxidative metabolism (Fox and Raichle, 1986; Fox et al., 1988; Raichle and Mintun, 2006). These observations have incited the debate about the importance of glycolytic versus oxidative metabolism in brain function (Pellerin et al., 2007; Pellerin and Magistretti, 2003; Chih and Roberts, 2003; Gladden, 2004; Shulman et al., 2004) since it was also found with MR spectroscopy that lactate is produced during functional activation (Prichard et al., 1991; Sappey-Marinier et al., 1992; Mangia et al., 2007).

Pellerin and Magistretti (1994) and Pellerin et al. (2007) put forward the hypothesis that glutamate released at the synapse is taken up into astrocytes, where it induces glycolysis leading to the production of lactate, which is then released and taken up by neurons and used as an energy substrate. This is also called the astrocyte–neuron lactate shuttle hypothesis (ANLSH). The competing view is that both neurons and astrocytes use glucose as their main substrate and lactate is produced when glycolysis exceeds the rate of oxidative metabolism, a situation that is both transient and potentially detrimental (Chih and Roberts, 2003). Astrocytes have many different functions in the brain and play an important role in neurovascular coupling and metabolism. They are a type of glial cell characterized by their star shape and until quite recently they were mostly considered filler material. However, they are involved in regulating homeostasis in the brain (Simard and Nedergaard, 2004) and providing energy to the neurons by supplying nutrients. Another function of astrocytes is the uptake of neurotransmitters released from nerve terminals, however, they can also release neuroactive agents themselves, like glutamate (Newman, 2003).

During activation, most of the energy used also seems to be accounted for by the needs of the Na^+/K^+-ATPase; Mata et al. (1980) found that glucose uptake upon stimulation is due to activity of the Na^+/K^+-ATPase. The pump is linked to glycolysis in different tissues and may be in the brain (see Ames (2000) for a review) but glycolysis by itself provides insufficient energy to feed the pump. However, it is possible that for fast and/or small increases like activation, neurons rely more on glycolysis while for sustained activity they rely on oxidative metabolism (Wong-Riley, 1989; Erecinska and Silver, 1994; Raichle and Mintun, 2006). In experiments by Fox et al. (1986, 1988) and others (Ito et al., 2005) $CMRO_2$ was not much elevated by functional activity but $CMRO_2$ increased with prolonged stimulation (Gjedde and Marrett, 2001; Mintun et al., 2002; Vlassenko et al., 2006b). Hence there may be a shift from initially more glycolytic metabolism towards more oxidative metabolism under sustained stimulation (Gjedde and Marrett, 2001; Mintun et al., 2002; Raichle and Mintun, 2006; Vlassenko et al., 2006a). Further studies are needed to clarify the contributions of glycolysis and oxidative metabolism during both baseline and activation.

But whether metabolism upon activation is primarily oxidative or glycolytic, or whether the nutrient that needs to be supplied is glucose or O_2 may not be most important to understand the BOLD response. What matters for the BOLD signal is what drives the signaling cascade that leads to the increase in blood flow, given that it is the blood flow response that is taken as an indicator of increases in neural activity. Hence, we need to look at neurovascular coupling and what is the trigger for the neurovascular response. Again, the processes related to synaptic function, synaptic transmission, restoration of electrical gradients and neurotransmitter recycling elicit functional blood flow increases (Lauritzen, 2005; Wang et al., 2005; Hoffmeyer et al., 2007; Iadecola and Nedergaard, 2007). Glutamatergic neurotransmission in particular leads to an increase in blood flow (Li and Iadecola, 1994; Mathiesen et al., 1998; Yang et al., 2003; Gsell et al., 2006; Hoffmeyer et al., 2007) although there are numerous vasoactive signaling molecules and multiple vasodilatory and constrictive mechanisms (Cauli et al., 2004; Iadecola, 2004; Iadecola and Nedergaard, 2007). Molecules involved in signaling pathways leading to vasodilation during activation are also associated more with synaptic signaling (Zhang and Wong-Riley, 1996; Yang et al., 2003; Hoffmeyer et al., 2007) than with action potentials or metabolic signals (Attwell and Iadecola, 2002; Iadecola, 2004) although the redox state during activation possibly plays a role through the cytosolic $NADH/NAD^+$ in astrocytes (Ido et al., 2001, 2004; Raichle and Mintun, 2006; Vlassenko et al., 2006b). Many neurovascular coupling processes are also mediated by astrocytes, because of their key location between capillaries and neurons

(Nedergaard et al., 2003). Their numerous processes contact the blood vessels and envelope synapses, and they are involved in the regulation of local blood flow, and hence are an important mediator of the BOLD signal. Their effect on the blood flow responses are again often triggered by neuronal glutamate release (Zonta et al., 2003; Koehler et al., 2006; Metea and Newman, 2006; Takano et al., 2006; Petzold et al., 2008; Schummers et al., 2008).

11.5.7 *The cortical circuit and the BOLD response*

We described earlier how the BOLD signal is associated more with input and local processing of cortical neurons than with their spiking output. But this opens further questions about what specific cortical processes and what properties of the cortical circuitry determine the BOLD response. For a better interpretation of fMRI it is essential that we understand which processes are represented in the BOLD response, and which are represented to a lesser extent, or not at all (Logothetis, 2008). In essence this boils down to a question about the transfer function: do all neural events contribute equally to the BOLD signal, or do some events contribute more than others?

Different cortical processes can and without doubt do have different contributions to the BOLD, LFP and MUA signals. The fMRI signal may also represent different aspects of neural processing that are not always observed with single-unit recording. For instance, a binocular rivalry stimulus showed robust functional activation in V1, while only a small fraction of the single-neuron's responses were modulated by the percept (Blake and Logothetis, 2002). Different processes, such as feedforward versus feedback processes, or stimulus driven or neuromodulatory processes, or for instance subcortical input versus cortico-cortical input or recurrent intracortical input, may affect these signals differently. It is not known whether the BOLD signal is more or less strongly weighted towards intracortical processing or cortico-cortical or thalamocortical processing. If we look at connectivity we find that most cortical connections are highly local. In contrast, subcortical input is rather weak in terms of number of synapses (Peters and Payne, 1993; Peters et al., 1994; Douglas and Martin, 2007), for instance thalamocortical input typically involves only about 10–20% of synapses (Douglas and Martin, 2007), and in V1 only ~5% (Peters and Payne, 1993; Peters et al., 1994).

Furthermore, different types of neurons may have different contributions to the BOLD signals as they do to the neural signals. For instance, interneurons are less visible in LFP and MUA signals than pyramidal neurons, but because they can have high firing rates they could possibly have substantial metabolic demands (Buzsaki et al., 2007) and a considerable effect on the BOLD response. Interneurons also have been shown to cause dilation and constriction of microvessels (Cauli et al., 2004). Another example is small cells that can also have higher firing rates and are more easily stimulated than large cells (Gur et al., 1999) and hence may have a substantial contribution to the BOLD response, although their contribution to the MUA and LFP is less.

Another question is how do excitation and inhibition contribute to the fMRI responses (Buzsaki et al., 2007; Logothetis, 2008) and is the BOLD signal mediated by specific neurotransmitters or receptors? We can ask the same question about the LFP and MUA because different types of neurotransmission may also not be equally represented in the recorded neural signals. Glutamatergic excitatory neurotransmission is commonly associated with BOLD and CBF responses (Gsell et al., 2006; Hoffmeyer et al., 2007). One reason is that these are the most common synapses in the cortex and they outnumber inhibitory synapses by about five to one (Braitenberg and Schüz, 1998). Glutamatergic excitatory synapses are also correlated with high cytochrome oxidase levels (Wong-Riley, 1989). However, although NMDA and AMPA receptor activity were both shown to contribute to the CBF and BOLD responses, in contrast to AMPA receptor activity, NMDA receptor activity did not contribute to the LFP (Mathiesen et al., 1998; Gsell et al., 2006; Hoffmeyer et al., 2007). NMDA receptor activity is linked to increased blood flow through the nitric oxide (NO) signaling pathway (Akgoren et al., 1994; Li and Iadecola, 1994) which mediates glutamate induced blood flow increases. For instance NO syntase knockout mice showed decreased activation-induced blood flow responses in the cerebellum (Yang et al., 2003). The effect of inhibitory neurotransmission on the blood flow response is much less clear (Buzsaki et al., 2007; Logothetis, 2008) and may be small (Waldvogel et al., 2000), possibly due to the lower number of inhibitory synapses. However a contribution of inhibitory neurotransmission to the BOLD signal in the cortex is not unlikely because GABAergic neurotransmission has been shown to account for ∼15% of the energy consumption (Patel et al., 2005; Hyder et al., 2006). Furthermore, inhibitory neurons can have high firing rates and their synapses are also associated with mitochondria and high CytOx levels (Wong-Riley, 1989). Inhibition has also been shown to increase 2-DG uptake (Ackermann et al., 1984; McCasland and Hibbard, 1997; Nudo and Masterton, 1986) and GABA was shown to induce vasodilation in hippocampal microvessels (Fergus and Lee, 1997). Finally, other neurotransmitters and neuromodulators are known to be vasoactive and may play a role in the BOLD response, for example acetylcholine, serotonin, dopamine, and noradrenaline (Attwell and Iadecola, 2002).

Hence, although it is without doubt true that the BOLD signal is related to neural processing, what cortical processes exactly it does and does not represent is still far from clear. For many studies knowing that activation is due to neural activity may be all that one needs to know. But if we want to go beyond that and interpret the BOLD findings in the context of these different neural processes it becomes increasingly important to understand better the neurovascular response and its relation to the underlying neural processes.

11.5.8 Perception and attention

Physiological signals tend to be higher when stimuli are perceived as opposed to when they are not and, intriguingly, in some regions like the motion processing area V5+/MT+, BOLD signal has appeared to reflect this more sensitively than electrophysiological measures (Zeki and Ffytche, 1998; Moutoussis et al., 2005; Moutoussis and Zeki, 2006). This phenomenon has been addressed most extensively using bistable percepts, such as in binocular rivalry, where the percept alternates spontaneously, despite constant physical stimulation.

Higher areas including V4, V5 and inferotemporal regions show both substantial BOLD as well as spiking modulations as a function of the percept (Logothetis and Schall, 1989; Moutoussis et al., 2005). However, LGN and V1/V2 have consistently revealed no or minimal modulation in spikes and multi-unit activity during bistable percepts, yet some limited modulations in LFPs (Lehky and Maunsell, 1996; Leopold and Logothetis, 1996; Gail et al., 2004; Wilke et al., 2006). BOLD signal modulations in human V1 and even in LGN on the other hand were in some (yet not all) experiments nearly as large during perceptual transitions as when the stimulus was changed physically (see e.g. Haynes et al., 2005; Wunderlich et al., 2005)). LFPs and BOLD may thus reflect perceptually modulated local processing in V1 that is captured less well in spiking output rates.

Indeed, recent laminar LFP recordings revealed perceptually modulated changes in membrane currents within the upper layers of V1 that barely affect pyramidal neuron spiking output (A. Maier (NIH/NIMH), personal communication). If properties of monkey and human LGN and V1 are similar, the above results reveal a prominent dissociation between BOLD signal and spiking activity obtained in a normal behavioral setting, potentially due to modulatory input to V1.

It may be no coincidence that a similar dissociation as in rivalry can be observed with attention, which is also associated with perceptual enhancement and with top-down modulation (Pessoa et al., 2003; Gilbert and Sigman, 2007; Yoshor et al., 2007). In V1, attention has usually revealed modest spike rate modulations (a notable exception though being e.g. line-segmentation tasks (Roelfsema et al., 1998; McAdams and Maunsell, 1999; Gilbert and Sigman, 2007)) and in LGN no signal change has been detected (Mehta et al., 2000; Bender and Youakim, 2001). These findings stand in contrast to human fMRI, reporting consistently strong attention effects in both V1 and LGN (O'Connor et al., 2002;

Pessoa et al., 2003). The discrepancy is unlikely to originate from species difference or task, as intracortical recordings of LFPs in human patients also revealed only modest attentional modulations, consistent with those obtained in monkey (Yoshor et al., 2007). If one considers that attention also affects other properties such as spatial integration and even attribute preference in V1, and potentially neural states in LGN, and that this is mediated by thalamocortical loops and by the release of neuromodulators such as acetylcholine, it would be surprising if there was *no* discrepancy between a mass-action signal such as BOLD and the more limited electrophysiological measures (Li et al., 2004; Roberts et al., 2005; McAlonan et al., 2006). The implication is that BOLD signals are not merely a low-resolution non-invasive version of multi-unit activity or LFPs, but that they constitute an independent, complementary measure. The BOLD signal may in many cases be correlated with other measures, but then in some important cases, it may not. The latter especially deserve more attention and investigation.

11.6 Conclusions

fMRI is without any doubt one of the most widely used tools for the investigation of brain function, and its non-invasive nature has opened an enormously growing field investigating the human brain. Compared to invasive electrophysiology, fMRI offers a very poor temporal resolution signal, but the advantage of whole-brain coverage. If there is any simple message to be drawn from the above complex relations between metabolism, synaptic activity, neuromodulation, spiking and BOLD signal, it is this: BOLD signal does not constitute a low-resolution version of neural spiking activity. Instead, it is a signal that cannot be grasped easily, as it is affected by numerous physiological mechanisms, some of which appear difficult to detect even using invasive recordings. In some circumstances fMRI appears even more sensitive than neural spiking activity, especially when it comes to experiments aimed at dissociating stimulus processing from conscious perception. Alterations in feedback and intensity of local processing may affect BOLD signal more than neural (output) spiking activity. Therefore, BOLD signal constitutes a formidable signal related to neural processing which, because of its distinct nature, can complement but in no way replace electrophysiological signals. There is still room for considerable advances in the MR technology, which will most likely provide much higher temporal and spatial resolutions than available at present. There is no doubt that this will pose new challenges to analysis techniques, in particular to those aimed at establishing the content, direction and augmentation of information flow across the brain volume, and to disentangle neural information from the numerous artifacts that, in part, may be directly related or caused by some of the neural activity, such as heart-rate, breathing and autonomous signals that also affect vascular responses.

There is no doubt that because of the complementary nature of fMRI, PET, EEG and invasive techniques, combinations of these may lead to breakthroughs both in the understanding of their respective contributions and in neuroscience.

References

Ackermann, R. F., Finch, D. M., Babb, T. L. and Engel, J., Jr. (1984). Increased glucose metabolism during long-duration recurrent inhibition of hippocampal pyramidal cells. *J. Neurosci.,* **4**, 251–264.

Adams, D. L., Sincich, L. C. and Horton, J. C. (2007). Complete pattern of ocular dominance columns in human primary visual cortex. *J. Neurosci.,* **27**, 10391–10403.

Akgoren, N., Fabricius, M. and Lauritzen, M. (1994). Importance of nitric oxide for local increases of blood flow in rat cerebellar cortex during electrical stimulation. *Proc. Natl. Acad. Sci. USA,* **91**, 5903–5907.

Ames, A., III (2000). CNS energy metabolism as related to function. *Brain Res. Rev.,* **34**, 42–68.

Attwell, D. and Iadecola, C. (2002). The neural basis of functional brain imaging signals. *Trends Neurosci.,* **25**, 621–625.

Attwell, D. and Laughlin, S. B. (2001). An energy budget for signaling in the grey matter of the brain. *J. Cereb. Blood Flow Metab.,* **21**, 1133–1145.

Bartels, A. and Zeki, S. (2003). Functional brain mapping during free viewing of natural scenes. *Human Brain Mapping,* **21**, 75–83.

Bartels, A. and Zeki, S. (2005a). Brain dynamics during natural viewing conditions – a new guide for mapping connectivity in vivo. *NeuroImage,* **24**, 339–349.

Bartels, A. and Zeki, S. (2005b). The chronoarchitecture of the cerebral cortex. *Philos. Trans. R. Soc. London, Ser. B,* **360**, 733–750.

Bartels, A., Zeki, S. and Logothetis, N. K. (2008a). Natural vision reveals regional specialization to local motion and to contrast-invariant, global flow in the human brain. *Cereb. Cortex,* **18**, 705–717.

Bartels, A., Logothetis, N. K. and Moutoussis, K. (2008b). fMRI and its interpretations: an illustration on directional selectivity in area V5/MT. *Trends Neurosci.,* **31**, 444–453.

Basar, E. (1980). *EEG-Brain Dynamics: Relation between EEG and Brain Evoked Potentials.* Amsterdam: Elsevier.

Belitski, A., Gretton, A., Magri, C., Murayama, Y., Montemurro, M. A., Logothetis, N. K. and Panzeri, S. (2008). Low-frequency local field potentials and spikes in primary visual cortex convey independent visual information. *J. Neurosci.,* **28**, 5696–5709.

Bender, D. B. and Youakim, M. (2001). Effect of attentive fixation in macaque thalamus and cortex. *J. Neurophysiol.,* **85**, 219–234.

Belliveau, J. W., Rosen, B. R., Kantor, H. L., Rzedzian, R. R., Kennedy, D. N., McKinstry, R. C., Vevea, J. M., Cohen, M. S., Pykett, I. L. and Brady, T. J. (1990). Functional cerebral imaging by susceptibility-contrast NMR. *Magn. Reson. Med.,* **14**, 538–546.

Belliveau, J. W., Kennedy, Jr., D. N., McKinstry, R. C., Buchbinder, B. R., Weisskoff, R. M., Cohen, M. S., Vevea, J. M., Brady, T. J. and Rosen, B. R. (1991). Functional mapping of the human visual cortex by magnetic resonance imaging. *Science,* **254**, 716–719.

Biswal, B., Yetkin, F. Z., Haughton, V. M. and Hyde, J. S. (1995). Functional connectivity in the motor cortex of resting human brain using echo-planar MRI. *Magn. Reson. Med.*, **34**, 537–541.

Blake, R. and Logothetis, N. K. (2002). Visual competition. *Nature Rev. Neurosci.*, **3**, 13–23.

Boxerman, J. L., Hamberg, L. M., Rosen, B. R. and Weisskoff, R. M. (1995). MR contrast due to intravascular magnetic susceptibility perturbations. *Magn. Reson. Med.*, **34**, 555–566.

Braitenberg, V. and Schüz, A. (1998). *Cortex: Statistics and Geometry of Neuronal Connectivity.* Berlin: Springer.

Buchwald, J. S. and Grover, F. S. (1970). Amplitudes of background fast activity characteristic of specific brain sites. *J. Neurophysiol.*, **33**, 148–159.

Buchwald, J. S., Halas, E. S. and Schramm, S. (1965). A comparison of multi-unit activity and EEG activity recorded from the same brain sites during behavioral conditioning. *Nature*, **205**, 1012–1014.

Buracas, G. T. and Boynton, G. M. (2002). Efficient design of event-related fMRI experiments using M-sequences. *NeuroImage*, **16**, 801–813.

Buxton, R. B. (2009). *Introduction to Functional Magnetic Resonance Imaging: Principles and Techniques* (2nd edition). Cambridge: Cambridge University Press.

Buxton, R. B., Uludag, K., Dubowitz, D. J. and Liu, T. T. (2004). Modeling the hemodynamic response to brain activation. *NeuroImage*, **23** (Suppl 1), S220–S233.

Buzsaki, G. and Chrobak, J. J. (1995). Temporal structure in spatially organized neuronal ensembles: a role for interneuronal networks. *Curr. Opinion Neurobiol.*, **5**, 504–510.

Buzsaki, G., Bickford, R. G., Ponomareff, G., Thal, L. J., Mandel, R. and Gage, F. H. (1988). Nucleus basalis and thalamic control of neocortical activity in the freely moving rat. *J. Neurosci.*, **8**, 4007–4026.

Buzsaki, G., Kaila, K. and Raichle, M. (2007). Inhibition and brain work. *Neuron*, **56**, 771–783.

Cauli, B., Tong, X. K., Rancillac, A., Serluca, N., Lambolez, B., Rossier, J. and Hamel, E. (2004). Cortical GABA interneurons in neurovascular coupling: relays for subcortical vasoactive pathways. *J. Neurosci.*, **24**, 8940–8949.

Chaigneau, E., Oheim, M., Audinat, E. and Charpak, S. (2003). Two-photon imaging of capillary blood flow in olfactory bulb glomeruli. *Proc. Natl. Acad. Sci. USA*, **100**, 13081–13086.

Cheng, K., Waggoner, R. A. and Tanaka, K. (2001). Human ocular dominance columns as revealed by high-field functional magnetic resonance imaging. *Neuron*, **32**, 359–374.

Chih, C. P., and Roberts, Jr., E. L., (2003). Energy substrates for neurons during neural activity: a critical review of the astrocyte-neuron lactate shuttle hypothesis. *J. Cereb. Blood Flow Metab.*, **23**, 1263–1281.

Cox, D. D. and Savoy, R. L. (2003). Functional magnetic resonance imaging (fMRI) "brain reading": detecting and classifying distributed patterns of fMRI activity in human visual cortex. *NeuroImage*, **19**, 261–270.

Dechent, P. and Frahm, J. (2000). Direct mapping of ocular dominance columns in human primary visual cortex. *NeuroReport*, **11**, 3247–3249.

Dennie, J., Mandeville, J. B., Boxerman, J. L., Packard, S. D., Rosen, B. R. and Weisskoff, R. M. (1998). NMR imaging of changes in vascular morphology due to tumor angiogenesis. *Magn. Reson. Med.*, **40**, 793–799.

Denys, K., VanDuffel, W., Fize, D., Nelissen, K., Peuskens, H., Van, E. D. and Orban, G. A. (2004). The processing of visual shape in the cerebral cortex of human and nonhuman primates: a functional magnetic resonance imaging study. *J. Neurosci.,* **24,** 2551–2565.

Disbrow, E. A., Slutsky, D. A., Roberts, T. P. and Krubitzer, L. A. (2000). Functional MRI at 1.5 Tesla: a comparison of the blood oxygenation level-dependent signal and electrophysiology. *Proc. Natl. Acad. Sci. USA,* **97,** 9718–9723.

Douglas, R. J. and Martin, K. A. (2007). Mapping the matrix: the ways of neocortex. *Neuron,* **56,** 226–238.

Duong, T. Q., Kim, D. S., Ugurbil, K. and Kim, S. G. (2001). Localized cerebral blood flow response at submillimeter columnar resolution. *Proc. Natl. Acad. Sci. USA,* **98,** 10904–10909.

Duong, T. Q., Yacoub, E., Adriany, G., Hu, X. P., Ugurbil, K. and Kim, S. G. (2003). Microvascular BOLD contribution at 4 and 7T in the human brain: gradient-echo and spin-echo fMRI with suppression of blood effects. *Magn. Reson. Medi.,* **49,** 1019–1027.

Duvernoy, H. M., Delon, S. and Vannson, J. L. (1981). Cortical blood vessels of the human brain. *Brain Res. Bull.,* **7,** 519–579.

Engel, A. K., Moll, C. K., Fried, I. and Ojemann, G. A. (2005). Invasive recordings from the human brain: clinical insights and beyond. *Nature. Rev. Neurosci.,* **6,** 35–47.

Erecinska, M. and Silver, I. A. (1994). Ions and energy in mammalian brain. *Prog. Neurobiol.,* **43,** 37–71.

Faraci, F. M. and Heistad, D. D. (1998). Regulation of the cerebral circulation: role of endothelium and potassium channels. *Physiol. Rev.,* **78,** 53–97.

Fergus, A. and Lee, K. S. (1997). GABAergic regulation of cerebral microvascular tone in the rat. *J. Cereb. Blood Flow Metab.,* **17,** 992–1003.

Fox, P. T. and Raichle, M. E. (1986). Focal physiological uncoupling of cerebral blood-flow and oxidative-metabolism during somatosensory stimulation in human-subjects. *Proc. Natl. Acad. Sci. USA,* **83,** 1140–1144.

Fox, P. T., Raichle, M. E., Mintun, M. A. and Dence, C. (1988). Nonoxidative glucose consumption during focal physiologic neural activity. *Science,* **241,** 462–464.

Friston, K. J., Frith, C. D., Liddle, P. F. and Frackowiak, R. S. J. (1993). Functional connectivity – the principal-component analysis of large (Pet) data sets. *J. Cereb. Blood Flow Metab.,* **13,** 5–14.

Friston, K. J., Holmes, A. P., Poline, J. B., Grasby, P. J., Williams, S. C., Frackowiak, R. S. and Turner, R. (1995). Analysis of fMRI time-series revisited. *NeuroImage,* **2,** 45–53.

Friston, K. J., Holmes, A., Poline, J. B., Price, C. J. and Frith, C. D. (1996). Detecting activations in PET and fMRI: levels of inference and power. *NeuroImage,* **4,** 223–235.

Friston, K. J., Buechel, C., Fink, G. R., Morris, J., Rolls, E. and Dolan, R. J. (1997). Psychophysiological and modulatory interactions in neuroimaging. *NeuroImage,* **6,** 218–229.

Friston, K. J., Harrison, L. and Penny, W. (2003). Dynamic causal modeling. *NeuroImage,* **19,** 1273–1302.

Friston, K. J., Ashburner, J. T., Kiebel, S. J., Nichols, T. E. and Penny, W. D. (editors). (2007). *Statistical Parametric Mapping.* London: Academic Press.

Fromm, G. H. and Bond, H. W. (1964). Slow changes in the electrocorticogram and the activity of cortical neurons. *Electroencephalogr. Clin. Neurophysiol.,* **17,** 520–523.

Fromm, G. H. and Bond, H. W. (1967). The relationship between neuron activity and cortical steady potentials. *Electroencephalogr. Clini. Neurophysiol.,* **22,** 159–166.

Gail, A., Brinksmeyer, H. J. and Eckhorn, R. (2004). Perception-related modulations of local field potential power and coherence in primary visual cortex of awake monkey during binocular rivalry. *Cereb. Cortex,* **14,** 300–313.

Gasser, H. S. and Grundfest, H. (1939). Axon diameters in relation to the spike dimensions and the conduction velocity in mammalian A fibers. *Am. J. Physiol.*, **127**, 393–414.

Genovese, C. R., Lazar, N. A. and Nichols, T. (2002). Thresholding of statistical maps in functional neuroimaging using the false discovery rate. *NeuroImage*, **15**, 870–878.

Gilbert, C. D. and Sigman, M. (2007). Brain states: top-down influences in sensory processing. *Neuron*, **54**, 677–696.

Gjedde, A. and Marrett, S. (2001). Glycolysis in neurons, not astrocytes, delays oxidative metabolism of human visual cortex during sustained checkerboard stimulation in vivo. *J. Cereb. Blood Flow Metab.*, **21**, 1384–1392.

Gladden, L. B. (2004). Lactate metabolism: a new paradigm for the third millennium. J. *Physiol.*, **558**, 5–30.

Goebel, R., Roebroeck, A., Kim, D. S. and Formisano, E. (2003). Investigating directed cortical interactions in time-resolved fMRI data using vector autoregressive modeling and Granger causality mapping. *Magn. Reson. Imaging*, **21**, 1251–1261.

Goense, J. B. and Logothetis, N. K. (2006). Laminar specificity in monkey V1 using high-resolution SE-fMRI. *Magn. Reson. Imaging*, **24**, 381–392.

Goense, J. B. and Logothetis, N. K. (2008). Neurophysiology of the BOLD fMRI signal in awake monkeys. *Curr. Biol.*, **18**, 631–640.

Goense, J. B. and Logothetis, N. K. (2010). Physiological basis of the BOLD signal. In: M. Ullsperger (editor), *Integrating EEG and fMRI: Recording, Analysis and Integration*. Oxford: Oxford University Press, pp. 21–46.

Goense, J. B., Zappe, A. C. and Logothetis, N. K. (2007). High-resolution fMRI of macaque V1. *Magn. Reson. Imaging,*. **25**, 740–747.

Gray, C. M., Maldonado, P. E., Wilson, M. and McNaughton, B. (1995). Tetrodes markedly improve the reliability and yield of multiple single-unit isolation from multi-unit recordings in cat striate cortex. *J. Neurosci. Methods*, **63**, 43–54.

Grill-Spector, K., Henson R. and Martin, A. (2006). Repetition and the brain: neural models of stimulus-specific effects. *Trends Cogn. Sci.*, **10**, 14–23.

Grover, F. S. and Buchwald, J. S. (1970). Correlation of cell size with amplitude of background fast activity in specific brain nuclei. *J. Neurophysiol.*, **33**, 160–171.

Gsell, W., Burke, M., Wiedermann, D., Bonvento, G., Silva, A. C., Dauphin, F., Buhrle, C., Hoehn, M. and Schwindt, W. (2006). Differential effects of NMDA and AMPA glutamate receptors on functional magnetic resonance imaging signals and evoked neuronal activity during forepaw stimulation of the rat. *J. Neurosci.*, **26**, 8409–8416.

Gur, M., Beylin, A. and Snodderly, D. M. (1999). Physiological properties of macaque V1 neurons are correlated with extracellular spike amplitude, duration, and polarity. *J. Neurophysiol.*, **82**, 1451–1464.

Haacke, M. E., Brown, R. W., Thompson, M. R. and Venkatesan, R. (1999). *Magnetic Resonance Imaging: Physical Principles and Sequence Design*. New York: Wiley-Liss.

Haacke, E. M., Lin, W. L., Hu, X. P. and Thulborn, K. (2001). A current perspective of the status of understanding BOLD imaging and its use in studying brain function: a summary of the workshop at the University of North Carolina in Chapel Hill, 26–28 October, 2000. *Nucl. Magn. Reson. Biomed.*, **14**, 384–388.

Hamel, E. (2006). Perivascular nerves and the regulation of cerebrovascular tone. *J. Appl. Physiol.*, **100**, 1059–1064.

Harada, Y. and Takahashi, T. (1983). The calcium component of the action potential in spinal motoneurones of the rat. *J. Physiol.*, **335**, 89–100.

Harel, N., Lin, J., Moeller, S., Ugurbil, K. and Yacoub, E. (2006a). Combined imaging-histological study of cortical laminar specificity of fMRI signals. *NeuroImage*, **29**, 879–887.

Harel, N., Ugurbil, K., Uludag, K. and Yacoub, E. (2006b). Frontiers of brain mapping using MRI. *J. Magn. Reson. Imaging*, **23**, 945–957.

Haynes, J. D. and Rees, G. (2006). Decoding mental states from brain activity in humans. *Rev. Neurosci.*, **7**, 523–534.

Haynes, J. D., Deichmann, R. and Rees, G. (2005). Eye-specific effects of binocular rivalry in the human lateral geniculate nucleus. *Nature*, **438**, 496–499.

Henze, D. A., Borhegyi, Z., Csicsvari, J., Mamiya, A., Harris, K. D. and Buzsaki, G. (2000). Intracellular features predicted by extracellular recordings in the hippocampus in vivo. *J. Neurophysiol.*, **84**, 390–400.

Hevner, R. F., Duff, R. S. and Wong-Riley, M. T. (1992). Coordination of ATP production and consumption in brain: parallel regulation of cytochrome oxidase and Na+, K(+)-ATPase. *Neurosci. Lett.*, **138**, 188–192.

Hirase, H., Creso, J., Singleton, M., Bartho, P. and Buzsaki, G. (2004). Two-photon imaging of brain pericytes in vivo using dextran-conjugated dyes. *Glia*, **46**, 95–100.

Hoffmeyer, H. W., Enager, P., Thomsen, K. J. and Lauritzen, M. J. (2007). Nonlinear neurovascular coupling in rat sensory cortex by activation of transcallosal fibers. *J. Cereb. Blood Flow Metab.*, **27**, 575–587.

Hopfinger, J. B., Buchel, C., Holmes, A. P. and Friston, K. J. (2000). A study of analysis parameters that influence the sensitivity of event- related fMRI analyses. *NeuroImage*, **11**, 326–333.

Huettel, S. A., Song, A. W. and McCarthy, G. (2008). *Functional Magnetic Resonance Imaging* (2nd edition). Sutherland, MA: Sinauer Associates.

Huk, A. C., Ress, D. and Heeger, D. J. (2001). Neuronal basis of the motion aftereffect reconsidered. *Neuron*, **32**, 161–172.

Hunt, C. (1951). The reflex activity of mammalian small-nerve fibers. *J. Physiol.*, **115**, 456–469.

Hyder, F., Patel, A. B., Gjedde, A., Rothman, D. L., Behar, K. L. and Shulman, R. G. (2006). Neuronal-glial glucose oxidation and glutamatergic-GABAergic function. *J. Cereb. Blood Flow Metab.*, **26**, 865–877.

Iadecola, C. (2004). Neurovascular regulation in the normal brain and in Alzheimer's disease. *Nature Rev. Neurosci.*, **5**, 347–360.

Iadecola, C. and Nedergaard, M. (2007). Glial regulation of the cerebral microvasculature. *Nature Neurosci.*, **10**, 1369–1376.

Ido, Y., Chang, K., Woolsey, T. A. and Williamson, J. R. (2001). NADH: sensor of blood flow need in brain, muscle, and other tissues. *FASEB J.* **15**, 1419–1421.

Ido, Y., Chang, K. and Williamson, J. R. (2004). NADH augments blood flow in physiologically activated retina and visual cortex. *Proc. Natl. Acad. Sci. USA*, **101**, 653–658.

Ito, H., Ibaraki, M., Kanno, I., Fukuda, H. and Miura, S. (2005). Changes in the arterial fraction of human cerebral blood volume during hypercapnia and hypocapnia measured by positron emission tomography. *J. Cereb. Blood Flow Metab.*, **25**, 852–857.

Jin, T., Zhao, F. and Kim, S. G. (2006). Sources of functional apparent diffusion coefficient changes investigated by diffusion-weighted spin-echo fMRI. *Magn. Reson. Med.*, **56**, 1283–1292.

Juergens, E., Eckhorn, R., Frien, A. and Woelbern, T. (1996). Restricted coupling range of fast oscillations in striate cortex of awake monkey. In: *Brain and Evolution*. Berlin: Thieme, p. 418.

Juergens, E., Guettler, A. and Eckhorn, R. (1999). Visual stimulation elicits locked and induced gamma oscillations in monkey intracortical- and EEG-potentials, but not in human EEG. *Exp. Brain Res.*, **129**, 247–259.

Kamitani, Y. and Tong, F. (2006). Decoding seen and attended motion directions from activity in the human visual cortex. *Curr. Biol.*, **16**, 1096–1102.

Kamondi, A., Acsady, L., Wang, X. J. and Buzsaki, G. (1998). Theta oscillations in somata and dendrites of hippocampal pyramidal cells in vivo: activity-dependent phase-precession of action potentials. *Hippocampus,* **8**, 244–261.

Kandel, A. and Buzsaki, G. (1997). Cellular-synaptic generation of sleep spindles, spike-and-wave discharges, and evoked thalamocortical responses in the neocortex of the rat. *J. Neurosci.*, **17**, 6783–6797.

Katzner, S., Nauhaus, I., Benucci, A., Bonin, V., Ringach, D. L. and Carandini, M. (2009). Local origin of field potentials in visual cortex. *Neuron*, **61**, 35–41.

Kennan, R. P., Zhong, J. and Gore, J. C. (1994). Intravascular susceptibility contrast mechanisms in tissues. *Magn. Reson. Med.*, **31**, 9–21.

Kennedy, C., Des Rosiers, M. H., Sakurada, O., Shinohara, M., Reivich, M., Jehle, J. W. and Sokoloff, L. (1976). Metabolic mapping of the primary visual system of the monkey by means of the autoradiographic [14C]deoxyglucose technique. *Proc. Natl. Acad. Sci. USA*, **73**, 4230–4234.

Kiebel, S. J., Poline, J. B., Friston, K. J., Holmes, A. P. and Worsley, K. J. (1999). Robust smoothness estimation in statistical parametric maps using standardized residuals from the general linear model. *NeuroImage*, **10**, 756–766.

Kim, T., Hendrich, K. S., Masamoto, K. and Kim, S. G. (2007). Arterial versus total blood volume changes during neural activity-induced cerebral blood flow change: implication for BOLD fMRI. *J. Cereb. Blood Flow Metab.*, **27**,1235–1247.

Kocsis, B., Bragin, A. and Buzsaki, G. (1999). Interdependence of multiple theta generators in the hippocampus: a partial coherence analysis. *J. Neurosci.*, **19**, 6200–6212.

Koehler, R. C., Gebremedhin, D. and Harder, D. R. (2006). Role of astrocytes in cerebrovascular regulation. *J. Appl. Physiol.*, **100**, 307–317.

Kohn, A. and Movshon, J. A. (2003). Neuronal adaptation to visual motion in area MT of the macaque. *Neuron*, **39**, 681–691.

Koyama, M., Hasegawa, I., Osada, T., Adachi, Y., Nakahara, K. and Miyashita, Y. (2004). Functional magnetic resonance imaging of macaque monkeys performing visually guided saccade tasks: comparison of cortical eye fields with humans. *Neuron*, **41**, 795–807.

Kreiman, G., Hung, C. P., Kraskov, A., Quiroga, R. Q., Poggio, T. and DiCarlo, J. J. (2006). Object selectivity of local field potentials and spikes in the macaque inferior temporal cortex. *Neuron*, **49**, 433–445.

Krekelberg, B., Boynton, G. M. and van Wezel, R. J. (2006). Adaptation: from single cells to BOLD signals. *Trends Neurosci.*, **29**, 250–256.

Kwong, K. K., Belliveau, J. W., Chesler, D. A., Goldberg, I. E., Weisskoff, R. M., Poncelet, B. P., Kennedy, D. N., Hoppel, B. E., Cohen, M. S. and Turner, R. (1992). Dynamic magnetic resonance imaging of human brain activity during primary sensory stimulation. *Proc. Natl. Acad. Sci. USA*, **89**, 5675–5679.

Lauritzen, M. (2005). Reading vascular changes in brain imaging: is dendritic calcium the key? *Nature Rev. Neurosci.*, **6**, 77–85.

Lauterbur, P. C. (1973). Image formation by induced local interactions – examples employing nuclear magnetic-resonance. *Nature*, **242**, 190–191.

Lauwers, F., Cassot, F., Lauwers-Cances, V., Puwanarajah, P. and Duvernoy, H. (2008). Morphometry of the human cerebral cortex microcirculation: general characteristics and space-related profiles. *NeuroImage*, **39**, 936–948.

Legatt, A. D., Arezzo, J. and Vaughan, H. G. J. (1980). Averaged multiple unit activity as an estimate of phasic changes in local neuronal activity: effects of volume-conducted potentials. *J. Neurosci. Methods*, **2**, 203–217.

Lehky, S. R. and Maunsell, J. H. (1996). No binocular rivalry in the LGN of alert macaque monkeys. *Vision Res.*, **36**, 1225–1234.

Lennie, P. (2003). The cost of cortical computation. *Curr. Biol.*, **13**, 493–497.

Leopold, D. A. and Logothetis, N. K. (1996). Activity changes in early visual cortex reflect monkeys' percepts during binocular rivalry. *Nature*, **379**, 549–553.

Li, J. and Iadecola, C. (1994). Nitric oxide and adenosine mediate vasodilation during functional activation in cerebellar cortex. *Neuropharmacology*, **33**, 1453–1461.

Li, W., Piech, V. and Gilbert, C. D. (2004). Perceptual learning and top-down influences in primary visual cortex. *Nature Neurosci.*, **7**, 651–657.

Liang, Z. P. and Lauterbur, P. C. (1999). *Principles of Magnetic Resonance Imaging: A Signal Processing Perspective.* New York: Wiley.

Lindsley, D. B. and Wicke, J. D. (1974). The electroencephalogram: autonomous electrical activity in man and animals. In: R. F. Thomson, and M. M. Patterson (editors), *Electroencephalography and Human Brain Potentials.* New York: Academic Press, pp. 3–83.

Lipton, M. L., Fu, K. M. G., Branch, C. A. and Schroeder, C. E. (2006). Ipsilateral hand input to area 3b revealed by converging hemodynamic and electrophysiological analyses in macaque monkeys. *J. Neurosci.*, **26**, 180–185.

Liu, T. T. (2004). Efficiency, power, and entropy in event-related fMRI with multiple trial types. Part II: design of experiments. *NeuroImage*, **21**, 401–413.

Liu, T. T. and Frank, L. R. (2004). Efficiency, power, and entropy in event-related FMRI with multiple trial types. Part I: theory. *NeuroImage*, **21**, 387–400.

Logothetis, N. K. (2003). MR imaging in the non-human primate: studies of function and of dynamic connectivity. *Curr. Opinion Neurobiolo.*, **13**, 630–642.

Logothetis, N. K. (2008). What we can do and what we cannot do with fMRI. *Nature*, **453**, 869–878.

Logothetis, N. K. and Schall, J. D. (1989). Neuronal correlates of subjective visual perception. *Science*, **245**, 761–763.

Logothetis, N. K., Guggenberger, H., Peled, S. and Pauls, J. (1999). Functional imaging of the monkey brain. *Nature Neurosci.*, **2**, 555–562.

Logothetis, N. K., Pauls, J., Augath, M., Trinath, T. and Oeltermann, A. (2001). Neurophysiological investigation of the basis of the fMRI signal. *Nature*, **412**, 150–157.

Logothetis, N. K., Merkle, H., Augath, M., Trinath, T. and Ugurbil, K. (2002). Ultra high-resolution fMRI in monkeys with implanted RF coils. *Neuron*, **35**, 227–242.

Logothetis, N. K., Kayser, C. and Oeltermann, A. (2007). In vivo measurement of cortical impedance spectrum in monkeys: implications for signal propagation. *Neuron*, **55**, 809–823.

Lorente, de Nó, R. (1947). Analysis of the distribution of action currents of nerve in volume conductors. In: *Studies from the Rockefeller Institute Medical Research*, Vol 132, *A Study of Nerve Physiology*, pp. 384–477.

Lowel, S., Freeman, B. and Singer, W. (1987). Topographic organization of the orientation column system in large flat-mounts of the cat visual cortex: a 2-deoxyglucose study. *J. Comput. Neurol.*, **255**, 401–415.

Lu, H., Zuo, Y., Gu, H., Waltz, J. A., Zhan, W., Scholl, C. A., Rea, W., Yang, Y. and Stein, E. A. (2007). Synchronized delta oscillations correlate with the resting-state functional MRI signal. *Proc. Natl. Acad. Sci. USA*, **104**, 18265–18269.

Maandag, N. J., Coman D., Sanganahalli, B. G., Herman, P., Smith, A. J., Blumenfeld, H., Shulman, R. G. and Hyder, F. (2007). Energetics of neuronal signaling and fMRI activity. *Proc. Natl. Acad. Sci. USA,* **104**, 20546–20551.

Mangia, S., Tkac, I., Gruetter, R., Van de Moortele, P. F., Maraviglia, B. and Ugurbil, K. (2007). Sustained neuronal activation raises oxidative metabolism to a new steady-state level: evidence from 1H NMR spectroscopy in the human visual cortex. *J. Cereb. Blood Flow Metab.,* **27**, 1055–1063.

Mansfield, P. (1977). Multi-planar image-formation using NMR spin echoes. *J. Phys. C,* **10**, L55–L58.

Mata, M., Fink, D. J., Gainer, H., Smith, C. B., Davidsen, L., Savaki, H., Schwartz, W. J. and Sokoloff, L. (1980). Activity-dependent energy metabolism in rat posterior pituitary primarily reflects sodium pump activity. *J. Neurochem.,* **34**, 213–215.

Mathiesen, C., Caesar, K., Akgoren, N. and Lauritzen, M. (1998). Modification of activity-dependent increases of cerebral blood flow by excitatory synaptic activity and spikes in rat cerebellar cortex. *J. Physiol.,* **512** (2), 555–566.

McAdams, C. J. and Maunsell, J. H. (1999). Effects of attention on orientation-tuning functions of single neurons in macaque cortical area V4. *J. Neurosci.,* **19**, 431–441.

McAlonan, K., Cavanaugh, J. and Wurtz, R. H. (2006). Attentional modulation of thalamic reticular neurons. *J. Neurosci.,* **26**, 4444–4450.

McCasland, J. S. and Hibbard, L. S. (1997). GABAergic neurons in barrel cortex show strong, whisker-dependent metabolic activation during normal behavior. *J. Neurosci.,* **17**, 5509–5527.

McKeown, M. J., Makeig, S., Brown, G. G., Jung, T. P., Kindermann, S. S., Bell, A. J. and Sejnowski, T. J. (1998). Analysis of fMRI data by blind separation into independent spatial components. *Human Brain Mapping,* **6**, 160–188.

Mehta, A. D., Ulbert, I. and Schroeder, C. E. (2000). Intermodal selective attention in monkeys. I: distribution and timing of effects across visual areas. *Cereb. Cortex,* **10**, 343–358.

Menon, R. S. and Goodyear, B. G. (1999). Submillimeter functional localization in human striate cortex using BOLD contrast at 4 Tesla: implications for the vascular point-spread function. *Magn. Reson. Medi.,* **41**, 230–235.

Metea, M. R. and Newman, E. A. (2006). Glial cells dilate and constrict blood vessels: a mechanism of neurovascular coupling. *J. Neurosci.,* **26**, 2862–2870.

Mintun, M. A., Vlassenko, A. G., Shulman, G. L. and Snyder, A. Z. (2002). Time-related increase of oxygen utilization in continuously activated human visual cortex. *NeuroImage,* **16**, 531–537.

Mitzdorf, U. (1985). Current source-density method and application in cat cerebral cortex: investigation of evoked potentials and EEG phenomena. *Physiol. Rev.,* **65**, 37–100.

Mitzdorf, U. (1987). Properties of the evoked potential generators: current source-density analysis of visually evoked potentials in the cat cortex. *Int. J. Neurosci.,* **33**, 33–59.

Moon, C. H., Fukuda, M., Park, S. H. and Kim, S. G. (2007). Neural interpretation of blood oxygenation level-dependent fMRI maps at submillimeter columnar resolution. *J. Neurosci.,* **27**, 6892–6902.

Moutoussis, K. and Zeki, S. (2006). Seeing invisible motion: a human FMRI study. *Curr. Biol.,* **16**, 574–579.

Moutoussis, K., Keliris, G., Kourtzi, Z. and Logothetis, N. (2005). A binocular rivalry study of motion perception in the human brain. *Vision Res.,* **45**, 2231–2243.

Mukamel, R., Gelbard, H., Arieli, A., Hasson, U., Fried, I. and Malach, R. (2005). Coupling between neuronal firing, field potentials, and fMRI in human auditory cortex. *Science*, **309**, 951–954.

Mulligan, S. J. and Macvicar, B. A. (2004). Calcium transients in astrocyte endfeet cause cerebrovascular constrictions. *Nature*, **431**, 195–199.

Nakada, T., Nabetani, A., Kabasawa, H., Nozaki A. and Matsuzawa, H. (2005). The passage to human MR microscopy: a progress report from Niigata on April 2005. *Magn. Reson. Med. Sci.*, **4**, 83–87.

Nakahara, K., Hayashi, T., Konishi, S. and Miyashita, Y. (2002). Functional MRI of macaque monkeys performing a cognitive set-shifting task. *Science*, **295**, 1532–1536.

Nauhaus, I., Busse, L., Carandini, M. and Ringach, D. L. (2009). Stimulus contrast modulates functional connectivity in visual cortex. *Nature Neurosci.*, **12**, 70–76.

Nedergaard, M., Ransom, B. and Goldman, S. A. (2003). New roles for astrocytes: redefining the functional architecture of the brain. *Trends Neurosci.*, **26**, 523–530.

Nelson, P. G. (1966). Interaction between spinal motoneurons of the cat. *J. Neurophysiol.*, **29**, 275–287.

Newman, E. A. (2003). New roles for astrocytes: regulation of synaptic transmission. *Trends Neurosci.*, **26**, 536–542.

Nicholson, C. and Llinas, R. (1971). Field potentials in the alligator cerebellum and theory of their relationship to Purkinje cell dendritic spikes. *J. Neurophysiol.*, **34**, 509–531.

Niessing, J., Ebisch, B., Schmidt, K. E., Niessing, M., Singer, W. and Galuske, R. A. (2005). Hemodynamic signals correlate tightly with synchronized gamma oscillations. *Science*, **309**, 948–951.

Nudo, R. J. and Masterton, R. B. (1986). Stimulation-induced [14C]2-deoxyglucose labeling of synaptic activity in the central auditory system. *J. Comput. Neurol.*, **245**, 553–565.

O'Connor, D. H., Fukui, M. M., Pinsk, M. A. and Kastner, S. (2002). Attention modulates responses in the human lateral geniculate nucleus. *Nature Neurosci.*, **5**, 1203–1209.

Oeltermann, A., Augath, M. A. and Logothetis, N. K. (2007). Simultaneous recording of neuronal signals and functional NMR imaging. *Magn. Reson. Imaging*, **25**, 760–774.

Ogawa, S., Lee, T. M., Kay, A. R. and Tank, D. W. (1990). Brain magnetic resonance imaging with contrast dependent on blood oxygenation. *Proc. Natl. Acad. Sci. USA*, **87**, 9868–9872.

Ogawa, S., Tank, D. W., Menon, R., Ellermann, J. M., Kim, S. G., Merkle, H and Ugurbil, K. (1992). Intrinsic signal changes accompanying sensory stimulation – functional brain mapping with magnetic-resonance-imaging. *Proc. Natl. Acad. Sci. USA*, **89**, 5951–5955.

Patel, A. B., de Graaf, R. A., Mason, G. F., Rothman, D. L., Shulman, R. G. and Behar, K. L. (2005). The contribution of GABA to glutamate/glutamine cycling and energy metabolism in the rat cortex in vivo. *Proc. Natl. Acad. Sci. USA*, **102**, 5588–5593.

Pellerin, L. and Magistretti, P. J. (1994). Glutamate uptake into astrocytes stimulates aerobic glycolysis: a mechanism coupling neuronal activity to glucose utilization. *Proc. Natl. Acad. Sci. USA*, **91**, 10625–10629.

Pellerin, L. and Magistretti, P. J. (2003). Food for thought: challenging the dogmas. *J. Cereb. Blood Flow Metab.*, **23**, 1282–1286.

Pellerin, L., Bouzier-Sore, A. K., Aubert, A., Serres, S., Merle, M., Costalat, R. and Magistretti, P. J. (2007). Activity-dependent regulation of energy metabolism by astrocytes: an update. *Glia*, **55**, 1251–1262.

Peppiatt, C. and Attwell, D. (2004). Neurobiology: feeding the brain. *Nature*, **431**, 137–138.

Pessoa, L., Kastner, S. and Ungerleider, L. G. (2003). Neuroimaging studies of attention: from modulation of sensory processing to top-down control. *J. Neurosci.,* **23**, 3990–3998.

Peters, A. and Payne, B. R. (1993). Numerical relationships between geniculocortical afferents and pyramidal cell modules in cat primary visual cortex. *Cereb. Cortex,* **3**, 69–78.

Peters, A., Payne, B. R. and Budd, J. (1994). A numerical analysis of the geniculocortical input to striate cortex in the monkey. *Cereb. Cortex,* **4**, 215–229.

Petzold, G. C., Albeanu, D. F., Sato, T. F. and Murthy, V. N. (2008). Coupling of neural activity to blood flow in olfactory glomeruli is mediated by astrocytic pathways. *Neuron,* **58**, 897–910.

Pfeuffer, J., Van de Moortele, P. F., Yacoub, E., Shmuel, A., Adriany, G., Andersen, P., Merkle, H., Garwood, M., Ugurbil, K. and Hu, X. P. (2002). Zoomed functional imaging in the human brain at 7 Tesla with simultaneous high spatial and high temporal resolution. *NeuroImage,* **17**, 272–286.

Pfeuffer, J., Merkle, H., Beyerlein, M., Steudel, T. and Logothetis, N. K. (2004). Anatomical and functional MR imaging in the macaque monkey using a vertical large-bore 7 Tesla setup. *Magn. Reson. Imaging,* **22**, 1343–1359.

Poline, J. B., Worsley, K. J., Evans, A. C. and Friston, K. J. (1997). Combining spatial extent and peak intensity to test for activations in functional imaging. *NeuroImage,* **5**, 83–96.

Prichard, J., Rothman, D., Novotny, E., Petroff, O., Kuwabara, T., Avison, M., Howseman, A., Hanstock, C. and Shulman, R. (1991). Lactate rise detected by 1H NMR in human visual cortex during physiologic stimulation. *Proc. Natl. Acad. Sci. USA,* **88**, 5829–5831.

Privman, E., Nir, Y., Kramer, U., Kipervasser, S., Andelman, F., Neufeld, M. Y., Mukamel, R., Yeshurun, Y., Fried, I. and Malach, R. (2007). Enhanced category tuning revealed by intracranial electroencephalograms in high-order human visual areas. *J. Neurosci.,* **27**, 6234–6242.

Raichle, M. E. and Mintun, M. A. (2006). Brain work and brain imaging. *Annu. Rev. Neurosci.,* **29**, 449–476.

Rajimehr, R., Young, J. C. and Tootell, R. B. (2009). An anterior temporal face patch in human cortex, predicted by macaque maps. *Proc. Natl. Acad. Sci. USA,* **106**, 1995–2000.

Rauch, A., Rainer, G. and Logothetis, N. K. (2008). The effect of a serotonin-induced dissociation between spiking and perisynaptic activity on BOLD functional MRI. *Proc. Natl. Acad. Sci. USA,* **105**, 6759–6764.

Riera, J. J., Schousboe, A., Waagepetersen, H. S., Howarth, C. and Hyder, F. (2008). The micro-architecture of the cerebral cortex: functional neuroimaging models and metabolism. *NeuroImage,* **40**, 1436–1459.

Roberts, M. J., Zinke, W., Guo, K., Robertson, R., McDonald, J. S. and Thiele, A. (2005). Acetylcholine dynamically controls spatial integration in marmoset primary visual cortex. *J. Neurophysiol,* **93**, 2062–2072.

Roelfsema, P. R., Lamme, V. A. and Spekreijse, H. (1998). Object-based attention in the primary visual cortex of the macaque monkey. *Nature,* **395**, 376–381.

Sanchez-Vives, M. V., Nowak, L. G. and McCormick, D. A. (2000). Membrane mechanisms underlying contrast adaptation in cat area 17 in vivo. *J. Neurosci.,* **20**, 4267–4285.

Sappey-Marinier, D., Calabrese, G., Fein, G., Hugg, J. W., Biggins, C. and Weiner, M. W. (1992). Effect of photic stimulation on human visual cortex lactate and phosphates using 1H and 31P magnetic resonance spectroscopy. *J. Cereb. Blood Flow Metab.*, **12**, 584–592.

Sawamura, H., Georgieva, S., Vogels, R., VanDuffel W. and Orban, G. A. (2005). Using functional magnetic resonance imaging to assess adaptation and size invariance of shape processing by humans and monkeys. *J. Neurosci.*, **25**, 4294–4306.

Sawamura, H., Orban, G. A. and Vogels, R. (2006). Selectivity of neuronal adaptation does not match response selectivity: a single-cell study of the FMRI adaptation paradigm. *Neuron,* **49**, 307–318.

Schummers, J., Yu, H. and Sur, M. (2008). Tuned responses of astrocytes and their influence on hemodynamic signals in the visual cortex. *Science*, **320**, 1638–1643.

Schwartz, W. J., Smith, C. B., Davidsen, L., Savaki, H., Sokoloff, L., Mata, M., Fink, D. J. and Gainer, H. (1979). Metabolic mapping of functional activity in the hypothalamo-neurohypophysial system of the rat. *Science*, **205**, 723–725.

Shmuel, A., Augath, M., Oeltermann, A. and Logothetis, N. K. (2006). Negative functional MRI response correlates with decreases in neuronal activity in monkey visual area V1. *Nature Neurosci.,* **9**, 569–577.

Shoham, S., O'Connor, D. H., Segev, R. (2006). How silent is the brain: is there a "dark matter" problem in neuroscience? *J. Comp. Physiol.*, **192**, 777–784.

Shulman, R. G., Rothman, D. L., Behar, K. L., Hyder, F. (2004). Energetic basis of brain activity: implications for neuroimaging. *Trends Neurosci.,* **27**, 489–495.

Silva, A. C. and Koretsky, A. P. (2002). Laminar specificity of functional MRI onset times during somatosensory stimulation in rat. *Proc. Natl. Acad. Sci. USA,* **99**, 15182–15187.

Simard, M. and Nedergaard, M. (2004). The neurobiology of glia in the context of water and ion homeostasis. *Neuroscience,* **129**, 877–896.

Smith, A. J., Blumenfeld, H., Behar, K. L., Rothman, D. L., Shulman, R. G. and Hyder, F. (2002). Cerebral energetics and spiking frequency: the neurophysiological basis of fMRI. *Proc. Natl. Acad. Sci. USA,* **99**, 10765–10770.

Sokoloff, L. (1960). The metabolism of the central nervous system in vivo. In: J. Field, H. W. Magoun and V. E. Hall (editors), *Handbook of Physiology–Neurophysiology*. Washington, DC: American Physiological Society, pp. 1843–1864.

Sokoloff, L. (1999). Energetics of functional activation in neural tissues. *Neurochem Res.*, **24**, 321–329.

Steriade, M. (1991). Alertness, quiet sleep, dreaming. In: *Cerebral Cortex*. New York: Plenum Press, pp. 279–357.

Steriade, M. and Hobson, J. (1976). Neuronal activity during the sleep-waking cycle. *Prog. Neurobiol.,* **6**, 155–376.

Stone, J. (1973). Sampling properties of microelectrodes assessed in the cat's retina. *J. Neurophysiol.,* **36**, 1071–1079.

Sun, P., Ueno, K., Waggoner, R. A., Gardner, J. L., Tanaka, K. and Cheng, K. (2007). A temporal frequency-dependent functional architecture in human V1 revealed by high-resolution fMRI. *Nature Neurosci.,* **10**, 1404–1406.

Szentagothai, J. (1978). The Ferrier Lecture, 1977. The neuron network of the cerebral cortex: a functional interpretation. *Proc. R. Soc. London, Sci. B,* **201**, 219–248.

Takano, T., Tian, G. F., Peng, W., Lou, N., Libionka, W., Han, X. and Nedergaard, M. (2006). Astrocyte-mediated control of cerebral blood flow. *Nature Neurosci.,* **9**, 260–267.

Talairach, J. and Tournoux, P. (1988). *Co-planar Stereotaxic Atlas of the Human Brain*. Stuttgart: Thieme.

Tolias, A. S., Keliris, G. A., Smirnakis, S. M. and Logothetis, N. K. (2005a). Neurons in macaque area V4 acquire directional tuning after adaptation to motion stimuli. *Nature Neurosci.*, **8**, 591–593.

Tolias, A. S., Sultan, F., Augath, M., Oeltermann, A., Tehovnik, E. J., Schiller, P. H. and Logothetis, N. K. (2005b). Mapping cortical activity elicited with electrical microstimulation using FMRI in the macaque. *Neuron*, **48**, 901–911.

Towe, A. L. and Harding, G. W. (1970). Extracellular microelectrode sampling bias. *Exp. Neurol.*, **29**, 366–381.

Ts'o, D. Y., Frostig, R. D., Lieke, E. E. and Grinvald, A. (1990). Functional organization of primate visual cortex revealed by high resolution optical imaging. *Science*, **249**, 417–420.

Tsao, D. Y., VanDuffel, W., Sasaki, Y., Fize, D., Knutsen, T. A., Mandeville, J. B., Wald, L. L., Dale, A. M., Rosen, B. R., Van Essen, D. C., Livingstone, M. S., Orban, G. A. and Tootell, R. B. (2003). Stereopsis activates V3A and caudal intraparietal areas in macaques and humans. *Neuron*, **39**, 555–568.

Tsao, D. Y., Freiwald, W. A., Tootell, R. B. and Livingstone, M. S. (2006). A cortical region consisting entirely of face-selective cells. *Science*, **311**, 670–674.

Tsao, D. Y., Moeller, S. and Freiwald, W. A. (2008). Comparing face patch systems in macaques and humans. *Proc. Natl. Acad. Sci. USA,* **105**, 19514–19519.

Ugurbil, K., Adriany, G., Andersen, P., Chen, W., Gruetter R., Hu, X. P., Merkle, H., Kim, D. S., Kim, S. G., Strupp, J., Zhu, X. H. and Ogawa, S. (2000). Magnetic resonance studies of brain function and neurochemistry. *Annu. Rev. Biomed. Eng.,* **2**, 633–660.

Ugurbil, K., Toth, L. and Kim, D. S. (2003). How accurate is magnetic resonance imaging of brain function? *Trends Neurosci.,* **26**, 108–114.

Ugurbil, K., Adriany, G., Akgun, C., Andersen, P., Chen, W., Garwood, M., Gruetter, R., Henry, P. G., Marjanska, M., Moeller, S., Van de Moortele, P. F., Pruessmann, K. P., Tkac, I., Vaughan, J. T., Wiesinger, F., Yacoub, E. and Zhu, X. H. (2006). High magnetic fields for imaging cerebral morphology, function, and biochemistry. In: P. M. Robitaille and L. J. Berliner (editors), *Ultra High Field Magnetic Resonance Imaging*. New York: Springer, pp. 285–342.

VanDuffel, W., Fize, D., Peuskens, H., Denys, K., Sunaert, S., Todd, J. T. and Orban, G. A. (2002). Extracting 3D from motion: differences in human and monkey intraparietal cortex. *Science*, **298**, 413–415.

Viswanathan, A. and Freeman, R. D. (2007). Neurometabolic coupling in cerebral cortex reflects synaptic more than spiking activity. *Nature Neurosci.*, **10**, 1308–1312.

Vlassenko, A. G., Rundle, M. M. and Mintun, M. A. (2006a). Human brain glucose metabolism may evolve during activation: findings from a modified FDG PET paradigm. *NeuroImage*, **33**, 1036–1041.

Vlassenko, A. G., Rundle, M. M., Raichle, M. E. and Mintun, M. A. (2006b). Regulation of blood flow in activated human brain by cytosolic NADH/NAD+ ratio. *Proc. Natl. Acad. Sci. USA,* **103**, 1964–1969.

Wager, T. D. and Nichols, T. E. (2003). Optimization of experimental design in fMRI: a general framework using a genetic algorithm. *NeuroImage*, **18**, 293–309.

Wald, L. L., Fischl, B. and Rosen, B. R. (2006). High-resolution and microscopic imaging at high field. In: P. M. Robitaille and L. J. Berliner (editors), *Ultra High Field Magnetic Resonance Imaging*. New York: Springer, pp. 343–371.

Waldvogel, D., Van, G. P., Muellbacher, W., Ziemann, U., Immisch, I. and Hallett, M. (2000). The relative metabolic demand of inhibition and excitation. *Nature*, **406**, 995–998.

Wang, H., Hitron, I. M., Iadecola, C. and Pickel, V. M. (2005). Synaptic and vascular associations of neurons containing cyclooxygenase-2 and nitric oxide synthase in rat somatosensory cortex. *Cereb. Cortex,* **15**, 1250–1260.

Weber, B., Keller, A. L., Reichold, J. and Logothetis, N. K. (2008). The microvascular system of the striate and extrastriate visual cortex of the macaque. *Cereb. Cortex,* **18**, 2318–2330.

Weisskoff, R. M., Zuo, C. S., Boxerman, J. L. and Rosen, B. R. (1994). Microscopic susceptibility variation and transverse relaxation: theory and experiment. *Magn. Reson. Med.,* **31**, 601–610.

Wilke, M., Logothetis, N. K. and Leopold, D. A. (2006). Local field potential reflects perceptual suppression in monkey visual cortex. *Proc. Natl. Acad. Sci. USA,* **103**, 17507–17512.

Wong-Riley, M. T. (1989). Cytochrome oxidase: an endogenous metabolic marker for neuronal activity. *Trends Neurosci.,* **12**, 94–101.

Woolrich, M. W., Ripley, B. D., Brady, M. and Smith, S. M. (2001). Temporal autocorrelation in univariate linear modeling of FMRI data. *NeuroImage,* **14**, 1370–1386.

Worsley, K. J. (2005). Spatial smoothing of autocorrelations to control the degrees of freedom in fMRI analysis. *NeuroImage,* **26**, 635–641.

Wu, E. X., Tang, H. and Jensen, J. H. (2004). Applications of ultrasmall superparamagnetic iron oxide contrast agents in the MR study of animal models. *Nucl. Magn. Reson. Biomed.,* **17**, 478–483.

Wunderlich, K., Schneider, K. A. and Kastner, S. (2005). Neural correlates of binocular rivalry in the human lateral geniculate nucleus. *Nature Neurosci.,* **8**, 1595–1602.

Xu, F., Liu, N., Kida, I., Rothman, D. L., Hyder, F. and Shepherd, G. M. (2003). Odor maps of aldehydes and esters revealed by functional MRI in the glomerular layer of the mouse olfactory bulb. *Proc. Natl. Acad. Sci. USA,* **100**, 11029–11034.

Yacoub, E., Shmuel, A., Pfeuffer, J., Van de Moortele, P. F., Adriany, G., Andersen, P, Vaughan, J. T., Merkle, H., Ugurbil, K. and Hu, X. P. (2001). Imaging brain function in humans at 7 Tesla. *Magn. Reson. Med.,* **45**, 588–594.

Yacoub E., Van de Moortele, P. F., Shmuel, A. and Ugurbil, K. (2005). Signal and noise characteristics of Hahn SE and GE BOLD fMRI at 7T in humans. *NeuroImage,* **24**, 738–750.

Yacoub, E., Ugurbil, K. and Harel, N. (2006). The spatial dependence of the poststimulus undershoot as revealed by high-resolution BOLD- and CBV-weighted fMRI. *J. Cereb. Blood Flow Metab.,* **26**, 634–644.

Yacoub, E., Shmuel, A., Logothetis, N. and Ugurbil, K. (2007). Robust detection of ocular dominance columns in humans using Hahn spin echo BOLD functional MRI at 7 Tesla. *NeuroImage,* **37**, 1161–1177.

Yacoub, E., Harel, N. and Ugurbil, K. (2008). High-field fMRI unveils orientation columns in humans. *Proc. Natl. Acad. Sci. USA,* **105**, 10607–10612.

Yang, G., Zhang, Y., Ross, M. E. and Iadecola, C. (2003). Attenuation of activity-induced increases in cerebellar blood flow in mice lacking neuronal nitric oxide synthase. *Am. J. Physiol. Heart. Circ. Physiol.,* **285**, H298–H304.

Yoshor, D., Ghose, G. M., Bosking, W. H., Sun, P. and Maunsell, J. H. (2007). Spatial attention does not strongly modulate neuronal responses in early human visual cortex. *J Neurosci.,* **27**, 13205–13209.

Zappe, A. C., Pfeuffer, J., Merkle, H., Logothetis, N. K. and Goense, J. B. (2008). The effect of labeling parameters on perfusion-based fMRI in nonhuman primates. *J. Cereb. Blood Flow Metab.,* **28**, 640–652.

Zeki, S. and Ffytche, D. (1998). The Riddoch syndrome: insights into the neurobiology of conscious vision. *Brain,* **121**, 25–45.

Zhang, C. and Wong-Riley, M. T. (1996). Do nitric oxide synthase, NMDA receptor subunit R1 and cytochrome oxidase co-localize in the rat central nervous system? *Brain Res.*, **729**, 205–215.

Zhao, F. Q., Wang, P. and Kim, S. G. (2004). Cortical depth-dependent gradient-echo and spin-echo BOLD fMRI at 9.4T. *Magn. Reson. Med.*, **51**, 518–524.

Zhao, F., Wang, P., Hendrich, K. and Kim, S. G. (2005). Spatial specificity of cerebral blood volume-weighted fMRI responses at columnar resolution. *NeuroImage*, **27**, 416–424.

Zhao, F., Wang, P., Hendrich, K., Ugurbil, K. and Kim, S. G. (2006). Cortical layer-dependent BOLD and CBV responses measured by spin-echo and gradient-echo fMRI: insights into hemodynamic regulation. *NeuroImage*, **30**, 1149–1160.

Zhao, F., Jin, T., Wang, P. and Kim, S. G. (2007). Improved spatial localization of post-stimulus BOLD undershoot relative to positive BOLD. *NeuroImage*, **34**, 1084–1092.

Zonta, M., Angulo, M. C., Gobbo, S., Rosengarten, B., Hossmann, K. A., Pozzan, T. and Carmignoto, G. (2003). Neuron-to-astrocyte signaling is central to the dynamic control of brain microcirculation. *Nature Neurosci.*, **6**, 43–50.

12

Perspectives

In the nineteenth century, Julius Bernstein invented an ingenious device called the "differential rheotome," a rotating wheel which could record the time course of action potentials (see Chapter 3). Since then, many sophisticated techniques have been introduced to measure correlates of neural activity: measurements of electricity produced by single neurons (Chapters 3 and 4) or multiple neurons (Chapters 5–7 and 9), measurements based on brain metabolism (Chapters 8 and 11) or on calcium dynamics (Chapter 10). These techniques are always more or less indirect measurements of neural activity, and they have diverse spatial and temporal resolutions, and spatial scales. Each chapter in this book has described the quantitative relationship between neural activity (e.g. membrane potential or synaptic activity) and the measured quantity, as it is currently understood. This effort serves two purposes: to give a better understanding and interpretation of the measurements, and to help enhance existing techniques or develop new ones. To conclude this book, the authors of all the chapters describe ongoing developments in their field, open questions to be addressed, and new emerging techniques.

12.1 Extracellular recording

K. H. Petersen, H. Lindén, A. M. Dale, G. T. Einevoll and T. Stieglitz

Substrate-integrated microelectrode arrays (MEAs) are planar arrays of microelectrodes used to record electrical activity in neuronal cell cultures or acute brain slices (Taketani and Baudray, 2006; Egert et al., 2010; Gross, 2010). While their history goes back to the 1970s, the rapid development of photolithographic techniques (stimulated by the needs of the computer industry) has now made prefabricated high-density MEA chips a popular research tool. A typical MEA may have 60 or so metal electrodes, each with a diameter of 10 to 30 μm, with an electrode spacing ranging from 50 up to a few hundred micrometers. Recently, chips with more than

10 000 electrodes and electrode spacings less than 20 μm have been made (Frey et al., 2009). Another development is perforated MEAs, i.e. MEAs with small holes in the insulating area between the electrodes, which allows for the use of suction to provide better contact between brain slices and the electrodes.

Typically the MEA is set up so that the electrodes can record electrical signals, and can also provide electrical stimulation to excite the neural tissue. A main application of MEAs has been to study spatiotemporal activity in neuronal ensembles in a simpler and better controlled setting than in vivo. They are also used for effective and rapid screening of pharmacological effects on neural systems.

The relationship between neural activity, i.e. transmembrane currents, and the potentials recorded by the electrodes is in principle the same as outlined for the in vivo situation in Chapter 4. However, additional difficulties in the estimation of neural activity arise from the more complicated macroscopic electric boundary conditions because the metal electrode, the insulating material between the electrodes, and the neural culture or slice (and saline covering the slice) have very different electrical conductivities. Thus it seems that the solution to the electrostatic forward problem must in general be based on solving Poisson's equation (see Equation (4.7)) using a finite-element model (FEM) scheme (Buitenweg et al., 2002; Frey et al., 2009).

12.2 Intracellular recording

R. Brette and A. Destexhe

Although electrodes and amplifiers are well established experimental devices, we can expect new developments in measuring techniques in the future, either in the way recordings are analyzed or in the way the experimental devices are controlled. We list below a few areas where new techniques might emerge in the future.

- Recording techniques using numerical models of neurons and/or of the experimental apparatus (e.g. electrodes), as were introduced recently for current clamp and dynamic clamp recordings (Brette et al., 2008). The challenge in single-electrode recordings is to estimate the electrode properties independently of neuronal properties, which is difficult in some situations (e.g. recording in axons or dendrites).
- Dynamic clamp techniques: dynamic clamp recordings consist in injecting a current that depends in real time on the measured potential, which poses specific technical problems (Brette et al., 2009), especially when modeling fast ionic channels (e.g. sodium channels).
- Single-trial conductance measurements: model-based and/or statistical techniques could be used to estimate the time course of synaptic conductances in

single trials. The difficulty is that two quantities (reversal potential and total conductance, or excitatory and inhibitory conductances) must be derived from a single measurement (membrane potential). Recent work suggests that this goal may be achievable (Pospischil et al., 2009).

12.3 Local field potentials

C. Bédard and A. Destexhe

A proper understanding and modeling of LFP signals will depend on solving a number of challenges in forthcoming years. First, we need more experimental data and theoretical efforts to understand the role of ionic diffusion. It was shown theoretically (Bédard and Destexhe, 2009) that taking into account ionic diffusion, the extracellular impedance varies as $1/\sqrt{\omega}$ ("Warburg" impedance), which gives $1/f$ in the power spectrum. This mechanism constitutes a plausible physical cause of the $1/f$ filtering. Indirect evidence also exists for such a filter, such as the mismatch between EEG and MEG power spectra found recently (Dehghani et al., 2010) and the transfer function between intracellular and extracellular potentials (Bédard et al., 2010). However, no direct evidence is presently available. The problem is that the exact weight of ionic diffusion will depend critically on the type and magnitude of extracellular currents. So, impedance measurements should be done with current amplitudes of the same order as physiological currents, and this represents a clear technical challenge.

Second, if the non-resistive aspect of the extracellular medium is confirmed, this will require a revision of several of the currently used methods (in addition to the modeling of extracellular potentials and spikes). One such method is the "current source density" (CSD) method, which is not valid if the medium is non-resistive. Ongoing theoretical work (Bédard and Destexhe, 2011) aims at providing CSD expressions which are valid for non-resistive media. Another method is the inverse source localization, from EEG or MEG. These methods are also based on resistive media, and will require substantial revisions if the medium is significantly non-resistive.

Third, the exclusive dipolar nature of the current sources is also questionable. The high-amplitude LFP signals in structures such as the thalamus (Contreras et al., 1996) are difficult to reconcile with the fact that these are "closed field" structures with few or no dipoles. There is a possibility that other configurations, such as monopoles or high-order multipoles, have a larger effect than is assumed by the canonical dipole model. This also constitutes an interesting aspect for exploration by future experiments and models.

Finally, the LFP signal becomes available at increasingly high densities and numbers of electrodes in an array, which will motivate the emergence of "LFP imaging" methods to map precisely the extracellular electrical activity in two or three dimensions. It will then become more and more important to account quantitatively for such signals, in order to reconstruct the underlying two-dimensional or three-dimensional maps of neuronal current sources.

12.4 EEG and MEG: forward modeling

J. C. De Munck, C. H. Wolters and M. Clerc

Various different approaches of the forward problem have been presented in a common mathematical framework (Chapter 6). For each method there is a trade-off between speed and accuracy and each method poses constraints regarding the underlying assumptions. Analytical methods can deal with anisotropy, but only for highly symmetric geometries, and BEMs can deal with arbitrary geometries, but only when all compartments are isotropic. The work of Kybic et al. (2005) can be considered as a breakthrough in the accuracy that can be achieved with BEM. Until now, only three-compartment isotropic (skin, skull, brain) realistically shaped BEM models seem to be practicable, but the importance of modeling skull inhomogeneity, the highly conducting CSF compartment and brain anisotropy on EEG source analysis has been demonstrated. Considering such a priori constraints, the FEM is the most flexible method, because it can deal with arbitrary geometry and anisotropy. However, routine application of the FEM in individual subjects is still hampered by the complexity of MRI segmentation required to set up detailed FEM head models.

Although the common framework in which different methods are presented would facilitate the choice of a "preferred" method, this choice is hampered by practical goals. In a fair comparison of methods, one should account for the fact that different methods have different setup times and, therefore, the effective speed depends on the number of forward computations required. The only thing one can say in this context is that analytical methods are only of theoretical interest, useful to determine the absolute accuracy of new numerical variants. The drawback thereof is, however, that these variants are only tested for highly symmetric conductors, such as concentric spheres. From our current experience, for three-compartment isotropic models, the symmetric BEM is the method of choice (Gramfort et al., 2011), while already in a four-compartment isotropic model with an additional CSF compartment, the FEM is advantageous because the error induced by neglecting the CSF compartment clearly exceeds the numerical errors (Vorwerk, 2011). However, further methodological improvements are possible for both BEM and FEM, so that further systematic comparisons of different variants in future studies are still necessary.

As we have seen, the variety of models available can incorporate different features of the head conductivity. In forward modeling, an important practical issue is therefore to determine the level of model detail that is sufficient for a given experimental situation. Further research that guides the user in this respect is strongly needed (Dannhauer et al., 2011).

12.5 EEG and MEG: source estimation

S. P. Ahlfors and M. S. Hämäläinen

Advances in MEG and EEG inverse modeling are expected to emerge from multimodal imaging, in which different types of functional and structural data are merged. These methods include fMRI, PEG, optical imaging, and TMS. Sometimes the data can be acquired simultaneously, thereby making possible the study of those brain events that cannot be fully repeated. Separate recordings, however, typically offer a higher signal to noise ratio. As a particular application of multimodal imaging, fMRI data can be used to suggest likely locations of MEG and EEG sources among all the possible locations that can explain the measured MEG and EEG signals equally well. A better understanding of the relationship between electrophysiological and hemodynamic signals emerging from experimental and modeling studies will facilitate the integration of the different types of non-invasive imaging data. Advances in the interpretation of MEG and EEG signals are also expected to result from the integration of source estimation with sophisticated signal processing methods. In particular, the inherent limitations of MEG and EEG inverse estimates should be taken into account throughout the analysis and interpretation of the data in relation to models of neural dynamics in studies of large-scale brain networks.

12.6 Intrinsic optical imaging

R. D. Frostig and C. H. Chen-Bee

The most exciting developments have to do with new methods of selective brain stimulation. Recent developments in optogenetic techniques enable researchers to excite or inhibit selected groups of neurons using light stimulation in the anesthetized or awake, behaving animals. These techniques expand the ability of researchers to probe the brain with a growing level of refinement at the level of its functional circuitry. A limitation of such techniques, however, is that they are invasive. New, non-invasive stimulation techniques seem therefore a promising complement to optogenetic techniques although they are not as refined. These include transcranial stimulation using weak ultrasound pulses (Tufail et al., 2010) and transcranial electric stimulation (Ozen et al., 2010), which have been successfully applied in rats.

12.7 Voltage-sensitive dye imaging

S. Chemla and F. Chavane

Optical imaging of voltage-sensitive dye (VSDI) is so far the only technique that allows a direct recording of neuronal activity with both a high spatial (few tens of millimeters) and temporal (1–10 ms) resolution over a large cortical area (about 2 cm^2). However, the Achilles heel of the technique lies in the multi-component origin of the underlying signal: the recorded fluorescence results from the integration of membrane depolarization of all stained cellular components over hundreds of cells (for one pixel). One striking example of this problem is the difficulty in predicting the contribution of inhibition to the overall signal. Indeed, the activity of both inhibitory and excitatory cells contributes positively to the signal, but inhibition causes a silent shunting of postsynaptic elements that will mostly be invisible to the signal.

The ideal technique does not exist yet and a combination of complementary recordings is one evident solution to overcome such limitations. However, we foresee that the combination of VSDI with modeling is another promising perspective for better interpretation and, hopefully, use of VSDI signals. Different approaches have recently been tackled, from biophysical models, which help us to understand better the origin of the signal, to neural field models, which help us to drive the possible neuronal interactions at the origin of the observed cortical dynamics. In the near future, we can envision that close interactions between physiologists and modelers will instigate predictions that can be tested experimentally, closing the loop of a virtuous cycle.

12.8 Calcium imaging

F. Helmchen

Several aspects of calcium indicator measurements and their interpretation are likely to gain more interest in the near future. First, more information is needed about the differences in calcium dynamics in the various cell types, for example, the many subtypes of GABAergic neurons found in the cortical microcircuits. This will be important in order to achieve a comprehensive reconstruction of activity patterns in local populations, taking all their diversity into account. Second, non-linear features of calcium indicators such as saturation will increasingly be considered to understand the relationship between electrical excitation, changes in calcium concentration, and fluorescence signals more completely. This is particularly true for newest generation GECIs, for which the binding curves, conformational changes, and interactions with other proteins can be complex. Third, consideration of kinetic properties of calcium indicators will become essential not only when analyzing calcium dynamics on the very fine spatial scale, in synaptic and dendritic structures, but also when starting to model the multiple parallel effects of calcium influx on the cytosolic protein networks. Finally, the combination of electrical recordings – be it juxtacellular or multi-electrode extracellular recordings – will remain a key technique for verifying the interpretation of calcium measurements in terms of neuronal spiking.

The principles of calcium imaging are rather independent of the microscope setup but the interpretation of measured calcium dynamics clearly depends on specific parameters of the imaging procedure such as sampling frequency, spatial resolution, and sensitivity. Implementations of calcium imaging may crucially depend on the specific imaging technology, which is obvious in the case of two-photon calcium imaging, because it is the only method that allows cellular imaging several hundred micrometers deep inside neural tissue (Helmchen and Denk, 2005; Kerr and Denk, 2008). Therefore, further advances in microscopy and labeling techniques may open entire new fields of application for calcium imaging. One promising direction is the development of three-dimensional laser scanning technologies, which enable comprehensive volumetric measurements from local cell populations and thus new insights into the fundamentals of neural network dynamics (Göbel and Helmchen, 2007; Göbel et al., 2007). The same holds true for imaging at higher speed, which is essential to capture rapid signal flow through neural networks (Grewe et al., 2010). Moreover, the ongoing improvement and further development of GECIs (Mank et al., 2008; Wallace et al., 2008) and their great potential for combination with other methods like viral tracing techniques (Granstedt et al., 2009), can be foreseen to revolutionize the analysis of neural circuits in the neurosciences. Soon it will be possible to monitor neural network activity repeatedly over days to weeks from the same cell population, which will open tremendous opportunities to study the formation, plasticity, and reconfiguration of neural circuits.

After forty years of calcium imaging we are still experiencing only the beginning of the use of this powerful method for the study of cellular function. Most of the principles of fluorescence measurements that were covered in Chapter 10 will remain valid so that we hope that young researchers have gained a solid foundation for exploring new research fields.

12.9 Functional magnetic resonance imaging

A. Bartels, J. Goense and N. Logothetis

Since the invention of fMRI, the recurrent theme in the development of hardware and of imaging sequences has always been "higher, faster, finer," with "higher" referring to the field strengths. While this is likely to stay true and to lead to improvements in the foreseeable future, it is also clear that the physiological mechanisms underlying the BOLD signal pose limits that may be difficult to overcome. Perhaps signals hitherto thought to be unreliable, such as the initial dip, will have to be re-examined in order to allow for neuroimaging to overcome this. While the higher signal provided by higher field strengths will allow improvements in spatial and temporal scales, crude practicalities such as heart-beat, breathing or body movements, susceptibility artifacts, and RF-inhomogeneity, cause artifacts or differences in the signal to noise ratio that become more pronounced at higher fields. Thus, new sequences that are ultra-fast or less sensitive to artifacts, or methods such as dynamic shimming that can correct for field distortions in real time, will hopefully provide solutions. At the same time, increasing spatial and temporal resolution will pose new challenges to analysis techniques, in particular to those aimed at establishing the content, direction and augmentation of information flow across the brain, and to disentangling neural information from the numerous artifacts that, in part, may be directly related or caused by some of the neural activity, such as heart-rate, breathing and autonomous signals that also affect vascular responses. There is no doubt that because of the complementary nature of fMRI, PET, EEG and invasive techniques, some combinations of these techniques may lead to breakthroughs both in their understanding and in neuroscience. In particular, traditional micro-stimulation techniques but also optical and optogenetic tools, or new contrast agents may, in combination with neuroimaging, vastly expand our understanding of functional circuitry. At the same time they may lead to invaluable clinical applications relying on controlled neurostimulation tailored not only with respect to anatomical position but also with respect to stimulation parameters to the individual case. There is no doubt that this relatively young discipline will benefit from some foreseeable, but hopefully also from more surprising and groundbreaking, innovations and insights.

References

Bédard, C. and Destexhe, A. (2009). Macroscopic models of local field potentials and the apparent $1/f$ noise in brain activity. *Biophys. J.*, **96** (7), 2589–2603.

Bédard, C. and Destexhe, A. (2011). A generalized theory for current-source density analysis in brain tissue. *Physi. Rev. E*, **84**, 041909.

Bédard, C., Rodrigues, S., Roy, N., Contreras, D. and Destexhe, A. (2010). Evidence for frequency-dependent extracellular impedance from the transfer function between extracellular and intracellular potentials. *J. Comput. Neurosci.*, **29** (3), 389–403.

Brette, R., Piwkowska, Z., Monier, C., Rudolph-Lilith, M., Fournier, J., Levy, M., Frégnac, Y., Bal, T. and Destexhe, A. (2008). High-resolution intracellular recordings using a real-time computational model of the electrode. *Neuron*, **59** (3), 379–391.

Brette, R., Piwkowska, Z., Monier, C., Francisco, J., Gonzalez, G., Frégnac, Y., Bal, T. and Destexhe, A. (2009). Dynamic clamp with high-resistance electrodes using active electrode compensation in vitro and in vivo, 30 pages.

Buitenweg, J. R., Rutten, W. L. C. and Marani, E. (2002). Modeled channel distributions explain extracellular recordings from cultured neurons sealed to microelectrodes. *IEEE Trans. Biomed. Eng.*, **49** (12), 1580–1590.

Contreras, D., Destexhe, A., Sejnowski, T. J. and Steriade, M. (1996). Control of spatiotemporal coherence of a thalamic oscillation by corticothalamic feedback. *Science*, **274** (5288), 771–774.

Dannhauer, M., Lanfer, B., Wolters, C. H. and Knösche, T. R. (2011). Modeling of the human skull in EEG source analysis. *Human Brain Mapping*, **32** (9), 1383-1399.

Dehghani, N., Bédard, C., Cash, S. S., Halgren, E. and Destexhe, A. (2010). Comparative power spectral analysis of simultaneous elecroencephalographic and magnetoencephalographic recordings in humans suggests non-resistive extracellular media. *J. Comput. Neurosci.*, **29** (3), 405–421.

Egert, U., Kindervater, R. and Stett, A. (2010). *Conference Proceedings of the 7th International Meeting of Substrate-Integrated Microelectrode Arrays. Reutlingen, Germany.* Stuttgart: BIOPRO Baden-Württemberg.

Frey, U., Egert, U., Heer, F., Hafizovic, S. and Hierlemann, A. (2009). Microelectronic system for high-resolution mapping of extracellular electric fields applied to brain slices. *Biosens. Bioelectron.*, **24** (7), 2191–2198.

Göbel, W. and Helmchen, F. (2007). New angles on neuronal dendrites in vivo. *J. Neurophysiol.*, **98** (6), 3770–3779.

Göbel, W., Kampa, B. M. and Helmchen, F. (2007). Imaging cellular network dynamics in three dimensions using fast 3D laser scanning. *Nature Methods*, **4** (1), 73–79.

Gramfort, A., Papadopoulo, T., Olivi, E. and Clerc, M. (2011). Forward field computation with OpenMEEG. *Comp. Intel. Neurosc.*, **2011**, 1–13.

Granstedt, A. E., Szpara, M. L., Kuhn, B., Wang, S. S. and Enquist, L. W. (2009). Fluorescence-based monitoring of in vivo neural activity using a circuit-tracing pseudorabies virus. *PloS One*, **4** (9), e6923.

Grewe, B. F., Langer, D., Kasper, H., Kampa, B. M. and Helmchen, F. (2010). High-speed in vivo calcium imaging reveals neuronal network activity with near-millisecond precision. *Nature Methods*, **7** (5), 399–405.

Gross, G. W. (2010). Multielectrode arrays. *Scholarpedia*, **3** (6), 5749.

Helmchen, F. and Denk, W. (2005). Deep tissue two-photon microscopy. *Nature Methods*, **2** (12), 932–940.

Kerr, J. N. D. and Denk, W. (2008). Imaging in vivo: watching the brain in action. *Nature Rev. Neurosci.*, **9** (3), 195–205.

Kybic, J., Clerc, M., Abboud, T., Faugeras, O., Keriven, R. and Papadopoulo, T. (2005). A common formalism for the integral formulations of the forward EEG problem. *IEEE Trans. Med. Imaging*, **24** (1), 12–28.

Mank, M., Santos, A. F., Direnberger, S., Mrsic-Flogel, T. D., Hofer, S. B., Stein, V., Hendel, T., Reiff, D. F., Levelt, C., Borst, A., Bonhoeffer, T., Hübener, M. and

Griesbeck, O. (2008). A genetically encoded calcium indicator for chronic in vivo two-photon imaging. *Nature Methods*, **5** (9), 805–811.

Ozen, S., Sirota, A., Belluscio, M. A., Anastassiou, C. A., Stark, E., Koch, C. and Buzsáki, G. (2010). Transcranial electric stimulation entrains cortical neuronal populations in rats. *J. Neurosci.*, **30** (34), 11476–11485.

Pospischil, M., Piwkowska, Z., Bal, T. and Destexhe, A. (2009). Extracting synaptic conductances from single membrane potential traces. *Neuroscience*, **158** (2), 545–552.

Taketani, M. and Baudray, M. (2006). *Advances in Network Electrophysiology Using Multi-Electrode Arrays*. New York: Springer.

Tufail, Y., Matyushov, A., Baldwin, N., Tauchmann, M. L., Georges, J., Yoshihiro, A., Tillery, S. I. H. and Tyler, W. J. (2010). Transcranial pulsed ultrasound stimulates intact brain circuits. *Neuron*, **66** (5), 681–694.

Vorwerk, J. (2011). Comparison of numerical approaches to the EEG forward problem. *Master thesis* in mathematics, Münster.

Wallace, D. J., zum Alten Borgloh, S. M., Astori, S., Yang, Y., Bausen, M., Kügler, S., Palmer, A. E., Tsien, R. Y., Sprengel, R., Kerr, J. N. D., Denk, W. and Hasan, M. T. (2008). Single-spike detection in vitro and in vivo with a genetic Ca^{2+} sensor. *Nature Methods*, **5** (9), 797–804.

d in the United States
& Taylor Publisher Services